International Studies in the History of Mathematics and its Teaching

Series Editors

Alexander Karp
Teachers College, Columbia University, New York, NY, USA

Gert Schubring
Universität Bielefeld, Bielefeld, Germany
Universidade Federal do Rio de Janeiro, Rio de Janeiro, Brazil

The International Studies in the History of Mathematics and its Teaching Series creates a platform for international collaboration in the exploration of the social history of mathematics education and its connections with the development of mathematics. The series offers broad perspectives on mathematics research and education, including contributions relating to the history of mathematics and mathematics education at all levels of study, school education, college education, mathematics teacher education, the development of research mathematics, the role of mathematicians in mathematics education, mathematics teachers' associations and periodicals.

The series seeks to inform mathematics educators, mathematicians, and historians about the political, social, and cultural constraints and achievements that influenced the development of mathematics and mathematics education. In so doing, it aims to overcome disconnected national cultural and social histories and establish common cross-cultural themes within the development of mathematics and mathematics instruction. However, at the core of these various perspectives, the question of how to best improve mathematics teaching and learning always remains the focal issue informing the series.

More information about this series at http://www.springer.com/series/15781

Évelyne Barbin • Marta Menghini • Klaus Volkert
Editors

Descriptive Geometry, The Spread of a Polytechnic Art

The Legacy of Gaspard Monge

Editors
Évelyne Barbin
UFR Sciences et Techniques
Nantes, France

Marta Menghini
Department of Mathematics
Sapienza–University of Rome
Rome, Italy

Klaus Volkert
University of Wuppertal
Wuppertal, Germany

ISSN 2524-8022   ISSN 2524-8030 (electronic)
International Studies in the History of Mathematics and its Teaching
ISBN 978-3-030-14810-2   ISBN 978-3-030-14808-9 (eBook)
https://doi.org/10.1007/978-3-030-14808-9

© Springer Nature Switzerland AG 2019
This work is subject to copyright. All rights are reserved by the Publisher, whether the whole or part of the material is concerned, specifically the rights of translation, reprinting, reuse of illustrations, recitation, broadcasting, reproduction on microfilms or in any other physical way, and transmission or information storage and retrieval, electronic adaptation, computer software, or by similar or dissimilar methodology now known or hereafter developed.
The use of general descriptive names, registered names, trademarks, service marks, etc. in this publication does not imply, even in the absence of a specific statement, that such names are exempt from the relevant protective laws and regulations and therefore free for general use.
The publisher, the authors, and the editors are safe to assume that the advice and information in this book are believed to be true and accurate at the date of publication. Neither the publisher nor the authors or the editors give a warranty, express or implied, with respect to the material contained herein or for any errors or omissions that may have been made. The publisher remains neutral with regard to jurisdictional claims in published maps and institutional affiliations.

This Springer imprint is published by the registered company Springer Nature Switzerland AG.
The registered company address is: Gewerbestrasse 11, 6330 Cham, Switzerland

Preface

The Purpose of the Book

The developments that followed the French Revolution of 1789 led to the birth of new kinds of schools and a new kind of mathematical teaching. In 1794–1795, two entirely novel institutions of higher education were founded in France. One of them was the *École normale de l'an III*, which was devoted to multistep teacher education for the new schools of the revolution. A second was the *École polytechnique* (initially *École des travaux publics*), which provided training for civil and military engineers and would serve as a model for many countries in the future.[1] One of the leading actors of the period was Gaspard Monge, whose lessons delivered to future teachers at the *École normale* were summarized in his well-known textbook *Géométrie descriptive* (1799). Monge was also influential in elaborating the initial conception of the *École polytechnique* and played a considerable role in it. The case of descriptive geometry is highly noteworthy: being taught publicly the first time at two institutions created in the revolutionary period, later it exemplified a novel impact of mathematics on society.

The military and civilian needs for well-prepared officers, engineers and administrators, which led to the creation of the *École polytechnique* and later of the *École Centrale des Arts et Manufactures*, were common to many countries. New types of schools were created everywhere like *Polytechnische Schulen* in German-speaking countries, *Politecnici* in Italy and military schools in the USA. They all aimed to address the needs of their students in scientific and technical studies. Descriptive geometry was an important subject in these new schools. New ways of understanding and teaching it were created in different places in response to the needs and teaching traditions of different countries and regions. For instance, in France—and in other countries—descriptive geometry was soon considered to be a new type of geometry and was taught to all students in the second part of the

---

[1] More on them in Chap. 1.

century. In other places, its interest remained restricted to its technical nature, and its dissemination was connected with the creation of military academies and analogous institutions.

This book intends to analyse how descriptive geometry spread through general, technical or artistic education and thus in "polytechnic" schools, in different countries around the world. Our purpose is to go beyond the analysis of the development of the subject in individual countries. We do so by gathering and confronting the local experiences in order to highlight the parallel processes and the connections among them. We also seek to understand how they may have been provided by the circulation between countries and persons or by publications related to our subject.

Depending on different regions, the period that is considered here ranges from approximately 1800 up to the 1900s and beyond. The titles of the contributions often refer to a country or a region. Of course, it is difficult to speak of countries in the nineteenth century in relation to German-speaking regions and also in relation to Italy before its unification—not to speak of the complicated situation in the Balkans. Similarly, there is a contribution about teaching in the Czech language although the country wasn't created until 1918. Moreover, the considered period corresponds to a political and colonial situation that influenced relations among countries (for instance, those between France, Greece and Italy at the beginning of the century and later between Portugal and Brazil), which makes attributing the dissemination process to one country even more difficult.

The process of the introduction of descriptive geometry is studied at different levels: the chapters may be devoted to the countries but also to specific towns and regions. The Austro-Hungarian Imperium is represented mainly by Vienna (Chaps. 11 and 12) and partly by Prague and Brno (Chap. 16) after the introduction of the Czech language. Russia is represented by a chief character, Charles Poitier in Saint Petersburg (Chap. 13). Latin America (Chap. 21) deals mainly with Brazil—where historical research on the subject is the most developed. Sometimes (although rarely), different chapters consider the same objects—it seems valuable because of the different context.

Not all countries involved in the process are included in this book: other countries like Canada, Sweden, Belgium and the Far East could be considered. Nevertheless, the 22 chapters of this book, written by different authors from 15 countries, offer a wide panorama and can also help to engage studies in the field concerning other countries or regions.

The content of the book is discussed below, but here, we need to make a note on some chapters because they address some special aims. The chapter on the Vienna School gives an account of the lives of the involved characters, of the academic life in this institution and of the work of the students (Chap. 12). In this way, the chapter addresses an aim to explore the academic world from another "human" perspective. Simultaneously, we aimed to take a closer look at the relations between projective and descriptive geometry. To *rise* descriptive geometry to a level of an exact science was, as the reader will see, a concern in many countries; in the work of Wilhelm Fiedler (Chap. 10) and also of Luigi

Cremona (Chap. 4), projective geometry was considered to be the theoretical basis for descriptive geometry—the connection between the two areas being established through the concept of central projection. Fiedler's and Cremona's point of view is placed in the context of "modern geometry" in the second part of the nineteenth century even if—in their case—attention was paid mainly to synthetic geometry. This "purist" trend influenced the development not only of descriptive geometry but of all geometry in some universities (e.g. Barcelona and Madrid, as shown in Chap. 6). The connection between projective and descriptive geometry epitomized in some way the relationship between pure and applied sciences and became also central in the relationship between the faculties of sciences and the schools for engineers and architects (Chaps. 4 and 6).

The Content

The chapters of the book are given in a geographical order, which begins from the "near" countries of Southern Europe (which also includes Greece) and moves to the more culturally independent countries of Central and Eastern Europe and then to England and lastly to non-European countries effected by colonization and decolonization (the Americas but also Egypt). The geographical order can broadly correspond to a chronological order referring to the introduction of the teaching of descriptive geometry.

The Initial Spread of Descriptive Geometry in Southern Europe The first part of the book concerns countries in Southern Europe: France with Italy and Spain, Portugal and Greece; four countries strongly influenced by France at the beginning of the nineteenth century before and during the Napoleonic wars. This influence resulted in the creation of new schools for engineers and military officers inspired not only by the *École polytechnique* but by French schools in general and in the importance given to French mathematics, especially to descriptive geometry. The textbook of Gaspard Monge (Chap. 1) was translated into Spanish, Italian and Portuguese between 1803 and 1812. Locally, people connected directly with France played an important role like Agustin de Betancourt in Spain (Chap. 5) and Ioannis Papadakis in Greece (Chap. 8). The chapters concerning these five countries indicate that interactions between them continued after this first period. Indeed, the teaching of descriptive geometry quickly spread to new kinds of schools and was taught in universities such as in Italy and in Portugal (Chaps. 3 and 7). During this period, the situation in France changed as descriptive geometry became a subject of preparatory grades for public schools (Chap. 2). In all the southern countries, the dissemination of descriptive geometry was followed by a lasting establishment of the subject in the curricula, in the course of which many books by French authors were translated.

Translations from French continued after the middle of the century when new conceptions concerning the teaching of descriptive geometry attracted new audiences, especially those related to secondary teaching. Such is the case of the textbooks by Antoine Amiot, translated into Italian in 1875 (Chap. 3), and by Paul-Louis Cirodde, translated into Spanish in 1865 (Chap. 5). Translations from French are more rare in Portugal, where authors wrote new textbooks with the help of different foreign textbooks, and in Greece, where it was not necessary to translate because students had to learn French.

In the last quarter of the century, the political context changed considerably; however, circulation between the five countries continued with the teaching of projective geometry and the importance given to schools for engineers or for people involved in other technical professions. Then, translations from languages other than French emerged, particularly with the textbook by Luigi Cremona, which was translated from Italian into French two years after its original publication in 1875, into German in 1882 and into English in 1885. In Southern Europe countries—as well as in other European countries—descriptive geometry at secondary education level came closer to the teaching of projective geometry, and it was finally included in projective geometry at the turn of the century. But it remained an independent subject for students in the arts and architecture (Chap. 4) and for candidates of *Grandes écoles* in France until the year 1970 (Chap. 2).

The Establishment of Descriptive Geometry in Central and Eastern Europe
Like in France, there had existed some tradition of teaching in technical schools in Central and Eastern Europe, although in general at a low level. Together with the dissemination of descriptive geometry, these schools were upgraded in their level and in their disciplinary range. For instance, in German-speaking countries, different kinds of technical secondary schools existed, which then evolved into polytechnics. Different kinds of *Realanstalten* could be, on the other hand, considered as the predecessors of the *Technische Hochschulen* (Chap. 9). The situation in German-speaking countries was indeed complicated: even after the creation of the so-called second German empire, the states were autonomous concerning teaching.

In Germany and in Central Europe, it was quite usual to find methods in descriptive geometry different from Monge's projections and a tendency developed towards using the methods of projective geometry in descriptive geometry (Chap. 10); we also find—in some other cases—a series of modifications to avoid perceived difficulties with respect to the French tradition, in particular when teaching to architects and engineers. In Austria, the emphasis was initially on empirical methods, and, therefore, the geometric methods needed for architects and engineers were taught in courses such as engineering or architectural drawing. Later, a need for progress in scientific and technical fields brought an autonomous expansion of the teaching of descriptive geometry at the university level (Chap. 11). Descriptive geometry was one of the main subjects also in Czech universities, but Monge's projection method was not included, and, instead, we find theories of curves and surfaces, collineations and projective and kinematic geometry; this was rather the Viennese influence. A new method for orthogonal projection was elaborated by Skuherský (Chap. 16). In Czech (real) secondary schools in the 1870s, the descriptive geometry exam was one of the obligatory parts of the examination system and was very difficult in comparison to other contemporary requirements in secondary education.

New colleges inspired by the *École polytechnique* appeared in Denmark in the 1830s. The Danish Polytechnic College deserves particular attention for its high academic ideals. Hielmlev's geometry of reality raised descriptive geometry to a new and higher scientific level (Chap. 15). Indeed, this approach to geometry focused on the accuracy with which one can execute geometrical constructions in a systematic (theoretical) way. In the Netherlands, the lack of well-trained engineers during the eighteenth century was seen as the cause of different problems in the country: mainly the lack of military success but also the management of, or lack of, flood controls. Political changes brought new laws in education leading to the renewal of primary education, the creation of a sort of "modern" curriculum for secondary school and the creation of schools for the training of engineers and of military schools. All these schools had descriptive geometry in their programmes. Descriptive geometry—also seen as a means to facilitate the study of analytic geometry and to train intellect and imagination (Chap. 14)—became part of the final examinations in secondary schools here too (HBS).

Interesting is the role that the teaching of descriptive geometry had in Serbia; it started in the 1850s and lasted till the twenty-first century. This longevity was due to an interest in practical mathematics and to the absence of a Euclidean tradition. Politically, Serbia was influenced by the Habsburgs more than by the Ottomans and mathematically by the geometry of Fiedler, and descriptive geometry was seen as an important method to study geometry and mathematics (Chap. 17). In contrast to Serbia, the existence of a "local" geometric tradition in England was a hindrance to the development of descriptive geometry. Due also to the wars between England and France, the translation of texts on descriptive geometry and its introduction into the educational system happened only after 1840s. Moreover, in this period, England was an important contributor to applied mathematics, and its interest in practical geometry brought to life approaches like Farish's "isometric perspective". It is interesting to note that polytechnic schools did not fully develop there until the twentieth century although the first such school was founded in 1838 (Chap. 18).

One of the rare cases of a direct transmission of descriptive geometry from France to the continental countries is represented by Russia; indeed, one of its leading characters, Charles Poitier, was invited to Russia as a former student of the *École polytechnique* and started there the teaching of descriptive geometry. Although closely aligned with Monge, in his works, he enriched the theory of surfaces and their study and applications, particularly in projects related to shipbuilding and the building of machines (Chap. 13).

Descriptive Geometry in America and Africa A very interesting feature of descriptive geometry is the fact that it soon reached many faraway regions. In the USA, the teaching of descriptive geometry became established relatively slowly and had a major link to that of geometry in general and was even seen as its continuation. Its teaching was influenced by France at the beginning of the nineteenth century. Authors generally introduced the method of projections in a simplified way to adapt it to the new readership until it established its place in emerging institutions as a graphic art for the training of engineers (Chap. 19). In Latin America, with particular reference to Rio de Janeiro, we can again point to a French influence: Lacroix seemed to be preferred to Monge even if both were considered very difficult. A new terminology was introduced for didactical reasons (Chap. 21). The translation of a book on descriptive geometry (Chap. 20) in Egypt marked the French influence on the learning of geometry there and was followed by the expansion of the teaching of descriptive geometry in engineering schools and military academies.

We include, at the end of this introduction, a timeline showing—for each considered country—the dates of three events: the year of the first translation of Monge's work, the year of the first original text on descriptive geometry and the year of the creation of a school in the spirit of the *École polytechnique*. The timeline refers to countries, but, as explained above, it is not always possible to speak of countries in the nineteenth century, in particular German-speaking countries, Italy or the Balkans. The timeline allows an overview of the spread of descriptive geometry in the world and its link with the creation of new higher schools. The timeline begins in 1794 when the *École polytechnique* was founded in France to give a general (polytechnic) formation for a meritocratically constituted elite composed of administrators, engineers and the military. The names given to the different higher schools created in each country give the first interesting information on the varieties of these kind of schools, especially when their purpose was restricted to military academies. It is interesting to examine the time interval separating the date of the first textbooks on descriptive geometry (translation or original publication) from the date of the creation of the schools in each country. This time interval is short in many European countries like Spain (1 year), Italy (4 years), Portugal (Brazil) (1 year), Germany (3 years) or England (3 years) showing that the teaching of descriptive geometry was very much considered as complementing the creation of such new schools. Our timeline finishes with the date of the translation of Gaspard Monge's textbook in German in 1900, but, at this late date, the book was considered as an historical source.

Elements for a Comparison We may say that the focus of this book is on three main points: the role of descriptive geometry as it appears from curricula and textbooks, the educational point of view and the institutions in which descriptive geometry was taught. Various common or diverging aspects emerge when following the development of our subject in the various countries covered. We propose some transversal analysis that can help the reader to gain an overview of the international role played by descriptive geometry in the training of a "polytechnic" audience:

1. *The question of the transfer of the ideas from Monge's school and the École polytechnique.* As shown above, whereas in some regions and countries the French influence was strong (Italy, Spain, Greece, but also Russia, USA etc.) through either early translations of or the direct use of French texts, in others, it was rather weak (such as the German-speaking part of Europe and, in general, in Central and Eastern Europe) due to different teaching traditions and also to different developments in mathematical research. An encounter with Monge's text in its different local guises is, anyway, part of almost all the chapters of our book.

2. *The nature of the (poly)technical teaching system at a higher level that was created in the various countries.* The École polytechnique was probably seldom copied as an institution (see Chap. 22), but it was often seen as a model and a reason to introduce a new teaching system in a country. The new schools were of different types and at different levels and had the aim of training a heterogeneous audience ranging from engineers to military and to secondary teachers. But, they all had a common denominator: the presence of descriptive geometry among the teaching subjects.
3. Linked to the previous point is *the role played by descriptive geometry in the different institutions*. This is apparent not only in the "modern" schools mentioned above but also in universities—e.g. faculties of science—and in secondary schools with a rather humanistic tradition such as gymnasia and *lycées*. Indeed, it was usual to find descriptive geometry taught in the French lycées or in Austrian, Portuguese and Italian universities, while such introduction remained restricted in many German-speaking areas to a second emerging type of higher education: the (poly)technical colleges (later: the technical universities). So, whereas descriptive geometry was often presented as a method useful to future engineers or architects, there were also attempts to integrate it into theoretical geometry, in particular—in the second half of the nineteenth century—with projective geometry (to use its current name). In addition, descriptive geometry opened the way not only to analytical geometry but also to other mathematical subjects from kinematics to conformal differential geometry (Chap. 11). On the other hand, the teaching of descriptive geometry often needed to be complemented by special courses on drawing in order to meet the exact needs of architects, engineers and also the military. So, descriptive geometry came to be considered as a theory of this practical field. England went its own way by replacing descriptive geometry—considered as being rather theoretical—by its own substitute for it. The same considerations lead us to reflect on the disappearance of descriptive geometry first from the secondary curricula—with the exception of art schools—(only a few countries, such as Austria, France, Serbia and Switzerland maintained descriptive geometry as a subject of scholarly technical teaching far into the twentieth century) and then from polytechnics and universities, except for faculties of architecture.
4. The place of *drawing within the educational role of descriptive geometry*. According to Monge, the fundamental role of descriptive geometry and geometric drawing in polytechnic schools is to foster "the learners intelligence by giving them the habit and the feeling for precision".[2] This point of view—along with the improvement of spatial representation—was also shared by Cremona and by others. To quote just one of the many statements about this point, Stojanović (1899, p. 29, see Chap. 17) stated that "descriptive geometry is the best way to study geometrical forms; it demands ability of representation and description

---

[2]Cited from Monge, Gaspard. 1839. *Géométrie descriptive: suivi d'une théorie des ombres et de la perspective*, Vol. II, n. V, p. 160. Also see Chap. 1.

and gives an opportunity to develop sharp spirit and rigorous thinking". In some countries, such as in early nineteenth-century Spain but also in Germany and the Netherlands, the emphasis on drawing within elementary technical education (in primary schools as well as in drawing schools and schools of arts and crafts, etc.) can be seen as preparing the way for the teaching of descriptive geometry. In Denmark, it was the work of the painter Eckersberg that introduced descriptive geometry through the teaching of linear perspective at the Royal Academy of Arts. In Russia, Sevastianov published an "application of descriptive geometry to drawing", which he considered a wonderful guide for studying the basics of drawing (Chap. 13). Moreover, the teaching of descriptive geometry was often accompanied by hours devoted to drawing and was conducted in specialized classrooms for drawing with the necessary equipment; this corresponds to a methodological change requiring autonomous activities by the students. In the second part of the century, this practice evolved into the reform of geometry teaching and in the introduction of new pedagogies that favoured student activity and led to increased students' spatial perception. On the other hand, in some institutions, the teaching of descriptive geometry became completely separated from geometric drawing, such as in French *lycées* (Chap. 2).

5. Many contributions evidence the *problem of translating the mathematical or technical terminology linked to descriptive geometry* and the creation of a national terminology. In German countries, there was even the question of the translation of the term "descriptive" (Chap. 9). We find a "philosophical" division of descriptive geometry into morphogenesis and iconography in the teaching of Tilscher in Prague (Chap. 16). In Chap. 17, a page of a dictionary of Serbian terms is shown with the German and the French translation. In Egypt, the role of translators was also important in transferring European mathematical knowledge and adapting it to the local context; this same role was also played by the Egyptian students sent to France who then became authors of the first Egyptian books on descriptive geometry, such as Ibrahim Ramadan (Chap. 20). In Russia, Sevastianov, the translator of the works by Poitier, created a Russian terminology, which is still in use today, while in Brazil (Chap. 21), a new terminology was created for didactical reasons. In England, some confusion about the "generating principle" of descriptive geometry gave the term "rabatting" a prominent role with respect to projection planes. This in turn gave more importance to the graphical operations to be performed rather than finding the real form of the considered object (Chap. 18). In general, the notion of *rabattement* comes up in many chapters, often linked to questions of translation and of interpretation of its meaning.

We thank all the researchers who collaborated with their work and their comments in the preparation of this work: Pierre Ageron, Kristín Bjarnadóttir, Konstantinos Chatzis, Leo Corry, Alessandra Fiocca, Jeremy Gray, Alexander Karp, Ladislav Kvasz, Jérôme Laurentin, Maria Rosa Massa, Guillaume Moussard, Philippe Nabonnand, Karen Parshall, Hélder Pinto, João Bosco Pitombeira de Carvalho, Leo Rogers, Erhard Scholz, Harm Jan Smid and Robert Wengel.

Moreover, we thank the institutions that supported our work: Centre International de Rencontres Mathématiques SMF (France), University of Wuppertal, University of Nantes, Sapienza University of Rome and GDR CNRS Histoire des mathématiques (France).

Nantes Cedex, France
Rome, Italy
Wuppertal, Germany

Évelyne Barbin
Marta Menghini
Klaus Volkert

Timeline

Year	First translation of a book on descriptive geometry	First original textbook	First creation of a "Polytechnic" School
1794			France (Paris): École polytechnique
1799		France: G. Monge	
1802			Spain (Madrid): Escuela de Caminos y Canales
1802			USA (New-York): West-Point Military Academy
1803	Spain: G. Monge (from French)		
1805	Italy: G. Monge (from French)		
1806			Czech Lands (Prague): Königlich-böhmische ständische Lehranstalt
1807		Italy: V. Flauti	
1811			Italy (Naples): Scuola di Ponti e Strade; Austria (Graz): Polytechnicum Portugal and Brazil (Rio de Janeiro): Academia Real Militar
1812	Portugal and Brazil: G. Monge (from French)		
1816		Russia: Ch. Potier	
1819		Spain: M. Zorraquin	
1821	Netherlands: S. Lacroix (from French)	USA: C. Crozet	Egypt (Cairo): Muhandishana
1825			Germany (Karlsruhe): Polytechnische Schule Karlsruhe
1828		Germany: G. Schreiber	Greece (Kitsi): Military School of Evelpides
1829			Denmark (Copenhagen): Polyteknisk Læreanstalt
1836		Denmark: L. S. Kellner	
1837	Egypt: É. Duchesne (from French)		Portugal (Lisbon): Escola Polytechnica; Portugal (Porto): Academia Polytechnica
1838			England (London): London Polytechnic
1840		Netherlands: H. Strootman	
1841		England: R. T. G. Hall	

(continued)

Year	First translation of a book on descriptive geometry	First original textbook	First creation of a "Polytechnic" School
1842			Netherlands (Delft): Koninklijke Akademie
1845	Chile: C. F. A. Leroy (from French)	Austria: J. Hönig	
1848			Colombia: Colégio Militar
1850			Serbia (Belgrade): Artillery School (from which the Military Academy grew later)
1852		Egypt: I. Ramadan	
1862		Czech Lands: D. Rysavy	
1873		Portugal: J. F. d'Assa Castel-Branco	
1874		Serbia: E. Josimovic	
1882	Serbia: G. Berger (from German)		
1883		Greece: M. Kanellopoulos; Greece: A. Apostolou	
1887	Greece: C. Leroy (from French)		
1900	Germany: G. Monge (from French)		

Contents

Part I First Spreading in Southern Europe

1. **Monge's Descriptive Geometry: His Lessons and the Teachings Given by Lacroix and Hachette** 3
   Évelyne Barbin

2. **Descriptive Geometry in France: Circulation, Transformation, Recognition (1795–1905)** 19
   Évelyne Barbin

3. **Descriptive Geometry in Italy in the Nineteenth Century: Spread, Popularization, Teaching** 39
   Roberto Scoth

4. **Luigi Cremona and Wilhelm Fiedler: The Link Between Descriptive and Projective Geometry in Technical Instruction** 57
   Marta Menghini

5. **Descriptive Geometry in the Nineteenth-Century Spain: From Monge to Cirodde** 69
   Elena Ausejo

6. **Descriptive Geometry in Spain as an Example of the Emergence of the Late Modern Outlook on the Relationship Between Science and Technology** 81
   Ana Millán Gasca

7. **Portuguese Textbooks on Descriptive Geometry** 97
   Eliana Manuel Pinho, José Carlos Santos, and João Pedro Xavier

8. **In Pursuit of Monge's Ideal: The Introduction of Descriptive Geometry in the Educational Institutions of Greece During the Nineteenth Century** 113
   Christine Phili

Part II Installation of Descriptive Geometry in Europe

9   A German Interpreting of Descriptive Geometry
    and Polytechnic 139
    Nadine Benstein

10 Otto Wilhelm Fiedler and the Synthesis of Projective and
    Descriptive Geometry 167
    Klaus Volkert

11 The Evolution of Descriptive Geometry in Austria 181
    Hellmuth Stachel

12 The Vienna School of Descriptive Geometry 197
    Christa Binder

13 At the Crossroads of Two Engineering Cultures,
    or an Unedited Story of the French Polytechnician
    Charles Potier's Descriptive Geometry Books in Russia 211
    Dmitri Gouzevitch, Irina Gouzevitch, and Nikolaj Eliseev

14 Engineering Studies and Secondary Education: Descriptive
    Geometry in the Netherlands (1820–1960) 233
    Jenneke Krüger

15 The Rise and Fall of Descriptive Geometry in Denmark 255
    Jesper Lützen

16 Descriptive Geometry in Czech Technical
    Universities Before 1939 275
    Vlasta Moravcová

17 The Love Affair with Descriptive Geometry:
    Its History in Serbia 295
    Katarina Jevtić-Novaković and Snezana Lawrence

18 Descriptive Geometry in England: Lost in Translation 313
    Snezana Lawrence

Part III Descriptive Geometry in America and Africa

19 Teaching Descriptive Geometry in the United States
    (1817–1915): Circulation Among Military Engineers, Scholars,
    and Draftsmen 339
    Thomas Preveraud

20 The Teaching of Descriptive Geometry in Egypt 359
    Pascal Crozet

| 21 | The Dissemination of Descriptive Geometry in Latin America .... | 377 |

Gert Schubring, Vinicius Mendes, and Thiago Oliveira

Part IV   Epilogue

| 22 | The Myth of the Polytechnic School .... | 403 |

Gert Schubring

Author Index 423

Subject Index 431

About the Authors

Elena Ausejo is professor of the History of Science at the University of Zaragoza (Spain). She has been editor of Llull, the *Journal of the Spanish Society for the History of Science and Technology* (ISSN 0210-8615) since 2004. Her research focuses on the social history of mathematics in Spain. Her most recent papers deal with commercial arithmetic in Renaissance Spain (https://doi.org/10.1007/978-3-319-12030-0_9) and with the introduction of infinitesimal calculus in Spain (https://doi.org/10.3989/asclepio.2015.31).

Évelyne Barbin was full professor of Epistemology and History of Sciences at the University of Nantes (France). Since 2014, she is professor emeritus and member of the Laboratory of Mathematics Jean Leray CNRS UMR 6629. Her main themes of research are mathematics and sciences in seventeenth and nineteenth century; mathematical proof in history; history of algorithms, instruments and machines; and history of mathematics teaching from the nineteenth to twentieth centuries. She worked in the IREMs (Institute for Research in Mathematics Education) since 1975, and she is member of the IREM of Nantes. As convenor of the IREM National Committee "Epistemology and History of Mathematic", she organized 20 national colloquia, 8 interdisciplinary summer universities on epistemology and history in mathematics teaching and the first European Summer University (ESU) on this theme in 1993. She was chair of 8 ESU from 1996 to 2014. She published about 135 papers and 30 books or editions.

Nadine Benstein obtained her Master of Education degree in English, Mathematics and Educational Sciences at the University of Wuppertal in 2015, where she has been doing her doctoral studies in the field of history of mathematics since then. Her main field of research is the teaching of descriptive geometry in the nineteenth century in Germany.

Christa Binder studied Mathematics at the University of Vienna. She received a promotion in 1971 with a work on uniform distribution modulo 1. She was assistant at the Institute of mathematics, University of Vienna, till 1981, scientific official (wissenschaftliche Beamtin) at the University of Technology of Vienna from 1981 till retirement in 2011 and organizer of 13 symposia on the history of mathematics. She also teaches regular courses on history of mathematics for more than 30 years.

Her main interests are history of mathematics in Austria, history of number theory and history of geometry.

Pascal Crozet is director of research in the CNRS. He is currently the director of the SPHERE laboratory (UMR 7219, CNRS, Paris-Diderot University, University Paris 1). He is also head of the Centre d'Histoire des Sciences et des Philosophies Arabes et Médiévales since 2009. His work falls into two distinct fields of research: first the study of classical Arabic mathematics, mainly between the ninth and eleventh centuries, and to a lesser extent the history of the modernization of the exact sciences in Arab countries in the nineteenth century. His work on Egypt is based on a long stay he made in Cairo between 1990 and 1995 as a researcher at the CEDEJ.

Nikolaj Eliseev is doctor in technical sciences (Russia) and lecturer of the chair "Descriptive geometry and graphic" of the State University of Ways of Communication (Saint Petersburg). His specialization includes machines and equipment for construction and road building, and he is a qualified mechanical engineer. He teaches 3D modelling of the railroad constructions, infographics, descriptive geometry and engineering infographics. He is also director of the Museum of the State University of Ways of Communication.

Ana Millán Gasca is professor of Mathematics at Roma Tre University (Rome, Italy). She has been managing editor of Llull (*Journal of the Spanish Society for the History of Science and Technology*) and is corresponding member of the Académie Internationale d'Histoire des Sciences. Her research subjects are the history of applied mathematics, the role of scientific and technological knowledge and of scientific communities in processes of modernization and historical aspects of elementary mathematics education. Her books include *Numeri e forme* (Zanichelli, 2016), *Fabbriche, sistemi, organizazzioni, Storia dell'ingegneria industriale* (Springer, 2006) and, with Giorgio Israel, *Pensare in matematica* (Zanichelli, 2012), *The World as a Mathematical Game: John von Neumann and Twentieth Century Science* (Birkhäuser 2009) and *The Biology of Numbers* (Bikhäuser, 2003).

Dmitri Gouzevitch is doctor in technical sciences (Russia) and received a master diploma in History and Civilization (Centre d'études des mondes russe, caucasien et centre européen, Paris), historian in the École des hautes études en sciences sociales de l'EHESS. Dmitri's Domains of research are the rise of the engineering profession in Russia (eighteenth and nineteenth centuries) and in Western Europe (technical training, professional organizations, production of knowledge at cognitive practices, role played by technological transfer and cultural exchanges). He has more than 250 publications.

Irina Gouzevitch is doctor in history of technology (PhD, University of Paris VIII, defended in 2001) and researcher in the Centre Maurice Halbwachs, École des hautes études en sciences sociales (Paris). Her domains of research are: transfer, acculturation and circulation of scientific and technical knowledge at the eighteenth and nineteenth centuries, with a particular regard on the engineering and engineers, technical training, communication networks, mobility of experts, identity questions, intercultural exchanges between Russia and Western Europe, technoscientific policies in Russia and Spain in the eighteenth century and comparative study. She has more than 200 publications in 6 languages.

Katarina Jevtić-Novaković is a professor of Descriptive Geometry and Urban Studies at the Higher Engineering and Geodesic School in Belgrade, Serbia. She is one of the organizers and an active member of the Serbian scholarly group under the name of monGeometria, meeting biannually to discuss education in the fields of geometry, graphics and visual communication. Katarina has published a book on descriptive geometry and perspective which has seen several editions and is used as a textbook across the country (Serbia).

Jenneke Krüger researches the history of mathematics and of mathematics education and is involved in the development of didactics for interdisciplinary science education in secondary schools. She studied Biology at Utrecht University (MSc) and Mathematics at Flinders University (Adelaide), Birkbeck College (London) and the University of Groningen. Her PhD research was on the influences on Dutch mathematics curricula for secondary education, starting from the seventeenth century (actors and factors behind the mathematics curriculum since 1600). She works as an independent researcher but also occasionally with the Freudenthal Institute for Science and Mathematics Education (Utrecht University, Netherlands).

Snezana Lawrence is a senior lecturer in Mathematics Education at Anglia Ruskin University in Cambridge, England. Snezana has published on the history of mathematics and its relationship to education, as well as on the history of geometry, its application to architecture and its perceptions in popular culture. She writes a column for *Mathematics Today*, the largest professional magazine for mathematicians in the UK. She has co-edited a book with Mark McCartney *Mathematicians and Their Gods*, published by the Oxford University Press. Snezana leads the nationwide teacher development programme for the Prince's Teaching Institute (UK). She is a keen swimmer.

Jesper Lützen is professor of history of mathematics at the Department of Mathematical Sciences of the University of Copenhagen. He is the author of a biography of Joseph Liouville, a book about Heinrich Hertz's mechanics and a book on the prehistory of the theory of distributions. He is now working on different aspects of the history of impossibility theorems about which he plans to write a book.

Vinicius Mendes is professor at the *Universidade Federal Fluminense*; he received his master's and doctorate degree from *Universidade Federal do Rio de Janeiro*. He is a researcher in History of Mathematics with special emphasis on History of Mathematics in Brazil. He has worked on the formation of prospective mathematics teachers and has published about mathematics history and mathematics teaching.

Marta Menghini is associate professor in the Department of Mathematics of the University of Rome Sapienza. She is the author of numerous published works in the fields of mathematics education, of history of mathematics and of the history of mathematics education.

She was in the Scientific Committee and chaired the Organizing Committee of the International Symposium held in Rome in March 2008 "The First Century of the International Commission on Mathematical Instruction. Reflecting and Shaping the World of Mathematics Education" and edited the volume published on this occasion by Enciclopedia Italiana. In 2012, she held a Regular Lecture at ICME 12 in Seoul. She wrote, with Évelyne Barbin, the chapter on "History of Teaching Geometry" in the *International Handbook on the History of Mathematics Education*. She was involved in the translation and edition of Felix Klein's third volume on *Elementary Mathematics from a Higher Standpoint*, which appeared in 2016.

Vlasta Moravcová, née Chmelíková, has taught mathematics and descriptive geometry at a gymnasium in Prague since 2005. She was awarded the doctoral degree in General Problems of Mathematics and Computer Science from Charles University in Prague in 2016. Currently, she also works as a senior assistant professor in the Department of Mathematics Education at Charles University in Prague, Faculty of Mathematics and Physics. In her research, she focuses on the history of descriptive geometry and mathematics education, particularly geometry education. Since 2016, she has been a member of the Union of Czech Mathematicians and Physicists.

Thiago Oliveira is Assistant Professor at the State University of Rio de Janeiro and at the Military School of Rio de Janeiro. He graduated in Mathematics at the Federal University of Rio de Janeiro (2002), obtained the Master's Degree in Science and Mathematics Teaching at the Federal Center for Technological Education Celso Suckow da Fonseca (2007) and the Ph.D. in History of Sciences and Techniques and Epistemology at the Federal University of Rio de Janeiro (2016). He has experience in mathematics, with emphasis on mathematics teaching and history of mathematics, working mainly in the following subjects: educational technologies applied to mathematics teaching, history of mathematics, descriptive geometry and perspective.

Christine Phili, a disciple of René Taton, taught mathematics and history of mathematics at the National Technical University of Athens. Her research interests mainly lie in the history of nineteenth-century mathematics. She is the author of articles in several journals including *Archives Internationales d'Histoire des Sciences, Bollettino di Storia delle Scienze Matematiche, Istorico-Mathematicheskii*

Issledovania and *Proceedings of the Academy of Athens*, and her recent publications include *Sovereignty and Mathematics* and the *Ancient Roots of Modern Mathematics*. Her book on Cyparissos Stéphanos (1857–1917) the *International and the Greek Mathematical Community* is forthcoming.

Eliana Manuel Pinho is a researcher and an artist. She has an education in physics, mathematics and fine arts. For two decades, she is dedicated to the research and teaching in mathematics, mathematics and architecture, and mathematics and fine arts. She currently maintains an artistic practice in sculpture, installation, drawing and engraving and does independent research by following an artistic and scientific approach.

Thomas Preveraud teaches mathematics and history of mathematics at Université Lille Nord de France. He is a member of the Laboratoire de mathématiques de Lens at Université d'Artois and an associated research member of the Centre de recherches en histoire internationale et atlantique at the University of Nantes. His research is turned towards mathematics circulations between France and the USA during the nineteenth century. He focuses on translations and adaptations of French mathematics textbooks and mathematical journal articles into American education, research and social area.

José Carlos Santos has a PhD in Mathematics (Université Paris VII-Denis Diderot) and is an associate professor at the Faculty of Sciences of the University of Porto. He made research in representation theory and published textbooks in complex analysis, number systems and Lie groups and Lie algebras.

Gert Schubring is a retired member of the Institut für Didaktik der Mathematik, a research institute at Bielefeld University, and at present is visiting professor at the Universidade Federal do Rio de Janeiro (Brazil). His research interests focus on the history of mathematics and the sciences in the eighteenth and nineteenth centuries and their systemic interrelation with social-cultural systems. One of his specializations is history of mathematics education. He has published several books, among which is *Conflicts Between Generalization, Rigor and Intuition: Number Concepts Underlying the Development of Analysis in seventeenth–nineteenth century France and Germany* (New York, 2005). He was chief-editor of the *International Journal for the History of Mathematics Education*, from 2006 to 2015.

Roberto Scoth read at the University of Cagliari, obtaining a degree in Mathematics and a PhD in History, Philosophy and Didactics of Sciences. From 2012 to 2016, he was an assistant professor in the Department of Mathematics and Computer Science at the University of Cagliari. He has participated in numerous congress and conferences and is a member of SISM (Italian Society of History of Mathematics). He is mainly interested in the history of the teaching of mathematics in Italy and the spread of mathematics in Sardinia in the seventeenth to nineteenth centuries. His publications include *Secondary School Mathematics Teaching from the Early Nineteenth Century to the Mid-Twentieth Century in Italy* (with Livia Giacardi);

Handbook on the History of Mathematics Education, Springer, Berlin, New York (2014): 201–228; and *"Higher Education, Dissemination and Spread of the Mathematical Sciences in Sardinia* (1720–1848)", *Historia Mathematica* 43, no. 2 (2016): 172–193.

Hellmuth Stachel was born in 1942 in Graz, Austria, and he graduated in 1965 with the secondary school teacher accreditation in "Mathematics" and "Descriptive Geometry" at the University of Technology (TU) in Graz. In 1969, he got a PhD in Mathematics at the University of Graz. His academic career started in 1966 at the Institute of Geometry, TU Graz. After his habilitation in 1971 for geometry at TU Graz, he was appointed in 1978 as full professor for Applied Geometry at the University for Mining in Leoben, Austria. In the years 1980–2011, he worked as full professor for geometry at the TU Vienna. In 1991, he was elected corresponding member of the Austrian Academy of Sciences. In 2010, he received a honorary doctorate from the Technische Universität Dresden, Germany. From 1996 till recent, he is the editor in chief of the *Journal for Geometry and Graphics*. His list of publications contains today 13 books and about 147 scientific articles.

Klaus Volkert is a full professor at the University of Wuppertal; he is lecturing on mathematics, didactics and history. His interests are mainly in the history of geometry and topology; recently he started a project on Wilhelm Fiedler. His is editor in chief of the journal *Mathematische Semesterberichte* (together with J. Steuding) and editor of the series *Mathematik im Kontext* (with David Rowe).

João Pedro Xavier has a PhD in Architecture (University of Porto) and is associate professor with habilitation at the Faculty of the Architecture of University of Porto. He worked in Álvaro Siza's office from 1986 to 1999. At the same time, he established his own practice as an architect. Xavier has always been interested in the relationship between architecture and mathematics, especially geometry (perspective and descriptive geometry). He published several books, works and papers on the subject and presented conferences and lectures. He is correspondent editor of the *Nexus Network Journal* and member of the executive board of Resdomus.

Part I
First Spreading in Southern Europe

Chapter 1
Monge's Descriptive Geometry: His Lessons and the Teachings Given by Lacroix and Hachette

Évelyne Barbin

Abstract After the French Revolution, the reorganization of teaching in France concerned schools for civil and military engineers, for teachers, and for workers. Gaspard Monge was involved in the projects to create these schools and he proposed what he called "descriptive geometry" for all of them. In 1794–1795, he gave two different courses, one in the *École normale de l'an III*, devoted to future teachers, and the other one in the *École centrale des travaux publics*, which will become the *École polytechnique*, devoted to engineers and the military. Monge's famous textbook *Géométrie descriptive* of 1799 is a transcription of the oral *Leçons* given in the *École normale*. We analyze the *Leçons*, taking into account the context of this school. Then we examine teaching given by students of Monge, Sylvestre-François Lacroix and Jean-Nicolas Pierre Hachette, in the contexts of the *École centrale des Quatre-Nations*, the *École polytechnique*, and the *Faculté des sciences de Paris*.

Keywords Descriptive geometry · Method of projections · Teaching of geometry · Secondary school · School for engineers · Gaspard Monge · Jean-Nicolas Pierre Hachette · Sylvestre-François Lacroix · *École polytechnique* · *École normale de l'an III* · *École centrale des Quatre-Nations* · *Faculté des sciences de Paris*

1 Monge's Descriptive Geometry: from a Method to Teaching (1785–1795)

In the historiography, the birth of descriptive geometry by Gaspard Monge has been linked with the *École polytechnique* and generally with the training of engineers (Deforge 1981; Sakarovitch 1998; Belhoste 2003). But, for a better

É. Barbin (✉)
UFR Sciences et Techniques, Nantes, France
e-mail: evelyne.barbin@wanadoo.fr

© Springer Nature Switzerland AG 2019
É. Barbin et al. (eds.), *Descriptive Geometry, The Spread of a Polytechnic Art*, International Studies in the History of Mathematics and its Teaching, https://doi.org/10.1007/978-3-030-14808-9_1

understanding of the *Géométrie descriptive* of 1799, a textbook that will become an international reference, it is necessary to situate descriptive geometry in Monge's aim on education in the years 1793–1795. Indeed, Monge developed the method of projections in the context of the *École royale du génie de Mézières* from 1764, but he introduced the term "descriptive geometry" for the first time in a text of 1793 on teaching in "secondary schools".

In the eighteenth century, mathematics and drawing were the most important teaching subjects in schools for engineers, like in the *École des ponts et chaussées*, created in 1747 (Michel 1981), and in the *École royale du génie de Mézières*, created in 1748 (Belhoste et al. 1990, pp. 53–109). The start of the method of projections was taught in this second school, around 1760, when its founder Nicolas de Chastillon considered that the teaching of Stonemasonry had to be taken as a basis for all subjects, which require vision in space, like the design of fortifications (Belhoste 1990). It consists of representing a solid by two projections on two perpendicular planes. Monge systematized this conception to unify and generalize practices of representations invented by artisans, architects, and engineers, when he became a teacher of mathematics in the *École royale du génie de Mézières*, but also teacher of drawing and stonemasonry (Brisson 1818; Dupin 1819; Arago 1854; Taton 1954; Bret 2007). In a paper of 1785, like Chastillon, Monge introduced the idea of projection to his students by means of shadows (Olivier 1847, pp. 26–35). He stated a "general problem", that is to find the shadow of a solid, when the direction of the rays of light are given and when the solid and the surface are given by their dimensions and positions. He explained that we have to construct projections on two perpendicular planes, one horizontal and the other vertical. In the first example, it is asked to determine the shadow of a cube given by its position on a horizontal plane: $BCDE$ and $IFGH$ are the horizontal and vertical projections of the cube, YA and ya are those of the ray of light, and KN is the "basis of elevation" (that will be named "line of ground" later) (Olivier 1847, p. 31) (Fig. 1.1). The problem and the reasoning are similar to those of Albrecht Dürer in his *Underweysung der messung* of 1538 where he introduced two planes where the shadows of the cube are represented: the lowered plane ef and the raised plane the $efgh$ (Dürer 1538, Fig. 52) (Fig. 1.2).

The term "descriptive geometry" occurred for the first time in September 1793 in a text titled "Object of the studies in the schools for artists and workers of various kinds", where Monge wrote:

> The order of knowledge, which is in question here, is founded on a particular three-dimensional geometry, for which a well-constructed treatise does not exist; for a rigorous purely descriptive geometry; and the purpose of which is to represent three-dimensional objects by two-dimensional drawings. This art is like a common language for heads of workshops who organize the works, and for workers who have to execute them (Taton 1992, p. 579).[1]

---

[1] All the translations of quotations are made by Évelyne Barbin.

1 Monge's Descriptive Geometry

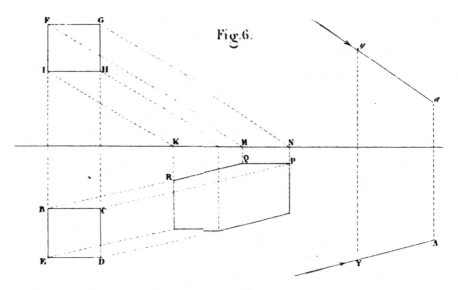

Fig. 1.1 The shadow of a cube in Monge's paper of 1785 (Olivier 1847, p. 31)

Then the term appears in 1794 in the projects of the *École normale de l'an III*, devoted to future teachers, and of the *École centrale des travaux publics*, which will become the *École polytechnique*, devoted to engineers and the military. The first lessons on "descriptive geometry" took place in the two schools in 1794–1795.

In the text of 28th of September 1794, "Developments on Teaching for the *École Centrale des Travaux Publics*", Monge presented the teaching of mathematics (Langins 1980). It was given over 2 years and it was divided into two parts, in such a way that the three-dimensional geometry received two treatments, by "analysis" and by "descriptive geometry", both taught by Monge (Table 1.1). Descriptive geometry had to occupy an important place: in the first year, "one will teach the general rules and the methods of descriptive geometry and, to make them familiar, the teachers will apply them to the drawings of stonemasonry, carpentry, to the rigorous determination of shadows in drawings, practice of linear perspective, [...] description of the main machines used in public works" (Langins 1980). On 6 days (among 10 days) of the first year, the students had a lesson of 1 h on descriptive geometry and made drawings during the following 5 h.

We only have the programme of Monge's *Leçons* given in 1794–1795 in the *École polytechnique*. From the preface of the *Traité de géométrie descriptive* of Jean Nicolas Pierre Hachette, who was his assistant, we know that Monge gave 24 lessons (Hachette 1822, pp. viii-x). The first lesson gave an exposition of the method of projections. Then, from *Leçon* 2 to *Leçon* 5, the topic was curved surfaces with their tangent and normal planes, the intersections of curved surfaces, the generations of curved surfaces and the curvatures. The applications of the method began with *Leçon* 6. There were three lessons on shadows, aerial perspective, linear perspective,

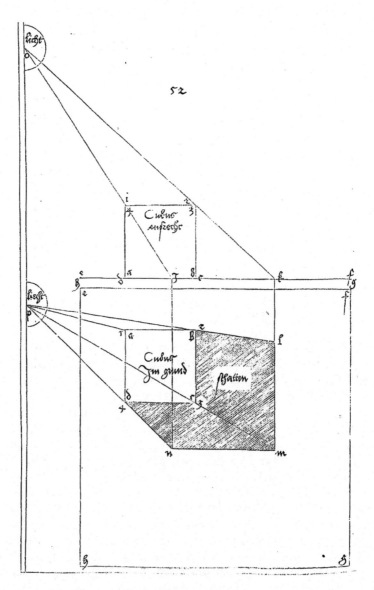

Fig. 1.2 The shadow of a cube by Dürer (1538, n. p., fig. 52)

three lessons on stereotomy, four lessons on art of carpentry, four lessons on topography, and four lessons on mechanisms. Monge edited his teaching on analysis in 1801 and in 1809, under the title of "analysis applied to geometry", which shows the importance for him to link these two domains (Monge 1801, 1809).

The spirit of the lessons given in the *École normale* was different because the audience consisted of future teachers, of the French *écoles normales* (normal schools) that trained all the teachers (Julia 2016, pp. 83–104), in order to make them able to teach a new geometry, and the context is not the same. Firstly, the

Table 1.1 Organization of mathematical teaching in the *École polytechnique* (1795)

Mathematics	Analysis	1st year	2nd year
		Three-dimensional geometry, mechanics, calculus for machines	
	Graphical description	Descriptive geometry	Stereotomy, architecture, fortifications
		Art of drawing	

public was not a little group of 20 students but it was an amphitheater of about 1400 students, more or less advanced in mathematics. Secondly, because of the historical context, the teaching had to be reduced to 4 months. Thirdly, the organizers of the *École normale* decided that the teaching had not to be read but improvised, in such a way that the *Leçons* were presented as an invention. The lessons were taken in stenography and then corrected by the professor. There were three kinds of sessions: lessons on sciences, lessons on "the science of teaching in itself", and discussions (Julia 2013). Fourthly, the teaching had to be "elementary" (Julia 2016, pp. 371–421).

In 1798, Monge left France for Rome and then for Egypt. The book *Géométrie descriptive* of 1799, prepared by Hachette, contains the oral teaching given by Monge without the structure in lessons and discussions and without the four last lessons given in this school (Monge 1799). In the fourth edition of the book, in 1820, Barnabé Brisson, a former student of the *École polytechnique* and engineer of *Ponts et Chaussées*, added a theory of shadows and perspective from lessons given by Monge (Monge 1820, pp. 137–188).

2 The *Leçons* in the *École normale de l'an III*

We analyze the *Leçons* given in the *École normale*, republished in 1992 (Monge 1992) and not the books of 1799 and 1820. In the "Programme" of the *Leçons*, Monge explained that descriptive geometry has two objects in view:

> The first one is to represent with exactness three-dimensional objects on drawings that have two dimensions only, and which can admit a rigorous definition. From this point of view, it is a necessary language to the man of genius who conceives a project, to those who have to supervise the carrying out, and finally to the craftsmen themselves who have to execute the different parts. The second object of descriptive geometry is to deduce from the exact description of bodies all that necessarily follows from their forms and respective positions. In this sense, it is a way to research truth; it offers perpetual examples of a passage way from known to unknown; and because it is always applied to objects open to the greatest obviousness, it is necessary to introduce it in national education (Monge 1992, pp. 305–306).

These two aims assign to descriptive geometry several kinds of audience and permit many possibilities for its presentation and its contents. But also, they install inside descriptive geometry the tension that exists between theoretical and applied mathematics.

In *Leçon* 1, Monge chose the order of invention and began with an inaugural problem, whose statement is simple: "how to determine the position of a point in space?". Monge examined four answers: the point can be determined by its distances to three points, to three straight lines, to three planes, and finally to two planes. So, he went from the simplest figures of geometry to others (from the points to the planes), but he noted that the last answer gives the simplest result. Indeed, the other answers need to introduce intersection of spheres and cylinders. In this manner, he motivated the "method of projections", which is introduced in *Leçon* 2 only. In *Leçon* 2, Monge introduced the projection of a straight line on a plane and then on two planes, to explain that these two projections completely determine the line, independently of the position of the "planes of projection". Moreover, as the artists who use the method are familiar with the horizontal plane and the direction of a plumb line, we suppose that the two planes are perpendicular. He added that, to have the drawings of the two projections on only one sheet of paper, the artists conceived that the vertical plane has to turn around its intersection with the horizontal plane, to descend upon the horizontal plane. So, Monge stressed the importance to draw this intersection precisely. Finally, he pointed out that this arrangement has the advantage to shorten the work on projections. Indeed, if a and a' are the projections of a point A, if the plane containing Aa and Aa' cuts the intersection in C, and if a'' is the point obtained when the vertical plane turns, then the three points a, C, a'' are aligned (Fig. 1.3).

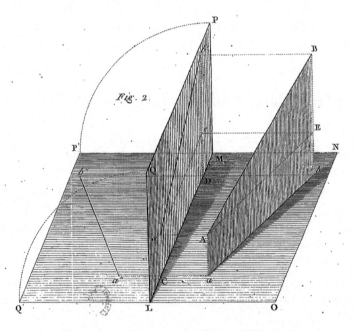

Fig. 1.3 The method of projection in Monge's *Leçons* (Monge 1798, plate I)

The two first discussions concern the "order of knowledge" and the idea of simplicity. In Monge's *Leçons*, there are 83 occurrences of the word "simple" and its derivatives, and 25 of them appear in these two first lessons (Barbin 2015). In the first discussion, a student affirmed that the definitions of the points, lines, and planes are not rigorous in the "ordinary elements of geometry" and he proposed a definition of the straight line using the notion of distance. Monge answered that we have to choose "the most simple and easiest property of this object" in its definition (Monge 1992, p. 319), but the definition has to be fertile also. In the second discussion, a student questioned about the necessity to confront difficulties with spheres, cylinders, and planes before coming to the definition of the projections. Monge answered that he had to show the simplicity of the method of projections and that he "did not want to lack an occasion to give a beautiful lesson of geometry" (Monge 1992, p. 322). So, the order of simplicity of figures—usually adopted when treating the elements of geometry—is not the best one. Monge explained that in geometry, curved surfaces have to be seen as classified in families, according to their own generations.

In *Leçon* 3, Monge presented the general "generation of surfaces", considered by him as "the complement of the method of projections", because the method of projections applied to surfaces needs a specific notion of surface, which is that every curved surface has to be considered as generated by the movement of a curved line. He explained that, for a given surface, we have to choose the simplest generating curve and to consider two generating curves, "as a long use taught to us" (Monge 1992, p. 328). As for the other chapters, he continued with problems that have two aims: to practice the method and to make progress. It is remarkable that in the "first general problem", Monge presented the intersection of two surfaces in the more general situation of any general surfaces, by a method that consists of intersecting the two surfaces by a system of horizontal planes. In the same manner, in the "second general problem" on the research of the tangent of the intersection of two curved surfaces, he stated generally that "the projection of the tangent of a curve of double curvature is itself a tangent to the projection of the curve, and its point of contact is the projection of that of the curve of double curvature" (Monge 1992, p. 371).

One important purpose of Monge is to show the union between descriptive geometry and analysis to students: "it would be desirable that these two sciences would be cultivated together" because descriptive geometry would bring its character of obviousness and analysis would bring its character of generality. He stressed the correspondence between descriptive geometry and analysis:

> To learn mathematics in the most advantageous way, it is necessary that the student becomes accustomed to feel the correspondence between the operations of analysis and those of geometry early; he has to be able to write all the movements that he can conceive in space by analysis, on the one hand, and to continually represent the moving spectacle of the analytical operations in space, on the other hand (Monge 1992, p. 367).

For instance, in *Leçon* 6 on the intersections of curved surfaces, Monge showed the analogy between the algebraic elimination and the operations by which one determines the intersection of surfaces. From his first paper on developable surfaces of 1771, his analytical works on surfaces are strongly linked to the method of projections and to the characteristics of the surfaces produced by machines (Dupin 1819, pp. 107–123). A major paper, on the classification of the curved surfaces by their way of generation, appeared in the *Mémoires de l'Académie des sciences de Turin* (1784–1785) (Monge 1784–1785). From this point of view, it is interesting to remark that problems on curvatures of spatial curves were treated by means of geometry in *Leçon* 8, with the introduction of a notion of pole of an arch (Fig. 1.4). But, in a paper of the Journal of the *École polytechnique* in 1795, Monge showed how the research of lines of curvature of an ellipsoid needs both analysis and geometry. He justified this research as a very useful tool for artists, for example, to divide a vault in voussoirs (Monge 1801, p. 148).

Since descriptive geometry has to become a principal part of the national education system, Monge gave some "useful" examples to show that this geometry can provide analysis in many questions. So, its teaching could replace ordinary geometry, but also analysis, in secondary schools. The first example is the "problem of the sphere", that is to find the center and the radius of a sphere for which four points in space are given. Monge wrote that the teaching of descriptive geometry for secondary schools should stop with *Leçon* 7. But, he continued in *Leçons* 8 and 9 with curvatures, evolutes, and involutes of curves. The three last lessons, published by Brisson in the 5th edition in 1827, concern the "useful things for craftsmen": the theory of shadows, considered as a complement of descriptive geometry rather than an application, and perspective.

The reading of the *Leçons* indicates that descriptive geometry requires newness with regard to the "ordinary elements of geometry" on at least two points: the priority given to the objects of space by "the method of projections" and the introduction of movement in geometry. The first point immediately needs to treat spatial geometry. The second one corresponds to a new way to associate geometry and analysis. Everywhere, Monge followed "an order of invention", which offers an important role to the resolution of problems in teaching. But above all, we have to remark that, in many places, he pointed out the relations between objects more than the objects themselves. This is the case for surfaces, which are classified by families according to their way of generation.

3 The Teachings of Descriptive Geometry by Lacroix and Hachette

Sylvestre-François Lacroix and Jean Nicolas Pierre Hachette were followers of Monge, they had been initiated to the method of projections early and they were his "assistants" for the *Leçons* in the *École normale*. Both were born in the beginning of the 1760s. Both edited textbooks considered by them as "Elements of descriptive geometry", in 1795 and in 1817. Their reading permits us to compare two ways of implementing Monge's educational project of a new geometry. The contexts of their

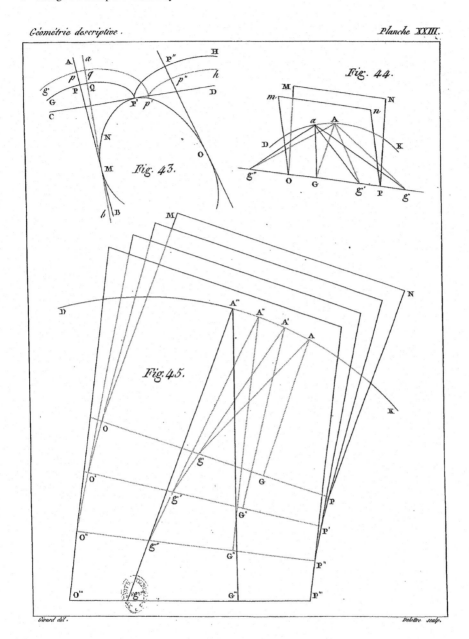

Fig. 1.4 Spatial curves in Monge's *Leçon* 8 (Monge 1798, plate XXIII)

writings were different: Lacroix was already an experienced teacher in many fields of mathematics (Schübring 1987; Ehrhardt 2009), while Hachette was interested in the theory and applications of descriptive geometry, he taught these subjects in the *École polytechnique* after Monge and he was author of many papers.

In 1795, Lacroix published a textbook *Essais de géométrie sur les plans et les surfaces courbes (ou Éléments de géometrie descriptive)*, the purpose of which was to present geometry in the framework of Monge's methods. Then, in 1799, he published his *Éléments de géométrie*, intended for the students of the *École Centrale des Quatre Nations* created in Paris in 1795 by the French Revolution to replace the ancient *Collèges*. In 1802, the Essais had been included in the *Compléments des éléments de géométrie*, in accordance with Lacroix's view of descriptive geometry as a sequel of geometry. The date of the publication of Lacroix's *Essais* incited historians to speak about "plagiarism" towards Monge (Belhoste 1992), but this word is incorrect because Lacroix was his former student, who quoted the master when he gave his solutions for problems (Lacroix 1795, p. 29). Moreover, Hachette wrote that Lacroix had written the major part of his book before 1795 (Hachette 1822, p. x). This can explain that the mention of "descriptive geometry" appeared in the subtitle as an addition, because, as we mentioned, the term appeared in 1793 and became public in 1794 only.

In the beginning of the preface, Lacroix expressed the correspondence between theorems of plane and spatial geometries and illustrated it with an inaugural problem, the "problem of the center of the sphere", also presented in Monge's *Leçons*. For him, this example shows that the "Elements of geometry" are incomplete and his textbook is a continuation of Euclid's Elements or of the "excellent treatise of geometry" of the "Citoyen Legendre" (Lacroix 1795, p. v). The aim is to present the "Elements" of a new geometry which follows the previous ones, useful for those who don't want a purely geometrical knowledge only but also want to apply it to the arts.

The textbook presented a sequence of problems ordered by deduction: to solve one problem needs to solve previous ones. The *Essais* are ordered in two parts—part I on the planes and the sphere and part II on the "generation of surfaces"—while Monge's *Leçons* considered many kinds of surfaces at once. Moreover, Lacroix followed an order of simplicity of figures, where he treated points and straight lines in a plane and then in a space, before the spheres. Contrary to Monge, he tackled part II with particular kinds of surfaces (conical surfaces, cylindrical surfaces, etc.) before coming to the general conception of surfaces. Thus, the spirit of generality of Monge is not very present. It seems that Lacroix had the project to treat the parallelism between geometry and analysis into his *Traité du calcul différentiel et du calcul intégral*, edited in 1797 after several years of preparation (Belhoste 1992, p. 568). Hachette was the "assistant" of Monge in the *École polytechnique* in 1794, where he became professor and developed an important activity until his dismissal in 1816 (De Sylvestre 1834). He was a teacher, but also an active mathematician. He wrote numerous papers on descriptive geometry and its applications (sometimes with Monge). He also wrote on machines and gears and he published his *Traité élémentaire des machines* in 1811. From 1804 to 1816, he published the *Correspondance sur l'École polytechnique* where he wrote historical papers on this school.

Hachette added a supplement to a new edition in 1811 of the *Géométrie descriptive of Monge* (Monge 1811), which had been published only in 1812 under the title *Supplément de la Géométrie descriptive de M. Monge* (Hachette 1812). At this period, Hachette was already professor of the *Faculté des sciences* de Paris. In this supplement, he studied the difficult problems on intersections of surfaces and tangents to surfaces. To treat this last problem, in 1817 he introduced what he will call "ruled surfaces" and the "synthetic method of tangents". The *Supplément* studied the general case of such surfaces generated by a mobile straight line, which leans on three given guiding curves named *directrix* (guiding line) (Hachette 1812, p. 1). A special case is the one of the hyperboloid of one sheet where the guiding curves are themselves straight lines. Hachette began to determine the plane tangent to the hyperboloid at a given point P. He called A, B, C three directrices of the mobile line, and A', B', C' three other lines of the hyperboloid leaned on the others, such that A' goes through P. Then he imagined A', B', C' as directrices and A, B, C as three positions of the mobile line, which leans on the new directrix. Consequently, if the plane spanned by P and the line B' cuts the line C' in point P', then the line PP' is on the hyperboloid and the plane determined by PP' and the line A' is the desired plane. To deduce the tangent plane of a ruled surface, Hachette explained that we can consider this surface as the envelope of a one-sheet hyperboloid whose three directrices are located in three planes that touch the surface in three points of a same straight line. The method of tangents had been praised by the mathematicians Arago and Legendre in a report at the *Académie royale des sciences* in Annales (1816, pp. 422–423).

When his *Éléments de géométrie à trois dimensions. Théorie des lignes et des surfaces courbes* appeared in 1817, Hachette taught descriptive geometry in the *Faculté des sciences* of Paris. He taught descriptive geometry in this *Faculté* until his death in 1834; after this date, his teaching was replaced by a teaching on probability (Bulletin Universitaire 1837, p. 68). His textbook contains two parts: a synthetic part, on the "Theory of lines and curved surfaces", and an algebraic part, on "Surfaces of second degree". In his foreword, the author explained that he wanted to bring the ancient and the modern geometries together and to write a continuation to the ancient geometry. His "geometry in three dimensions" includes descriptive geometry "to give a new growth to the arts, which are the principal source of the public prosperity" (Hachette 1817, p. viii). The preliminaries of the *Éléments* end with what Hachette called "the operations of descriptive geometry":

> It is very remarkable that the number of the graphical constructions of descriptive geometry is reduced to two: 1° to construct the distance of two points from which we have the projections; 2° to construct the point of intersection of a plane, which goes through three given points, and a straight line that goes through two given points. It is as well as all the operations on the numbers can be reduced to the practice of the common rules of the Arithmetic (Hachette 1817, p. 10).

This order of problems is linked with the practice of drawing, because when one has to construct a problem, it is better to decompose it into simple operations than to review a long sequence of problems. Moreover, Hachette's *Éléments* are presented as a sequence of short numbered paragraphs organized around drawings for better understanding the generation and chain of ideas. So, in their Elements, both Lacroix and Hachette adopted an "order of invention". In particular, we note the presence of an inaugural problem to motivate the method of projections. But the orders of problems are not the same: the will of generality guided Monge's order, the deductive way organized Lacroix's one, while Hachette isolated simple problems, into which the others can be composed (Barbin 2015).

Like Monge and contrary to Lacroix, Hachette introduced surfaces with the general conception of a surface as generated by the motion of a mobile curve: "if we consider a surface as the locus of a mobile curve, of which the form is constant or variable, the law of the motion of this curve determines those curves which link the points of the surface between them. This mobile curve is called the generator of the surface" (Hachette 1817, p. 11). From this point of view, we have to be attentive to the titles of the textbooks of Lacroix and Hachette: the first one indicates "planes and curved surfaces" and the second one "curved lines and surfaces". Hachette examined the surfaces of revolution, the developable surfaces, and the ruled surfaces. He completed the synthetic method by deducing the tangent plane to any surface or curve (Hachette 1817, pp. 58–62). Examples are given in the *Second supplément de la géométrie descriptive* of 1818, like the one for the ellipse (Fig. 1.5) (Hachette 1818, pp. 4–6). The notion of ruled surfaces will be deepened in the *Géométrie descriptive. Traité des surfaces réglées of Gaston Dascheau* (Dascheau 1828).

Hachette's *Éléments* can be associated with his teaching in the *Faculté des sciences de Paris*, as mentioned on the cover of his textbook (Hachette 1817), while his *Traité de géométrie descriptive* of 1822 corresponds to his teaching in the *École polytechnique* (Barbin 2015). The *Traité* contains applications to shadows, perspective, and stereotomy. It begins with a history of the method of projections, linked with the practices of engineers, and an account of Monge's life and his lessons in the *École polytechnique*. In the last sentence of his long preface, Hachette wrote that "he hopes that this book will help to propagate a fertile doctrine in its applications, which is the fruit of 20 years of working in the School, whose celebrity proves the services given to arts and sciences" (Hachette 1822, p. xix). The textbook contains a part on "pure descriptive geometry", with a rational part and a technical part, which is the art of representing points and lines of the space on paper. Hachette stressed the construction and the quality of drawings and, in an appendix, he gave instructions on the drawing for cutting stones. He wrote in the preface that, immediately after the lesson on descriptive geometry. Students of the *École polytechnique* made drawings with the help of models in wood for the timber structure or in plaster for the cutting of stones. In the *Traité*, Hachette gave a general definition of surfaces, as generated by the motion of a curve, which is considered as a main feature of the new geometry:

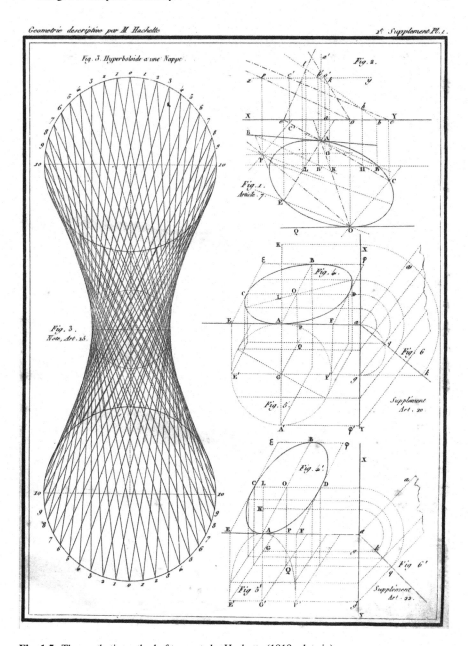

Fig. 1.5 The synthetic method of tangents by Hachette (1818, plate iv)

> The geometers express the nature of a surface by a relation between the three distances of any point of this surface to three given orthogonal planes, and this relation determines the position of the surface in relation with this planes. In descriptive geometry, we consider a surface as the locus of a mobile curve, of which the form is constant or variable and given at each instant; the law of the motion of this curve determines the form and the position of the surface: the mobile curve is called the generator of the surface (Hachette 1822, p. 23).

Then he examined surfaces of revolution, developable surfaces, and ruled surfaces.

4 Conclusion: Two Purposes, Two Kinds of Schools, One New Geometry

As Monge wrote in the "Programme" of his *Leçons*, descriptive geometry has two purposes: the first one is to represent three-dimensional objects by drawings in two dimensions, the second one is to research truth in geometry. These two purposes had their place in national education: the first one is essential for workers and the second one concerns all the students (Monge 1992, pp. 305–306). The construction of problems can be seen as essential for these two purposes, as part of their activity for engineers and workers, and as an educational task for beginners. But, from the beginning, each purpose of descriptive geometry had been favored by each of the two different institutions where Monge taught. The first purpose is inscribed in the *École polytechnique*, for preparatory teaching with various applications to train engineers, and the second one in the *École normale*, for a new teaching of geometry to enable the opening of the new system of public education and to train future teachers of schools.

Moreover, from the *Leçons* for the *École normale*, it appears that descriptive geometry is a new geometry, which can be learned without its applications for engineers. Lacroix and Hachette used this way in their "Elements". But Hachette kept Monge's first purpose well in his mind. So, the role of drawings in his teaching explains that he did not adopt the same order for problems as Lacroix. Also, the treatment of surfaces was different for the two authors. Hachette conceived a synthetic method of tangents, which is typical of Monge's idea of generating surfaces. Easier examples of using descriptive geometry to solve geometrical problems will be given soon, like in the *Annales de Gergonne* of 1816, where Coste has shown, for instance, how to describe a parabola that touches four straight lines (Coste 1816).

With regard to the reception of Monge's descriptive geometry, an important historical fact is that only his Lessons in the *École normale* had been edited, while his name is more associated with the *École polytechnique*. In the context of the *École normale*, the oral teaching of descriptive geometry was followed by pedagogical and philosophical conceptions, which will be associated with this new domain (Barbin 2015). Firstly, this teaching adopts an "order of invention" which consists of posing problems and then to introduce tools to solve them. Secondly, it presents geometric objects, not in the order of simplicity, starting from those of in the plane to those in space, but immediately the second. Thirdly, it expresses a will for generalization, especially for the "generation of surfaces". Fourthly, it promotes the union of descriptive geometry and analysis. The alternative left to the successors will be to keep or to leave these conceptions.

References

Arago, Louis. 1854. Bibliographie de Gaspard Monge. *Mémoires de l'Académie des Sciences de l'Institut de Paris* 24: xxiii–lxxiii.
Arago, Louis, Legendre, Adrien-Marie. 1816. Mémoire de M. Hachette, sur la théorie des lignes et des surfaces courbes. Mémoire de l'Académie royale des sciences, tome I. Paris: Firmin Didot.
Barbin, Évelyne. 2015. Descriptive geometry in France: history of the elementation of a method (1795–1865). *International Journal for the History of Mathematics Education* 10: 39–81.
Belhoste, Bruno. 1990. Du dessin d'ingénieur à la géométrie descriptive. In *Extenso* (June): 103–155.
———. 1992. Sylvestre-François Lacroix et la géométrie descriptive. In *L'école normale de l'an III. Leçons de mathématiques,* ed. Jean Dhombres, 564–568. Paris: Dunod.
———. 2003. *La formation d'une technocratie*. Paris: Belin.
Belhoste, Bruno, Antoine Picon, and Joël Sakarovitch. 1990. Les exercices dans les Écoles de l'ancien régime. *Histoire de l'Éducation* 46: 53–109.
Bret, Patrice. 2007. Les études sur Monge. *Bulletin de la Sabix* 41: 39–44.
Brisson, Barnabé. 1818. *Notice historique sur Gaspard Monge*. Paris: Plancher.
Bulletin Universitaire. 1837. Volume IV. Paris: Imprimerie Royale.
Coste, M. 1816. Géométrie descriptive. Application de la méthode des projections à la resolution de quelques problèmes de géométrie plane. *Annales de mathématiques pures et appliquées* 7: 304–310.
Dascheau, Gaston. 1828. *Géométrie descriptive. Traité des surfaces réglées*. Paris: Bachelier.
Deforge, Yves. 1981. *Le graphisme technique, son histoire et son enseignement*. Seysell: Champ-Vallon.
De Sylvestre, Antoine-Isaac. 1834. *Discours prononcé le 18 janvier 1834*. Paris: Huzard.
Dupin, Charles. 1819. *Essai historique sur les services et les travaux scientifiques de Gaspard Monge*. Paris: Bachelier.
Dürer, Albrecht. 1538. *Underweysung der messung mit dem Zirckel und richtscheyt*. Nuremberg: Formschneider.
Ehrhardt, Caroline. 2009. L'identité sociale d'un mathématicien et enseignant. *Histoire de l'éducation* 123: 5–43.
Hachette, Jean Nicolas. 1812. *Second supplément de la géométrie descriptive*. Paris: Klostermann.
———. 1817. *Éléments de géométrie à trois dimensions*. Paris: Courcier.
———. 1818. *Supplément de la géométrie descriptive par M. Hachette*. Paris: Firmin Didot.
———. 1822. *Traité de géométrie descriptive*. Paris: Corby.
Julia, Dominique. 2013. L'École normale de l'an III et 'l'art d'enseigner': les séances de débats. *La Révolution Française* 4: 1–20.
Julia, Dominique. 2016. *L'école normale de l'an III. Une institution révolutionnaire et ses éléves*, ed. Dominique Julia. Paris: Éditions rue d'Ulm.
Lacroix, Sylvestre-François. 1795. *Essais de géométrie sur les plans et les surfaces courbes (ou Éléments de géométrie descriptive)*. Paris: Fuchs.
Langins, Janis. 1980. Sur la première organisation de l'École polytechnique. Texte de l'arrêté du 6 frimaire an III. *Revue d'histoire des sciences* 33: 289–313.
Michel, Jean. 1981. Le patrimoine documentaire de l'école nationale des ponts et chaussées. *Annales des Ponts et Chaussées* 2: 25–31.

Monge, Gaspard. 1795. *Sur les lignes de courbure de la surface de l'Ellipsoïde, Journal de l'École polytechnique* 2: 145–165.

——. 1798. *Géométrie descriptive, Leçons données aux Écoles normales, l'an III de la République.* Paris: Baudoin.

——. 1799. *Géométrie descriptive.* Paris: Beaudoin.

——. 1801. *Feuilles d'analyse appliquée à la géométrie à l'usage de l'École polytechnique.* Paris: Beaudoin.

——. 1809. *Application de l'analyse à la géométrie à l'usage de l'École impériale polytechnique.* Paris: Bernard.

——. 1811. *Géométrie descriptive par G. Monge avec un supplément par M. Hachette.* Paris: Klostermann fils.

——. 1820. *Géométrie descriptive. Quatriéme édition augmentée d'une théorie des ombres et de la perspective.* Paris: Bachelier.

——. 1992. Leçons. In *L'École normale de l'an III. Leçons de mathématiques*, ed. Jean Dhombres, 305–453. Paris: Dunod.

Olivier, Théodore. 1847. *Applications de la géométrie descriptive.* Paris: Carilian-Goeury et Dalmont.

Sakarovitch, Joël. 1998. *Épures d'architecture.* Berlin: Birkhäuser.

Schübring, Gert. 1987. On the Methodology of Analysing textbooks: Lacroix as Textbook Author. For the Learning of Mathematics 7: 41âĂŤ51.

Taton, René. 1954. *L'œuvre scientifique de Monge.* Paris: PUF.

——. 1992. Un projet d'organisation d'écoles secondaires pour artisans et ouvriers, préparé par Monge en septembre 1793. In *L'École normale de l'an III. Leçons de mathématiques*, ed. Jean Dhombres, 574–582. Paris: Dunod.

Chapter 2
Descriptive Geometry in France: Circulation, Transformation, Recognition (1795–1905)

Évelyne Barbin

Abstract Descriptive geometry had been taught by Monge in 1794–1795 in two schools: the *École polytechnique*, intended for future engineers and officers, and the *École normale*, intended for future teachers. Monge's two proposals were preparatory teaching for various applications, but also a new teaching of geometry, which could follow (or replace) the ordinary Elements of geometry. In this chapter, our main goal is to examine the future of these two proposals in France. Indeed, the spirit of the first lessons given by Monge changed at the same time that descriptive geometry underwent a considerable dissemination into all French education and society. In relation to that dissemination, we examine the circulation of knowledge towards artists, craftsmen, and engineers. We investigate teaching given in the preparatory grades for the entrance examination to the *École polytechnique*, to the *École centrale des arts et manufactures*, and to other schools, and we analyze the process that led to teaching descriptive geometry at secondary level. We also examine the role of descriptive geometry in the teaching of geometry in the end of the nineteenth century.

Keywords Descriptive geometry · Projective geometry · Teaching of geometry · Teaching of methods · *Rebatment* · Rotation · Secondary school · School for engineers · Gaspard Monge · Jean-Nicolas Pierre Hachette · Joseph Adhémar · Théodore Olivier · Eugène Rouché · Louis Léger Vallée · *École polytechnique* · *École des ponts et chaussées* · *École centrale des arts et manufactures*

É. Barbin (✉)
UFR Sciences et Techniques, Nantes, France
e-mail: evelyne.barbin@wanadoo.fr

© Springer Nature Switzerland AG 2019
É. Barbin et al. (eds.), *Descriptive Geometry, The Spread of a Polytechnic Art*,
International Studies in the History of Mathematics and its Teaching,
https://doi.org/10.1007/978-3-030-14808-9_2

1 The Spreading of Descriptive Geometry into Preparatory Grades (1813–1833)

In 1794–1795, Monge's first lessons on "descriptive geometry" took place in two schools. We have the text of those for the *École normale de l'an III*, but we only have the program for the *École centrale des travaux publics* (the future *École polytechnique*). The starting point of the famous *Géométrie descriptive* of 1799 (Monge 1799) was a transcription of oral teaching given in this *École normale* (Monge 1992), without the structure in lessons and without the four last lessons (Barbin, Chap. 1, this volume). The purpose of the lessons in the *École normale* was to present a new geometry in the spirit of this school (Barbin 2015b). Firstly, lessons adopted an "order of invention" which consisted in giving problems before introducing tools to solve them. So, *Leçon* 1 began with an inaugural problem: "how to determine the position of a point in space?" before introducing the method of projections in *Leçon* 2. Secondly, they presented geometric objects, not in the order of simplicity, from plane to space, but starting with the latter immediately. Thirdly, they expressed a will of generalization, especially with the general "generation of surfaces" in *Leçon* 3, where a curved surface was defined as generated by the movement of a curved line. Fourthly, they promote the union of descriptive geometry and analysis.

Jean Nicolas Pierre Hachette was the "assistant" of Monge in the *École polytechnique* in 1794 and he replaced him as professor of descriptive geometry until his dismissal in 1816. His *Traité de géométrie descriptive* contained applications to shadows, perspective, and stereotomy (Hachette 1822). He gave great importance to the general definition of surfaces introduced by Monge, as generated by the motion of a curve, and he introduced the notion of ruled surfaces (Barbin, Chap. 1, this volume). Charles-François Leroy succeeded Hachette. In 1834 he edited his *Traité de géométrie descriptive* in which he criticized Monge's textbook, because it did not offer numerous and various examples, and it did not give clear drawings (Leroy 1834, p. v). He did not adopt Monge's order: he began with problems on straight lines and planes and continued with the trihedral angle. He introduced the general generation of surfaces, but the problems on tangent planes only concern cylinders and cones. His textbook met considerable success with 15 editions from 1834 to 1910 (with additions of Émile Martelet from the fourth edition of 1855). The contents (developable surfaces, curvatures, etc.) renders clear that the students had to be familiar with descriptive geometry before they entered the *École polytechnique*.

Indeed, teaching descriptive geometry rapidly decreased in the *École polytechnique* but appeared in the entrance program to this school in 1813. From its creation, there existed an examination to enter the *École polytechnique*, which had more and more candidates over the years. The *Conseil de perfectionnement* defined the entrance program and chose the examiners every year, but most of them remained for a long time. In 1804, the first examiners (like Monge or Jean-Baptiste Biot) had been replaced by former students of the *École polytechnique*, like Louis-Benjamin Francœur and Charles-Louis Dinet, who were also teachers in *Lycées*. From 1810, examiners began to ask questions outside the entrance program, especially in descriptive geometry. As a result, in 1813, the *Conseil de perfectionnement* decided that "the candidates would be questioned on the first six lessons of descriptive geometry concerning the straight line and the plane; and that they would construct, with the compass and the ruler, at a given scale, one figure of the elements of geometry which will be indicated by the examiner"[1] (Fourcy 1828, p. 320).

The introduction of descriptive geometry in the entrance program led to the publication of many textbooks explicitly devoted to the candidates to the *École polytechnique*, but also to the other schools of government (Military school of Saint-Cyr, Naval school of Brest, Forestry school). These candidates were students in upper grades of *Lycées* or private *Collèges*, named "special mathematics" and created in the beginning of the century, some in provinces but most in Paris (Belhoste 2001). Throughout the century, many authors of textbooks on descriptive geometry taught in these schools, they were former students of the *École polytechnique* and sometimes examiners. They constituted a network of Parisian authors, who wrote collections of textbooks on all the parts of mathematics (Barbin 2015a).

In 1828, Émile Duchesne wrote his *Éléments de géométrie descriptive*, a short and elementary textbook devoted to candidates for the *École polytechnique* and other schools. The later authors will be more prolix. Among the textbooks on descriptive geometry for the entrance examination to *École polytechnique* in the years 1820–1840, the most famous was the *Traité de géométrie descriptive* of Louis Lefébure de Fourcy, edited in 1830. The author was a teacher in the *Collège Royal Saint-Louis*, a former student of the *École polytechnique*, and an examiner during more than 30 years. This longevity explains the success of the textbook, which had been republished eight times until 1881. The first volume contains 295 pages in a small format and the second one around 150 pages of figures (Lefébure de Fourcy 1830). It began with an inaugural problem, which was to find the center of a sphere circumscribed to a triangular pyramid, and it continued with a list of problems, mixed with theorems. The order was far away from the one of Monge's lessons, with three parts: straight line and plane, curved surfaces and tangent planes, curved lines and their tangents. Lefébure de Fourcy gave Monge's general conception of a surface of revolution but he added that it was not useful for the applications. The part named "Exercises" shows that the textbook was a tool for training the students to prepare examinations. There were not any applications of descriptive geometry in the textbook, although the author considered it as "complete". The existence of new students, schools, and teachers was at the origin of a new conception of descriptive geometry, oriented not by problems but by formatted examination exercises (Barbin 2015b).

---

[1] All the translations of quotations are made by Évelyne Barbin.

2 Descriptive Geometry as a Part or as a Sequel of a Geometry Teaching (1812–1844)

Independently of the entrance into the *École polytechnique*, some teachers of *Lycées* proposed to introduce descriptive geometry in secondary school, as a part of geometry or as a sequel to ordinary geometry. In some sense, they followed the "Elements of descriptive geometry" edited in 1795 and 1817 by Monge's students, Sylvestre-François Lacroix (Lacroix 1795) and Jean Nicolas Pierre Hachette (Barbin, Chap. 1, this volume). The fact that some examiners or former students of the *École* became teachers of *Lycées* is an important factor for this dissemination of descriptive geometry. As in the case of Jean-Guillaume Garnier, who was an examiner of the *École*, an assistant of Lagrange until 1802, and then a teacher at the *Lycée* of Rouen. His *Éléments de géométrie* of 1812 contained a part on descriptive geometry composed of "preliminary notions" and five problems. For him, these notions "constitute a natural sequel to plane geometry, and introduce at the same time space geometry, in other words descriptive geometry" (Garnier 1812, p. 258).

Antoine-André-Louis Reynaud, who entered in the *École* in 1796, became a teacher in a *Lycée* in 1800 and an examiner of the *École* in 1809. As soon as 1812, he introduced a part with around 80 pages entitled "Elements of descriptive geometry" into his *Notes sur la géométrie*, which followed a new edition of Bézout's *Cours de mathématiques*. He wrote: "The principal purpose of Descriptive Geometry is to provide the means to exactly represent bodies in a plane. Scholars and artists invented more or less ingenious methods to solve this problem and thanks to research we reached to give constructions with the degree of simplicity that we have to make known" (Bézout and Reynaud 1812, p. 130). It is remarkable that he gave many theorems on projections of a point, a straight line, and a curve on only one plane, before he treated the case of two planes of projections, and finally the better case, where these planes are perpendicular. He ended by showing the simplicity of the solution of problems in space geometry using descriptive geometry. In *Problèmes et développements sur diverses parties des mathématiques*, written with Jean-Marie Duhamel in 1823, the authors did not introduce descriptive geometry but used the notion of projection. On the contrary, the *Théorèmes et problèmes de géométrie* of the "Baron Reynaud" of 1833 contained an important part on descriptive geometry, similar to the one of the *Notes* of 1812. At this period, Reynaud was still the examiner and his textbook was now intended for candidates to enter the *École polytechnique* and the other schools of government (Reynaud 1833).

But we have to remark that Olry Terquem, in his *Manuel de géométrie* of 1829, written to present the "writings of contemporary geometers" to beginners, did not mention Monge and the descriptive geometry and preferred to introduce the theories of projections and polars of Poncelet and Gergonne to teach conics (Terquem 1829, pp. 347–350). He entered the *École polytechnique* in 1801, was a "répétiteur" of this school and a teacher at the *Lycée* of Mayence. Later, in 1842, he created mathematical journals with Camille Gerono: *Nouvelles Annales mathématiques* and *Journal des candidats aux Écoles polytechnique et normale*. In papers of these journals, he appeared an attentive commentator on the teaching of descriptive geometry (Barbin 2015b).

Hyppolite Sonnet was assistant of mechanics in the *École centrale des arts et manufactures* created in 1829 (see further) and author of many textbooks. In 1839 he edited his *Géométrie théorique et pratique*, intended for the *Écoles normales primaire*, where the teachers of primary schools and industrial schools were trained. The edition of 1848 was also intended for special teaching given in the *Facultés*. It contained applications of geometry to drawing, architecture, perspective, and "the first elements of descriptive geometry", which covered 25 pages. Sonnet motivated this geometry with three inaugural problems on trihedrals, which had been solved in the plane just before. From the beginning of his textbook, he introduced the *rebatment* and its properties, as a direct method to solve problems, like Adhémar in 1832 (see further). A chapter concerned "the curved surfaces in general, and the cylindrical and conical surfaces in particular" (Sonnet 1848, pp. 251–264). The general surfaces were divided into ruled surfaces and surfaces of revolution with applications of cylindrical surfaces to the vaults, and of conical surfaces to a machine. Paul-Louis Cirodde was also author of many textbooks on various matters. He introduced "elementary notions on descriptive geometry" in the second edition of his *Leçons de géométrie* of 1844, in a period when he taught in the *Collège Royal Henri IV*. It essentially consisted in a collection of 26 problems and their solutions, those on surfaces concerning cylinders, conics, and spheres only (Cirrodde 1844, pp. 50–61).

3 Descriptive Geometry for Civil Engineers and Artists (1819–1841)

The teaching of descriptive geometry spread quickly in schools for engineers and technicians. For instance, Gabriel Gascheau was a teacher at the *École des arts et métiers* de Macon when in 1828 he edited his short *Géométrie descriptive. Traité des surfaces réglées* (Gascheau 1828), in which he introduced the ruled surfaces of Hachette (Barbin, Chap. 1, this volume).

Until the Revolution, the *École des ponts et chaussées* was a school for engineers, organized around lessons given by engineers and projects made by the students. In 1795, it became a school of application of the *École polytechnique*. Three professors were appointed in civil architecture, mechanics applied to the art of construction, and cutting of stones (Michel 1981). The third one was Léon Bruyère, a former student of this school, architect and engineer, who attended the lectures of Monge. Joseph Mathieu Sganzin started as teacher of descriptive geometry in 1797. In 1825–1826, Barnabé Brisson, a teacher of construction, who reedited Monge's textbook in 1820, promoted descriptive geometry in the school (Picon 1989). Two authors, related to this school, introduced novelties into the teaching of descriptive geometry, often adopted by their successors: Vallée and Adhémar.

Louis Léger Vallée was a former student of the *École polytechnique* and an engineer of the *École des Ponts et Chaussées*. He edited his *Traité de géométrie descriptive* intended for artists in 1819, where he compared constructions in a plane and in space with two problems: (1) to construct the center of a circle where three points are given in a plane; (2) to construct the center of a sphere four points of which are given in space. He defined the notion of orthogonal projection and the two planes of projection, then he explained that, before solving problems, we have to examine the representations of a point, of a straight line, and of a plane in their most "remarkable positions" (Vallée 1819, p. 10) and he gave them in a set of figures (Fig. 2.1). The first problems only concerned points, straight lines, and planes. In the same manner, he began to study the projections of curves before those of surfaces. He showed the usefulness of the notion of tangent to a curve, since the projection

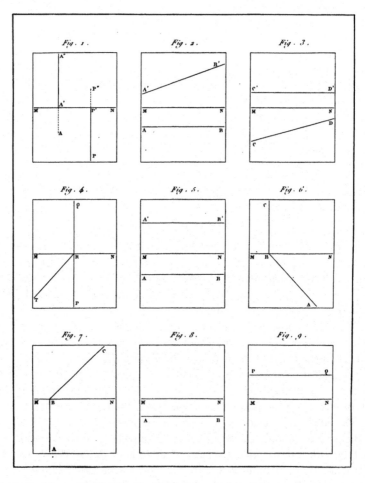

Fig. 2.1 Table of "traces" in Vallée's *Traité de géometrie descriptive* (n. p.)

of the tangent is tangent to the projection of the curve. Thus, Vallée introduced two orders, different from Monge's one, which will often be used by the successors: (1) a decomposition of the projections according to the simplicity of the figures, those of a point, a straight line, and a plane, with their different "traces" on the planes of projection and their problems; (2) a study of the projection of curves, with their "traces" and problems on tangents. As in Monge's *Leçons*, the part on surfaces began with their general conception, as generated by the motion of a curve, but it continued with cylindrical and conical surfaces, surfaces of revolution, warped and envelope surfaces, tangent planes, and then intersections of surfaces. Vallée used the word "rabattement" (rebatment), introduced by Charles Potier in 1817 (Barbin 2015b), to name the result of the motion of the vertical plane to the horizontal position, but here the notion was also introduced to solve problems on angles by turning a plane around a straight line (on Potier, see Gouzevitch, Chap. 13, this volume).

In 1821 Vallée wrote his *Traité de la science du dessin* devoted to artists, in which he introduced descriptive geometry, and not only perspective, contrary to other authors, like the teacher of drawings Thénot (1834). For Vallée, "descriptive geometry is an indispensable help to deeply penetrate all that concerns the mechanism of the eye. [...] As it furnishes the means to rigorously represent points, lines and surfaces defined with exactness; we conceive that it has to serve as a basis of the science of drawing" (Vallée 1821, pp. viii–ix). To represent an object and its shadow, he introduced two planes of projection, one horizontal (its plane) and one vertical (its cutting), and he deleted the ground line, for instance, for a niche (Fig. 2.2). The second edition of the book contained the enthusiastic support of Joseph Fourier, Gaspard de Prony, and François Arago in the name of the Royal academy of sciences.

Joseph Adhémar was a private teacher of mathematics and a prolific author of textbooks on descriptive geometry and its applications, intended for beginners, craftsmen, and civil engineers (Barbin 2015b). His textbooks were published by Carillan-Gœury, the bookseller of the *Corps des Ponts et Chaussées* and the *Corps des Mines*. In his *Cours de géométrie descriptive* of 1823, like Vallée, he decomposed the projections of a point, a straight line, and a plane. In 1832, he edited a collection of textbooks named *Cours de Mathématiques à l'usage de l'ingénieur civil*, and the first one is devoted to descriptive geometry. Like Vallée, he introduced a study of projections of curves before coming to surfaces. In another collection, named Cours de *Mathématiques à l'usage des architectes, ingénieurs civils, etc.*, he treated the applications of descriptive geometry: cutting of stones (1834), shadows (1840), frames (1849), and bridges (1853).

One novelty introduced by Adhémar in 1832 concerned the choice of the planes of projection. He stressed: "always, in the applications, we will have to choose the system of coordinate planes or auxiliary planes on which the projections will be the simplest. And as long as we change nothing in the data and their related position, the generality of the question will remain complete". He added: "the choice of the planes of projection is one of the most essential parts of the solution of problems" (Adhémar 1832, p. 158). Later, in his *Traité de géométrie descriptive* edited in 1841,

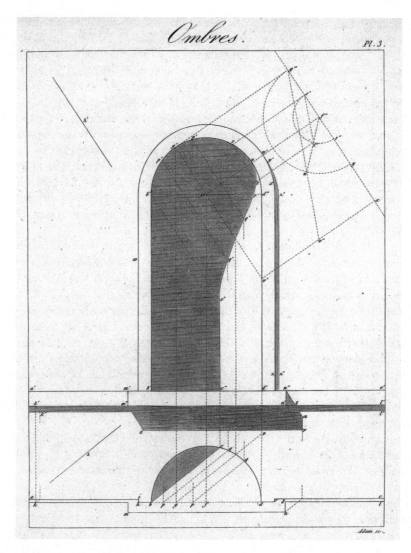

Fig. 2.2 Drawing of a niche in Vallée's *Traité de la science du dessin* (plate 3)

he stressed the advantages of using "auxiliary planes" for making a construction easier and he introduced a special chapter of 12 pages entitled "Rebatments". For him, this operation corresponded to the "transformation of coordinates" in algebraic analysis. He used rebatments to construct several problems and he used the properties of the rotations without making them explicit. He wrote:

> To have the true magnitude of a portion of a straight line, which joins two points, we have to turn this line until it becomes parallel to one of the planes of projection. This operation, named *rebatment*, served in the solution of several previous questions. The importance that

the *rebatments* have to acquire in the further applications of descriptive geometry has to engage us to present some general considerations on this kind of operation right now. [...] When we make a *rebatment*, each point describes a circle in space whose center is on the axis of rotation, and whose radius is the distance of this axis to the turning point (Adhémar 1832, pp. 44–46).

Another novelty was introduced in 1841, with the role given to cylindrical surfaces. Adhémar did not begin with the general conception of surfaces and neither with ruled surfaces. He started with the cylindrical surfaces, not as the simplest surfaces but as "the first and most essential curved surfaces" (Adhémar 1841, p. 134). For him, cylindrical, conical, and spherical surfaces are "the essential basis of almost all the combinations of the industry" (Adhémar 1841, p. 264). He defined a cylindrical surface as generated by a straight line that moves in parallel to itself, following a curve named the generative curve of the cylinder. He used this surface to show that "to construct a tangent to a given curve at a given point, it is sufficient to construct the two tangents of the projections of the curve at the projections of the given point. These lines are the projections of the tangent to the curve" (Adhémar 1841, p. 143). In this way, he stressed a general study of the projection on one plane only.

Notions élémentaires de géométrie descriptive, edited by M. F. Amadieu in 1838, was the first textbook for the candidates to the *École polytechnique* that took into account the novelties of Adhémar. The author was a former student of the *École militaire de Saint-Cyr* and he was a teacher at the *Lycée* of Versailles (near Paris). The textbook was small, with only 110 pages: it contained "preliminary notions" with theorems, two pages on drawings and a list of "problems to solve". It followed Vallée's textbook with the drawings of the nine projections of a straight line depending of its different positions with regard to the planes of projection, etc. In the part called "Resolution of problems by the method of rebatments", the properties of the motion of rotation are used implicitly (Amadieu 1838, pp. 19–20). In the part named "Changing the vertical plane", the problems of the change of planes are treated systematically: given the two projections of a point to find the projection on a new vertical plane, then the same problem for a straight line and for a plane.

4 Descriptive Geometry in the *École centrale des arts et manufactures* (1829–1853)

The *École centrale des arts et manufactures* was created in 1829 to train civil engineers and managers for industries and to develop the applications of the new sciences. It was an initiative of an industrialist man, Alphone Lavallée, with the chemist Jean-Baptiste Dumas, the physicist Eugène Péclet, and the mathematician Théodore Olivier. Lavallée was a shareholder of the journal Le Globe, created in 1824 to spread the saint-simonian doctrine, which granted a major role to the engineers in society (Comberousse 1879). Olivier was a former student of the

École polytechnique (1811) and of the *École d'artillerie* of Mézières. He was a student of Monge and Hachette and he remained friendly with Hachette. In 1851, he wrote, about the creation of the *École centrale des arts et métiers*, that, from 1816, the students of the *École polytechnique* received the same teaching as those of the *École normale*, so many of them preferred to become teachers than engineers, which means "philosophers" than "workers" (Olivier 1851, pp. xiii–xxiii). For him, many textbooks were edited for the entrance examination to the *École polytechnique* because of a "thirst of lucre" but without making any progress in science. He concluded that the industry needed civil engineers, who were not trained anymore in this school. At this time, courses for technicians were given in the *Conservatoire National des Arts et Métiers*, created in 1794, and in some towns, like in Metz by Poncelet (Fox 1992), while *Écoles centrales* in Châlons, Angers and Saint-Étienne prepared supervisors. For Olivier, the purpose of the creation of the *École centrale des arts et manufactures* was to "recreate the ancient *école centrale des travaux publics* like Monge had conceived it" to train engineers and not "scholars" (Olivier 1851, p. xx). Four sciences had to be taught: geometry, mechanics, physics, and chemistry. The teaching of mechanics contained a part on analytical geometry and analysis. The teaching of geometry was reduced to descriptive geometry and occupied a great part of the timetable, with 2 h every Tuesday and Saturday morning in first year, only on Tuesday in second year (Comberousse 1879, p. 46). During the 3 years of studies, many hours were devoted to drawings.

As a mathematician, Olivier was author of works on geometry, descriptive geometry, and the mechanics of gears. His thesis in 1834 concerned the geometrical study of curves and surfaces of second order and the applications of gears. He had been professor of descriptive geometry in *École centrale des arts et manufactures* from 1829 until his death in 1853, and also in the *Conservatoire National des Arts et Métiers*. In this latter school, in order to help students to understand ruled surfaces, he designed concrete "models", which had been manufactured by Fabre de Lagrange in 1872. Made out of threads, some of them were static (like those imagined by Monge) or some were articulated. For instance, a cylinder could be transformed in a hyperboloid, and then to a cone (Sakarovitch 1994, pp. 332–333). Olivier wrote many textbooks on descriptive geometry, like his *Cours de géométrie descriptive* in two volumes (1843), *Développements de géométrie descriptive* (1843), *Compléments de géométrie descriptive* (1845), *Applications de la géométrie descriptive* (1846).

In his *Cours de géométrie descriptive*, Olivier began with an inaugural problem, not easily solved by ordinary geometry, which is to fix the direction of the perpendicular of a plane passing through a given point. Then, he transformed two ideas of Vallée and Adhémar into two methods: the "method of point, straight line and plane" and the "method of changes". He wrote about the first method: "as soon as we will know how to represent a point, a right line and a plane by the method of projections [...], we will know descriptive geometry" (Olivier 1843, p. vi). The different possible "traces" of a point or a line are called "alphabet". He justified the "method of changes", meaning changes of planes of projection or rotations of figures around an axis, by remarking that a figure drawn on the planes of projection can be very complicated, but difficulties "will disappear by a suitable choice of planes of projection; we can also keep the same planes and change the position of the figure, this last operation is always made by turning the figure around on an axis" (Olivier 1843, p. 2). He introduced the operation of rebatment: "if we make a plane turn around its intersection with another plane, until it meets with this one, we say that we rebat ('rabattre') the first plane on the second. This operation is frequently used in descriptive geometry" (Olivier 1843, pp. 18–19). For him, this operation is identical to Euler's formula employed in analysis to find an equation of the section of a solid. The properties of what he called "rotation" are clearly expressed, and are used systematically. Like Monge, Olivier defined the surfaces in general and insisted on problem solving. He treated developable surfaces, tangent planes to conical and cylindrical surfaces, envelope surfaces and surfaces of revolution.

It is interesting to compare the contents of Olivier's textbooks with manuscripts of his students, preserved in the archives of the school. Every student had two notebooks, one for lessons, where the figures were made freehand, and another one for exact drawings. This last notebook was constituted by a list of problems, and many of them concerned motions. For instance, in 1839–1840, problem 12 asked to carry a given cone parallel to itself in such a manner that its summit reaches a given point (Fig. 2.3). In a notebook of 1845–1846, many problems concerning rotations

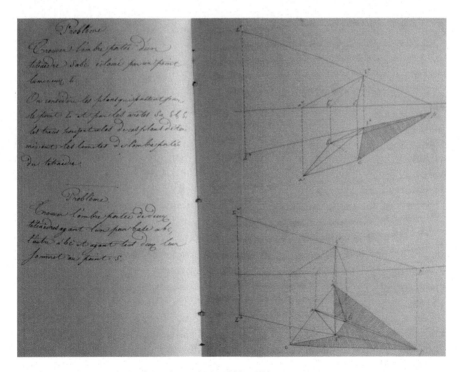

Fig. 2.3 From a notebook of drawings (OLI1 1839–1840)

were studied for themselves. Olivier's interest in rotations came from descriptive geometry but also from his study of gears. The notebook contained problems of application with drawings of carpentry, stones, and stairs.

5 From the Preparatory Grades to Secondary Schools (1847–1869)

Olivier's "method of changes" was criticized by Émile Martelet in the 4th edition of Leroy's *Traité*. For him, the method led to complicated constructions, but it could be a good subject for exercises (Leroy 1855, p. 394). In the 2nd edition of the *Traité élémentaire de géométrie descriptive* of Eugène Catalan and Henri Charette de Lafrémoire, Lafrémoire wrote: "during these recent years, the auxiliary projections were excessively recommended. It was believed that we had to recourse to their use in every circumstance, and in all the problems of descriptive geometry [...]. A new thing needed a new name: the Method of Planes of Projection was invented!" (Lafrémoire de et al. 1852, p. 120). The controversy continued in the *Nouvelles Annales de mathématiques* from 1851 to 1856 (Barbin 2015b).

Jules de La Gournerie also criticized Olivier's method. He was a former student of the *École polytechnique* and of the *École des ponts et chaussées*, and he succeeded to Leroy as professor of descriptive geometry in the first school. In the preface of his *Traité de géométrie descriptive* of 1860, he defined descriptive geometry as the "abstract science of the drawing line" (La Gournerie 1860, p. v) and he wrote four pages against Olivier's method, considering that it was not suitable for applications and was not new—Abraham Bosse used it in 1643 and he was not approved, while Monge did not use it in his drawings (La Gournerie 1860, p. viii). He quoted Hachette, Leroy, Vallée, and Fourcy. Like Vallée and contrary to Monge, he began treating the simplest elements, straight lines, and planes, and the simplest surfaces, cylinders, and cones. The first volume contained chapters on the "quoted geometry", used in topography, and on the axonometric perspective. The second volume (1862) and the third volume (1864) gave a complete study on surfaces and their curvatures, using analysis, presented recent results and gave applications on shadows. La Gournerie also gave lessons to the *Conservatoire des arts et métiers*, to promote descriptive geometry among workers and artists.

Despite this negative advice, Olivier's method had been adopted in many textbooks for candidates to the *École polytechnique* and to other schools of government. As soon as 1847, Bertaux-Levillain, a former student of the *École polytechnique* and a teacher, wrote in his *Éléments de géométrie descriptive* that he "followed the movement shown by a learned professor, M. Olivier": "it seemed to me that it was useful to put into the hands of students a textbook where [...]they could find the notions of change of planes of projection, of rebatment of a plane on another in the beginning, notions that one rejects in the middle of the course wrongly, because they are indispensable for properly understanding the solution of almost all the problems" (Bertaux-Levillain 1847, p. xiv). In 1852, Henri Édouard Tresca, a former student of the *École polytechnique* and a teacher of mechanics in the *Conservatoire des arts et métiers*, edited a *Traité élémentaire de géométrie descriptive*, "written in accordance with the textbooks and the lessons of Th. Olivier". The textbook began with the method of the projections of "point, straight line, plane", then it gave the representations of the projections of curves, of cylindrical and conical surfaces and surfaces of revolution. It presented two methods required to solve problems: "the general method of changes of planes" and "the general method of motions of rotations", which replaced the method of rebatment (Tresca 1852, p. 95). Here, the idea of rotation introduced by Olivier to define the changes of planes was enlarged to become a notion, that was the basis of a general method. Tresca gave what he called "the rules" to execute motions of rotations of one or several points.

In 1853, Antoine Amiot edited his *Leçons nouvelles de géométrie descriptive*, intended for the students of the preparatory grades to the *École polytechnique*, but also for the *École normale supérieure*, which trained future teachers. He was a former student of this latter school, a teacher at the *Lycée Saint-Louis* and the *École des beaux-arts*. The textbook was a small textbook of 190 pages only (without figures), which belonged to the collection of textbooks written by Amiot. It did not begin with an inaugural problem. In some chapters, there were theorems followed by a list of problems given as exercises, like in many textbooks of geometry of this period. Other chapters were composed of problems, like in Monge's. It began with the decomposition of the projections of point, straight line, and plane. Then Amiot introduced what he called "transformations of projections", which are the changes of plane of Olivier, and the rotations, which are systematically used to solve problems. He wrote: "it is M. Th. Olivier who gave a scientific character to the ideas expounded in this chapter under a particular title, and making it a basis of a method of resolution of questions in space geometry" (Amiot 1853, pp. 25–26). Like Monge, he defined general surfaces, as generated by the motion of a line, which meets one or many fixed guiding lines called "directrices". He continued with tangent planes to cylindrical surfaces and their problems, solved with the theorem on the projection of a tangent: "the tangent at a point M of any curve generally has, for orthogonal or oblique projection on any plane, a tangent to the projection of the curve, in the projection m of the point M" (Amiot 1853, p. 150).

In 1865, descriptive geometry became an autonomous part of secondary school teaching at the "elementary mathematics" level (students aged 17–18 years), which prepared the students for the *Baccalauréat*, the military school of Saint-Cyr, and the naval school of Brest. The program was close to Amiot's textbook, with projections of point, line, and plane, method of rebatments, projections of prisms, pyramids, cylinders, cones, plane intersections of polyhedrons (Belhoste 1995, p. 404). A typical textbook was the *Éléments de géométrie descriptive*, edited in 1869 by Charles Briot and Charles Vacquant, two former students of the *École normale supérieure* and teachers at the *Lycée Saint-Louis* and at the *Lycée Henri IV*. In the beginning of the textbook, the authors explained that an ordinary drawing is not sufficient to have exact lengths, but there exists the method of projections. They gave the figures of the "traces" of the projections of a point, considered as "an alphabet" and they proposed problems on point, straight line, and plane. The chapter entitled "Methods in descriptive geometry" contains three methods, namely rebatments, rotations, changes of planes. The following chapter began with the projection of a curve and with the theorem on the projection of the tangent to a curve (Briot and Vacquant 1869, p. 74). The last two chapters treated the projections of polyhedrons and their intersections. Consequently, the projection and its properties appeared as the major subject of the problems.

6 Descriptive Geometry and Modern Geometry at the Turn of the Century

The teachings of geometry and descriptive geometry became closer at the end of the nineteenth century. It was the result of two movements. On one hand, from the 1860s, the teaching of descriptive geometry widened to allow other ways of representing space objects and took the study of projections as a preliminary. On the other hand, from the 1870s, authors proposed to enlarge the teaching of geometry to "modern geometry" where the projections play an important role, as well as methods coming from descriptive geometry, like rotations. Indeed, the accent on the introduction of methods to solve problems of descriptive geometry and the tendency to understand these methods as subjects for teaching led to abstract notions, which can be fruitful for solving any kind of geometrical problems (Moussard 2012, 2015; Chevalarias 2014).

An interesting actor and witness of the connection between these two movements is Eugène Rouché, who taught descriptive geometry in the *École centrale des arts et manufactures*, from 1867 to 1888, and edited *Éléments de géométrie descriptive* for "special secondary school teaching" in 1875. In 1872, 30 years after the first edition, he reedited Olivier's *Cours de géométrie descriptive*. As he was a student of Jules de La Gournerie in the *École polytechnique*, we can interpret that as a choice between two teaching methods. He is also an author with Charles de Comberousse of an important *Traité de géométrie élémentaire* in 1866, which had been a book of initiation to modern geometry among teachers in France and had an impact for the teaching of mathematics (Barbin 2012). On that date, the two authors were teachers of preparatory grades in the famous *Lycée Charlemagne and Collège Chaptal* in Paris. Charles de Comberousse was a former student of the *École centrale* and he taught kinematics and applied mechanics in this school from 1862 on.

Rouché did not edit his lessons of descriptive geometry given in the *École centrale* but the archives of this school contain many series of students' notebooks. In 1871–1872, the first lesson began with definitions of projections of points, straight lines, then continued with the notions of sheaf of rays and anharmonic ratio, in the spirit of Poncelet's geometry. The problem of the section of a cone was treated with projections on three planes (Fig. 2.4). Thus, the conception of descriptive geometry

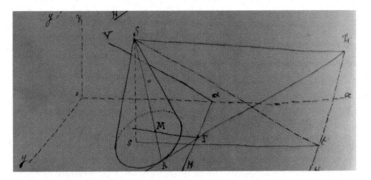

Fig. 2.4 A section of a cone in a notebook (ROU14 1871–1872, n. p.)

became larger than that of Monge. Eight years later, in 1879–1880, the presentation of descriptive geometry indicated: "it treats firstly, of geometrical notions, secondly, of various modes of graphical representations, thirdly, of stereotomy or cutting stones. The first part is disseminated in the middle of the others, so, as one goes along, the need for new theories appear" (ROU01, n. p.). The first lesson began with conical projections, and the first figure examined the case of the projection of a curve and its tangent, quite like many problems that considered one plane of projection only. Lessons included perspective and many applications, like carpentry or stereotomy, with lessons on vaults. Like Monge, Hachette, Vallée, Adhémar, Leroy, or Olivier, Rouché considered descriptive geometry for its applications: in 1893 with Charles Brisse, who was a professor in the *École centrale* and in the *École des Beaux-arts*, he edited a textbook entitled *Coupe des pierres* (cutting stones) (Rouché and Brisse 1893).

The *Éléments de géométrie descriptive* of Rouché, edited in 1875, were intended for the new "special secondary school teaching", which granted considerable importance to sciences and their applications, unlike "classical teaching". The program of 1866 introduced a part on descriptive geometry in the 3rd year (students aged 14–15 years), with the use of Olivier's models of the "Conservatoire des arts et métiers" (Belhoste 1995, p. 432). Rouché's textbook began with projective notions and their properties. Chapter VII, entitled "Rebatments and rotations", presented the two methods. The method of rotations was considered as a principle of solution, with the rotation of a point around a vertical axis, around a perpendicular axis to the vertical plane, then the rotations of a straight line, of a plane, etc. Here rotations were taught for themselves. As for his lessons in the *École centrale*, Rouché put forward general notions, which could be part of geometry teaching.

The *Traité de géométrie élémentaire* of Rouché and Comberousse was edited about 15 times from 1866 until 1935. The first edition was composed of 776 pages with two parts, plane geometry and space geometry. The historical introduction attached considerable value to Monge's descriptive geometry and Poncelet's theory of projections. The notion of rotation and its properties were used throughout the textbook, to study the similitude of polygons but also the homothetic figures in space. The notion of rebatment also appeared, especially in the third edition where the ellipse was defined as the orthogonal projection of a circle (Rouché and Comebrousse 1873, pp. 320–330). In the following editions, rotation would be considered as a method for solving problems (Rouché and Comebrousse 1900, p. 264). In 1866, the notion of projection was introduced in the beginning of the part on geometry in space, it served, for instance, to introduce the notion of angle of a plane and a line and the shorter distance of two lines. It was proven that "the projection of a straight line AB on a plane is a straight line" (Rouché and Comebrousse 1866, p. 347). So, projection was considered as an abstract notion: properties were not considered as perfectly obvious but needed proofs. After having given the theorems on projection, the authors extended the notions of projection and perspective by considering many kinds of projections. The curves and usual surfaces were defined and studied in the part on space geometry. For instance, it was proven that the intersection of a circular cone by a plane is an ellipse, a hyperbola, or a parabola. Here, the notion of projection became central in a textbook of geometry (Barbin 2015a).

In 1864, Amédée Mannheim succeeded Jules de La Gournerie to teach descriptive geometry in the *École polytechnique*. He was a former student of this school and of the *École de Metz* before to become an officer. It is important to remark that his *Cours de géométrie descriptive de l'École polytechnique*, edited in 1881, began with various ways to represent bodies, like shadows, projections, conical perspective, axonometric perspective. It contained also a part on kinematics and its applications to descriptive geometry. Mannheim defended an important theoretical part with a quotation of Gabriel Lamé who considered that the principal utility of studies in the *École polytechnique* was "to exercise reasoning" (Mannheim 1880, p. ix). In two papers of 1882 and in the second edition of his textbook, Mannheim proposed to delete the ground line when it was not useful, to follow the habit of the engineers (Barbin 2015a). Ernest Lebon adopted this in the third edition of his *Traité de géométrie descriptive* intended for the level of "elementary mathematics" (Lebon 1901). He was a teacher at the *Lycée Charlemagne* and author of many papers on descriptive geometry. His textbook contained the three methods of descriptive geometry in this order: rotation, change of plane, rebatment. In 1891, the program of "modern secondary school teaching", which replaced "special secondary school teaching", introduced teaching of descriptive geometry for the "second and first levels" (students aged 15–17 years) (Belhoste 1995, pp. 543, 546). Lebon wrote a *Géométrie descriptive* intended to these students in 1891, where he presented the method of change only (Lebon 1891, pp. 57–64).

The reform of secondary school teaching in 1902–1905 strengthened the proximity of geometry and descriptive geometry teaching. Firstly, teaching of plane geometry was not separated from the one in space. Secondly, the teaching of geometry contained drawings, projections, and perspective, and thus many notions of descriptive geometry. Thirdly, the teaching of descriptive geometry began in the first grade (student aged 16–17 years), with projections, representations of point, line and plane on two planes of projection, rebatment of a plane on the horizontal plane, and the change of the vertical plane. The teaching in the last grade treated rebatments, change of a plane of projection, and rotation around a perpendicular axis of a plane of projection. The teaching of descriptive geometry was teaching of methods with exercises and problems, but without concrete applications. This continued with the next reform of 1912, and a typical example is the *Cours de géométrie descriptive* of F. G.-M. (Frère Gabriel-Marie) in 1917. The "method of projection" occupied half the lessons for the first grade and the "theory of change of the horizontal plane", rebatment of a plane and rotations the third of those of the next grade (F. G.-M 1917). Many textbooks for preparatory grades were edited in this period, when, despite the criticisms of professors of the *Faculté des sciences* and of the *École polytechnique*, the teaching remained theoretical (Barbin 2015a). It is the case with the *Cours de géométrie descriptive* edited many times by Xavier Antomari for the candidates to the "great schools", which means *École polytechnique*, *École normale supérieure*, and *École centrale*. It was an impressive textbook of 641 pages, where the author quoted Rouché's method to determine the intersections between a surface and a straight line (Antomari 1910, pp. 486–491). As a result of the use of the methods, many problems of construction turned to examining the case of a projection on only one plane.

7 Conclusion: The Two Purposes of Monge and Their Historical Futures

The process of transformation of Monge's conceptions is linked with the emergence of two institutions. The first one is the preparatory grades for entrance into the *École polytechnique*, where students are prepared to answer exercises, far away from applications or drawings. The second one is the *École centrale des arts et manufactures*, created to train engineers for the industries. In some sense, these different teaching methods of descriptive geometry converge after the Cours of Olivier given in the *École centrale*, which was itself inherited from the textbooks for artists, craftsmen, and engineers written by Vallée and Adhémar. Indeed and maybe paradoxically, the introduction of "methods" by Olivier seemed equally suitable to train engineers for industries, "workers" who need applications, and to help students for examinations. In a long period, from 1815 until 1900, professors of the *École polytechnique* did not play a major role in the spreading of descriptive geometry. But, there existed a lineage of actors of changes, all linked with this school, who recognized their predecessor as a master: Monge, Hachette, Vallée, Olivier, and Rouché.

These five men promoted the applications of descriptive geometry. So, despite what the mathematician Carlo Bourlet wrote in 1906 (Sakarovitch 1998, p. 345), the first purpose of Monge was perennial in this period. Indeed, during the second part of the nineteenth century, descriptive geometry penetrated the world of craftsmen and technicians, as we can verify by comparing the first edition of 1844 of the *Nouveau manuel complet de la coupe des pierres* (cutting stones) written by the architect C.-J. Toussaint with its edition by F. Fromholt in 1902. The first book quoted "the kind of method created by Desargues" and gave a short explanation on elementary descriptive geometry (Toussaint 1844, pp. 9–10). While, in his edition, Fromholt went further by introducing and making intelligible descriptive geometry to a simple worker, because this geometry "will teach him cutting of stones with considerable ease" (Toussaint 1902, p. 19).

The second purpose of Monge by teaching descriptive geometry at secondary level (Taton 1992) had been granted in 1865, even if it was far from his conceptions defended in the *École normale*. This teaching met projective geometry, inherited from his student Poncelet, at the end of the century. Thus, between 1902 and 1962, the teaching of descriptive geometry and geometry became closer at each reform. Finally, in 1962, descriptive geometry became only a part of the program of geometry in the last grade of the *Lycées*, from which it disappeared in 1966. It remained a subject for teaching in preparatory levels until it disappeared from the entrance examination to the "great schools" and schools of engineers in 1970.

References

Sources

OLI, *Archives of the Library of the École Centrale des arts et manufactures*.
ROU, *Archives of the Library of the École Centrale des arts et manufactures*.

Publications

Adhémar, Joseph. 1832. *Cours de mathématiques à l'usage de l'ingénieur civil. Géométrie descriptive*. Paris: Carilian-Gœury.
———. 1841. *Traité de géométrie descriptive*. Paris: Carilian-Gœury, Bachelier, Mathias.
Amadieu, M.F. 1838. *Notions élémentaires de géométrie descriptive*. Paris: Bachelier.
Amiot, Antoine. 1853. *Leçons nouvelles de géométrie descriptive*. Paris: Guiraudet & Jouas.
Antomari, Xavier. 1910. *Cours de géométrie descriptive* (5th ed.). Paris: Vuibert.
Barbin, Évelyne. 2012. Teaching of conics in 19th and 20th centuries : On the conditions of changing (1854–1997). In *Proceedings of the Second International Conference on the History of Mathematics Education*, ed. Kristin Bjarnadottir, Fulvia Furinghetti, Johan Pritz, Gert Schubring, 44–59. Lisbon: Universidade Nova.
———. 2015a. Top-down: the role of the classes préparatoires aux grandes Écoles in the French teaching of descriptive geometry (1840–1910). In *Proceedings of the Third Conference on the History of Mathematics Education*, ed. Kristin Bjarnadottir, Fulvia Furinghetti, Johan Pritz, Gert Schubring, 49–64. Uppsala: Uppsala University.
———. 2015b. Descriptive geometry in France: history of the elementation of a method (1795–1865). *International Journal for the History of Mathematics Education* 10: 39–81.
Belhoste, Bruno. 1995. *Les sciences dans l'enseignement secondaire français*. Textes officiels. Paris: INRP.
———. 2001. La préparation aux grandes écoles scientifiques au XIXe siècle: établissements publics et institutions privées. *Histoire de l'éducation* 90: 101–130.

Bertaux-Levillain. 1847. *Éléments de géométrie descriptive*. Paris: Carilan-Gœury & Valmont.
Bézout, Étienne, and Antoine-André-Louis Reynaud. 1812. *Cours de mathématique seconde partie. Notes sur la géométrie*. Paris: Courcier.
Briot, Charles, and Charles Vacquant. 1869. *Éléments de géométrie descriptive*. Paris: Hachette.
Chevalarias, Nathalie. 2014. Changes in the teaching of similarity in France: from similar triangles to transformations (1845–1910). *International Journal for the History of Mathematics Education* 9: 1–32.
Cirrodde, Paul-Louis. 1844. *Leçons de géométrie suivies de notions élémentaires de géométrie descriptive* (2d ed.). Paris: Hachette.
Comberousse, Charles de. 1879. *Histoire de l'École centrale arts et manufactures depuis sa fondation jusqu'à ce jour*. Paris: Gauthier-Villars.
F. G.-M. 1917. *Cours de géométrie descriptive*. Tours: Mame & Fils. Paris: de Gigord.
Fourcy, Ambroise. 1828. *Histoire de l'École polytechnique*. Paris: École polytechnique.
Fox, Robert. 1992. *The Culture of Science in France, 1700–1900*. Aldershot: Ashgate.
Garnier, Jean-Guillaume. 1812. *Éléments de géométrie*. Paris: Béchet.
Gascheau, Gabriel. 1828. *Géométrie descriptive. Traité des surfaces réglées*. Paris : Bachelier.
Hachette, Jean Nicolas. 1822. *Traité de géométrie descriptive*. Paris: Corby.
La Gournerie de, Jules. 1860. *Traité de géométrie descriptive*, vol. I. Paris: Mallet-Bachelier.
Lacroix, Sylvestre-François. 1795. *Essais de géométrie sur les plans et les surfaces courbes (ou Éléments de géométrie descriptive)*. Paris: Fuchs.
Lafrémoire de, Henri Charette, and Eugène Catalan. 1852, *Traité élémentaire de géométrie descriptive*. Paris: Carilian-Gœury et Dalmont.
Lebon, Ernest. 1891. *Géométrie descriptive pour l'enseignement secondaire moderne*. Paris: Delalain.
———. 1901. *Traité de géométrie descriptive et géométrie côtée*. Paris: Delalain.
Lefébure de Fourcy, Louis. 1830. *Traité de géométrie descriptive*. Paris: Bachelier.
Leroy, Charles-François. 1834. *Traité de géométrie descriptive*. Paris: Carillan-Gœury & Anselin.
———. 1855. *Traité de géométrie descriptive*. Paris: Mallet-Bachelier.
Mannheim, Amédée. 1880. *Cours de géométrie descriptive de l'École polytechnique*. Paris: Gauthier-Villars.
Michel, Jean. 1981. Le patrimoine documentaire de l'école nationale des ponts et chaussées. *Annales des Ponts et Chaussées* 2: 25–31.
Monge, Gaspard. 1799. *Géométrie descriptive*. Paris: Beaudoin.
———. 1992. Leçons. In *L'École normale de l'an III. Leçons de mathématiques*, ed. Jean Dhombres, 305–453. Paris: Dunod.
Moussard, Guillaume. 2012. The notion of method in 19th century French geometry teaching : three textbooks. In *Proceedings of the Second International Conference on the History of Mathematics Education*, ed. Kristin Bjarnadottir, Fulvia Furinghetti, Johan Pritz, Gert Schubring, 333–350. Lisbon: Universidade Nova.
———. 2015. *Les modifications des notions de problème et de méthode dans l'enseignement des mathématiques aux XIXe et XXe siècles*. Nantes: Thesis.
Olivier, Théodore. 1843. *Cours de géométrie descriptive*. Paris: Carilian-Gœury & Dalmont.
———. 1851. *Mémoires de géométrie descriptive théorique et appliquée*. Paris: Carilian-Gœury & Dalmont.
Picon, Antoine. 1989. Les ingénieurs et la mathématisation. L'exemple du génie civil et de la construction. *Revue d'histoire des sciences* 42: 155–172.
Reynaud, Antoine-André-Louis. 1833. *Théorèmes et problèmes de géométrie*. Paris: Bachelier.
Rouché, Eugène. 1875. *Éléments de géométrie descriptive*. Paris: Delagrave.
Rouché, Eugène, and Charles Brisse. 1893. *Coupe des pierres*. Paris: Baudry & Cie.
Rouché, Eugène, and Charles de Comebrousse. 1866. *Traité de géométrie élémentaire* (1st ed.). Paris: Gauthier-Villars.
———. 1873. *Traité de géométrie élémentaire* (3rd ed.). Paris: Gauthier-Villars.
———. 1900. *Traité de géométrie élémentaire* (7th ed.). Paris: Gauthier-Villars.

Sakarovitch, Joël. 1994. Théodore Olivier, Professeur de géométrie descriptive. In *Les professeurs du Conservatoire national des arts et métiers*, 1794–1955, ed. Claire Fontanon and André Grelon, 326–335. Paris: INRP/CNAM.
——. 1998. *Épures d'architecture*. Berlin: Birkhäuser.
Sonnet, Hypollite. 1848. *Géométrie théorique et pratique*. Paris: Hachette.
Taton, René. 1992. Un projet d'organisation d'écoles secondaires pour artisans et ouvriers, préparé par Monge en septembre 1793. In *L'École normale de l'an III. Leçons de mathématiques*, ed. Jean Dhombres, 574–582. Paris: Dunod.
Terquem, Olry. 1829. *Manuel de géométrie ou exposition élémentaire des principes de cette science*. Paris: Roret.
Thénot. 1834. *Traité de perspective pratique*. Paris: chez l'auteur.
Toussaint, T. C. 1844. *Nouveau manuel complet de la coupe des pierres*. Paris: Roret.
Toussaint, T. C., Fromholt, F. 1902. *Nouveau manuel complet de la coupe des pierres*. Paris: Mulo.
Tresca, Henri Édouard. 1852. *Traité élémentaire de géométrie descriptive*. Paris: Hachette.
Vallée, Louis Léger. 1819. *Traité de géométrie descriptive*. Paris: Courcier.
——. 1821. *Traité de la science du dessin*. Paris: Courcier.

Chapter 3
Descriptive Geometry in Italy in the Nineteenth Century: Spread, Popularization, Teaching

Roberto Scoth

Abstract The spread of descriptive geometry in Italy began in the years of the French occupation (1796–1813) and continued during the Restoration, mainly through military schools and universities. At the same times, the first technical-professional schools (public and private) were founded in Italy. Elementary descriptive geometry (theoretical and/or graphical) was taught in many of these schools, which played a key role in the popularization of this subject. In the first decades of the nineteenth century, Italy, after France, "was the European country that, before all others, had provided original contributions in the field of descriptive geometry" (Loria. 1921. Storia della Geometria descrittiva. Milano: Hoepli. [p. vii]). After unification and the Founding of the Kingdom of Italy (1861), a new model for universities and secondary education was created. The tradition that began at the time of the French invasion continued in the universities even in the united Italy. In contrast, after a major attempt to reform secondary schools in 1871, the teaching of descriptive geometry was gradually neglected.

This chapter focuses attention on the main aspects of the spread, popularization, and teaching of descriptive geometry in Italy during the nineteenth century. In the first part, we consider the era before unification (1796–1861). In the second part, we analyze the period after unification (1861–1923) with particular regard to the lack of interest shown in teaching descriptive geometry at secondary education level.

Keywords Descriptive geometry · Projective geometry · School for engineers · University · Secondary school · Technical education · Teaching of geometry · Casati law · Filippo Corridi · Luigi Cremona · Vincenzo Flauti · Giuseppe Tramontini

---

R. Scoth (✉)
University of Cagliari, Cagliari, Italy
e-mail: biscoth@libero.it

1 Descriptive Geometry in Italy Before Unification

1.1 The Historical Background

During the first French invasion between 1796 and 1799, the Italian territory was divided into numerous satellite states. The three largest of these were the Cisalpine Republic, situated in the north, the Roman Republic, which included Rome and the territories of the Church, and the Neapolitan Republic, situated in the south.

From 1800 to 1813 there was the second phase of the French invasion, during which the Italian territory was partly annexed by France and partly divided into new satellite states, each with its own laws but without any real political autonomy. The most extensive territorial states, as well as the most strategic for French politics, were the Italian Republic in the north, which became the Kingdom of Italy in 1805, and the Kingdom of Naples in the south.

After the defeat of Napoleon, the Congress of Vienna sanctioned the division of the Italian peninsula into seven principal states. The only ones that retained their autonomy from the Habsburg Empire were the Kingdom of Sardinia, ruled by the Savoy, and the Papal State, which was returned to the Pope. The other principal states were the Kingdom of Lombardy-Venetia, which was directly dependent on the Austrian Emperor Francis I, and the Kingdom of the Two Sicilies, which was created in 1816 after the union of the former Kingdoms of Naples and Sicily and was returned to the Bourbons.

This political division lasted until 1859 when the unification process began.

1.2 Descriptive Geometry and Higher Education in the Napoleonic Period and During the Restoration

Descriptive geometry was introduced to Italy via the new cultural channels created by the French during the first phase of their invasion. Monge sojourned twice in Italy between 1796 and 1798 to carry out the tasks related to the office of the Commissioner of the French Republic. At the start of the peace negotiations with the Austrians (1797), Napoleon requested the presence of Monge in French headquarters and was able to converse with him about various topics, including descriptive geometry (Fiocca 1992, p. 196).

It can be assumed that the presence of Monge in Italy somehow facilitated the spread of descriptive geometry. In fact, just 2 years after the French invasion began, and just 3 years after the same Monge had held his famous lectures at the *École normale* and the *École centrale des travaux publics*, descriptive geometry was introduced to Italy. In 1798, a first course was opened at the *Scuola di Artiglieria e Genio* in Modena (Cisalpine Republic), and in 1799, another course was opened in Naples, at the *Scuole teoriche di Artiglieria e Genio*, founded to replace the old military schools of the Bourbon period. Simultaneously with the start of French reformist activity, new plans for education (which remained uncompleted) were prepared in 1798, which included the creation of two engineering schools in the cities of Milan and Rome with courses in descriptive geometry. At the beginning of the nineteenth century, with the introduction of the new French educational models and the creation of the *Lycées*, the main Italian universities were reformed, and a descriptive geometry course was inaugurated for the first time at the University of Naples in 1806.

In Italy, alongside the profession of military engineer, the profession of civil engineer began to take shape in the early decades of the nineteenth century. In fact, during the French invasion, a process began that was consolidated during the Restoration and which transformed the traditional architect (who operated independently and acquired skills via training and experience) to a more modern professional belonging to special corps that were responsible for the design, construction, and maintenance of roads, canals, ports, and rivers as in the French model of the *Corps des ponts et chaussées*. This process favored the creation of the first courses for civil engineers, in which teaching was focused on subjects such as geodesy, hydraulics, structural engineering, stereotomy, and the branches of mathematics related to them like differential calculus, mechanics, and descriptive geometry. In Italy, however, the training of civil engineers took place almost exclusively in the universities, and in the period preceding unification, only two schools for engineers in Naples and Rome were founded. In Naples, the *Scuola del Corpo Reale di Ponti e Strade* was created by the French in 1811. It was re-founded by the Bourbons during the Restoration and became one of the leading engineering schools in the country after the unification of Italy. In Rome, after the first unfinished projects of 1798, the *Scuola d'ingegneri pontifici* was opened in 1817 and in 1826 was integrated into the University.

The spread of descriptive geometry throughout Italian universities took place at different times. In some states, the discipline was introduced in the first years following the Restoration, for example, in Turin (Kingdom of Sardinia). In 1801, the French devised a reform plan for the University of Turin that included the teaching of descriptive geometry, but this plan remained unrealized (Conte and Giacardi 1990, pp. 285 ff.). Monge's discipline was taught at this university only in 1824,[1] after a course in descriptive geometry had been established at the *Regia Accademia Militare* in 1816.

In other universities such as those in Pisa (Grand Duchy of Tuscany), Pavia, and Padua (Kingdom of Lombardy-Venetia), the introduction of descriptive geometry took place a few decades later. Overall, according to Fiocca (1992, p. 188), it can be said that the widespread dissemination of descriptive geometry in Italy was realized between 1798 and 1840. In 1840s, the discipline even transferred to Sardinia, the Italian region with the lowest school attendance rate and in which the Savoy had implemented policies that had slowed cultural development (Scoth 2016, p. 186).

---

[1] During the Napoleonic occupation, many Piedmontese students were recruited to attend the *École polytechnique* in Paris (see Id., pp. 289–296).

1.3 The Emergence of Technical and Professional Education and the Popularization of Descriptive Geometry

In Italy, the development of technical and scientific teaching at the pre-university level began in the early decades of the nineteenth century with the creation of a number of educational institutions that, for the first time, attributed a "popular" character to the subjects and removed them from the hegemony of military schools and universities. There are basically two reasons for this development: the growth of the economy and modern industry that required a radically different education from the one linked to the eighteenth-century models, and the orientation of Italian society towards greater popular culture. In those decades, technical and professional education in Italy was very varied and took place in many different schools, supported by patrons and entrepreneurs. The best initiatives were linked to the industrialization of northern Italy and the modernization of agriculture. There were, however, numerous examples also in the south of both schools for artisans and agricultural schools. Moreover, some nautical schools were established in several coastal cities of the peninsula and on Sicily and Sardinia.

In 1830s and 1840s, there were the first attempts to "popularize" descriptive geometry and to teach it in these new schools. One of the most famous examples was that of the *Società d'incoraggiamento d'arti e mestieri* in Milan, founded in 1838 on the initiative of a group of intellectuals and entrepreneurs, where a descriptive geometry course held by Giuseppe Colombo, one of the most famous Italian engineers of those times, was introduced. Other famous technical and professional schools were opened in 1838 in the Kingdom of Sardinia, for example, those of the *Società per l'avanzamento delle arti, dei mestieri e dell'agricoltura* in Biella, and the *Istituto di arti e mestieri* in Novara. In these schools, descriptive geometry was part of the teaching of solid geometry and had as its goal, above all, its application to mechanical and construction drawing (MAIC 1862, pp. 409 ff.).

Alongside the *Scuole di arti e mestieri* (Schools for Arts and Crafts), which were at the elementary school level, some secondary technical schools were created in Italy from public initiatives in the nineteenth century. In general, private initiative was faster than political initiatives in interpreting the needs of the local economy but, in the middle of the century and with the increase in industrial development, the creation of public technical schools as an alternative to grammar schools became could no longer be delayed, and all governments enacted special measures to create new courses of study. In Lombardy-Venetia, the *Scuole Reali* (similar to the Austrian and German *Realschulen*) were founded in Venice (1841) and Milan (1851). In Turin in 1852, the Municipal Technical Schools and the Royal Technical Institute, a school that depended directly on the Ministry of Education, were founded. In all these schools, descriptive geometry applied to the drawing of orthogonal projections was taught (Scoth 2008, pp. 33–34). In Florence, a chair of descriptive geometry at the *Istituto Tecnico*, a high-level institution was set up in 1853.

Overall, it may be concluded that in the mid-nineteenth century, especially in northern Italy, descriptive geometry was widely taught in secondary technical schools and primary schools for artisans.

1.4 The Italian Treatises of Descriptive Geometry in the First Half of the Nineteenth Century

In the first half of the nineteenth century, the introduction and the spread of descriptive geometry in Italy was supported by an original local production of treatises and manuals. In this period, almost 40 new texts, reprints, and translations (wholly or partly devoted to the descriptive geometry and its applications) were published.

The lectures given at the *École normale de l'an III* were translated into Italian by Carlo Lauberg, a Neapolitan philosopher and revolutionary who moved to the Cisalpine Republic for political reasons. These translations (*Lezioni ad uso delle scuole normali di Francia raccolte per mezzo dei Stenografi e rivedute dai Professori*, Milano, Raffaele Netti, 1798) are very rare, and only the first two volumes are known. They contain lessons by Garat, Sicard, Berthollet, Vandermonde, Bauche de la Neuville, Mentelle, La Harpe, Volney, Daubenton, and also five mathematics lessons by Laplace, four physics lessons by Hauy, and the first three lessons of descriptive geometry by Monge (Pepe 2003, pp. 332–333).

The first Italian translation of Monge's *Géométrie descriptive* (Placci 1805) was published in Bologna by Giuseppe Placci, a former student of the Military School in Modena. This version, which contains some notes added by the translator, probably had a low circulation (Fiocca 1992, p. 205). In fact a second translation by the mathematician Filippo Corridi was printed in Florence in 1838 (Corridi 1838), with the subtitle "first Italian edition" (Fig. 3.1).

Instead, the first Italian treatise of descriptive geometry (Flauti 1807) was published in Naples by the mathematician Vincenzo Flauti for the students of the *Scuole teoriche di Artiglieria e Genio*. This treatise, as the author explains, follows the classic setting of Monge's work although it contains new examples and different construction methods (Id., pp. 212 ff.). A second work (Flauti 1815, and subsequent eds.) contains a personal re-working of descriptive geometry. The originality lies mostly in the fact that the language used was that of the ancient Greek geometers with references to *Book of the Euclid's Data* (Id., pp. 219–220). Flauti, in fact, was an exponent of the so-called *Scuola sintetica napoletana* (Neapolitan synthetic school), namely a then philosophical movement that favored the model of Euclid's Geometry in contrast to analytical methods (see also Mazzotti 1998, pp. 675 ff.).

One of the descriptive geometry teachers of the *Scuola di Artiglieria e Genio* in Modena, Giuseppe Tramontini, published a treatise (Tramontini 1811) that shows several important aspects (Fiocca 1992, pp. 199 ff.):

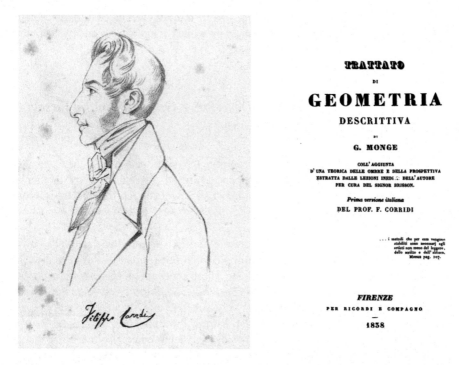

Fig. 3.1 Mathematician Filippo Corridi, translator of the most famous Italian edition of Monge's *Géométrie descriptive*

- for the first time, he introduced the technique of the change of projection planes in the most general form, making use of non-orthogonal planes between them[2];
- in some cases, he used three projection planes since in general there is no bijective correspondence between all the points of an object and their representation on two orthogonal planes;
- the second section of his work was given to applications, in particular prospective and theory of shadows.[3] This section was necessary because school courses in Modena were divided into a theoretical and an applied part such as those of Monge at the *École Centrale des Travaux Publics*.

The Military school in Modena, in spite of its short life (it was closed in 1815), gave rise to an important tradition in the field of descriptive geometry. Another

---

[2]In particular in chapter IV of his work (see also Loria 1921, p. 191 ff.; Torelli 1875; Fiocca 1992, pp. 202–203). The question of the change of the projection planes (the *Méthode d'Olivier*) was also studied by Francesco Paolo Tucci, a lecturer at the School for engineers in Naples Tucci (1823). In France, this topic was considered a few years later by Adhémar and Olivier (Barbin, Chap. 2, this volume).

[3]The treatise of Tramontini was published before Monge's lessons about the perspective and the theory of shadows was printed in France by Bernabé Brisson.

former student of this school, Carlo Sereni, was a professor for almost 50 years of this discipline in the *Scuola d'ingegneri pontifici* in Rome and published two treatises, which were very popular in Italy, one theoretical and one applied (Sereni 1826 and subsequent eds.; Sereni 1846).

Quintino Sella, one of the most famous statesmen in the united Italy and teacher of geometry at the Royal Technical Institute in Turin, was the first to introduce isometric drawing to Italy (Càndito 2003, pp. 62 ff.). He published a pamphlet on the principal methods of representation, which contain his lessons from 1856 (Sella 1856). Half of this treatise is dedicated to axonometric drawing and the methods proposed by William Farish (see also Lawrence, Chap. 17, this volume) and perfected by Julius Weisbach.

Among the various works printed before the unification of Italy, it is worth mentioning (Bellavitis 1851), which was written by Giusto Bellavitis, professor at the University of Padua. In this text, and for the first time in Italy, there is reference to modern theories of central projection.

Overall, we can say that in almost all pre-unification Italian states the diffusion of descriptive geometry was supported by the production of original local treatises.[4] Works on this subject were not published only in Piedmont. At the University of Turin in 1830, manual written by Sereni for engineering schools of Rome was adopted (Fiocca 1992, p. 222).

Currently, there are no studies regarding the transfer and circulation of treatises of descriptive geometry in Italy at that time, and so it is not possible to make definitive assertions. It is likely that the production of Italian texts hindered the translation of foreign treatises. In fact, in addition to Monge's *Géométrie descriptive*, only three other works of descriptive geometry translated into Italian in the first half of the nineteenth century are known.[5]

Another important aspect is that of didactic transposition of descriptive geometry in "schools of arts and crafts". The model of the *Conservatoire des arts et métiers* had a major impact, especially in northern Italy, thanks to the activity of disclosure made by Italian scientists that were sent to France (and to other advanced nations) to study the processes of industrialization and thanks to the existence of various technical and scientific information magazines. The *Géométrie et méchanique des arts et métiers et des beaux-arts*, the work summarizing the lessons given

---

[4]Since the 1840s, the treaties written by the professor of descriptive geometry were used at the University of Pavia in the Kingdom of Lombardy-Venetia. (Carlo Pasi, *Sunto di lezioni di geometria descrittiva*, 1843; *Saggio di applicazioni della geometria descrittiva*, 1844 and subsequent eds.)

[5]Two of these were the works of Sylvestre-François Lacroix, *Saggio di Geometria riguardante le superficie piane e curve o sia Elementi della Geometria descrittiva di S. F. Lacroix. Prima traduzione italiana fatta sopra la terza edizione francese*, 1829, and Charles-François-Antoine Leroy, *Trattato di geometria descrittiva con una collezione di disegni. Prima versione dal francese con note di Salvatore D'Ayala e Paolo Tucci*, 1838. The third was a treaty of Georg Schaffnit adopted in the Polytecnic of Wien: *Scienza Geometrica delle Costruzioni ovvero Geometria Descrittiva. Versione dal tedesco con illustrazioni e aggiunte di Vincenzo Tuzzi*, 1841.

to the working classes by Charles Dupin, contained an entire volume devoted to elementary geometry and its applications, including descriptive geometry. The work was twice translated Italian[6] and was an important reference for elementary technical teaching. Following the example of the famous text by Dupin, similar books were printed in Italy, which contained, in more or less expanded form, an elementary discussion of descriptive geometry.[7]

2 Descriptive Geometry in Italy After Unification

2.1 A New Educational and Academic Model

The most crucial phase in the unification of Italy began in 1859 with the annexation of Lombardy to the Kingdom of Sardinia and ended in 1861 with the annexation of the remaining territories and the birth of the Kingdom of Italy.[8] At that time, the Italian educational system was based on the 1859 Casati law (named after the Minister Gabrio Casati), created to reform education in the Kingdom of Sardinia and was subsequently extended to other areas of the new Kingdom of Italy. The educational system was divided into: Elementary, Secondary (classical, technical, and normal), and Superior (university).

New *Facoltà di Fisica, Matematica e Scienze naturali* were created in the universities. Moreover, two engineering schools were opened, called respectively *Scuola d'applicazione per ingegneri* and *Istituto tecnico superiore* in Turin and Milan. Only students who had done two first years of mathematical studies in universities were allowed into these schools. Consequently, the training of engineers was divided into two phases: a first phase of mathematical studies in the Faculty of Science (which also included descriptive geometry) and a second phase of application studies in the engineering schools. Even when schools for engineers were gradually established in other major Italian cities (Palermo, Naples, Bologna, Rome) this structure was maintained and remained unchanged for the rest of the nineteenth century.[9]

---

[6]Anon. (actually Antonio Cioci), *Geometria e meccanica delle arti, dei mestieri e delle belle arti, ad uso degli artisti, e direttori d'officine e manifatture, del Barone Carlo Dupin*, 1829, and Giacomo Laderchi, *Geometria e meccanica delle arti e mestieri e delle belle arti del Barone Carlo Dupin. Prima versione italiana*, 1829–1830.

[7]Luigi Poletti, *Geometria applicata alle arti belle e alle arti meccaniche*, 1829; Giovanni Alessandro Majocchi, *Manuale di geometria per le arti e pei mestieri*, 1832.

[8]Venice was annexed in 1866 and the city of Rome in 1870.

[9]The courses for architects and engineers were reformed during the Fascist rule (see also Menghini, Chap. 4, this volume).

Under the Casati law, Secondary classical (humanistic) instruction consisted in the *Ginnasio-Liceo* axis. Technical instruction constituted a separate sector and was divided into two levels called *Scuola tecnica* and *Istituto tecnico*, respectively. The *Scuola tecnica* started after elementary school; the *Istituto tecnico* started after *Scuola tecnica* and was divided into different sections, one of which—called *Sezione Fisico Matematica (SFM)*—was dedicated to mathematics and scientific studies and, unlike the others, provided access to university. Descriptive geometry was never taught in the *Ginnasio-Liceo*. It was instead taught in the *SFM* and in the two new sections (industrial and for surveyors) created after 1859. In fact, the regulations, the structure, and the syllabi of technical institutes were changed eight times between 1859 and 1891 (Scoth 2010). Moreover, descriptive geometry was taught in nautical schools during the entire nineteenth century.

2.2 *Descriptive Geometry in the Universities and Engineering Schools*

As regards higher education, one of the effects of the Casati law was that descriptive geometry was taught only at the beginning of university mathematics courses (Fig. 3.2) except for the engineering schools where its applications were taught

Anno Primo

Primo Semestre	*Secondo Semestre*
Introduzione al calcolo	Come nel primo Semestre meno la Geometria descrittiva
Fisica	
Geometria descrittiva	
Disegno	

Anno Secondo

Calcolo differenziale ed integrale	Come nel primo Semestre meno la Geometria descrittiva
Geometria descrittiva	
Fisica	
Disegno	

Anno Terzo

Meccanica razionale	Come nel primo Semestre
Geodesia	
Chimica	
Disegno	

13. Il Professore di disegno sarà coadiuvato da Assistenti, a cui potrà essere affidato l'insegnamento di qualche parte del disegno. Questi saranno in tal numero che a ciascun di loro ed al Professore tocchino circa trenta studenti.

Gli studi del disegno avranno per oggetto più specialmente l'*architettura*, le *macchine* e la *topografia*.

14. Gli esami orali speciali che gli studenti di Matematica debbono sostenere per conseguire la laurea, appaiono nel seguente specchio:

N.° d' ordine	ESAMI	Durata dell' Esame
1	Introduzione al calcolo	45′
2	Calcolo differenziale ed integrale	45′
3	Meccanica razionale	45′
4	Geodesia	30′
5	Fisica sperimentale	30′
6	Chimica	30′
7	Geometria descrittiva	30′
8	Disegno	30′

15. Prima degli esami orali di Geometria descrittiva e Disegno, si eseguiranno due componimenti, l'uno sulla Geometria descrittiva, e l'altro sul Disegno.

I temi di questi componimenti saranno tratti a sorte da quelli proposti ne' programmi degli esami; si faranno a porte chiuse in una sessione non maggiore di sei ore per la Geometria descrittiva, ed in due sessioni ciascuna non maggiore di sei ore pel Disegno.

16. I componimenti, di cui all'articolo precedente, saranno sottoposti alla Commissione deputata all'esame orale; della Commissione per l'esame di disegno farà parte l'Assistente che vegliò il candidato nell'esecuzione del lavoro.

Il candidato presenterà pure alla Commissione per l'esame di disegno i lavori fatti durante il corso.

17. Gli esami speciali debbono farsi entro un quinquennio comin-

Fig. 3.2 A timetable of the mathematical courses in Italian Universities (1860). Descriptive geometry was mandatory in the first 2 years (Decreto 1860, pp. 2174–2175)

ORARIO DELLA SCUOLA D'APPLICAZIONE

1ª CLASSE

	8–9	9–10	10–11	11–12	12–1	1–2	2–3	3–4	4–5
Lunedì	Meccanica razionale	Applicazioni di geometria descrittiva			Statica grafica		Disegno di statica grafica		
Martedì	Disegno di applicazioni di geometria descrittiva				Mineralog.ª e geologia applicate		Chimica docimastica		
Mercoledì	Meccanica razionale	Geodesia			Statica grafica		Disegno di statica grafica		
Giovedì	Meccanica razionale	Applicazioni di geometria descrittiva			Mineralog.ª e geologia applicate		Chimica docimastica		
Venerdì	Disegno di applicazioni di geometria descrittiva				Statica grafica		Disegno di statica grafica		
Sabato	Meccanica razionale	Geodesia			Mineralog.ª e geologia applicate		Chimica docimastica		

Fig. 3.3 Subjects and lesson timetables in the Engineering School of Rome (1877/78). Descriptive geometry was limited to the applications (Programma 1877, p. 20)

exclusively (Fig. 3.3). This situation also concerned other branches of mathematics and was a consequence of the fact that engineering schools were not considered high school alternatives to university faculties but as university post-training schools. In these conditions, the teaching of descriptive geometry in universities was assimilated into other branches of pure mathematics. The courses were almost always held by mathematicians who were more interested in investigating the properties of three-dimensional geometric bodies rather than developing the drawing of orthogonal projections and practical applications. Luigi Cremona, one of the most famous Italian mathematicians of the time, who taught descriptive geometry at the University of Bologna, wrote: "I think I am unfit to teach descriptive geometry; I am not a designer, drawing figures makes me uncomfortable and I am not an expert in applications" (Gatto 1996, p. 26).

Besides Cremona, many famous Italian mathematicians were professors of descriptive geometry, for example, Gino Fano, Gino Loria, and later Beppo Levi, Federigo Enriques and Francesco Severi.

In 1875, a compulsory course in projective geometry separate from that of descriptive geometry was introduced into the mathematical degree courses at Italian universities. Research into projective geometry had many development prospects, and mathematicians followed them with great attention. At that time in Italian engineering schools, the teaching of graphic statics was introduced, and this made preparatory teaching of projective geometry at universities even more important. The introduction of graphic statics had already been tested at the end of 1860s in the *Istituto tecnico superiore* in Milan, where the director (the mathematician Francesco Brioschi) and Cremona had introduced new methods devised by Karl Culmann and

Theodor Reye at the Polytechnic of Zurich (see also Benstein, Chap. 9, this volume) into Italy.

In schools for engineers, applications of descriptive geometry were focused on classic topics: contour lines, perspective, shadows, stereotomy, and photogrammetry. In some cases, applications were taught in the mathematics faculty but were generally limited to the more purely geometrical topics, perspective, theory of shadows, or gnomonic projections.

New textbooks of descriptive geometry were produced in Italy in the second half of the nineteenth century and the first decades of the 20th. As well as those written by Fano (1910) and Loria (1909) the books written by Ferdinando Aschieri, professor of descriptive geometry at the University of Padoa (Aschieri 1884) and by Federigo Enriques (1902) were quite successful. All these texts were reprinted and revised several times. Those of Loria and Fano, in particular, considered the two methods of central projection and orthogonal projection, following a trend that had developed since the introduction of projective geometry course at Italian universities.

2.3 Descriptive Geometry in Secondary Technical Education: the SFM

After the Casati law, the first syllabi for technical institutes were introduced in 1860. Descriptive geometry was taught in the *SFM* and was a part of the course in pure and applied mathematics.[10] These syllabi were changed a few years later. In fact, a law of 1861 shifted the technical institutes from the Ministry of Public Education to Ministry of Industry. Consequently, technical schools remained isolated from the rest of the school environment and the regulations were changed several times in search of an optimal structure.

In 1865, the mathematics course was modified and descriptive geometry became an independent subject. The syllabus was divided (according to the teaching tradition that dated back to the time of Monge) into a theoretical part and a more practical one, which also included stereotomy. The changes of the projection planes and the rotations were added to the theoretical part.

In 1871, the *SFM* was transformed into a specific preparatory course for higher education in engineering. The mathematics syllabi were expanded and modernized by Cremona and projective geometry was introduced (see Menghini, Chap. 4, this volume; Menghini 2006), and the parallel course in descriptive geometry was assigned to a professor other than mathematics. This reform, however, was

---

[10] All mathematics syllabi cited in this chapter are available at http://www.associazionesubalpinamathesis.it/storia-insegnamento/provvedimenti-legislativi/.

unsuccessful for various reasons (Scoth 2011, pp. 276 ff.), but partly because the curricula were far too innovative.

In 1876, technical institutes were back under the control of the Ministry of Education. In the *SFM*, mathematics syllabi were gradually reduced in 1876, in 1885, and in 1891, and the teaching of descriptive geometry experienced a steady decline.

In 1885, projective geometry was removed and descriptive geometry, in the context of orthogonal projections, was absorbed into the mathematics course. In 1891, the syllabus was restricted to the representation of points, lines, and planes and to a limited portion on the representation of solids. In the early twentieth century, descriptive geometry took up such an irrelevant role in teaching to the point that two famous mathematicians, Enriques and Severi, proposed eliminating it from the teaching programs of *SFM* "because it is taught more effectively in the Universities, after a good introductory course of projective geometry" (Enriques et al. 1903, p. 55).

This situation had important repercussions on the production of elementary treatises. In contrast with the important production of other mathematics treatises for secondary schools, only a dozen textbooks on descriptive geometry were published in Italy between 1860 and 1900.[11] Among the most important examples are the text for adoption in the *Istituto tecnico* in Florence (Peri 1869) (Fig. 3.4), the Italian translation (reprinted five times) of *Leçons nouvelles de Géométrie descriptive* by the French mathematician Antoine Amiot[12] (Mazzitelli 1875), and the treatise written by Salvatore Ortu Carboni (1894–1895), a mathematician who was very active in the Italian debate on the teaching of mathematics in secondary schools.[13] Even the handbooks of geometry that had a section devoted to descriptive geometry were very few, and this overall shortage forced teachers and students in some cases to use university treatises or old texts from the first half of the nineteenth century that were unsuitable for elementary teaching.

---

[11] The list is available at http://www.associazionesubalpinamathesis.it/storia-insegnamento/libri-di-testo/#1513285344390-8eed2f07-ed22.

[12] About Amiot see also (Barbin, Chap. 2, this volume).

[13] Ortu Carboni was one of the few Italian mathematicians who at the time had dealt with the problems of secondary teaching of descriptive geometry. He believed in the educational value of this discipline and during his speech at the congress of the Association of Italian mathematics teachers (the *Mathesis*) in 1901, he said: "Even in Italy descriptive geometry should become a subject of elementary mathematics courses. It helps to broaden and deepen geometric ideas, to arouse interest in the examination of figures, to understand many simple representations of everyday life and to present to people a broader view of the territories of mathematics. These few reasons are enough to show how important it is." (Ortu Carboni 1902, p. 110).

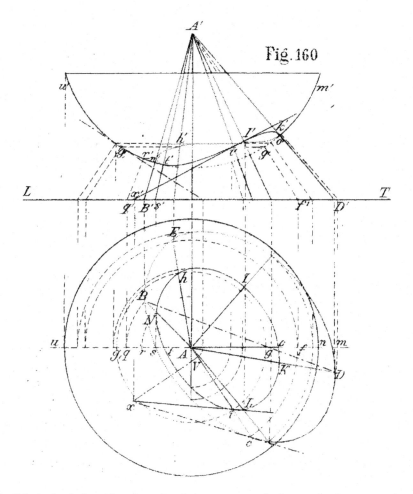

Fig. 3.4 A classical problem from descriptive geometry: the intersection between conic and spherical surfaces (Peri 1869, vol. II p. XVI)

2.4 The Teaching of Descriptive Geometry in Italian Secondary Schools After Unification: A Case Study

At this point it is worth asking a question: why did Italy, the country that first imported the tradition of Monge, that had many scholars and has produced new ideas and books in the field of descriptive geometry, gradually neglect secondary teaching of this discipline after unification? It is not easy to give an exact answer. We should first of all observe that the decline of descriptive geometry since the second half of the nineteenth century has not been a specifically Italian trend. As you can read in this volume, descriptive geometry in secondary education at that time did not adequately developed also in other European countries. In the case of

Italy, however, we need to consider particular factors that may have led to this state of affairs and make some preliminary comments.

- After unification, the establishment of effective secondary technical education was a difficult problem in Italy. The Casati law was designed for a small country like the Kingdom of Sardinia, but the speed with which unification took place forced the authorities to extend it throughout Italy. Therefore, this law was not the best solution to harmonize the different school systems of the pre-unification states. Moreover, according to the ideology of the time, humanistic studies were privileged and technical education was considered less important (Giacardi and Scoth 2014, p. 217). In Italy, humanistic education was financed directly by the state, while technical education was financed by the municipalities that did not have sufficient resources for scientific labs, materials, and skilled teachers. For this reason, technical education suffered significantly in those years.
- Another problem was that of creating an efficient model of public technical education. In Italy, there were no previous examples to refer to, and, for many years, governments proceeded with uncertainty by looking at the experience of the main European countries. For this reason, there were constant changes of syllabi and regulations, and the technical institutes were placed under the supervision of the Ministry of Industry.

Other considerations relate more specifically to the teaching of mathematics.

- At that time, the teaching of mathematics in Italian secondary schools was characterized by a rigorist approach, excessive use of deductive reasoning, dogmatism, and recourse to mnemonic study.
- In general, the syllabi were poorly designed, not finalized to inter-disciplinarity and in some cases over-extended.
- The methods used to recruit and train teachers were inadequate. Only a small number of them had a degree in mathematics or related disciplines, and, especially in the lower grade levels, they were recruited without possessing the necessary qualifications.

Moreover, the syllabi were often superficiality designed and without taking account of educational needs. Those of 1865 for the *SFM*, for example, anticipated the teaching of drawing of orthogonal projections by 1 or 2 years compared to that of descriptive geometry.[14] Those of 1871 proved to be too demanding for of the secondary technical education in those years and for the preparation of teachers (Scoth 2011, p. 276).

The insufficient number of hours devoted to the study of mathematics in relation to the content of the curriculum forced teachers reduce the hours taught. In technical institutes, descriptive geometry was the most frequently neglected

---

[14] We should consider that in Italian technical institutes the method of orthogonal projections was taught also in courses of drawing that included a part of artistic and a part of industrial drawing. This teaching, however, was detached from that of descriptive geometry and was often assigned to teachers who came from the world of art and had no competence in the field of industrial drawing (MAIC 1869, p. 180).

discipline. Besides, a survey done at the beginning of the twentieth century by *Mathesis* revealed that it was not either taught, it was reduced to a few concepts with some drawings by more than half of the teachers interviewed (Scorza 1911, p. 74).

These conditions not only certainly had an impact on the teaching of descriptive geometry but also especially on its training value as a means to develop spatial intuition. Furthermore, the didactic importance of descriptive geometry for the study of three-dimensional Euclidean geometry was not considered,[15] and this pedagogical conception was the fruit of the rigorist and rational trend in Italian teaching. Consequently, descriptive geometry was always seen as an atypical science which was located half way between pure mathematics and engineering sciences and was only taught in technical institutes and nautical schools, where it was thought to be useful for applications. The structure of the technical institutes, however, did not allow an optimal teaching of descriptive geometry. In the *SFM*, where the scientific subjects were predominant, this discipline was considered redundant because it had no links to applicative topics, and for these reasons, mathematicians like Enriques and Severi requested its abolition. In other sections of the technical institutes, a utilitarian purpose prevailed and descriptive geometry was reduced to a simple graphic exercise.

After the fascist reform of education (1923), all scientific teaching was weakened in Italian secondary schools. Technical institutes were reformed and the *SFM* was closed. Descriptive geometry was taught only in industrial and nautical schools and was restricted only to the drawing of orthogonal projections.

References

Aschieri, Ferdinando. 1884. *Geometria projettiva e descrittiva*, vol. II (Geometria descrittiva). Milano: Hoepli.
Bellavitis, Giusto. 1851. *Lezioni di geometria descrittiva. Con note contenenti i principii della geometria superiore ossia di derivazione, e parecchie regole per la misura delle aree e dei volumi*. Padova: Tipografia del Seminario.
Càndito, Cristina. 2003. *Le proiezioni assonometriche. Dalla prospettiva isometrica all'individuazione dei fondamenti del disegno assonometrico*. Firenze: Alinea.
Conte, Antonio, and Livia Giacardi. 1990. La matematica a Torino. In *Ville de Turin*, ed. Giuseppe Bracco, 1798–1814, vol. II , 281–329. Torino: Archivio Storico.
Corridi, Filippo. 1838. *Trattato di Geometria Descrittiva di G. Monge coll'aggiunta d'una teorica delle ombre e della prospettiva estratta dalle lezioni inedite dell'autore per cura del Sig. Brisson. Prima versione italiana*. Firenze: Ricordi e Compagno.

---

[15]This aspect was highlighted during the congress of the *Mathesis* in 1919, organized to standardize the programs of the former Austrian schools assigned to Italy after World War I. In Austria descriptive geometry "had got a lot of attention on the part of school authorities", while in Italy "it had been kept away from middle school because it was not considered necessary to provide students with a good education on conception of space" (Nordio 1920, p. 40).

Decreto. 1860. Decreto Luogotenenziale 7/11/1860 n. 4403 che approva il regolamento per la Facoltà di Scienze matematiche, fisiche e naturali. *Raccolta degli Atti del Governo di Sua Maestà il Re di Sardegna* 28: 2169–2187.

Enriques, Federigo. 1902. *Lezioni di geometria descrittiva*. Bologna: Zanichelli.

Enriques, Federigo, Francesco Severi, and Antonio Conti. 1903. Relazione sul tema: Estensione e limiti dell'insegnamento della matematica, in ciascuno dei due gradi, inferiore e superiore, delle Scuole Medie. *Il Bollettino di Matematica 2* 3–4: 50–56.

Fano, Gino. 1910. *Lezioni di geometria descrittiva date nel R. Politecnico di Torino*. Torino: Paravia.

Fiocca, Alessandra. 1992. La geometria descrittiva in Italia (1798–1838). *Bollettino di Storia delle Scienze Matematiche 12* 2: 187–249.

Flauti, Vincenzo. 1807. *Elementi di geometria descrittiva*. Roma: Salvioni.

——. 1815. *Geometria di sito sul piano, e nello spazio*. Napoli: Stamperia della Società Tipografica.

Gatto, Romano. 1996. Lettere di Luigi Cremona a Enrico Betti (1860–1890). In *Per la corrispondenza dei matematici italiani. La corrispondenza di Luigi Cremona (1830–1903)*, ed. Marta Menghini, vol. III, 7–90. Quaderni P.RI.ST.EM., n. 9. Milano: Università Bocconi.

Giacardi, Livia, and Roberto Scoth. 2014. Secondary school mathematics teaching from the early ninetheent century to the mid-twentieth century in Italy. In *Handbook on the History of Mathematics Education*, eds. Alexander Karp, Gert Schubring, 201–228. New York: Springer.

Loria, Gino. 1909. *Metodi di Geometria descrittiva*. Milano: Hoepli.

——. 1921. *Storia della Geometria descrittiva*. Milano: Hoepli.

MAIC (Ministero di Agricoltura, Industria e Commercio). 1862. *Relazione del ministro di Agricoltura, industria e commercio (Pepoli) sopra gli Istituti tecnici, le Scuole di arti e mestieri, le Scuole di nautica, le Scuole delle miniere e le Scuole agrarie. Presentata alla Camera dei Deputati nella tornata del 4 luglio 1862*. Torino: Tipografia della Camera dei Deputati.

MAIC. 1869. *Gl'Istituti Tecnici in Italia*. Firenze: Tipografia di G. Barbèra.

Mazzitelli, Domenico. 1875. *Nuove lezioni di geometria descrittiva riordinate ed accresciute di applicazioni alle ombre e del metodo dei piani quotati da A. Chevillard, Professore di prospettiva alla Scuola imperiale di belle arti. Traduzione eseguita sulla terza edizione francese*. Napoli: Pellerano.

Mazzotti, Massimo. 1998. The geometers of God. *Isis* 89: 674–701.

Menghini, Marta. 2006. The role of projective geometry in Italian education and institutions at the end of the 19th century. *International Journal for the History of Mathematics Education* 1 (1): 35–55.

Nordio, Attilio. 1920. Sull'insegnamento della geometria descrittiva. *Bollettino della Mathesis* 1–4 (12): 32–41.

Ortu Carboni, Salvatore. 1894–1895. *Geometria descrittiva elementare ed alcune sue applicazioni. Vol. I (Problemi fondamentali e metodi della rappresentazione), vol. II (Taglio delle pietre e dei legnami, esercizi; appendici)*. Torino-Roma-Milano-Firenze-Napoli: Paravia.

——. 1902. L'insegnamento della matematica nelle Scuole e negli Istituti tecnici. In *Atti del 2° Congresso dei professori di matematica delle scuole secondarie tenuto in Livorno nei giorni 17–22 agosto 1901 ad iniziativa dell'Associazione Mathesis*, 69–127. Livorno: Giusti.

Pepe, Luigi. 2003. Matematica e matematici nell'Italia repubblicana (1796–1799). In *Universalismo e nazionalità nell'esperienza del giacobinismo italiano*, eds. Luigi Lotti, Rosario Villari, 323–337. Roma-Bari: Laterza.

Peri, Giuseppe. 1869. *Corso elementare di geometria descrittiva. Libri tre seguiti seguiti da un'appendice sul metodo delle projezioni quotate. Vol. I (Testo), vol. II (Atlante)*. Pistoia: Niccolai e Quarteroni.

Placci, Giuseppe. 1805. *Trattato elementare di geometria descrittiva. Tomo I che contiene le lezioni di geometria descrittiva di Gaspard Monge tradotto dal francese con note*. Bologna: Masi.

Programma. 1877. *Programma della R. Scuola d'Applicazione per gli Ingegneri in Roma per l'Anno Scolastico 1877–78*. Roma: Salviucci.

Scorza, Gaetano. 1911. L'insegnamento della matematica nelle Scuole e negli Istituti tecnici. *Bollettino della Mathesis* 3: 49–80.

Scoth, Roberto. 2008. Gli insegnamenti matematici nella legge Casati: il caso della Geometria Descrittiva. *L'educazione Matematica* 3 (29): 32–42.

———. 2010. La matematica negli istituti tecnici italiani. Analisi storica dei programmi d'insegnamento (1859–1891). Supplemento a *L'educazione Matematica* 2 (31).

———. 2011. I programmi di matematica per gli istituti tecnici italiani del 1871: ricadute didattiche di un progetto avveniristico. *Annali di Storia dell'educazione e delle istituzioni scolastiche* 18: 259–283.

———. 2016. Higher education, dissemination and spread of the mathematical sciences in Sardinia (1720–1848). *Historia Mathematica* 43: 172–193.

Sella, Quintino. 1856. *Sui princìpi geometrici del disegno e specialmente dell'axonometrico. Dalle lezioni di geometria applicata alle arti dette nel R. Istituto Tecnico di Torino*. Torino.

Sereni, Carlo. 1826. *Trattato di geometria descrittiva*. Roma: De Romanis.

———. 1846. *Applicazioni di geometria descrittiva*. Roma: Salviucci.

Torelli, Gabriele. 1875. Notizie storiche relative alla teoria delle trasformazioni in geometria descrittiva. *Giornale di Matematiche* 13: 352–355.

Tramontini, Giuseppe. 1811. *Delle projezioni grafiche e delle loro principali applicazioni. Trattato teorico-pratico ad uso della Reale scuola militare del Genio, e dell'Artiglieria come ancora di tutti i giovani architetti, ed ingegneri civili*. Modena: Società Tipografica.

Tucci, Francesco Paolo. 1823. Su la permutazione de' piani di projezione in Descrittiva. *Biblioteca analitica* 1: 129–137.

Chapter 4
Luigi Cremona and Wilhelm Fiedler: The Link Between Descriptive and Projective Geometry in Technical Instruction

Marta Menghini

Abstract This paper considers Luigi Cremona's and Wilhelm Fiedler's outlook on technical instruction at school and university level, their vision about the educational role of descriptive geometry and its relation to Monge's original conception. Like Cremona, Fiedler sees a symbiosis between descriptive and projective geometry via the fundamental idea of central projection. The link between projective and descriptive geometry plays a double role: an educational one due to the graphical aspects of the two disciplines and a conceptual one due to the connection of theory to practice. Thus, projective and descriptive geometry contribute to form a class of scientifically educated people, and the link between them epitomizes—in the opinion of Cremona—the link between pure mathematics and its applications. According to Fiedler, the main scope of the teaching of descriptive geometry is the scientific construction and development of "Raumanschauung", as stated in a paper published in the Italian journal *Giornale di Matematiche*. The textbooks by Fiedler (1874) and Cremona (1873) were used in Italy to develop the geometry programs for the *sezione fisico matematica* (physics and mathematics section) within technical secondary instruction. While the relation between projective and descriptive geometry—and, thus, between pure and applied mathematics—had a short life at secondary school level in Italy, at the turn of the century there was a new expansion at university level due to the important role that the mathematicians had in the creation of the Faculty of Architecture.

Keywords Descriptive geometry · Projective geometry · Luigi Cremona · Wilhelm Fiedler · History of technical education

M. Menghini (✉)
Department of Mathematics, Sapienza–University of Rome, Rome, Italy
e-mail: marta.menghini@uniroma1.it

1 Introduction

In the 1870s, soon after the unification of Italy, the new country established its educational system based on different previous experiences (Scoth, Chap. 3, this volume). These were crucial years for technical instruction, which was particularly influenced by French and German models. The new programs for the *sezione fisico matematica* (physics and mathematics section) of the *Istituti tecnici* (technical institutes that are Italian technical secondary schools), set up by Luigi Cremona in 1871 represented the apex of the teaching of descriptive and projective geometry in Italy (Menghini 2006). In 1873, Cremona published his book on projective geometry to accompany the programs of the physics-mathematics section. In the same year, Cremona contributed to the creation of the "Citadel of Science" in Rome, a new university site that had to "physically" highlight the link between mathematics and its applications and, in particular, the link between projective and descriptive geometry. Furthermore, Wilhelm Fiedler's book on *descriptive geometry* was translated into Italian in 1874 and was used, alongside Cremona's text, in technical institutes.

The political unification of Italy in 1861 led to various areas of Italian mathematics becoming integrated into the context of European research. The most eminent Italian mathematicians were involved in bringing Italy back to the forefront of international developments in the fields of science and economics. Of significance in this context were geometric studies of a synthetic nature, which achieved their greatest development in the school of geometry headed by Luigi Cremona, whose extensive correspondence (Israel 2017) shows the many links with foreign researchers. Cremona anchored his work to the classical school of projective geometry, with particular attention given to the ramifications of the work by Jean Victor Poncelet and Michel Chasles in France and by Christian von Staudt, Julius Plücker, August F. Möbius, Jakob Steiner and Alfred Clebsch in Germany.

Among the letters to Cremona, there are 33 letters sent by Wilhelm Fiedler between 1862 and 1888 (Knobloch and Reich 2017). The correspondence mostly concerns the exchange of publications, but we also find references to the roots of the two geometers (René Descartes, Girard Desargues, Brook Taylor, Johann H. Lambert, Gaspard Monge, Poncelet, Möbius, Steiner, Chasles, von Staudt, Plücker) and to authors that both appreciated (such as George Salmon, Carl Culmann, Clebsch, Felix Klein and Theodor Reye). Very often Fiedler confirms his interest, and he praises Cremona's book and the simple way in which Cremona introduces the topics in the teaching of descriptive geometry in a letter written at the beginning of 1873. Furthermore, he praises Italian technical education.

With the arrival of Cremona in Rome in 1873, an interesting project was initiated based on the experiences of the Polytechnical Schools in Northern Italy. The project transferred part of the structures and professors of the Faculty of Science from the old 'Sapienza' to a new site at San Pietro in Vincoli. Thus, a sort of 'citadel of science' was established, in which, along with other disciplines, all teaching of a mathematical nature was brought together in a newly founded autonomous

Institute of Mathematics. In the new setting, one could find beside the School for Engineers, the School of Mathematics, the Library and the School of Drawing and Architecture. All these schools were part of the University of Rome. The position that mathematics should occupy in science is clearly reflected in this "physical" arrangement.

Again, this decision reflected the close ties between aspects of a theoretical nature and those of a 'concrete' nature[1]—linked to the practice of drawing—in the ideas of that time and in those of Cremona in particular, who headed the "School for Engineers" until his death in 1903 (after Cremona's death, another mathematician,[2] Valentino Cerruti, headed the school). So, the 'mathematical school' of the Faculty of Science could soon appreciate the relationship between applied and pure research, where the studies in descriptive geometry represented a clear example of the tradition inspired by Monge and by the French Polytechnics schools.

As shown in many papers of this volume, the *École* influenced the development of technical schools in many countries in nineteenth century Europe (also see Schubring 1989; Barbin and Menghini 2014). Concerning the *Politecnici* in Italy, there was a direct influence due to the French occupation (1796–1813); the situation was maintained during the Restoration (see Scoth, Chap. 3, this volume), and descriptive geometry spread widely in the Italian states, thanks to the military schools and the universities, which preserved the syllabuses of the French era.

The key role played by Monge is not only to be considered for its institutional influence but also for its influence on aspects of a didactic nature concerning the teaching of geometric disciplines in the nascent technical secondary schools. The fundamental role given to descriptive geometry and to geometric drawing in such schools fostered the learners' intelligence by giving them the habit and the feeling for precision (Monge 1839, Vol. II, n. V).

These aspects were taken up again by Luigi Cremona in the preface to the programs of 1871, where he stressed the educational role of descriptive geometry in secondary education and the exactness of its methods. Cremona never neglected the applications of geometry and attributed considerable importance to drawing and thereby preserved Monge's ancient conception of this doctrine.

Cremona proposed a theoretical introduction of a projective environment to address the issues of 'graphic' character in his courses on static graphics at the Institute of Higher Technical Education in Milan and in his courses at the University of Bologna. He introduced some fundamental views of higher geometry in his lecture course on descriptive geometry held in Bologna from 1861 to 1867. In a note from 1865, written with the pseudonym (anagram) Marco Uglieni, Cremona explained the solution to some graphic problems of central projection proposed in a booklet by Brook Taylor (see Lawrence, Chap. 18, this volume), following the methods used later by Wilhelm Fiedler in descriptive geometry.

---

[1] Some of the considerations contained in this section, as well as in the last section, were developed with L. Dell'Aglio (see Dell'Aglio et al. 2001).

[2] Actually, both Cremona and Cerruti were engineers.

So, with reference to teaching, the position of Cremona is characterized by the interchange of theoretical and graphical methods. This view—which assigns a central role to projective geometry and recognizes the 'empirical' importance of a graphic discussion of its principles—is essentially classic.

2 Projective and Descriptive Geometry in Secondary Schools: Cremona and Fiedler

2.1 Cremona's Projective Geometry

The teaching of geometry in the *Istituti tecnici*, the Italian secondary schools for technical instruction (Menghini 2006) was reformed in 1871. The reform presumed the explicit introduction of the fundamental principles of projective geometry, which was deemed to be a necessary theoretical preamble to the study of descriptive geometry. The aim was to form a course focusing on projective geometry, which was able to compete with its counterpart in the humanities.

The main points of the programs are summarized here. Geometry includes:

- the theory of projections of geometric forms (projective ranges and pencils, cross ratio, complete quadrilateral) and its application to the graphical solution of the problems of first and second degree and to the construction of the curves of the second order, seen as projections of the circle (this requires: projective ranges in a circle and self-corresponding elements of superposed forms);
- the theory of involution (conjugate points with respect to a circle);
- the duality principle in the plane;
- elements of stereometry and the graphic construction of the barycentres of plane figures.

Cremona's *Elementi di Geometria Projettiva*, published in 1873, was written to accompany the part of the syllabus concerning projective geometry. The book, which—as Cremona himself states in the introduction—owes much to Poncelet, also had great success outside of Italy and had numerous translations. But, although the book was written for Italian secondary schools, it was to be adopted mostly at university level outside Italy.

From this work, the conviction clearly emerges of the importance of inserting theoretical topics concerning projective geometry within the curriculum of future engineers—from the secondary technical institutes to the Polytechnics—even when the central purpose was related to drawing. In the introduction to the book, Cremona claims that the methods of projective geometry could, one day, solve the problem of teaching geometry also in classical instruction. He writes that he applied the methods of projective geometry to the teaching of descriptive geometry when he was at Bologna University and that he followed the methods suggested by Fiedler.

In his book, Cremona makes reference to the "affine" formulation of the theorems, which considers projections of figures from a plane onto a parallel plane, thus using parallelism and points at infinity. For example, he introduces homothetic (similar) triangles as a significant case of homological (perspective) triangles or parallelograms as a particular case of quadrilaterals; he uses length and sign of a segment, together with similitude, to prove the invariance of cross ratio in accordance with Moebius' barycentric calculus rather than basing it on the complete quadrilateral as von Staudt and Reye did.

After establishing the fundamental concepts of space, such as surface, line, point, straight line and plane, Cremona introduces the important concept of collineation. Like Fiedler, Cremona starts from central projection, but, differently from Fiedler, he considers only the graphic (projective) properties.

So he considers a centre of projection O and a figure made of points and lines *ABCabc*... The rays that connect the centre with the points or the lines of the given figure (which form straight lines and planes) are cut by a plane called the picture plane (Fig. 4.1, left).

Then he gives some properties: for instance, if the figure *ABCabc*... lies on a plane, we can have a one to one correspondence between the picture plane and the plane of the figure by introducing the points at infinity. The image of a point at infinity is a vanishing point. Cremona can as such introduce the Desargues theorem and homology in space. The Desargues theorem on the plane is obtained by applying the construction of Fig. 4.1 (left) twice: first from σ to σ' and then from σ' to another triangle on σ (Fig. 4.1, right).

The focal properties of conic sections, mentioned in the programs of 1871, were to have been covered in the second volume, but this volume was never published because the syllabuses were reduced in 1876.

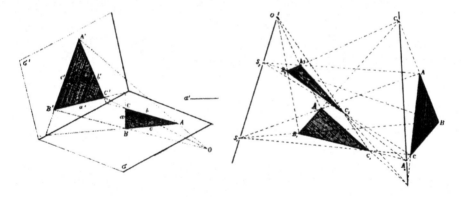

Fig. 4.1 From table I of figures in Cremona (1873)

2.2 Fiedler's Descriptive Geometry

The section from the same programs for the physics-mathematics class of 1871 concerning the teaching of descriptive geometry states that the teacher should start from central projection and from the projective properties of figures and should handle the theory of collinearities, of affinities and of similarities, paying attention to homology, up to the construction of intersections of surfaces of the second degree. The teachers of mathematics and descriptive geometry should cooperate, as both are concerned with the projection of geometric figures (Ministero di agricoltura 1871, pp. 52–63).

The Italian translation of the book on descriptive geometry by Wilhelm Fiedler (Fiedler 1874) appeared in 1874.[3] Although it was written for the *Technische Hochschulen*, which were near to university level in Germany (see Benstein, Chap. 9, this volume), the Italian edition was explicitly translated and adapted for use at the secondary school level in Italian technical institutes.

It was certainly appropriate to have Fiedler's book alongside Cremona's "Projective Geometry" in the parallel course on descriptive geometry at the technical institutes. According to Fiedler, the main scope of the teaching of descriptive geometry is the scientific construction and development of "Raumanschauung" (space-intuition). Fiedler reinforced this point of view in a paper translated and published in the *Giornale di Matematiche*.[4] Fiedler sees a complete symbiosis between descriptive and projective geometry and maintains that by starting from central projection, which corresponds to the process of viewing, we can develop the fundamental parts of projective geometry in a natural and complete way (Fiedler 1878, p. 248). He feels supported by the Swiss pedagogue and educator Johann Heinrich Pestalozzi, who argued that teaching must start from intuition. Fiedler sees these strategies as the best method for the reform of geometry teaching at all levels (in this respect, he shares Monge's opinion about the role of descriptive geometry in rethinking secondary education). Moreover, great importance is given to the parallel development of plane and solid geometry, to the duality principle and to motion and geometrical transformations.

According to Fiedler, projective geometry allows us to start from a few fundamental relations to construct all geometry. History shows, in his opinion, that these fundamental relations are substantially linked to the methods of descriptive

---

[3] The original title "Die darstellende Geometrie in organischer Verbindung mit der Geometrie der Lage" (1871) was simply translated as "Trattato di Geometria descrittiva". In the preface of his *Elementi di Geometria Projettiva*, Cremona considers 'Geometrie der Lage' (geometria di posizione/geometry of position), to be equivalent to projective geometry (geometria projettiva).

[4] This journal was founded in 1863 in Napoli by Giuseppe Battaglini (with his colleagues Vincenzo Janni and Nicola Trudi) and was directed by him until 1893. In 1894, the journal changed its name to become *Il giornale di matematica di Battaglini*. The journal was addressed principally to "young scholars of the Italian universities to serve as a connection between university lectures and other academic issues". It was much appreciated by people interested in mathematics education (Furinghetti 2017).

geometry and to the fact that Poncelet used the methods of perspective, already developed by Desargues, as basis for his work. Fiedler also refers to the 'elementary forms' of Jacob Steiner and to the search for the projective properties of figures by means of projections. He also refers to Möbius when looking at congruences, similarities and affinities as particular collineations, which can be completely understood starting from central projection. Thus, the whole of geometry must become descriptive (Fiedler 1878, pp. 243–248). The representation methods are a premise to 'motion', which can be organically introduced in geometry from the beginning; thereby, geometry is the science of comparing figures that are created one from another by means of representation.

In a letter to the editor of the journal the *Giornale di Matematiche* (Giuseppe Battaglini), which is annexed to his paper, Fiedler explains his view on geometry teaching, proposing a chronology of the arguments to be developed:

- START. Intuitive geometric teaching. Drawings from models made by bars and the building of nets. Geometric drawing.
- BEGINNING OF THEORY: Deductions and definitions with many combinations of forms. Exercises on definitions and their use. Central projection for simple forms in plane and in space. Duality principle.
- GEOMETRY AS THE SCIENCE OF COMPARISON. Comparison of figures that are mapped on to each other. Visual parallel rays: congruence, reflection (in a line/plane) and affinity. Trigonometry and Cartesian coordinates together with descriptive geometry (Monge). Visual concurring rays: similarity, reflection in a point, collinearity and involution.
- GENERAL IDEA OF PROJECTIVITY AND EXTENSION THROUGH NON-REAL ELEMENTS. Projective coordinates and algebraic and geometric treatment of the different forms. That is synthetic and analytic teaching of geometry of position.

In his textbook of 1874, Fiedler explains that his aim is to deduce from the methods of representation all the elements that are necessary for the study of the properties of figures. In the first part, he indicates those methods by which one passes from a given figure in space to another figure in the plane or in space. He looks for the properties common to the two figures and for those properties that do not change in the given representation. The simplest and most natural method of representation is central projection. So he starts from this (see Volkert, Chap. 10, this volume).

In the third section of Fiedler's book, the author presents homology in three dimensions, which allows representing any figure given in space, and only the final section deals with Monge's method and axonometry. All the representation methods are deduced from central projection. In fact, with rays starting from a finite centre, Fiedler obtains similarity, central symmetry, central collineations and involution; with rays having the same direction, he obtains congruence, axial and plane symmetry and affinity. Finally, Fiedler presents Monge's descriptive geometry.

Particularly in Italy, Fiedler's merit is recognized as he realized the "fusion" between projective and descriptive geometry, framed the method of central projection

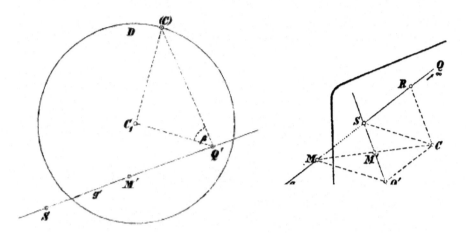

Fig. 4.2 From pages 8 and 9 of Fiedler (1874). The rounded curve on the right delimits a plane in the space

in descriptive geometry and realized a complete 1–1 correspondence between plane and space (Comesatti 1937–1938; Loria 1921). Let us see how Fiedler establishes this correspondence at the beginning of his work. To this purpose, metric relations are of course necessary. He considers the orthogonal projection of a centre C onto the plane of the drawing, C_1. We call d the distance CC_1. Given C_1 and d, the position of the centre C of projection is determined once an orientation has been fixed (Fig. 4.2, left; (C) is not on the projection plane).

If a line g does not pass through the centre C, its image on the picture plane is given by the line passing through the point S (intersection of g with the picture plane) and its vanishing point Q' (that is the intersection with the picture plane of the line through C parallel to g) (Fig. 4.2, left). The image of a point M cannot be given only by the intersection M' of the ray CM with the picture plane because any other point of this line has the same image. So, a point has to be characterized by two elements: M' and the image g' of a line g passing through it (Fig. 4.2, right is in three dimensions and the projection plane is represented). Therefore, given C_1, d, M' and g', we can find univocally the point M that has M' as an image.

The syllabuses of 1871 undoubtedly covered a great deal of ground. Indeed, the original aim was to prepare students in the physics-mathematics courses for direct entry to the School for Engineers without having to attend a 2 years preparatory course. In the end, this was not permitted, and a new reform took place in 1876 based on proposals from teacher councils. The aim of mathematics teaching was still that of enhancing the faculties of the mind while acquiring notions which are fundamental for further studies at university, but the syllabuses were highly reduced. The teaching of projective and descriptive geometry was combined and appeared only in the fourth year. After 2 years of plane geometry and 1 year of solid geometry and trigonometry, the study of projective geometry was whittled down to the study of the projective ranges and pencils and of the harmonic properties and projective relationships in a circle. Descriptive geometry was restricted to orthogonal and central projections, which were taught together with congruences, similarities, affinities and perspective collinearities. In the following years, the teaching of these subjects was further reduced and only some elements of technical drawing remained. Some of the motivations for this reduction are described by Scoth (see Scoth, Chap. 3, this volume).

As a consequence, the idea of deriving descriptive geometry from projective geometry also became lost in the following decades, and, from about 1880, textbooks for technical institutes only dealt with orthogonal projections, while texts by Cremona and Fiedler were used in universities.

3 Theory and Applications in Italy: A Slow Separation

The "classic" point of view of Cremona was shared by the majority of the geometers who taught in Italian universities towards the end of the nineteenth century. For instance, Federigo Enriques wrote

> The fact of considering all the methods of representation from a single point of view (i.e. as falling within the method of the projections) is due to the influence of projective geometry over descriptive [...]; moreover, projective geometry returned to descriptive geometry – in a more general and fruitful form – those principles which originated from the latter (Enriques 1894–1895, p. 5).

In that period, the principal Italian books on descriptive geometry at university level contained a list of the methods of representation of space onto a plane. As in most books, Enriques starts from central projection, beginning with graphic issues and then moving on to metric ones. Only in the second chapter does he "move" the centre of projection to infinity and treat orthogonal (Monge's) projections, again starting from graphic issues and then passing to metric ones. In the third chapter, he covers axonometry. This order allows him, from Enriques's point of view, to underline the passage from pure to applied mathematics. This order of exposition changed in the following decades.

In fact, it is possible to make a clear distinction between nineteenth century mathematical thought, which is characterized by a general symbiosis between theory and application, and that of the twentieth century, which shows a clear division between the two aspects, especially from a foundational point of view. In geometry, this position was to lead to the complete separation of the teaching of projective geometry from that of descriptive geometry and, at university level, to the complete exclusion of descriptive geometry from faculties other than for engineering and architecture. In universities, this separation was slower than in technical institutes (Scoth, Chap. 3, this volume).

The relationship between mathematicians and architects in the first decades of 1900s is of particular interest.

In 1921, the introduction of mathematical studies in the curriculum of every architect represented one of the main innovations of the Superior School for Architecture. In 1935, the foundation of the Faculty of Architecture in Rome saw the substantial participation of mathematicians in its establishment. Right from its opening, courses on descriptive geometry, mathematical analysis and applications of descriptive geometry were taught to architecture students by some of the leading exponents of the Roman mathematics community of that period, such as Enrico Bompiani, Francesco Severi and Ugo Amaldi (Dell'Aglio et al. 2001).

For example, the assignment to Severi (to 'His Excellency' Severi, Rector of the University from 1923 to 1925) of the teaching of 'Applications of Descriptive Geometry', which seems rather marginal, at least from today's point of view, is outstanding, in particular, if compared to the one provided in parallel by Bompiani.

How do we explain such an intense participation of Italian mathematicians in the birth of the Faculty of Architecture? Although the answer to this question also involves aspects of a global nature—related to the political weight and image assigned to the Faculty of Architecture during Fascism—we have mainly to consider a conceptual point of view, again linked to the general issue of the evolution of the teaching of geometry in faculties of a technical nature.

In the Faculty of Science, we can indeed notice a gradual weakening of the presence of descriptive geometry as an autonomous course; this corresponds precisely to a separation between the theoretical aspects of the subject and its applications. In 1910, Italian law provided for two courses: analytic geometry and projective and descriptive geometry with drawing. For a certain period in Rome, and also in other cities, Cremona's idea was still followed, namely having the teaching of analytic and projective geometry and the teaching of descriptive geometry with drawing. So, for a period of time, some weight was still given to descriptive geometry, but the end was nigh (in fact, the teaching of descriptive geometry disappeared from the course in mathematics only in 1960 after the separation of the first 2 years common to mathematics and engineering).

The climax of the separation took place in 1935: the birth of the Faculty of Architecture created a sudden expansion and autonomy of descriptive geometry and drawing, in contrast with what took place in the Faculty of Sciences.

In the Faculty of Architecture, the teaching of descriptive geometry and drawing still represented a challenge for Bompiani and Severi. It returned to its origins in that it re-proposed the classical problem of the relationship between concrete and theoretical aspects in geometry, and particularly in projective geometry. It is not surprising that these authors tended to re-propose some aspects of the discussion on geometry teaching present at the beginning of the twentieth century.

In the prefaces of the textbooks of that period, such as those by Amaldi or Bompiani, authors always refer to the derivation of descriptive geometry from painting and perspectivity (which was not done by Cremona). Nevertheless, the order of the presentation is changed: even if central projection is still treated, the treatises start from Monge's orthogonal projections, surely the most important element for future architects. But the link to projective geometry is always present: epitomic is Bompiani's reference to the classic harmony between descriptive and projective geometry (Bompiani 1942). In the Introduction to his *Lessons of descriptive geometry* he again underlines the didactical role of descriptive geometry. And, quoting Monge, he claims:

> [...] it seems that we can still – and even more – repeat with Monge that descriptive geometry is a 'means to search for the truth'. More modestly, we can assert with certainty that it is a necessary link in the chain that leads to secure understanding of higher geometrical truths (Bompiani 1942, pp. 7–8).

References

Barbin, Evelyne, and Marta Menghini. 2014. History of teaching geometry. In *Handbook on the History of Mathematics Education* ed. Alexander Karp and Gert Schubring, 473–472. New York: Springer.

Bompiani, Enrico. 1942. *Geometra Descrittiva*. Roma: R. Pioda.

Comesatti, Annibale. 1937–1938. Geometria descrittiva ed applicazioni. In *Enciclopedia delle Matematiche Elementari e complementi* ed. Luigi Berzolari, Giulio Vivanti, Duilio Gigli, vol. II, 307–375. Milano: Ulrico Hoepli.

Cremona, Luigi. 1873. *Elementi di geometria projettiva*. Roma: G. B. Paravia.

Dell'Aglio, Luca, Michele Emmer, and Marta Menghini. 2001. Le relazioni tra matematici e architetti nei primi decenni della Facoltà di Architettura: aspetti didattici, scientifici e istituzionali. In *La Facoltà di Architettura dell'Università di Roma La Sapienza, dalle origini, al duemila* ed. Vittorio Franchetti Pardo, 55–72. Roma: Gangemi Editore.

Enriques Federigo. 1894–1895. *Lezioni di Geometria Descrittiva*, Bologna: Regia Università.

Fiedler, Wilhelm. 1874. *Trattato di geometria descrittiva*, tradotto da Antonio Sayno e Ernesto Padova – Versione migliorata coi consigli e le osservazioni dell'Autore e liberamente eseguita per meglio adattarla all'insegnamento negli istituti tecnici del Regno d'Italia. Firenze: Successori Le Monnier. Original edition: 1871, *Die darstellende Geometrie in organischer Verbindung mit der Geometrie der Lage*. Leipzig: Teubner.

———. 1878. Sulla riforma dell'insegnamento geometrico. Seguita da tre lettere inedite dell'autore. *Giornale di Matematiche* XVI: 243–255.

Furinghetti, Fulvia. 2017. The mathematical journals for teachers and the shaping of teachers' professional identity in post-unity Italy. In *"Dig where you stand 4". Proceedings of the fourth International Conference on the History of Mathematics Education* eds. Kristín Bjarnadóttir, Fulvia Furinghetti, Marta Menghini, Johan Prytz, and Gert Schubring, 103–118. Rome: Edizioni Nuova Cultura.

Israel, Giorgio (ed). 2017. *De Diversis Artibus, Correspondence of Luigi Cremona (1830–1903)*. Turnhout: Brepols.

Knobloch, Eberhard, and Karin Reich (eds). 2017. 50. Letters from Wilhelm Fiedler (1862–1888). In *De Diversis Artibus, Correspondence of Luigi Cremona (1830–1903)* ed. Giorgio Israel, 637–697. Turnhout: Brepols.

Loria, Gino. 1921. *Storia della geometria descrittiva dalle origini sino ai giorni nostri*. Torino: Ulrico Hoepli.

Menghini, Marta. 2006. The Role of Projective Geometry in Italian Education and Institutions at the End of the 19th Century. *International Journal for The History of Mathematics Education* 1: 35–55.

Ministero di agricoltura, industria e commercio. 1871. *Ordinamenti degli istituti tecnici*. Firenze: Claudiana.

Monge, Gaspard. 1839. *Géométrie descriptive: suivi d'une theorie des ombres set de la perspective*; extraite des papiers de l'auteur par M. Brisson. Bruxelles: Societe belge de librairie.

Schubring, Gert. 1989. Pure and applied mathematics in divergent institutional settings in Germany: The role and impact of Felix Klein. In *The History of Modern Mathematics* eds. D. Rowe and J. McCleary, (V. II), 171–220. San Diego, London: Academic Press.

Uglieni, Marco (Luigi Cremona). 1865. I principi della prospettiva lineare secondo Taylor, *Giornale di matematiche* 3: 338–343.

Chapter 5
Descriptive Geometry in the Nineteenth-Century Spain: From Monge to Cirodde

Elena Ausejo

Abstract A reliable Spanish translation of Gaspard Monge's *Géométrie Descriptive* (1799) was published as early as 1803 for the training of building engineers at the recently established School of Roads and Waterways (1802), an institution conceived following the model of the French school of *Ponts et Chaussées*. The early introduction of descriptive geometry as a basic subject in the syllabus of the first engineering school marked a trend that expanded and consolidated with the development of the institutional framework for higher technical education in Spain throughout the nineteenth century. Descriptive geometry was included in the syllabi of engineers, artillerymen, architects, and navy officers, and also in the Master's Degree in Mathematics and Physics at the new Faculties of Science (1857). French influences continued beyond Monge in different ways: Leroy's, Olivier's, La Gournerie's, Adhémar's, and Cirodde's works on descriptive geometry were used either as textbooks or as selected sources in the production of original textbooks, even if only Olivier's *Cours de géométrie descriptive* and Cirodde's *Leçons de géométrie suivies de notions élémentaires de géométrie descriptive* were translated into Spanish. In the last quarter of the nineteenth century, civil engineers gradually began to transfer descriptive geometry from their syllabi to the highly competitive entrance examination, while university mathematicians evolved towards new developments in geometry.

Keywords Descriptive geometry · Higher technical education · Textbooks · Spain · Nineteenth century · Gaspard Monge · Mariano Zorraquín · Charles François Antoine Leroy · José Bielsa y Ciprián · Théodore Olivier · Joseph-Alphonse Adhémar · Jules de la Gournerie · Luis Felipe Alix · Ángel Rodríguez Arroquía · Paul-Louis Cirodde · Pedro Pedraza y Cabrera

---

E. Ausejo (✉)
University of Zaragoza, Zaragoza, Spain
e-mail: ichs@unizar.es

© Springer Nature Switzerland AG 2019
É. Barbin et al. (eds.), *Descriptive Geometry, The Spread of a Polytechnic Art*, International Studies in the History of Mathematics and its Teaching, https://doi.org/10.1007/978-3-030-14808-9_5

1 Introduction and Expansion of Descriptive Geometry in Spain

A reliable Spanish translation of Gaspard Monge's *Géométrie Descriptive* (1799) (Barbin, Chap. 1, this volume) was published as early as 1803 (Monge 1803) for the training of building engineers at the recently established *Escuela de Caminos y Canales* (School of Roads and Waterways, Madrid 1802), an institution conceived following the model of the French school of *Ponts et Chaussées*. Both this pioneering translation and the school were the result of the work of the Spanish engineers Agustín de Betancourt (1758–1824) and José María Lanz (1764–1839). From the very beginning they were under the influence of Monge and Hachette at the *École polytechnique*. Further results of this collaboration were the translation into Spanish of Francœur's *Traité de mécanique élémentaire* (Francœur 1803) and the publication of Lanz and Betancourt's *Essai sur la Composition des Machines* (Lanz et al. 1808) by the *École polytechnique* (Campo y Francés 1996).

The introduction of descriptive geometry as a basic subject in the syllabus of the first engineering school marked a trend that expanded and consolidated with the development of the institutional framework for higher education of civil engineers in new schools for mining (1835), forestry (1846), mechanical (1850), and agricultural engineering (1855)—all four in Madrid.

From 1819, descriptive geometry was also taught at the military College of Artillery (Segovia) and at the military Academy of Engineers (Alcalá de Henares). Later on in 1857, descriptive geometry was included in the syllabus of the Master's Degree in Mathematics and Physics at the new Faculties of Science.

In the last quarter of the nineteenth century, civil engineers gradually began to transfer mathematical disciplines—among them, descriptive geometry—from their syllabi to the highly competitive entrance examination, while the Naval School (1869) included descriptive geometry in its entrance examination and later in its syllabus (1895).[1]

This paper focuses on the teaching of descriptive geometry in the nineteenth-century Spain through the textbooks used for the purpose, while Millán Gasca will analyze the case of descriptive geometry in Spain as an example of the emergence of the late modern European outlook on the relationship between pure science and technology (Gasca, Chap. 6, this volume).

---

[1] For a detailed study of mathematics in the syllabi of civil engineers in the nineteenth-century Spain see Martínez García (2004), of mathematics education at military schools see Velamazán (1994), for the Navy see Comas (2015). For an all-embracing approach to mathematics in higher technical education in the nineteenth-century Spain, see Vea and Velamazán (2011).

2 Textbooks on Descriptive Geometry in the Nineteenth-Century Spain

2.1 Monge's Influence

The School of Roads and Waterways (1802) and the military academies of Artillery and Engineers were closed during the War of Independence against Napoleon (1808–1814). His defeat in Spain enthroned Ferdinand VII for an absolutist 6-year period (1814–1820), which was followed by a constitutional 3-year period (1820–1823) and then by a totally repressive decade (1823–1833), which ended only with the King's death in 1833. Both Ferdinand VII's absolutist reigns were periods of deep mistrust towards anything related to the Enlightenment: intellectuality, education, and scientific and technical development. Thus, the pre-war French-styled reforms program in the field of public education came to an almost total standstill.

In this context, the School of Roads and Waterways was kept closed after the war ended in 1814. A second short-lived school ran during the Liberal Triennium (1820–1823), the third and definitive school being founded in 1834. The leading role that descriptive geometry played in the syllabi of all three schools extended Monge's influence throughout the first half of the nineteenth century (Rumeu de Armas 1980, pp. 396–397, 451–453, 490–492).

However, descriptive geometry was taught at the College of Artillery and at the Military Academy of Engineers, which reopened after the war. The latter, which before had attempted to draft textbooks in Spanish for cadets since its foundation in 1803, successfully resumed this task. Colonel Mariano Zorraquín—a military engineer who during the war was imprisoned and then deported to France in 1814— was assigned on his return to the newly reopened military Academy of Engineers, where the 1816 syllabus included *Analytic geometry and geometric analysis* as a subject in the first year (Velamazán 1994, p. 276).

Zorraquín, who was appointed professor of descriptive geometry, found that the independent teaching of analytic and descriptive geometry at the Academy— as two separate subjects and by two different teachers during the same academic period—was time-consuming because the repetitive exposition of problems and solutions was unavoidable. As an alternative, he proposed a combined presentation of analytic and descriptive geometry given that they were connected, the former being the symbolic written and algebraic expression of geometry and the latter its graphic translation and representation. In Zorraquín's view, the long childhood of geometry came to an end with the application of algebra by Zorraquín (1819, p. 2). He acknowledged the role of descriptive geometry in the training of engineers not only because of the applications developed by Monge and his followers but also as a means to develop visual representation without previous education in mathematics. However, the power of analytic geometry was required in higher geometry. Consequently, both descriptive and analytic geometry should be taught making the reciprocal connection between analytic and space operations clear, as Monge wanted (Zorraquín 1819, pp. X–XIII, 88–89; Monge 1803, p. 11).

This was Zorraquín's main purpose when writing his *Analytical-descriptive geometry*, which became the official textbook at the Academy after an internal peer review reporting on the originality of the book, the high-quality and modernity of its sources, and the convenient integration of descriptive methods into analytic geometry (Zorraquín 1819, pp. XV–XIX). Actually, Zorraquín acknowledged having taken all the essential contents from the works by Monge, Lacroix, Biot, Puissant, Hachette, Garnier, and Boucharlat (Zorraquín 1819, p. IX) although only Monge, Lacroix, and Hachette were more precisely referenced in the book (Zorraquín 1819, pp. X, XII, 145, 347, 382, 420, 440, 463) together with Lagrange and Laplace (Zorraquín 1819, pp. 139, 142, 156, 158). He also referred to papers published in the *Correspondance sur l'École polytechnique* (Zorraquín 1819, pp. 420, 423, 445).

Zorraquín rejected using geometrical perspective for representation since magnitudes were not preserved (Zorraquín 1819, pp. 84–85). Alternatively, descriptive geometry provided the rules for the two-dimensional representation of space objects, whose properties could be deduced from this exact representation, and showed the meaning of analytic expressions. In this sense, descriptive and analytic geometry were one single science: any system and movement of points, lines, and surfaces in space could be expressed by analytic operations—their results representing the generated objects; reciprocally, any analytic operation of three variables expressed a combination in space. Consequently, Zorraquín's book aimed at developing the abilities to write the analytic expression of any combination or movement in space and easily represent the systems expressed by analytic expressions—again referring to Monge (Zorraquín 1819, pp. 87–89; Monge 1803, p. 50).

The book started with analytic geometry, with a first part—titled "Determinate analysis"—on constructions and problems using first and second degree equations—in one single chapter. On considering the sign of quantities in analytic geometry at the end of this part, Zorraquín quoted Carnot's *Géométrie de position* in order to introduce his ideas on correlative systems for the better understanding and representation of the geometric meaning of the roots of equations (Zorraquín 1819, pp. 26–27, 46–48; Carnot 1803, pp. 20–21, 46–47, 53–54; Velamazán 1993, pp. 592–594). The second part of Zorraquín's book titled "Indeterminate analysis" was divided into four chapters and was devoted to Cartesian geometry. Descriptive geometry was introduced at the end of the first chapter on first degree equations in two variables in order to deal with points and straight lines in space. It was mostly used to work with straight lines in space by means of the equations of the resulting straight lines in the projection planes,[2] but descriptive analysis was often also directly applied to the discussion of the analytic solutions in the projection planes in order to solve the problem in space, as in Zorraquín (1819, pp. 95–96).

---

[2]For a detailed example of how Zorraquín combined analytic and descriptive geometry, including the analysis of how Monge's results were used, see Velamazán (1993, pp. 596–602), Zorraquín (1819, pp. 96–103), and Monge (1803, p. 20).

All chapters but the fifth included some problems: nine in the first chapter, nine in the second, four in the third, and ten in the fourth. However, no practice of descriptive geometry was included in this book because Zorraquín was planning to publish a collection of the detailed constructions and examples of problems he actually used in teaching in a separate volume (Zorraquín 1819, p. XIII). This volume was never published as Zorraquín was elected deputy during the constitutional 3-year period (1820–1823) and died on 27 April 1823 fighting against the French army mobilized by the Bourbon King of France Louis XVIII to help the Spanish Royalists restore King Ferdinand VII of Spain to the absolute power of which he had been deprived during the Liberal Triennium.

Even so, Zorraquín's influence survived him since he was replaced by Captain José García Otero—his assistant teacher at the Academy and the author of the figures in Zorraquín's published and unpublished books (Zorraquín 1819, pp. IX, XIII) (Fig. 5.1)—until the King closed all military academies and colleges in 1823. In 1825, a new Royal Military General College opened for the training of officers, where descriptive geometry followed Monge's teachings (Monge 1803), while the application of algebra to geometry was taught according to Zorraquín's textbook (Zorraquín 1819; Velamazán 1994, p. 158). As for García Otero, expunged in 1824, he retired from the Army in 1828 and developed a successful career as an architect and civil engineer once the liberals took power after Ferdinand VII's death in 1833. Significantly enough, he was chosen to teach geometry at the definitive School of Roads and Waterways—founded in 1834, a position he left in 1836 to take on higher responsibilities (Necrología 1856).

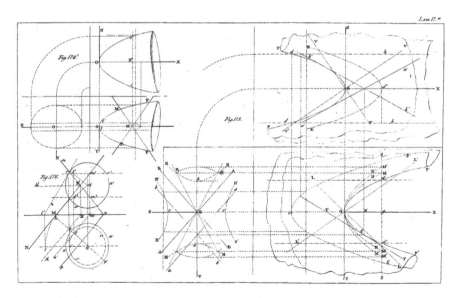

Fig. 5.1 García Otero's figures in Zorraquín (1819, plate 17)

Yet, this was not the case at the Royal Conservatory of Arts, founded in 1824 following the model of the French *Conservatoire Royale des Arts et Métiers*. Juan López de Peñalver—a teacher at the first and second schools of Roads and Waterways and a former colleague of Betancourt at the *École des Ponts et Chaussées* in Paris and at the Royal Cabinet of Machines in Madrid—was appointed director. He was instructed to follow, translate, and publish both Dupin's *Programme d'un cours de géométrie et de mécanique appliquée aux arts* (1824) (Dupin 1827) and *Géométrie et mécanique des arts et métiers et des beaux-arts* (1825–1826) (Dupin 1830–1835).[3]

In the second half of the century, José Jiménez y Baz referred to Monge, Lacroix and Zorraquín for further learning in his brief exposition of descriptive geometry as part of a textbook also including plane trigonometry and topography, a work adapted to the syllabus of the College of Infantry and to be used by infantry officers. According to the author, the 20 problems following the seven page introduction sufficed to easily use descriptive geometry as applied to topography and fortification (Jiménez y Baz 1857, pp. 5–6, 34).

Finally, Zorraquín's textbook was also included in the official lists of textbooks to be used at the Faculties of Science from 1847 to 1866, but only in analytic geometry.

2.2 Further French Influence[4]

Ferdinand VII's death finally meant the arrival to power of the liberals since the Regent Queen had to seek their support in order to preserve the right to the throne of her only daughter from the claims of the conservative partisans of the King's brother. Furthermore, although the dynastic conflict entailed three civil wars throughout the nineteenth century, at least the civilian institutional framework for upper scientific education began to settle. The 1857 General Law on Public Education regulated the resulting education system (Ausejo 2014).

As for descriptive geometry, new French textbooks were considered between 1848 and 1875, namely Leroy's *Traité de géométrie descriptive* (1842), Olivier's *Cours de géométrie descriptive* (1843–1844), and Adhémar's *Traité de géométrie descriptive* (1841). Leroy's *Traité* was the first to be used in Spain at the Preparatory School for Engineers and Architects (1848–1855), but all three were recommended at the schools for mechanical engineering which started in 1851,[5] mining engineering (1859–1860, 1870–1871), forestry engineering (1870–1871, 1873–1874), and

---

[3]Dupin's third volume on dynamic did not appear in his *Programme*. It was not translated into Spanish although it is not clear whether intentionally or because of López de Peñalver death in 1835.

[4]On the contents and uses of the French textbooks referred to in this chapter see Barbin (2015) and Barbin (Chap. 2, this volume).

[5]Together with Monge (1803).

at the School of Roads and Waterways (1870–1871). Only Olivier's *Cours* and Adhémar's *Traité* were recommended at the School of Architecture (1861–1864) and in 1867 also Vallée (1819).

As for Faculties of Science, only Vallés's *Traité* covered the 1858–68 decade together with Olivier's (1858–1867), Leroy's (1858–1861, 1867–1868), and Adhémar's (1866–1868) treatises. Lefebure de Fourcy (1830–1831) was also used between 1861 and 1867.

Oddly enough, none of the abovementioned teaching institutions ever produced translations or appropriations of any of these French authors. Only long afterwards, the first volume of Olivier's Course (Olivier 1872–1873) was translated into Spanish by Urbano Mas Abad (Olivier 1879). José Antonio Elizalde (1821–1875), professor of descriptive geometry at the Central University of Madrid from 1859, published the first part of his course as late as 1873 (Elizalde 1873–1878, vol. 1). The second part was edited posthumously (Elizalde 1873–1878, vol. 2). This lack of teaching resources might be connected to private tuition—at home or in preparatory academies, an activity that enabled the teaching staff to supplement their low salaries (Ausejo 2006, pp. 7–8). Publication of textbooks was fostered and rewarded only in military teaching institutions, so that seven textbooks on descriptive geometry were produced by the Spanish military between 1846 and 1875. Among these, Bielsa (1846, 1857), Rodríguez Arroquía (1850, 1865), and Alix (1866) were essentially original works based on a careful selection of different French sources.

Bielsa's treatise in six books (Bielsa 1846) was extracted from the second edition of Leroy's nine-book treatise (Leroy 1842)[6]; Leroy's fifth, sixth, eighth, and ninth books were omitted—the latter containing dimensioned drawings (*plans cotés*). Instead, Bielsa devoted his sixth book to the theory of shadows, so as to show, in his words, the immediate practical application of descriptive geometry (Bielsa 1846, p. 11). For this purpose, he used the third and fourth books of Cloquet's elementary treatise on perspective (Cloquet 1823, pp. 155–164, 175–177, 208–209, 211, 217–219, 222–235). It is also worth mentioning that Bielsa was a pioneer in translating into Spanish the French *rabattement* to *superposición* (Leroy 1842, p. 108; Bielsa 1846, p. 96). Later on, this word was replaced by *rebatimiento*, the literal translation introduced by Manuel María Barbery (Cirodde 1858, p. 412; Cirodde 1865, p. 412; Alix 1866, p. 87).

Four years after the publication of Bielsa's textbook for artillerymen,[7] Ángel Rodríguez Arroquía, a captain of engineering, published a textbook on dimension drawing. It was designed as a complement to descriptive geometry dealing with irregular surfaces in the practice of topography and fortification. The introduction presented a precise, well-balanced discussion of the practical advantages of skipping the graphical representation of the vertical projection by using one single—horizontal—projection plane together with the numerical value of the horizontal

---

[6]Bielsa was also familiar with Olivier's works (Bielsa 1846, pp. 105–106)

[7]Between 1869 and 1877, the first two chapters of this book were also recommended for preparing for the entrance examination to the Naval School (Comas 2015, p. 88).

projection line (Rodríguez Arroquía 1850, pp. 5–8). As for contents, the book consisted of three chapters in seventy six pages in octavo, where contents were mainly introduced with examples or explained through up to 45 problems.

We know for certain that this new method was included in the training of military engineers since they published a second edition of Rodríguez Arroquía's book (Rodríguez Arroquía 1865). The second edition of Bielsa's textbook—including a new book on dimension drawing (Bielsa 1857, pp. 186–205)—proves that it was also adopted by artillerymen. In the next decade, Artillery Captain Luis Felipe Alix produced a new elementary treatise on descriptive geometry, perspective, and theory of shadows for the training of artillerymen. His aim was to arrange a new presentation on the essentials of the subject collected from modern works (Alix 1866, Advertencia). The book focused on problem-solving in three parts—lines and planes, curves and surfaces, and perspective and shadows. Dimension drawing was considered at the end of the second part; axonometric perspective was introduced for the first time in Spain (Alix 1866, 3rd Part, pp. 21–33) (Fig. 5.2), and the theory of shadows was extracted from Adhémar's Traité (Adhémar 1840; Alix 1866, 3rd Part, pp. 40, 62). He also acknowledged La Gournerie (1860–1864) as his main source (Alix 1866, 1st Part, p. 105, 2nd Part, pp. 66, 179) and Olivier for notation and rotations (Alix 1866, 1st Part, pp. 13, 41, 46).

Infantry Commander Lozano y Ascarza (1866) followed Jiménez y Baz (1857) in his eleven lessons textbook on descriptive geometry—eight lessons in thirty nine pages—and dimension drawing—three lessons in twenty four pages. This was a problem-solving textbook—nine out of eleven lessons—adapted to the syllabus of the College of Infantry and designed as a preliminary to topography and fortification.

Finally, it should be noted that basic training in descriptive geometry was available in Spain from 1858 when Carlos Bailly-Baillière[8] published the Spanish translation of Cirodde's *Leçons de géométrie suivies de notions élémentaires de géométrie descriptive* (Cirodde 1858) with great success; the third edition (Cirodde 1865)—corrected, annotated, and enlarged by the Spanish translator Manuel María Barbery—was reprinted nineteen times until 1902.

3 Descriptive Geometry in the Late Nineteenth-Century Spain

Military men produced eleven new textbooks on descriptive geometry in the last quarter of the nineteenth century (Velamazán 1993, pp. 589, 612, 617–619). Of these, the most widely used was *Lessons on descriptive geometry* by Captain of Engineering Pedro Pedraza y Cabrera (Pedraza and Ortega 1880; Pedraza 1879).

---

[8]Charles Bailly added the family name of his uncles—the French publishers Baillière—to his own name in 1848 when he opened his publishing house in Madrid.

5 Descriptive Geometry in the Nineteenth-Century Spain

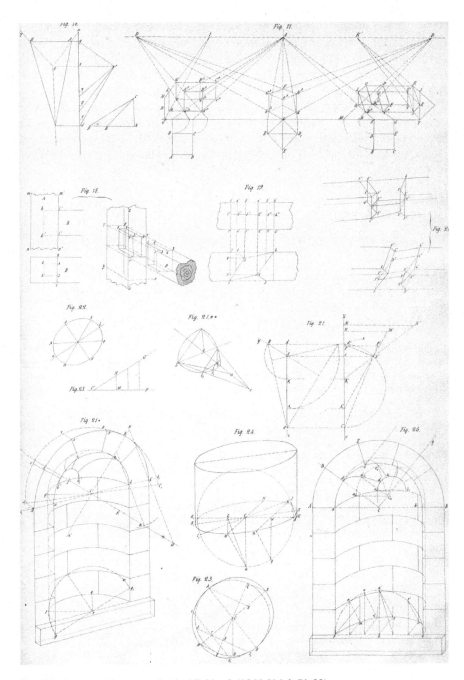

Fig. 5.2 Axonometric perspective in Alix' book (1866, Vol. 2, Pl. 38)

In the meantime, civil engineers were gradually transferring mathematical disciplines—among them descriptive geometry—from their syllabi to the highly competitive entrance examination. In this context, Elizalde's course was recommended in order to prepare for this test—as well as the treatises by Leroy and Adhémar, and even displaced French authors at the Faculties of Science.

From 1876 onwards, Eduardo Torroja, Elizalde's successor as professor of descriptive geometry at the Central University of Madrid, started to evolve towards new developments in geometry (Torroja 1884).[9] However, descriptive geometry was included in the 1900 syllabus of the Master's Degree in Mathematics at the Faculties of Science.

Acknowledgement This work is financed by the research project HAR2015-70985-P (MINECO/FEDER, UE).

References

Adhémar, Joseph. 1840. *Traité des ombres*. Paris: Bachelier.
———. 1841. *Traité de géométrie descriptive* (2nd ed.). Paris: Carilian-Goeury et V. Dalmont.
Alix, Luis Felipe. 1866. *Tratado elemental de geometría descriptiva, seguido de unas ligeras nociones sobre perspectiva y sombras*, 2 vols. Valencia: Imprenta de Ferrer de Orga.
Ausejo, Elena. 2006. Quarrels of a marriage of convenience: On the history of mathematics education for engineers in Spain. *International Journal for the History of Mathematics Education* 2 (1): 1–3. http://journals.tc-library.org/index.php/hist_math_ed/article/view/276/235
———. 2014. Mathematics education in Spain and Portugal. 1. Spain. In *Handbook on the History of Mathematics Education*, ed. Alexander Karp and Gert Schubring, 284–291. New York, NY: Springer. https://doi.org/10.1007/978-1-4614-9155-2_14
Barbin, Evelyne. 2015. Descriptive geometry in France: History of Élémentation of a method (1795–1865). *International Journal for the History of Mathematics Education* 10 (2): 39–81.
Bielsa, José. 1846. *Tratado elemental de geometría descriptiva y sombras, para el uso de los Caballeros Cadetes de Artillería de Segovia, extractado de las obras de Mr. Le-Roy y Mr. Cloquet*, 2 vols. Segovia: Imprenta de los Sobrinos de Espinosa.
Bielsa y Ciprián, José. 1857. *Tratado de geometría descriptiva, sombras, topográfico y sistema de acotaciones*, 2 vols (2nd ed.). Segovia: Imprenta de los Sobrinos de Espinosa.
Campo y Francés, Ángel del. 1996. La descriptiva de Monge en la Escuela de Caminos. In *Geometría descriptiva*, ed. Gaspard Monge, 9–54. Madrid: Colegio de Ingenieros de Caminos, Canales y Puertos, Facsimile ed. of (Monge 1803).
Carnot, Lazare Nicolas Marguérite. 1803. *Géométrie de position*. Paris: Chez Duprat, de l'imprimerie de Crapelet.
Cirodde, Paul-Louis. 1858. *Lecciones de geometría con algunas nociones de la descriptiva. Revisada y arreglada por Alfredo y Ernesto Cirodde. Traducida por Manuel María Barbery*. Madrid: Cárlos Bailly-Bailliere.
———. 1865. *Lecciones de geometría con algunas nociones de la descriptiva escritas en francés por P. L. Cirodde. Revisada y arreglada á los últimos programas oficiales de aquella nacion por Alfredo y Ernesto Cirodde. Traducida de la última edicion francesa por Manuel María*

---

[9]On this process, see Millán (1991).

Barbery. Tercera tirada, corregida, anotada y adicionada por el traductor. Madrid: Carlos Bailly-Bailliere.

Cloquet, Jean-Baptiste. 1823. *Nouveau traité élémentaire de perspective á l'usage des artistes.* Paris: Bachelier.

Comas Roqueta, Joaquín. 2015. *La enseñanza de las matemáticas en la armada española en el siglo XIX.* Zaragoza: Universidad de Zaragoza. https://zaguan.unizar.es/record/32763/files/TESIS-2015-085.pdf

Dupin, Carlos. 1827. *Programa de un Curso de Geometría y Mecánica, aplicadas á las Artes, para uso de los artesanos, y de los maestros y demas personas que dirigen talleres ó fábricas. Explicado en el Real Conservatorio de Artes y Oficios de Paris por el baron Carlos Dupin, miembro del Instituto en la Academia Real de las Ciencias, oficial superior del cuerpo de ingenieros de la marina, oficial de la legion de honor y caballero de San Luis. Traducido del frances por D. Juan Lopez de Peñalver y La Torre. De orden superior.* Madrid: Imprenta Real.

———. 1830–1835. *Geometría y Mecánica de las Artes y Oficios y de las Bellas Artes. Curso Normal para el uso de los artistas y menestrales, y de los maestros y veedores de los talleres y fábricas. Esplicado en el Conservatorio Real de Artes y Oficios por el baron Carlos Dupin, miembro del Instituto (Academia de Ciencias), oficial superior del cuerpo de ingenieros de marina, oficial de la legión de honor y caballero de San Luis. Traducido al castellano de orden del rey nuestro señor, que esta en gloria, por don Juan López Peñalver de la Torre, del consejo de S. M. y su secretario honorario*, 2 vols. Madrid: Imprenta de Don José del Collado.

Elizalde, José Antonio. 1873–1878. *Curso de geometría descriptiva. Primera parte - Texto: Del punto, de la recta, del plano y de sus combinaciones. Segunda parte - Texto: Generación y representación de las curvas y superficies en general*, 2 vols. Madrid: Imprenta y Fundición de Manuel Tello.

Francœur, Louis-Benjamin. 1803. *Tratado de mecánica elemental, para los discípulos de la Escuela Politécnica de Paris, ordenado según los métodos de R. Prony. Traducido al castellano para el uso de los estudios de la Inspección General de Caminos.* Madrid: Imprenta Real.

Jiménez y Baz, José. 1857. *Elementos de geometría descriptiva, trigonometría rectilínea y topografía para uso de los oficiales de Infantería y de los caballeros cadetes del cuarto semestre de estudios del colegio de dicha arma.* Toledo: Imprenta de José de Cea.

La Gournerie, Jules de. 1860–1864. *Traité de géométrie descriptive*, 2 vols. Paris: Mallet-Bachelier.

Lanz, José María de, Agustin de Betancourt, and Jean-Nicolas-Pierre Hachette. 1808. *École impériale polytechnique. Programme du cours élémentaire des machines, pour l'an 1808, par M. Hachette. Essai sur la composition des machines, par MM. Lanz et Bétancourt.* Paris: Impr. Impériale.

Lefebure de Fourcy, Louis-Étienne. 1830–1831. *Traité de géométrie descriptive, précédé d'une introduction qui renferme la théorie du plan et de la ligne droite considérée dans l'espace*, 2 vols. Paris: Bachelier.

Leroy, Charles-François-Antoine. 1842. *Traité de géométrie descriptive, suivi de la méthode des plans cotés et de la théorie des engrenages cylindriques et coniques, avec une collection d'épures*, 2 vols (2nd ed.). Paris: Bachelier.

Lozano y Ascarza, Antonio. 1866. *Lecciones fundamentales de geometría descriptiva como estudio preliminar necesario para el de la topografía y fortificación.* Toledo: Imprenta de Ricardo Romero.

Martínez García, Mª Ángeles. 2004. *Las matemáticas en la ingeniería: las matemáticas en los planes de estudios de los ingenieros civiles en España en el siglo XIX.* Zaragoza: Seminario de Historia de la Ciencia y de la Técnica de Aragón, Facultad de Ciencias, Universidad de Zaragoza.

Millán, Ana. 1991. Los estudios de geometría superior en España en el siglo XIX. LLULL, *Revista de la Sociedad Española de Historia de las Ciencias y de las Técnicas* 14: 117–186. https://dialnet.unirioja.es/descarga/articulo/62085.pdf.

Monge, Gaspard. 1799. *Géométrie descriptive. Leçons données aux Écoles normales, l'an 3 de la République.* Paris: impr. de Baudouin.

———. 1803. *Geometría descriptiva. Lecciones dadas en las Escuelas Normales en el año tercero de la República, por Gaspar Monge, del Instituto Nacional. Traducidas al castellano para el uso de los estudios de la Inspección General de Caminos*. Madrid: Imprenta Real.
Necrología. 1856. *Revista de Obras Públicas* 4 (18): 210–212.
Olivier, Théodore. 1843–1844. *Cours de géométrie descriptive*, 2 vols. Paris: Carilian-Goeury et Vve Dalmont.
———. 1872–1873. *Cours de géométrie descriptive. 3e édition revue et annotée par M. Eugène Rouché*, 2 vols. Paris: Dunod.
———. 1879. *Curso de geometría descriptiva: del punto, de la recta y del plano*. Madrid: Guirnalda.
Pedraza y Cabrera, Pedro. 1879. *Lecciones de geometría descriptiva. Superficies en general*. Madrid: Imprenta del Memorial de Ingenieros.
Pedraza y Cabrera, Pedro, and Miguel Ortega y Sala. 1880. *Lecciones de geometría descriptiva. Rectas y planos*, 2 vols. Madrid: Imprenta del Memorial de Ingenieros.
Rodríguez Arroquía, Ángel. 1850. *Complemento a la geometría descriptiva: empleo de un solo plano de proyección valiéndose del sistema de acotaciones para servir de aplicación de los principios generales de la ciencia a las superficies irregulares, y como preliminar a la topografía y a la desenfilada de las obras de fortificación*. Madrid: Boix Mayor y Cia.
———. 1865. *Complemento a la geometría descriptiva: empleo de un solo plano de proyección valiéndose del sistema de acotaciones para servir de aplicación de los principios generales de la ciencia a las superficies irregulares, y como preliminar a la topografía y a la desenfilada de las obras de fortificación*. Madrid: Imprenta del Memorial de Ingenieros.
Rumeu de Armas, Antonio. 1980. *Ciencia y tecnología en la España ilustrada. La Escuela de Caminos y Canales*. Madrid: Turner.
Torroja, Eduardo. 1884. *Programa y Resúmen de las lecciones de geometría descriptiva esplicadas en la Universidad Central por el Catedrático D. Eduardo Torroja. Tomo Primero (Con diez láminas)*. Madrid: año 1884. Manuscript text.
Vallée, Louis Léger. 1819. *Traité de la géométrie descriptive*, 2 vols. Paris: Vve Courcier.
Vea, Fernando, and Mª Ángeles Velamazán. 2011. La formación matemática en la ingeniería. In *Técnica e Ingeniería en España, Vol. VI El Ochocientos. De los lenguajes al patrimonio*, ed. Manuel Silva Suárez, 299–344. Zaragoza: Real Academia de Ingeniería, Institución 'Fernando el Católico', Prensas Universitarias de Zaragoza.
Velamazán, Mª Ángeles. 1993. Nuevos datos sobre los estudios de geometría superior en España en el siglo XIX: la aportación militar. LLULL, *Revista de la Sociedad Española de Historia de las Ciencias y de las Técnicas* 16: 587–620. https://dialnet.unirioja.es/descarga/articulo/62122.pdf
———. 1994. *La enseñanza de las matemáticas en las Academias militares en España en el siglo XIX*. Zaragoza: Seminario de Historia de la Ciencia y de la Técnica de Aragón, Facultad de Ciencias, Universidad de Zaragoza.
Zorraquín, Mariano. 1819. *Geometría Analítica-Descriptiva*. Alcalá: Imprenta Manuel Amigo.

Chapter 6
Descriptive Geometry in Spain as an Example of the Emergence of the Late Modern Outlook on the Relationship Between Science and Technology

Ana Millán Gasca

Abstract The teaching of descriptive geometry in Spain spread at the pace of the development of technical education. Despite the awareness of the need of innovation, and the foundation of local schools, until the end of the nineteenth century the State did not organize a network of schools for workers and artisans–technicians. Instead, the efforts were concentrated on the schools for engineers, with two kinds of difficulties: the lack of students, so that the only stable schools were at Madrid (while there was a great demand for elementary and middle technical education); and the relatively unbalanced relationship with the Faculty of Sciences, and specially the mathematical department, which was considered as preparatory for the school of architecture and the several engineering schools. The contribution by Culmann to further development of the graphical instruments of the engineer and the new didactical approach to descriptive geometry by Bellavitis, Fiedler, and Cremona were considered with interest by scholars in Barcelona and Madrid, although this led to an attention being paid almost exclusively to synthetic geometry. As a consequence, descriptive geometry had a key role in the curriculum of the master in mathematics. Around 1900, this became a hindrance to the development of the mathematical level of information in the country, and to the actual diffusion of the ethos of research.

Keywords Luigi Cremona · José de Echegaray · Zoel García de Galdeano · Julio Rey Pastor · Eduardo Torroja Caballé · Santiago Mundi i Girò · Synthetic geometry · Technical education

---

A. Millán Gasca (✉)
Department of Education, Roma Tre University, Rome, Italy
e-mail: anamaria.millangasca@uniroma3.it

1 The Teaching of Descriptive Geometry: Was It a Factor of Scientific Development or a Blockage to It in Nineteenth Century Spain?

> We note, moreover, that there is a dividing line between the theoretical and the practical mathematician, that is, descriptive geometry and mechanics. From this common area the mathematician rises towards the theory and the engineer descends towards the material applications. (García de Galdeano 1908, p. 1) (all the translations are by the author)

Zoel García de Galdeano (1846–1924), professor of mathematics at the university of Zaragoza, put descriptive geometry on the borderline between mathematics and engineering in his address to his colleagues in the newly founded Spanish Association for the Advancement of Science in 1908. In the equilibrium between the cultivation of science and the development of industry, following the nineteenth century European trend, García de Galdeano pinpointed a key for national progress, and the education of young mathematicians should have been planned in this framework. But the new (1900) curriculum of mathematics students in the Spanish Faculty of sciences (*ciencias exactas*, exact sciences) included a course in descriptive geometry in the fourth and last year, after courses in euclidean geometry, analytical geometry, and synthetic projective geometry (*géométrie de position*), in spite of the astonishing explosion of mathematical branches in the final decades of the nineteenth century that had little or no place in the training of professional mathematicians in Spain.

This choice deserves a historical explanation, because it has been identified as an obstacle to the development of original mathematical research in the country, and the slow development of free scientific research in its turn is generally considered a hindrance to the modernization and Europeanization of Spain (Ausejo and Hormigón 2002; Ausejo et al. 1990). In fact, at the beginning of the 1960s, the analysis of Julio Rey Pastor's (1888–1962) crucial contribution to the development of mathematical research in the twentieth century in Spain and Latin America had to take into account the starting point of his career in the early years of the century. He met at the University of Zaragoza García Galdeano (an intellectual marginalized in the province) and was then doctoral student of Eduardo Torroja y Caballé (1847–1918, a powerful and recognized professor of descriptive geometry at the University of Madrid), on a problem of synthetic projective geometry regarding the development of Christian von Staudt's and Ernst Kötter's work (Millán Gasca 1990a). In 1962, in the *Revista Matemática Hispanoamericana*, Enrique Vidal Abascal (1908–1994) recalled Rey Pastor's words in a 1956 speech at the Academy of Sciences:

> At the end of the nineteenth century we took a giant leap forward with the introduction of Staudt, more studied here than in Germany, but geometry got straightened out following the analytical direction, and both Cremona and Torroja, and those of us who followed them, again remained out of the riverbed.

And Vidal Abascal commented:

> This "out of the riverbed", we had to say it with harsh sincerity to honor the legacy of the egregious master, phased out the Spanish mathematics for another 50 years. Even today

those years, still recent, when *descriptive geometry was one of the most important subjects in the master curriculum*, and graphical and synthetic methods were studied in Spain with the same effort and enthusiasm that our ancestors put, for example, into crossing the Andes, weight down our most acknowledged mathematicians. (Vidal Abascal 1962, p. 119, my emphasis; see Souto Salorio et al. 2016).

So the teaching of descriptive geometry appears as a crucial historiographical issue in order to understand the evolution in the period between the eventful years of the 1868 Revolution, 1874 I Republic and 1875 Bourbon Monarchic Restoration, and the II Republic (1931–1939), and particularly in the decades before the turn of the century: was it a period in which Spanish culture managed to cope with the challenge of mathematical modernity—as regards the circulation of mathematical knowledge and the assimilation of the ethos of research (Turner 1971)—or was it a period of stagnation, due to the encumbrance of a narrow, utilitarian view of mathematics? Was the interest in descriptive geometry helpful or was it the source of a deep misunderstanding regarding the modern idea of university and the mission of a university professor originating in Prussian culture and increasingly assimilated in Europe?

2 Descriptive Geometry and the Education of Technicians at Several Levels

The *Dictionnaire des mathématiques appliquées* (1867) by Hippolite Sonnet (see Barbin Chap. 2, this volume) shows the great range of mathematical concepts and techniques available in the professions and in the crafts in 1860s; secondary mathematical education was spreading and an alternative to Euclid's *Elements* was an important concern (Barbin and Menghini 2013); geometry had been given great impulse by the new ideas brought by projective geometry to the application-oriented tradition of geometry, and Luigi Cremona's contributions to the theory of curves and theory of surfaces had again shown the possibility of using purely geometric methods in advanced research. Descriptive geometry was at the crossroads of all these developments, and its cultivation in Spain, marked by the influence of French mathematics (see Ausejo, Chap. 5, this volume), had been a factor of improvement of the level of technical training and information.

The diffusion of the teaching of descriptive geometry in Spain in the early nineteenth century gave momentum to the preceding efforts to introduce drawing in the training of workers–artisans–technicians. As much as in France, and consistent with the original intentions of Monge himself, graphical representation was seen as a mathematical language of technology, and linear drawing (*dibujo lineal*), as separated from traditional drawing (*dibujo de adorno, dibujo de figura*), marked the increasing distance between the artisan/technician (*artes y oficios*) and the artist (*bellas artes*).[1] Drawing, together with basic arithmetic and the natural sciences, should become the core of a new way to educate technicians as an alternative to the traditional on-the-job training (apprenticeship), and such an innovation in education and training was seen as a potential impulse to industrialization and modernization.[2] Descriptive geometry as a banner of "mathematical exactness" when representing objects was even upheld by the Spanish government in 1862 in the face of the complaints of drawing teachers as to the examination requirements:

> The more frequent case in this kind of drawing is to represent objects in projection and the hypothesis of a teacher of linear drawing is not acceptable without a knowledge of descriptive geometry and even linear perspective [...] (quoted in Bermúdez Abellán (2005), p. 293)

In the early nineteenth century Spain, the development of elementary technical education (drawing schools, *escuelas de aplicación*, schools of arts and crafts) was the result of many local initiatives carried on in continuity with the original Enlightenment spirit, combined with an increasing influence of the national idea, and many recent studies on these schools show a panorama of great dynamism.[3] The overall cultural-political framework was actually present in government documents, such as the Royal decree of September 1850 under Queen Isabel II: for the prosperity and wealth of the homeland not only classical, literary, and scientific institutions were needed, but also schools where "those who devote themselves to industrial careers could find all the needed instruction to stand out in the arts, or to become perfect chemists and skilled mechanics". Elementary industrial education

---

[1] See Bermúdez Abellán (2005). The late eighteenth century, enlightened idea of "patriotic" schools of drawing has its main sources in the writings by Campomanes (see note 3). On the French evolution, see Laurent (1999) and D'Enfert (2003): a "rational drawing" was seen as an ally of "useful arts" in the new "institutions of industrial art" (Laurent 1999, pp. 312–313). See also Ashwin (1981) and Efland (1990). The consolidation of this evolution is shown by the presence of descriptive geometry in a booklet published in 1879 by the architect Eugène Viollet-le-Duc (1814–1879) in the Hetzel series for children; in Spain see the booklet for women Ferrer de Pertegás (1897). In fact, linear drawing and geometry had a crucial role in the outlook on modern primary education of pioneering scholars such as Louis-Benjamin Francœur (1773–1849) and Johann Pestalozzi (1746–1827): primary education and elementary technical education were the hard-core of the education of the working classes in the age of liberalism (Millán Gasca 2016).

[2] In the 1832 lessons at the Royal Conservatory of Arts, which represent this cultural project at its best, together with Charles Dupin's books in their Spanish translation, Francœur's book on linear drawing and Louis Léger Vallée's (1784–1864) treatise on descriptive geometry were used (Ramón Teijelo 2011, p. 109; on Vallée's treatise see Chap. 1, on the contribution of Francœur to the invention of linear drawing see D'Enfert 2003). This evolution has a prehistory: on technical drawing in the 16th–18th centuries, see Cámara Muñoz (2016).

[3] The references are too many to be included here, as they are mostly monographical studies on single institutions that had different origins and evolution in different local contexts. See, for example, Blanes Nadal et al. (2002–2003) (where national regulations are described), Guijarro Salvador (2009), and their bibliographical references. Pedro Rodríguez de Campomanes (1723–1802) was the Spanish politician and scholar whose views on popular education for the development of industry inspired the movement of the "patriotic schools of drawing" (this was the name of the school at Murcia; see Rodríguez de Campomanes 1975).

was assigned to secondary schools, over 4 years: the preparatory year included knowledge and drawing of elementary geometrical figures, linear drawing (together with figure and ornament drawing), and elements of descriptive geometry; further teaching of descriptive geometry was included in the 3-year extension schools for pupils starting at 14 years old.

In the 1857 Law of Education, basic geometry, linear drawing, and surveying were included in higher primary school; figure and linear drawing were included in application studies in secondary schools (*institutos*); descriptive geometry was part of the curriculum of nautical professional schools, the school of architecture and every engineering school (denoted as "higher schools", *escuelas superiores*), and in the newly established Faculty of Sciences in the university. In Chap. 5 the diffusion of the teaching of descriptive geometry in engineering schools (and in military academies), the use of French texts and the available Spanish treatises is analyzed. The convenience of descriptive geometry for all the various degrees of responsibility and competence in the technical activity was acknowledged in the political and cultural milieu which focused on the link between industrial development and knowledge and education. There is a striking list of 13 kinds of technicians—from architects and engineers to carpenters and masons—listed on the title page of the 1865 treatise on descriptive geometry and its applications by the Catalan teacher Baltasar Cardona y Escarrabill (1828–1868), secretary of the physico-mathematical section of the Academy of Sciences of Barcelona. Nevertheless, there was no ordered organization of a state network of institutions for the education of artisans, workers, surveyors, and so on (school of arts and crafts) until the 1880s–1890s.

Descriptive geometry instead became one of the *atouts* of the newly founded Faculty of Sciences, whose potential for the modernization of the country was closely linked to its role of preparatory school for the higher technical schools: the state supported this role in article 79 of the 1857 law of Education[4] and in subsequent regulations, to the disappointment of the engineers themselves, led by José de Echegaray y Eizaguirre (1832–1886). This state of affairs was aggravated by the chronic lack of funding of education in an economy that was longing for modernity (Fusi and Palafox 1996), and by the fluctuating interest in education and in science of successive cabinets, due to the resistance to the consolidation of the liberal state: pure science was considered at best a luxury that the country could not actually afford, at the worst an enemy of religion, diverting young people towards secularism. The main faculties of sciences were at Madrid and Barcelona, while in the provinces there was no chance of obtaining a master in sciences, but only to study such subjects that were part of the professional studies in the university (Medicine, Pharmacy).

---

[4]"The regulations of the higher and professional Schools will establish the subjects of secondary education and of the faculty of sciences that had to be proved by an exam verified by the Schools themselves of those who aim to enter them" (Law of public instruction, September 9, 1857, art. 79).

3 Beyond the "Old and Paltry Forms of Descriptive Geometry": The Search for *Bildung* in Technical Education and the Diffusion of the Ethos of Research in the Universities

Echegaray, a liberal senator who took part in the 1868 Revolution, acting in the government, shared the views that Luigi Cremona (1830–1903) helped to propagate throughout Europe: science as the religion of the century, and the need to develop both technical and liberal education (Millán Gasca 2012). In the first aspect both were influenced by French culture. The second aspect combined the spirit of the Prussian educational reforms with a vision of technical education as *Bildung*, consistent with the one developed at the Zurich federal polytechnical school founded in 1855, which carried on and enlarged Monge's vision of the education of learned technicians through descriptive geometry. The idea of wrapping up technical education in the *Bildung* neohumanistic ideal was developed in the Swiss institution by Carl Culmann (1821–1881) through the projective foundation of graphical methods of statics, to which Cremona gave a crucial contribution (Scholz 1994). Cremona was in touch with Culmann and with Wilhelm Fiedler (1832–1912), who developed a projective foundation of descriptive geometry,[5] and he had appreciated a similar approach in the *Lezioni di geometria descrittiva* (1851) by Giusto Bellavitis (1803–1880) (Cremona 1873, pp. iv–v.). Beyond the didactical-mathematical details, Cremona shared the overall underlying *Bildung* approach, and he even aimed to extend it to the elementary technical education, "exposing projective geometry, either on its own or together with ordinary geometry, in an increasingly simple, more elementary form, more accessible to even mediocre intellects" and informing descriptive geometry to the methods of projective geometry, so that students could understand the application of modern geometry (*Geometrie der Lage, geometria superiore*) to technical drawing-design, as he wrote in his 1873 treatise on elements of projective geometry for technical secondary schools (Cremona 1873, pp. vi–vii; see Millán Gasca 2011).

As a matter of fact, the nineteenth century evolution of the geometrical tools of the engineer and the architect (drawing and representation, the training of spatial vision, design, and "the art of avoiding calculation") had an influence both on the cultural background of these professionals or civil servants and on the evolution of mathematics and the "mathematical world". Cremona's vision of the transformation of technical education by geometry had as a counterpart a theoretical interest in developing "pure" geometrical synthetic research: this combination of *vita activa* and *vita contemplativa* was shared by a rich network of mathematicians in the second half of the nineteenth century (Hormigón and Millán 1992; Millán

---

[5]Cremona's contacts with the Polytechnic of Zurich were intense, including Theodor Reye (1838–1919) and other scholars (Knobloch 2013). On the connection between Fiedler and Cremona, see Chap. 4.

Gasca 1990b). As pointed out by Israel (2017), the international correspondence of Cremona, who himself was a leading figure in this context, will help in the future to better appreciate this evolution, which is part of the constitution of the "Europe of sciences" (Blay and Nicolaïdis 2001), the European space combining national scientific development and international dialogue and networking. As Ivor Grattan-Guinness (2002) pointed out, France was a main actor at least until the mid 1870s, but the interplay among the nations contributing to research—and to shared values and cultural projects, we may add—became increasingly complex, something which can be considered as a factor of dynamism and vitality. Now, descriptive geometry treatises were among the main support of French influence among mathematicians, while new ideas able to renew the Mongian tradition came from the German speaking area and from Italy.[6]

Let us view this evolution from the Spanish prospective. Until the mid 1870s French textbooks of elementary and descriptive geometry had a wide circulation, also through reelaborations by Spanish engineers, as has been shown in Chap. 5.[7] The scholars taking part in the 1875–1876 competitive examination for a chair of descriptive geometry[8]—which was finally obtained by Eduardo Torroja—asked for the books by Adhémar, Leroy, Vallée, and Olivier, used in the Spanish engineering schools—but also for La Gournerie's treatise, and Libre-Irmand Bardin's (1794–1867) and Jacques Babinet's (1794–1872), in addition to the books by José Bielsa for artillerymen (see Ausejo, Chap. 5, this volume) and José Antonio Elizalde, the former holder of the professorship (Elizalde 1873–1878). Nevertheless, the French education system had graded descriptive geometry as elementary, propædeutic knowledge (see Barbin, Chap. 1, this volume), while in Spain the legal regulations established descriptive geometry as a main subject—as a polytechnic subject—in the mathematics branch of the Faculty of Sciences. In his Ph.D. graduation speech at the university of Madrid in 1861 Elizalde placed on the same level infinitesimal calculus and descriptive geometry in modern mathematics. Moreover, among the Faculties of sciences, the chair existed only in Madrid, and only Barcelona and Zaragoza obtained a professorship of descriptive geometry at the end of the century.

Echegaray, in a series of papers published in the scientific journal *Revista de los progresos de las ciencias exactas, físicas y naturales*, attempted to impose projective geometry as a higher, theoretical, geometry beyond descriptive geometry, following Michel Chasles (Echegaray 1867). He intended to mark the difference between

---

[6]Letters to Cremona from Gaston Darboux (1842–1917) and even from the publisher, secretary of the Société amicale des anciens éléves de l'École polytechnique, Jean-Albert Gauthier-Villars (1828–1898) show the bitter feelings aroused at having lost the leadership in the geometrical area linked to applications, a trend that—after all—had had its origins in France.

[7]This circulation of French books was part of the greater influence of French culture in Spain, of the prestige of French science, and of the fact that many influential Spanish scholars in the early nineteenth century, such as José Mariano Vallejo (1779–1846) and Juan de Cortázar (1809–1873), had had direct contacts with Parisian institutions.

[8]The file is conserved at the Spanish National Archive (Archivo General de la Administración, see references).

applied technical knowledge and theoretical research, trying to widen mathematical knowledge, which should have been the mission of the faculty of sciences. At Barcelona, the industrial capital of Spain, there was an interest in Culmann's and Cremona's development of graphical methods in engineering and architecture, as can be seen in the reports presented to the local Academy of Sciences and Arts in the 1880s. This prompted a new interest in synthetic projective geometry (*geometría de posición*). In 1883 Santiago Mundi i Giró (1842–1915), professor of analytical geometry since 1881, presented an essay on the evolution of synthetic projective geometry and its connections with analytical geometry, descriptive geometry, and its application to mechanics, and in 1884 a book on the subject (reproducing handwriting, edited by four of his students) following Mundi's 1883–1884 lessons at the University of Barcelona (based on Reye's and Antonio Favaro's treatises) was published:

> The publication of these lessons will benefit not only the students of our University, but also all those who need to devote themselves to the study of Culmann's graphostatics, whose more natural bases are graphical calculus and Staudt's geometry. (Mundí 1884, pp. 7–8)

The same idea was followed by the industrial engineer Carlos Maria de Moÿ, who in 1888, after a journey to Rome where he met Cremona and his collaborators at the Engineering school, published a book on projective geometry as the first volume of a treatise on graphical statics. Nevertheless, the mathematical milieu in Barcelona—and its eventual influence outside Catalonia—was damaged by a harsh controversy regarding growing abstraction and the departure from intuition and the physical world of geometry, in which the radical Mundí was opposed to a conservative group to which the architect Josep Domènech i Estapà (1858–1917), who in 1895 won the competitive examination in Descriptive geometry, belonged.[9]

At Madrid, Torroja decided rather to transform his descriptive geometry course into a course on synthetic projective geometry. The influence of Torroja was greater also because from 1900 he taught also the Ph.D. course on higher studies in geometry. A Catalan from Tarragona, the son of a primary teacher, Torroja was one of the first generation of students educated under the 1857 Law of Education, master (1866) of and Ph.D. (1873) in mathematics at Madrid, and he also graduated from the school of architecture in 1869. He taught descriptive geometry at Madrid in 1869–1973 as a replacement of Elizalde, collaborating also at the National Astronomic Observatory and as an external examiner at the school of architecture. In the syllabus written by Torroja for the above-mentioned 1875 competitive examination, he examined the split between technical and scientific-liberal education, and proposed to widen the approach under the name of descriptive geometry:

> in it [descriptive geometry] also the study of the general properties of lines and surfaces and even those specific to any of them is also included, from a purely geometrical point of view and excluding algebraic considerations almost completely; and proceeding this way,

---

[9]Doménech y Estapá 1898. See Millán (1991) and the bibliography herein. García de Galdeano's treatises included projective geometry as well as differential geometry.

not only is the instrument perfected, and the mind and even the hand become accustomed to its use, but the way is paved towards pursuing without setback the important applications of it.[10]

In 1879 he published in the series "Anales de la construcción y de la industria" a book on the quite new axonometric perspective, the interest of which for the representation of machines and buildings and in crystallography he underlined, comparing it to Monge's method (Torroja 1879). His references included Fiedler's (1875) treatise and the 4th edition (1876) of Karl Wilhelm (Pohlke 1860) Pohlke's (1810–1876) treatise on descriptive geometry for the Berlin Bauakademie, La Gournerie's treatise, and the *Corso teorico-pratico ed elementare di disegno assonometrico applicato specialmente alle macchine* (1861) by Agostino Cavallero (1833–1885).[11] The first sentence of the book was a quite broad definition of descriptive geometry:

> The immediate object of Descriptive Geometry is the determination of figures by means of other simpler or more appropriate ones regarding the questions under study. The examination of the existing relationships between the former and the latter, the investigations of the properties of the former by means of the corresponding ones in the latter, and the resolution of problems regarding the former through constructions performed on the latter, offer a wide and fertile field cultivated with great success by modern Geometry (Torroja 1879, p. 1).

In his 1884 work *Program and synopsis of the lessons on descriptive geometry*, Torroja developed the program presented in 1875. He supported the need to study the geometrical basis of the representation systems used in the applications: in his view, the contents of descriptive geometry as established in France should be given up, and Fiedler's approach should be followed. Therefore, his course actually became a course in projective geometry, or higher geometry, following the synthetic approach of Staudt:

> There will be people who will find strange the turn give to this course in Descriptive geometry which breaks, completely, with the traditions invariably followed in our homeland, and even in the neighbouring Republic whose progress we usually imitate, and not always as a wise move. [...]
>
> In some French treatises and in the Spanish by my dear master D. José Antonio Elizalde axonometrical and dimensional drawing methods are included, even if less importance is given to them than to Monge's; but, in order to find the method of the conic projection (foundation of linear perspective independently of that put forward by Monge) one needs to resort to treatises published in Germany and even Belgian ones. This system of conic projection is the one which has the highest scientific character and offers more elements that can give a solid basis in the field of descriptive geometry to the theories of curves and of surfaces. That is why we find in Germany the first essays of a truly scientific systematization

---

[10] *Oposiciones a la cátedra de Geometría descriptiva vacante en la Universidad Central. Programa presentado para las mismas por D. Eduardo Torroja.* November 1875, manuscript, see unpublished sources in the references.

[11] Axonometric representation was applied to crystallography by Quintino Sella. Ernesto Bellavitis published a treatise on axonometric drawing which he taught at Padua University in 1876. On machine drawing in Spain see Zulueta Pérez (2011).

of Descriptive geometry, removing it from the narrow framework that usually circulates among us (Torroja 1884, p. ix)

This work was praised by the State education national council in a report in 1892 for having broken the "old and paltry forms of descriptive geometry" (an example of these "paltry forms" is shown in Fig. 6.1) adopting new, broader ones more suitable for "the fair aspirations of pure science".[12] And in 1893, during Torroja's reception at the Royal Academy of Sciences, the forestry engineer Francisco de Paula Arillaga y de Garro (1846–1920) underlined that the university had not the freedom needed to enlarge the scope of higher education, and that in order to introduce modern projective geometry it had been necessary to give a wider interpretation to the name "descriptive geometry" (Torroja 1893, p. 80). In some ways, the new definition of descriptive geometry was for Torroja a new definition of geometry tout court, a new approach to geometry from the point of view of a mathematician-architect interested in the 2D representation of figures: although this idea stemmed from technical knowledge (see Chap. 4), a theoretical point of view focused on geometrical transformations was progressing also in Spain.

4 Final Remarks

An answer to the question formulated above is that the study and teaching of descriptive geometry, even if mainly aimed at training skilled technicians, encouraged the interest in geometric research in Spain. The admiration for geometrical "purism" in Italy and Germany led to undue emphasis on synthetic methods, which prevented the diffusion of the whole array of European geometric research. But at the same time it helped the spreading of a new university ethos of research, even if this modern mission of the university was marred by the overwhelming attention paid to technical education. In fact, the situation improved radically in the 1920s and 1930s, thanks to the grants and institutions of the Council for Advanced Education and Scientific Research (*Junta para ampliación de estudios e investigaciones científicas*), founded in 1907 and chaired by the Nobel prize for Medicine Santiago Ramón y Cajal. José Gabriel Álvarez Ude (1876–1958), professor of descriptive geometry at Zaragoza from 1902, and Julio Rey Pastor, both Ph.D. students of Torroja's, had the opportunity to study in Germany and in 1915 founded with José María Plans (1878–1934 a professor of Mathematical physics) the Mathematical Laboratory and Seminar of the Council. The issue of the European circulation of scientific information was oftentimes emphasized by Rey Pastor, and in particular regarding the graphical methods and geometry in the late nineteenth century; in his

---

[12] The report is included in Torroja's personal file at the National Archive (Archivo general de la Administración; see unpublished sources in the references).

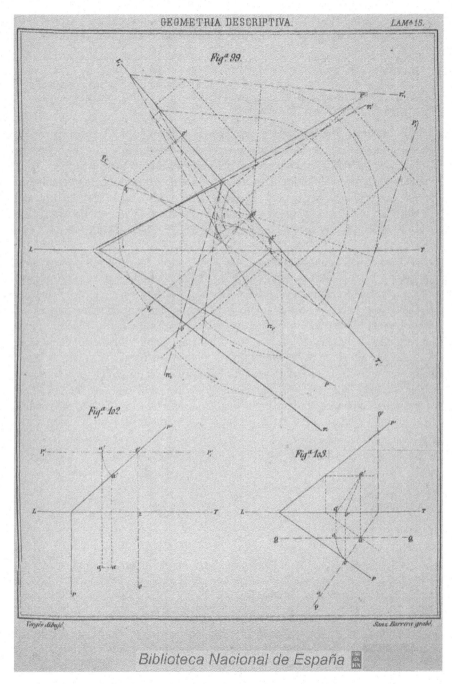

Fig. 6.1 Two figures from the book by Esteban Vergés Galofre. We read the name of the draftsman (the author himself) and the engraver Sanz Barrera (Vergés Galofre 1895, Volume 2, plates, corresponding to the chapter VII on "rebatimientos"). From the Biblioteca Digital Hispánica (Madrid, Spain)

1918 prologue to the Montevidean professor Federico García Martínez (born 1875) *Lecciones de Cálculo gráfico, Estática gráfica y aplicaciones*, Rey Pastor wrote[13]:

> The successful applications of Staudt's methods, that allowed Culmann to organize Graphical statics, has not yet found great acceptance in French books, for a long time the sole source of enquiry among people speaking Castilian; this excessive predilection, which hindered a knowledge of Italian and German sources, explains the scarce diffusion of Graphical statics. (García Martínez 1918, p. 2)

In Henkel (1926) Álvarez Ude published a translation of a German treatise on graphical statics by Otto Henkel. Moreover, together with Rey Pastor they published a translation of Moritz Pasch's treatise on modern geometry, and sent many young graduate students of the Laboratory to Italy in order to study the approach of the Italian school of algebraic geometry (Ausejo and Millán Gasca 1989). Significantly enough, Torroja's sons were an outstanding structural engineer (Eduardo) (Levi et al. 2000) and a pioneer of photogrammetry (José María) (Muro Morales et al. 2002).

Therefore, the example of Spain can help gain a better understanding of the emergence of the late modern European outlook on the relationship between pure science and technology (Nicolaïdis and Chatzis 2000), analyzing the case of descriptive geometry in this framework: the tension between pure science and technology is still present in the European Union scientific research policy. As we have seen, in order to take off, the late nineteenth century mathematical research in Spain needed to consider descriptive geometry no longer as a proper branch of mathematics. This evolution is considered as a paradox by some architecture scholars today: "the task of maintaining alive this noble and ancient science" is today "entrusted only to architects and engineers, as mathematicians have long since given it up" (Migliari 2014, p. 23; see also Carlevaris et al. 2012; Gómez Acosta 2013).

References

Sources

National Spanish Archive, Archivo General de la Administración, Alcalá de Henares (Spain):

Legajo (file) 1470–76 (E. y C.), Eduardo Torroja.
Legajo 5401–41 1875–76 public competitive examination for the professorship in descriptive geometry, University of Madrid, winner Eduardo Torroja.
Legajo 1445–43, 1875–76 public competitive examination for the professorship in descriptive geometry, University of Madrid, winner Eduardo Torroja: manuscript *Oposiciones a la cátedra*

---

[13] Rey Pastor had a great influence on the development of mathematical research in Latin America. He met García Martínez on his first visit to Argentina in 1917. (see Ortiz and Otero 2001, García Martínez 1918). The prologue has been republished in (Rey Pastor 1988, vol. 2.).

de Geometría descriptiva vacante en la Universidad Central. Programa presentado para las mismas por D. Eduardo Torroja, November 1875.
Legajo 5403–36 (E.C.), 1883 call for a public competitive examination for the professorship in descriptive geometry, University of La Habana.
Legajo 5404–60 (E.C.), 1895 public competitive examination for the professorship in descriptive geometry, University of Barcelona, winner Josep Domènech i Estapà.
Legajo 5407–4 (E.C.), 1902 public competitive examination for the professorship in descriptive geometry, University of Zaragoza, winner José Gabriel Álvarez Ude.

Publications

Ashwin, Clive. 1981. *Drawing and Education in German-speaking Europe 1800–1900*. Ann Arbor: UMI Research Press.
Ausejo, Elena, and Ana Millán Gasca. 1989. La organización de la investigación matemática en España en el primer tercio del siglo XX: el Laboratorio y Seminario matemático de la Junta para Ampliación de Estudios (1915–1938). *Llull, Revista de la Sociedad Española de Historia de las Ciencias y de las Técnicas* 12 (23): 261–308.
Ausejo, Elena, and Mariano Hormigón. 2002. Spanish initiatives to bring mathematics in Spain into the international mainstream. In *Mathematics Unbound: The Evolution of an International Mathematical Research Community 1800–1945*, ed. Karen H. Parshall and Adrian C. Rice, 45–60. Providence, RI: American Mathematical Society/London Mathematical Society.
Ausejo, Elena, Mariano Hormigón, Ana Millán, and María Ángeles Velamazán. 1990. Mathematics and engineering at the rise of capitalism in Spain. In *The Interaction Between Technology and Science*, ed. Bart Gremmen, 199–209. Wageningen: Wageningen Agricultural University.
Barbin, Evelyne, and Marta Menghini. 2013. History of teaching geometry. In Handbook on the History of Mathematics Education, ed. Alexander Karp and Gert Schubring, 473–457. Berlin: Springer.
Bermúdez Abellán, José. 2005. *Génesis y evolución del Dibujo como disciplina básica en la Segunda Enseñanza (1836–1936)*. Ph.D. Thesis, Murcia, University of Murcia.
Blanes Nadal, Georgina, Carlos Millán Verdú, and Rafael Sebastiá Alcaraz. 2002–2003. El origen de la Escuela de Artes y Oficios de Alcoy, 1886–1888. *Quaderns d'Història de l'Enginyeria* 5: 85–97.
Blay, Michel, and Efthymios Nicolaïdis. 2001. *L'Europe des sciences. Constitution d'un espace scientifique*. Paris: Seuil.
Cámara Muñoz, Alicia, ed. 2016. *Draughtsman engineers serving the Spanish monarchy in the sixteenth to eighteenth centuries*. Madrid: Fundación Juanelo Turriano.
Cardona y Escarrabill, Baltasar. 1865. *Tratado de geometría descriptiva y de sus principales aplicaciones al dibujo de projectos, sombras, perspectiva, cortes de piedra, de madera y de hierro, etc. para uso de los señores ingenieros, arquitectos, maestros de obras, directores de caminos vecinales, ayudantes de obras públicas, aparejadores, delineantes, agrimensores, maquinistas, contra-maestres, carpinteros, albañiles, canteros y demás carreras profesionales y de aplicación*, 2 vols. Barcelona (Establecimiento tipográfico de Jaime Jepús).
Carlevaris, Laura, Laura De Carlo, and Riccardo Migliari, eds. 2012. *Attualità della geometria descrittiva*. Roma: Gangemi Editore.
Cremona, Luigi. 1873. Elementi di geometria projettiva ad uso degli istituti tecnici del Regno di Italia. Roma: G. B. Paravia.
de Moÿ, Carlos María. 1888. *Estática gráfica. Primera parte: Geometría Proyectiva*. Barcelona.
D'Enfert, Renaud. 2003. *L'enseignement du dessin en France. Figure humaine et dessin géométrique (1750–1850)*. Paris: Belin.

Doménech y Estapá, José. 1898. *Programa de las lecciones de geometría descriptiva explicadas en la Facultad de Ciencias de la Universidad de Barcelona, por el catedrático de la expresada asignatura*, Barcelona (Imprenta de la Casa provincial de caridad).

Efland, Arthur D. 1990. *A history of art education: Intellectual and social currents in teaching the visual arts*, New York and London: Teachers College Press, Columbia University. Fiedler, Wilhelm. 1867. "Das Polytechnikum in Mailand (Augsburger)". Allgemeine Zeitung, 24.2, Nr. 55: 894–895. 1871. Die darstellende Geometrie: ein Grundriß für Vorlesungen an Technischen Hochschulen und zum Selbststudium. Leipzig: Teubner.

García Martínez, Federico. 1918. *Lecciones de cálculo grafico, estática gráfica y aplicaciones*. Montevideo.

Gómez Acosta, Juan. 2013. *La estática gráfica instrumento para el conocimiento estructural intuitivo y el diseño de los arquitectos*. Valencia: Universitat Politècnica de València.

Echegaray, José. 1867. *Introducción a la geometría superior*. Madrid: Imprenta y Librería de D. Eusebio Aguado.

Elizalde, José Antonio. 1861. *Resumen histórico de los progresos de las matemáticas desde los tiempos más remoto hasta nuestros días. Discurso leído en la Universidad Central por Don José Antonio Elizalde, catedrático de Geometría descriptiva de la misma, en el acto de recibir la investidura de Doctor en la Facultad de Ciencias, Sección de Ciencias Esactas*. Madrid.

———. 1873–1878. *Curso de geometría descriptiva*, 2 vols. Madrid.

Ferrer de Pertegás, Francisca. 1897. *Elementos de geometría plana y descriptiva y nociones de dibujo con aplicación a las labores de la maestra*. Valencia (Imprenta Gombau, Vicent y Masiá).

Fusi, Juan Pablo, and Jordi Palafox. 1996. *España 1808–1936: el desafío de la modernidad*, Madrid, Espasa Calpe.

García de Galdeano, Zoel. 1908. Algunas observaciones pedagógicas acerca de la Matemática. In *Memorias presentadas por Don Zoel García de Galdeano. Actas del Congreso de Zaragoza de la Asociación Española para el Progreso de las Ciencias*, 1–4.

Grattan-Guinness, Ivor. 2002. The end of dominance: The diffusion of French mathematics elsewhere, 1820–1870. In *Mathematics Unbound: The Evolution of an International Mathematical Research Community, 1800–1945*, ed. Karen Parshall, and Adam Rice, 17–44. Providence, RI: American Mathematical Society.

Gropp, Harald. 2004. Reseaux réguliers or regular graphs. Georges Brunel as a French pioneer in graph theory. *Discrete Mathematics* 276 (1): 219–227.

Guijarro Salvador, Pablo. 2009. La enseñanza del dibujo en Tudela durante el siglo XIX. *Principe de Viana* 246: 67–104.

Henkel, Otto. 1926. *Estática gráfica* (Spanish translation by José G. Álvarez Ude. 4th edition, Barcelona: Labor, 1953)

Hormigón, Mariano, and Ana Millán. 1992. Projective geometry and applications in the second half of the nineteenth century. *Archives Internationales d'Histoire des Sciences* 42 (129): 269–289.

Israel, Giorgio. 2017. Luigi Cremona. In *Correspondence of Luigi Cremona (1830–1903)*, ed. Giorgio Israel, vol. 1, 33–58. Turnhout: Brepols.

Knobloch, Eberhard. 2013. Luigi Cremona and his German correspondents. In *Unpublished lecture, International Conference Mathematical Schools and National Identity (16th–20th cent.), Turin, October 10–12, 2013*.

Laurent, Stéphane. 1999. *Les Arts appliqués en France : Genése d'un enseignement*. Paris: Editions CTHS.

Levi, Franco, Mario Alberto Chiorino, Chiara Bertolini Cestari. 2000. *Eduardo Torroja. From the Philosophy of Structures to the Art and Science of Building*. Milano: Franco Angeli.

Migliari, Riccardo. 2014. La geometria descrittiva dalla tradizione all'innovazione. In *La geometria descrittiva dalla tradizione*, ed. Cesare Cundari, and Riccardo Migliari, 15–26. Roma: Aracne Editrice.

Millán Gasca, Ana. 1990a. *Estudio de la obra de Julio Rey Pastor en el contexto del desarrollo histórico de la geometría proyectiva*. Unpublished Ph.D. Thesis. Zaragoza: University of Zaragoza, Spain.
———. 1990b. Methods of synthetic geometry in the second half of the second century. In *International Symposium Structures in Mathematical Theories (San Sebastián, 25–29 September 1990)*, ed. A. Díaz et al., 283–288. San Sebastián, Universidad del País Vasco.
———. 1991. Los estudios de geometría superior en España en el siglo XIX. Llull. *Revista de la Sociedad Española de Historia de las Ciencias y de las Técnicas* 14 (26): 117–186.
———. 2011. Mathematicians and the nation in the second half of the nineteenth century as reflected in the Luigi Cremona correspondence. *Science in Context* 24 (1): 43–72.
———. 2012. La matematica nella "sfida della modernitá" della Spagna liberale e il ruolo del modello italiano. In *Europa matematica e Risorgimento Italiano*, ed. Luigi Pepe, 81–90. Bologna: CLUEB.
———. 2016. Mathematics and children's minds: The role of geometry in the European tradition from Pestalozzi to Laisant. *Archives internationales d'histoire des sciences* 65 (175): 759–775.
Mundì i Girò, Santiago. 1884. *Apuntes de geometría de la posición tomados de las explicaciones del Dr. D. Santiago Mundí, catedrático de dicha asignatura en esta Universidad*, por D. Julio Enamorado, D. Arturo Ydrach, D. Luis Cuello y D. Arturo Vidal. Barcelona (selfpublished treatise); accessible in Biblioteca Digital Hispánica (Spanish National Library, http://bdh.bne.es/bnesearch/detalle/bdh0000123752)
Muro Morales, José Ignacio, Luis Urteaga, and Francesco Nadal. 2002. La fotogrametría terrestre en España (1914–1958). *Investigaciones geográficas* 27: 151–172.
Nicolaïdis, Efthymios, and Kostas Chatzis, eds. 2000. *Science, Technology, and the 19th Century State*. Athens: Institut de Recherches Néohelleniques.
Ortiz, Eduardo, and Mario Otero. 2001. Removiendo el ambiente: la visita de Einstein a Uruguay en 1925. *Mathesis* 1 (1): 1–35.
Pohlke, Karl W. 1860. *Darstellende Geometrie, zunächst für den Gebrauch bei den Vorträgen an der Königlichen Bau-Akademie und dem Königlichen Gewerbe-Institut zu Berlin*. Gaertner: Berlin (4th ed. 1876).
Ramón Teijelo, Pío-Javier. 2011. *El Real Conservatorio de Artes (1824–1887): Un intento de fomento e innovación industrial en la España del XIX*. Ph.D. Thesis, Universitat Autònoma de Barcelona.
Rey Pastor, Julio. 1988. *The Works of Julio Rey Pastor*, ed. Eduardo L. Ortiz, 8 vols. London: The Humboldt Society.
Rodríguez de Campomanes, Pedro. 1975 (1874 and 1875). *Discurso sobre el fomento de la Industria Popular. Discurso sobre la educación popular de los artesanos y su fomento*, ed. John Reeder. Madrid: Istituto de Estudios Fiscales, Ministerio de Hacienda.
Souto Salorio, María José, and Tarrío Tobar, Ana Dorotea. 2016. Enrique Vidal Abascal (1908–1994). Un renacentista en el siglo XX. *La Gaceta de la Real Sociedad Matemática Española* 19 (2): 385–406.
Scholz, Erhard. 1994. Graphical statics. In *Companion Encyclopedia of the History and Philosophy of the Mathematical Sciences*, ed. Ivor Grattan-Guinness, vol. 2, 987–993. London: Routledge.
Torroja, Edoardo. 1879. *Axonometría o perspectiva axonométrica: sistema general de representación geométrica que comprende, como casos particulares, las perspectivas caballera y militar, la proyección isográfica y otros varios*. Madrid (Imp. de Fortanet).
———. 1884. *Geometría descriptiva. Programa y resumen de las lecciones de Geometría descriptiva esplicadas en la Universidad Central*, tomo I. Madrid (reproduction of handwriting).
———. 1893. *Discursos leídos ante la Real Academia de Ciencias Exactas, Físicas y Naturales en la recepción pública del Sr. D. Eduardo Torroja y Caballé el día 29 de Junio de 1893*. Madrid: Real Academia de Ciencias Exactas, Físicas y Naturales.
Turner, Roy Steven. 1971. The growth of professional research in Prussia, 1818 to 1848: Causes and context. *Historical Studies in the Physical Sciences* 3: 137–182

Vergés Galofre, Esteban. 1895. *Elementos de geometría descriptiva. Primera parte, Puntos, rectas y planos*. 2 vols. Barcelona: Imprenta de la Casa provincial de Caridad. *Biblioteca Digital Hispánica*. http://www.bne.es/en/Catalogos/BibliotecaDigitalHispanica/Inicio/

Vidal Abascal, Enrique. 1962. El profesor Rey Pastor. *Revista matemática hispano-americana* 21 (2): 116–120.

Viollet-le-Duc, Eugène-Emmanuel. 1879. *Histoire d'un dessinateur, comment on apprend á dessiner. Texte et dessins par Viollet le Duc*. Paris: J. Hetzel

Zulueta Pérez, Patricia. 2011. El dibujo de máquinas: sistematización de un lenguaje gráfico. In *Técnica e ingeniería en España, vol. VI. El Ochocientos. De los lenguajes al patrimonio*, ed. Manuel Silva Suárez, 213–251. Zaragoza: Real Academia de Ingeniería, Institución "Fernando el Católico", Prensas Universitarias de Zaragoza.

Chapter 7
Portuguese Textbooks on Descriptive Geometry

Eliana Manuel Pinho, José Carlos Santos, and João Pedro Xavier

Abstract The introduction of descriptive geometry in Portuguese did not begin in Portugal itself but in what had so far been the outskirts of the Empire: in Brazil. There, a commented translation of Monge's treatise was published in 1812, but it was decades later before a descriptive geometry textbook was first published in Portugal. Furthermore most of the first descriptive geometry booklets and textbooks published in Portugal were written for secondary school students, and the publication of a comprehensive descriptive geometry textbook intended for higher education had to wait until the end of the nineteenth century. This chapter will describe the evolution of the teaching of descriptive geometry in Portugal up to the end of the nineteenth century.

Keywords Castel-Branco · Achilles Machado · Shiappa Monteiro · Descriptive geometry · Polytechnic school · Portugal · Mota Pegado · *Universidade de Coimbra* · *Escola Politécnica de Lisboa* · *Academia Politécnica do Porto*

1 Introduction

Portugal was invaded by France in 1807. As a consequence the entire royal family of Portugal, including Queen Maria I (1734–1816) and her son Prince Regent (later King) João VI (1767–1826) escaped to Brazil. King João VI returned to Portugal in

E. M. Pinho
Independent Researcher, Porto, Portugal

J. C. Santos
Faculty of Sciences, University of Porto, Porto, Portugal
e-mail: jcsantos@fc.up.pt

J. P. Xavier (✉)
University of Porto, Faculty of Architecture, Porto, Portugal
e-mail: jxavier@arq.up.pt

1821, and the next year Brazil—from where Portugal had been reigned after the end of the Napoleonic invasions—became an independent country.

Before 1807, Brazil had been kept underdeveloped as a colony by the Portuguese crown. For instance, printing presses had not been allowed and were imported only by the royal fleet in 1808. Now, however, all efforts were undertaken to develop this huge country and to establish an educational system and to promote a proper economic and technological system. The first signs of this technical transformation had been the *Real Academia de Artilharia, Fortificação e Desenho* founded in 1792 following the Real Academia de Artilharia, Fortificação e Desenho founded in 1790 in Lisbon. Now, an ambitious institution to educate military and civil engineers for Brazil was created, the *Real Academia Militar* in 1811. It was there that descriptive geometry was taught for the first time in the Portuguese Empire (see Schubring, Chap. 21, this volume). The first professor was an army officer, Captain José Victorino dos Santos e Souza, who published in 1812 a translation of Monge's textbook (Souza 1812) with the title *Elementos de Geometria Descritiva; com aplicações às artes* (Elements of descriptive geometry; with applications to the arts). Its subtitle emphasized the objective of forming the students of the military institution: "Para uso dos alunos da Real Academia Militar". It was not only a translation, though as it had a lengthy introduction, in which the author explained the aim of descriptive geometry. Here is what he wrote about the subject of the book (in a single sentence):

> The goal of descriptive geometry is not only to lead us to a large number of theorems, to give us the solution to an infinite number of problems deduced from the properties possessed by points, lines and planes when we consider them placed in any way in space with respect to each other, and in reference to three dimensions, of which the planar geometry problems, which are usually dealt with in the first approaches to this subject, are nothing but particular cases, but also to a way of representing on a paper the shapes of bodies from nature, or from art, which, as we know, have three dimensions, that is to transmit these descriptions to those instructed in the conventions and methods of this geometry, or to deduce from this representation new truths about their properties, their shapes and their dimensions, in various senses.[1] (Souza 1812, pp. 206–207)

And he emphasized that his desire in publishing the translation was to help raise knowledge of the sciences and of the fine arts in this new world (see Schubring, Chap. 21, this volume).[2] The introduction also mentions Sylvestre-François Lacroix's *Essais de Géométrie sur les plans et les surfaces courbes ou Éléments de Géométrie Descriptive* (Lacroix 1795), which was published in 1795, that is 4 years before Monge's *Géométrie Descriptive* (Monge 1799) (see Barbin, Chap. 1, this volume).

King João VI returned to Portugal in 1821, and Brazil became an independent country the year after.

---

[1] Translated by the authors.

[2] In fact, the translation of Monge's textbook had no impact in Portugal. For instance, the book is held by neither the National Library in Lisbon nor the library of the Academia de Marinha in Lisbon.

2 Descriptive Geometry at the University of Coimbra

At that time, the only university in Portugal was the University of Coimbra, which was founded in 1290 and had a Faculty of Mathematics from 1772 onwards. At the beginning of the nineteenth century, Coimbra actively fought for its monopoly as the only Portuguese university (Carvalho 1986, p. 567), and new universities (in Lisbon and Porto) would only be established in 1911. Due to the political and social turbulence of this period, the university could not open in some academic years, in particular between 1831 and 1834 when the civil war ended, and lacked teachers until 1837 (Freire 1872, pp. 61–63). The necessary adjustments to the new achievements in mathematical theories were made only after this date.

According to Gino Loria (Loria 1921, p. 402), the French mathematicians whose treatises on descriptive geometry had the strongest influence in Portugal were Olivier (1843), Fourcy (1830), and Leroy (1834) (on these textbooks, see Barbin, Chap. 2, this volume). After 1821, the first Portuguese mathematician to write a book on descriptive geometry was Rodrigo Ribeiro de Sousa Pinto (1811–1891). He was a professor of mathematics and astronomy at the University of Coimbra (and director of the Astronomical Observatory of Coimbra) and it was there that he published the book *Complementos da Geometria Descriptiva de M. de Fourcy* (Pinto 1853). However, as the author himself states (and as the title suggests), the book is not a standalone textbook. It is mostly a collection of observations requiring the original treatise (Fourcy 1830) to be comprehensible. However, this book would have been a major reference for the teaching of descriptive geometry at the University of Coimbra. According to Sousa Pinto (besides de Fourcy's treatise) his knowledge of descriptive geometry was based on Monge's textbook and a little-known booklet by Gascheau (1828) (Barbin, Chap. 2, this volume).

A course on descriptive geometry was created in 1840 as part of the degree in Mathematical Science (Freire 1872, pp. 64–69). It was titled "Descriptive geometry, geodesy and architecture", and it was given in the 4th year of a five-year degree, and the textbook adopted was de Fourcy's treatise (Fourcy 1830). Descriptive geometry was successively merged with other subjects in 1844 and 1852 and acquired the status of a discipline on its own in 1861 with the name "Descriptive geometry; applications to stereotomy, to perspective and to shadow theory". The book *Complementos da Geometria Descriptiva de M. de Fourcy* (Pinto 1853) was adopted in 1853 and, according to (Freire 1872), Leroy's treatise on Stereotomy became the adopted textbook in 1872 (Leroy 1870).

3 The First Standalone Descriptive Geometry Textbook

At last, a descriptive geometry booklet written by an engineer (and army officer), José Frederico d'Assa Castel-Branco (1836–1912), was published in 1873 (Castel-Branco 1873). It was published in Goa, which is currently part of India, but which was part of the Portuguese State of India until 1961. It was a very short work: it was only 47 pages long (including 3 pages containing exercises), and it had no introduction (although Monge is mentioned on the first page), bibliography or diagrams. It was written for the students of the *Instituto Profissional de Nova-Goa*, an institution of higher learning created in 1871. It so happens that of all the Portuguese colonies of that time, Goa was the one in which the school system was most developed (Pery 1875, pp. 383–384). We found no evidence of its having been used for teaching descriptive geometry in mainland Portugal.

4 The Establishment of Polytechnic Schools and of Industrial Schools

Two schools with the word "polytechnic" in their name were created in Portugal in 1837, the *Escola Polytechnica de Lisboa* and *Academia Polytechnica do Porto*. In fact, both of them were replacements for other pre-existing schools. At the political level, this was mainly due to the action of the Prime Minister Manuel da Silva Passos (better known as Passos Manuel), who felt the need to have more and better educated engineers in Portugal. In spite of the fact that his time in office as Prime Minister was less than a year, his government had a strong impact on higher education in Portugal, mainly because of the founding of these two schools. The presence of the word "polytechnic" in their names suggests that they were modelled on the *École polytechnique*, but it is more likely that they were based upon the *École Centrale* (Basto 1937, p. 152), which had been founded in 1829 by, among others, Théodore Olivier (Barbin, Chap. 2, this volume). This is in fact rather natural since the *École polytechnique* was mainly meant to prepare its students for public service, whereas the training of students at the *École Centrale* was concerned with work in industry (Schubring, Chap. 22, this volume). The first director of the *Escola Polytechnica de Lisboa* was José Feliciano da Silva Costa (1798–1866), a former student of the École des Ponts et Chaussées (Pereira 2009).[3]

Unlike the *École Centrale*, which was a private school until 1857, the Portuguese *Polytechnic* schools were created by the central government and, as a consequence, depended heavily on the changing mood of whoever was in power. They soon became underfunded, as Passos Manuel himself acknowledged in 1857 (Santos 2011, p. 75).

The *Academia Polytechnica do Porto* had a course on descriptive geometry right from its foundation (Pinto 2012) and adopted the textbook by de Fourcy (Santos 2011, p. 70). The first chair from 1838 until 1850 was an engineer, General José Victorino Damásio (1807–1875) (Delgado 1877, p. 9), a remarkable teacher of engineering but with no published texts on descriptive geometry (Pinto 2012, p. 142). The *Traité de géométrie descriptive* by Leroy (1834) was the adopted textbook for almost all of the last quarter of the nineteenth century (Pinto 2012, pp. 218–223). Although the course on descriptive geometry had existed since the

---

[3]For more about the influence of the *École des Ponts et Chaussées* on the *Escola Polytechnica de Lisboa*, see (Matos 2013).

foundation of the school, the first explicit references to its content relate to the school year of 1879–1880 due to the start of an annual publication, *Annuario da Academia Polytechnica do Porto*. Topics concerning the straight line and the plane were dealt with in a mathematic course, whereas the "descriptive geometry of the curves and tangent planes and of the curves with their tangents" were part of another course. The adopted textbook for both courses was de Fourcy's textbook, from which the exercises and problems solved in class were picked.

The next number of the *Annuario* contained the first known syllabus of descriptive geometry for the *Academia Polytechnica do Porto*. This syllabus was designed in 1880 by the Academic Council and is divided into two parts, the first with 12 lectures and the second with 9 lectures, described as follows (Anonymous 1880–1881, 1886–1887):

- Study of notable curves, especially the helix, epicycloids and involutes of the circle.
- Aim of descriptive geometry. Several projection systems. Representation of points, straight lines, and planes.
- Representation by a single projection with elevation. Problems.
- Intersection of two straight lines and the angle between them. To make a straight line pass through two given points.
- To make a plane pass through three given points. Intersection of a straight line and a plane, angle between them. Intersection of two planes.
- Distance between a point and a straight line and between two straight lines: shortest distance.
- Representation of two curved surfaces. Second degree surfaces. Tangent planes.
- Construction of the plane that is tangent to the cylinder and to the cone, given a point of the contact generatrix.
- Construction of the plane that is tangent to the ellipsoid of revolution or tri-axial, given the contact point.
- Construction of the plane that is tangent to the ellipsoid and that passes through a point outside the ellipsoid. Contact curve of the ellipsoid and of the circumscribed cone.
- Intersections of a plane and a cylinder, a cone, an ellipsoid.
- Intersection of a cylinder or cone with an ellipsoid.
- Intersection of two ellipsoids.

Second part (nine lectures):

- On the generation of surfaces.
- On developable surfaces. Developable helicoid.
- On skew surfaces.
- On the hyperboloid of one sheet.
- On the hyperbolic paraboloid.
- On the skew helicoid.
- Evolutes and involutes. Spherical involute.
- Curvature of surfaces. Lines of curvature of an ellipsoid.
- The torus and its tangent planes.

At the *Escola Polytechnica de Lisboa*, descriptive geometry started to be taught only in 1860, 6 years after such a course was proposed (Sequeira 1937, p. 5). The syllabus of that course was (in a summarized version):

- Aims of descriptive geometry.
- Representation of points, straight lines, and planes on two orthogonal planes.
- Rabatment of the vertical projection plane.
- Notation.
- Representation of polyhedra.
- To find the length of a straight line segment.
- Change of the projection planes.
- Rotation of a shape around an axis.
- Intersection of planes, straight lines and planes.
- To find planes that are perpendicular to given straight lines, or planes, under several conditions.
- Angles between straight lines and planes and study of several particular cases.
- Problems concerning the definition of straight lines and planes given different sets of data concerning angles, distances, points, projections, etc.
- Solid angles and resolution of several particular problems. General considerations concerning curved lines, tangent asymptotes, singular points, curvature, etc.
- To find tangents and their contact points.
- General definition of a surface.
- Curves on a surface and their tangents.
- Tangent plane and normal.
- The cylinder and the cone, their generatrices, apparent contours, projections, nets, tangent planes and sections.
- Several problems concerning the determination of generatrices and tangent planes, given different sets of conditions.
- General problem of cone and cylinder planar sections.
- Surfaces of revolution, axes, tangent planes and sections.
- Main properties of the hyperboloid of revolution of one sheet, sections and its asymptotic cone.
- The intersection of curved surfaces.
- Problems concerning curves with double curvature.
- Several special cases of the intersection of cones, cylinders and other surfaces of revolution.
- The helix, its projection, generatrices, tangents and subtangents.
- The developable helicoid, its tangent plane, sections and nets.
- The epicycloid.
- Definition of ruled surfaces.
- Considerations concerning the properties of developable and skew ruled surfaces.
- Definition of conoids and skew surfaces of the second degree.
- The hyperbolic paraboloid and the hyperboloid of one sheet and their double generation by straight lines, tangent planes, planar sections, and other properties.
- Particular conditions for the intersection of skew surfaces.

Fig. 7.1 Luiz Porfírio da Mota Pegado (1831–1903) (Sequeira 1937, pp. 6–7)

- Tangent planes and normals to skew surfaces.
- Study of several skew surfaces: conoid, bias passé, skew helicoid, screw surface, groin vault.
- Problems concerning planes that are tangent to several surfaces, in particular to two or three spheres, or to a sphere and a cone.

The first teacher was a former student of the Military Academy and of the *Escola Polytechnica de Lisboa*, General Luiz Porfírio da Mota Pegado (1831–1903), who taught there until 1902 (Fig. 7.1). Near the end of this long period, he published a textbook on descriptive geometry (Pegado 1899).

Before examining Mota Pegado's textbook, it should be noted that he published original research (although not in descriptive geometry) in two Portuguese scientific journals: *Jornal de Sciencias Mathematicas* and *Physicas e Naturais* and *Jornal de Sciencias Mathematicas e Astronomicas*. We shall have more to say about the second one below.

Fortunately for us, Mota Pegado not only tells us which works influenced his textbook, but he also tells us from where he learned descriptive geometry. He learned it from two textbooks. Not surprisingly, one of them was José Victorino dos Santos e Souza's translation of Monge's textbook (Souza 1812). The other one was a booklet that he described this way: it was only 28 pages long, it was divided into 39 sections and it mentioned 32 figures, which were absent from the copy that he studied from (and he did not know whether they had ever existed). Mota Pegado tells his readers that, according to an old rumour, the author was José Monteiro Feio (1787–1884), a former teacher at the *Escola Polytechnica de Lisboa*. He complains about the fact that no other Portuguese textbooks existed from which one could learn descriptive geometry. Therefore, he was presumably unaware of the existence of the booklet by Castel-Branco (Castel-Branco 1873).

Although he had been teaching descriptive geometry since 1861, Mota Pegado kept in touch with developments in the discipline, as reflected in his book. He

Fig. 7.2 Alfredo Augusto Schiappa Monteiro de Carvalho (1838–1919) (Sequeira 1937, pp. 10–11)

quotes an article by Amédée Mannheim published in 1882 as well as the edition of Mannheim's textbook (Mannheim 1886) published in 1886 and the third edition of Félix Chomé's textbook (Chomé 1898). From these texts, Mota Pegado extracted the idea of teaching descriptive geometry without using the concept of ground line. Mota Pegado states that he was influenced by La Gournerie's textbook (Gournerie 1860) in the context of the interaction between descriptive geometry and projective geometry, and he also mentions two textbooks written by Michel Chasles: his geometry textbook (Chasles 1880) and his textbook on the history of geometry (Chasles 1875). It is perhaps for this reason that Mota Pegado's textbook starts with a 42-page long chapter about modern geometry. Finally, he also mentions textbooks by Luigi Cremona (Menghini, Chap. 4, this volume), G. F. Monteverde and Ferdinando Aschier; all of them published no more than 20 years before the publication of his textbook.

General Alfredo Augusto Schiappa Monteiro de Carvalho (usually known as Schiappa Monteiro) was an assistant to Mota Pegado from 1870 until Mota Pegado retired in 1903, and then he replaced him until his own retirement in 1911 (Fig. 7.2). Unlike Mota Pegado, Schiappa Monteiro did publish articles about descriptive geometry. Most of these, if not all, were published in French in the *Jornal de Sciencias Mathematicas e Astronomicas*. These articles were usually short notes about very specific aspects of descriptive geometry. The descriptive geometry textbooks that he cites are all by French authors: Gournerie (1860), Fourcy (1830) (he cites the 1842 edition) and (Poncelet 1865–1866).

Schiappa Monteiro was one of the two people that Mota Pegado thanked concerning the writing of his textbook.

The *Jornal de Sciencias Mathematicas e Astronomicas*, a mathematical journal, was founded in 1877 by the Portuguese mathematician Francisco Gomes Teixeira (1851–1933). The first issue of the journal had a short list of topics in applied mathematics that were considered for publication, and one of them was stereotomy. As we saw above, it did indeed publish articles on that subject.

Gomes Teixeira was a professor at the University of Coimbra and then a professor at the *Academia Polytechnica do Porto*, where he taught descriptive geometry for a short period although he published nothing about it. From 1911 on, he was a professor and the first rector of the University of Porto, and he was a towering figure in the Portuguese mathematical community in the late 19th and early twentieth century.

Minister Fontes Pereira de Melo (1819–1887), a military engineer who studied at the *Escola Polytechnica de Lisboa*, had a good understanding of industrial and scientific matters, as opposed to the vast majority of politicians, who had a background in law (Carvalho 1986, p. 587). Fontes Pereira de Melo proceeded with several reforms in factories and public infrastructures. The need for professional expertise in industrial arts became apparent, and two new institutions were created by a decree published in the Diério do Governo on December 30, 1852 (pp. 864–870)—the *Escola Industrial do Porto* and the *Instituto Industrial de Lisboa*. These were very important in the spread of descriptive geometry in Portugal. With their focus on the education of industrial workers and craftsmen, descriptive geometry was established from the beginning as a fundamental discipline, articulated with linear drawing and machine drawing. These institutions had a course on descriptive geometry from their foundation in 1852. The first syllabus that we found for the industrial school in Porto was from 1887, when the course is designated as "Descriptive geometry and stereotomy—topography". This follows on from a decree in 1886 that reformulates industrial and commercial teaching in Portugal and is very similar to the syllabus from the Lisbon institute dating back to 1872. This syllabus, apart from the topography section, has two parts: the first one that we list below and the second part called "Applications of descriptive geometry" with the subsections shadows and wash drawing, machine drawing,[4] stone cutting and wood cutting.

The first part of the descriptive geometry syllabus begins with some preliminaries followed by the sections "Method of orthogonal projections" with six topics and "Method of projections with elevation" with two topics (Anonymous 1886):

- Introduction.
- Aims and usefulness of descriptive geometry. Projections: several methods.
- Method of orthogonal projections.

  (i) Preliminary notions:

  1. Representation of points
  2. Representation of lines in general
  3. Representation of straight lines
  4. Representation of the planes[5]

---

[4]In the Porto school, machine drawing is not included and is replaced by linear perspective. However, the study of screws, gears and serpentines appears in the first part of the course in the Porto school syllabus.

[5]In the Porto school syllabus, the notion of rabatment of planes is added here.

(ii) Problems concerning straight lines and planes:

1. Intersection of straight lines and planes
2. Straight lines and planes defined by different conditions
3. Perpendicular straight lines and planes
4. Angles of straight lines and planes

(iii) 1. Representation of right and oblique prisms
2. Representation of right and oblique pyramid;
3. Representation of frustums of prisms and pyramids

(iv) Curved surfaces:

1. Generation of surfaces and their graphic representation
2. Tangent planes in general, apparent contours, normal
3. On the several kinds of surfaces and their main properties

(v) Problems concerning tangent planes:

1. Tangent plane through a given point on the surface
2. Tangent plane parallel to a given straight line

(vi) Intersection of surfaces:

1. General principles
2. Planar sections of curved surfaces
3. Intersection of curved surfaces

– Method of projections with elevation.

(i) Preliminary notions:

1. Representation of points;
2. Representation of lines in general
3. Representation of straight lines
4. Representation of the planes
5. Representation of curved surfaces and especially of the terrain surface

(ii) Problems:

1. Problems concerning straight lines
2. Problems concerning planes

5 Descriptive Geometry for Secondary School Teaching

Let us now see which other descriptive geometry books were written by Portuguese authors in the nineteenth century.

Already in 1840, Maurício José Sendim (1790–1870), a painter and a teacher at the Casa Pia (which was, and still is, an educational institution dedicated to helping youngsters at risk of social exclusion or without parental support) announced the publication of such a book. It was to be part of a series of booklets published under the general title *O Estudante de Desenho e Pintura* (The student of drawing and painting). Unfortunately, although other booklets in the collection were published, the one about descriptive geometry never was Rodrigues (2001) and Pimenta (2003).

In 1878, a descriptive geometry booklet written by António Augusto Gonçalves (1848–1932) was published (Gonçalves 1878). António Augusto Gonçalves studied at the Universidade de Coimbra, but he never graduated. In spite of that, after having been a teacher of drawing in several schools for many years, he became professor of drawing at the University de Coimbra in 1902 (Rodrigues 1992). His book is only 39 pages of text followed by 6 pages of diagrams. It is divided into 93 very short sections; many of which consist of a single sentence. It has no bibliography, references to other authors or practical applications of what it teaches. The word "rebatimento" (which is the Portuguese word for "rabatment") is already mentioned in it. The booklet has a short two-page introduction, in which the author claims that the contents of the book follow the official syllabus. That is quite likely to be the secondary school syllabus, although he doesn't state that explicitly. Indeed, descriptive geometry was taught in secondary schools in Portugal around 1880. In this period, the syllabus for the *Lyceus* (secondary schools) had descriptive geometry in the 6th year, but it only addressed elementary notions related to points, straight lines and planes (Anonymous 1880, 1886). More precisely, the students were exposed to the following topics:

- Method of projections
- Representation of points, straight lines and planes
- Problems concerning straight lines and planes
- Intersection between a straight line and a plane
- Orthogonal straight lines and planes
- Method of rabatments
- Angle between two straight lines; angle between two planes
- Rotation around a vertical axis

By the end of the nineteenth century, descriptive geometry was no longer an independent course although it was still taught in drawing classes. More details concerning the teaching of descriptive geometry in secondary schools can be found in Palaré (2013).

Another descriptive geometry book published in 1883, which was explicitly meant for secondary schools, was *Elementos de Geometria no Espaço e de Geometria Descriptiva para uso nos Lyceus*, written by two army officers and teachers of Physics at the *Escola Polytechnica de Lisboa* (Vidal and Almeida 1883). This was an expanded edition of an earlier book, which did not mention descriptive geometry at all. Actually, the part of the book which deals with descriptive geometry is very short; of the 150 pages, only 27 are about descriptive geometry, and the rest of the book is about geometry in space.

Finally, an anonymous booklet (Machado 1885) about descriptive geometry was published in 1885 (Fig. 7.3). It was part of a collection called *Biblioteca do Povo e das Escolas* (Library for the People and the Schools), which was published both in

Fig. 7.3 Cover of *Geometria Descriptiva* (Machado 1885, cover)

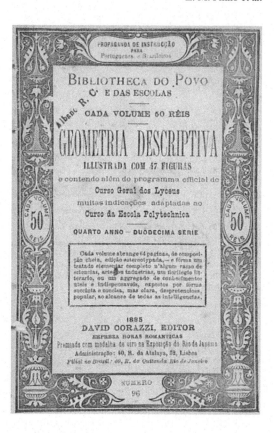

Portugal and in Brazil. This collection was very wide-ranging with booklets about many varied subjects such as history, biology, gymnastics, geography, agriculture, medicine and so on. In particular, there were three other booklets on geometry, two of which (about plane geometry and space geometry) were written by Carlos Adolfo Marques Leitão (1855–1938); the third one (about spherical geometry, as well as spherical trigonometry) was written by Rodolphe Guimarães (1866–1918). Although the collection had several books published anonymously, in most cases the names of the authors were provided. The author of this booklet was Achilles Alfredo da Silveira Machado (1862–1942), who, at the time he wrote this text, had just graduated from the *Escola Polytechnica*, where he must have been a student of Mota Pegado. He published two other booklets in the collection (under his name) about gunpowder and railways. We know that he is the author of this booklet from two sources; the editor of the collection says so in the foreword of its twelfth series[6] and in (Anonymous 1942) it is stated that Achilles Machado published a book on descriptive geometry for the general public.

---

[6]Personal communication from Nabo (2012).

The long title of the booklet states that not only its contents follow the official secondary schools syllabus but also that part of its contents correspond to what is taught at the *Escola Polytechnica*. In spite of being only 58 pages long, the author spends some time explaining what descriptive geometry is. He states that it "can be seen both as an art and a science" and that it was "only after the works of the famous Monge at the end of the last century that descriptive geometry started to be seen as a science". Also, he describes descriptive geometry as "the science which allows us to represent objects placed in space precisely in a plane, and to solve, using drawings on the plane, the problems related to these objects". Therefore, even outside the academic world, by this time the idea that descriptive geometry is a useful and important science was firmly established in Portugal.

6 Conclusions

In Portugal, as far as universities and similar institutions were concerned, the teaching of descriptive geometry, followed closely the way that this subject was being taught in France, mainly at the *École polytechnique* and the *École Centrale*. This, together with the fact that knowledge of French was expected from students at institutions of higher learning in Portugal, is possibly the reason why it took so long before a Portuguese textbook on descriptive geometry was published. However, it was felt that there was a need for descriptive geometry at the secondary school level, in part in the context of drawing. So, in Portugal, books on descriptive geometry at the secondary school level were published before they were published at the university level.

Acknowledgements The authors wish to thank Patrícia Costa (Museum of the *Instituto Superior de Engenharia do Porto*) and Roger Picken (*Instituto Superior Técnico*) for their valuable help.

José Carlos Santos was partially supported by CMUP (UID/MAT/00144/2013), which is funded by FCT (Portugal) with national (MEC) and European structural funds (FEDER), under the partnership agreement PT2020

References

Anonymous. 1886. *Novos Programmas do Curso dos Lyceus a que se refere a Portaria de 19 de Novembro de 1886*. Coimbra: Livraria J. Mesquita.
——. 1880. *Programmas approvados por Decreto de 14 de Outubro de 1880 para o ensino nos Institutos Secundarios*. Porto and Braga: Livraria Internacional de Ernesto Chardron.
——. 1880–1881. Resoluções do Conselho Academico. *Annuario da Academia Polytechnica do Porto*. A. 4: 1886–1887.
Anonymous. Achilles Machado. 1942. *Revista de Chimica Pura e Aplicada*, série III: 81–89.
Basto, Artur de Magalhãis. 1937. *Memória histórica da Academia Politécnica do Porto*. Porto: Universidade do Porto.

Carvalho, Rómulo de. 1986. *História do Ensino em Portugal*. Lisboa: Fundação Calouste Gulbenkian.
Castel-Branco, José Frederico d'Assa. 1873. *Elementos de Geometria Descriptiva: Lições coordenadas para uso dos alumnos do Instituto Profissional de Nova-Goa*. Pangim, Goa: Imprensa Nacional.
Chasles, Michel. 1875. *Aperçu historique sur l'origine et le développement des méthodes en Géométrie, particulièrement de celles qui se rapportent á la géométrie moderne*. 2ème édition. Paris: Gauthier-Villars.
——. 1880. *Cours de Géométrie Supérieure* (2nd ed.). Paris: Gauthier-Villars.
Chomé, Félix. 1898. *Cours de Géométrie Descriptive de l'École Militaire*, (3rd ed.), vol. I. Paris: Gauthier-Villars.
Delgado, Joaquim Filipe Nery. 1877. *Elogio historico de José Victorino Damasio*. Lisboa: Imprensa Nacional.
Fourcy, Louis Lefébure de. 1830. *Traité de Géométrie Descriptive*. Paris: Bachelier.
Freire, Francisco de Castro. 1872. *Memoria historica da Faculdade de Mathematica nos cem annos decorridos desde a reforma da Universidade em 1772 até o presente*. Coimbra: Imprensa da Universidade de Coimbra.
Gascheau, Gabriel. 1828. *Géométrie descriptive: traité des surfaces réglées*. Paris: Bachelier.
Gonçalves, António Augusto. 1878. *Brevíssima noção elementar sobre o método das projecções ortogonais*. Coimbra: Imprenssa da Universidade de Coimbra.
Gournerie, Jules de La. 1860. *Traité de Géométrie Descriptive*. Paris: Mallet-Bachelier.
Lacroix, Sylvestre-François. 1795. *Essais de Géométrie sur les plans et les surfaces courbes (ou Éléments de Géométrie Descriptive)*. Paris: Fuchs.
Leroy, Charles-François-Antoine. 1870. *Traité de stéréotomie: comprenant les applications de la géométrie descriptive á la théorie des ombres, la perspective linéaire, la gnomonique, la coupe des pierres et la charpente*. Paris: Gauthier-Villars.
Leroy, Charles-François-Antoine. 1834. *Traité de Géométrie Descriptive*. Paris: Carilian-Gœry.
Loria, Gino. 1921. *Storia della geometria descrittiva dalle origini sino ai giorni nostri*. Milan: Ulrico Hoepli.
Machado, Achilles Alfredo da Silveira. 1885. *Geometria Descriptiva: illustrada com 47 figuras e contendo além do programma official do Curso Geral dos Lyceus muitas indicações adaptadas ao Curso da Escola Polytechnica*. Editado por David Corazzi. Lisboa: Empreza Horas Romanticas.
Mannheim, Amédée. 1886. *Cours de Géométrie Descriptive de l'École Polytechnique* (2nd ed.). Paris: Gauthier-Villars.
Matos, Ana Cardoso de. 2013. The influence of the École des Ponts et Chaussées of Paris on the Lisbon polytechnic school (1836–1860). *Journal of History of Science and Technology*, 7 13–35.
Monge, Gaspard. 1799. *Géométrie Descriptive*. Paris: Baudouin.
Nabo, Olímpia de Jesus de Bastos Mourato. 2012. *Educação e Difusão da Ciência em Portugal: A "Bibliotheca do Povo e das Escolas" no Contexto das Edições Populares do Século XIX*. Tese de mestrado. Portalegre: Instituto Politécnico de Portalegre/Escola Superior de Educação de Portalegre.
Olivier, Théodore. 1843. *Cours de Géométrie Descriptive*. Paris: Carilian-Gœury et Vor. Dalmont.
Palaré, Odete Rodrigues. 2013. *Geometria descritiva: História e didática – Novas perspetivas*. Tese de doutoramento. Lisboa: Universidade de Lisboa.
Pegado, Luiz Porfírio da Mota. 1899. *Curso de Geometria Descriptiva da Escola Polytechnica*. Lisboa: Academia Real das Sciencias.
Pereira, Pilar de Lurdes Alagoinha. 2009. *École Polytechnique de Paris versus Escola Politécnica de Lisboa*. XXIIIe Congrés International d'histoire des Sciences et des Techniques. Budapeste. http://hdl.handle.net/10451/173
Pery, Gerardo A. 1875. *Geographia e estatistica geral de Portugal e colonias*. Lisboa: Imprensa Nacional.

Pimenta, Joaquim Alberto Borges. 2003. *Desenho: Manuais do século XIX de autores portugueses*, vol. I. Master's Thesis, Porto: Faculdade de Letras da Universidade do Porto.

Pinto, Hélder Bruno Miranda. 2012. *A Matemática na Academia Politécnica do Porto*. Tese de doutoramento, Universidade de Lisboa, Lisboa.

Pinto, Rodrigo Ribeiro de Sousa. 1853. *Complementos da Geometria Descriptiva de Lefébure de Fourcy*. Coimbra: Universidade de Coimbra.

Poncelet, Jean-Victor. 1865–1866. *Traité des propriétés projectives des figures* (2nd ed.). Paris: Gauthier-Villars.

Rodrigues, Carlos Telo. 2001. *Maurício José Sendim: Professor e Litógrafo (1790–1870)*, vol. I. Master's Thesis, Porto: Faculdade de Letras da Universidade do Porto.

Rodrigues, Manuel Augusto. 1992. *Memoria professorum Universitatis Conimbrigensis II: 1772–1937*. Coimbra: Arquivo da Universidade de Coimbra.

Santos, Cândido dos. 2011. *História da Universidade do Porto*. Porto: Editorial UP.

Sequeira, Luiz Guilherme Borges de. 1937. *Escola Politécnica de Lisboa: A cadeira de Geometria Descritiva e os seus professores*. Lisboa: Faculdade de Ciências de Lisboa.

Souza, José Victorino dos Santos e. 1812. *Elementos de Geometria Descriptiva; com aplicações às artes*. Rio de Janeiro: Imprensa Regia.

Vidal, António Augusto de Pina, and Carlos Augusto Morais de Almeida. 1883. *Elementos de Geometria no Espaço e de Geometria Descriptiva para uso dos Lyceus* (4th ed.). Lisboa: Typ. da Academia Real das Sciências.

Chapter 8
In Pursuit of Monge's Ideal: The Introduction of Descriptive Geometry in the Educational Institutions of Greece During the Nineteenth Century

Christine Phili

Abstract From 1824, during his teaching Ioannis Carandinos introduced descriptive geometry, as well as Monge's classical treatise in the Ionian Academy. A few years later in independent Greece, the newly founded Military School, inspired by the model of the *École polytechnique*, included in the curriculum the teaching of descriptive geometry from Monge's book. Konstantine Negris, a former student at the *École polytechnique*, tried to diffuse the spirit and methods of Monge during his period at the University of Athens. In the Polytechnic School of Athens, Monge's treatise was also adopted in the teaching of descriptive geometry as a useful tool for the instruction of craftsmen and engineers. Moreover, the translation of Louis-Benjamin Francœur's book on linear drawing, as well as that of Jean-Pierre Thénot's on perspective, gave a considerable impulse in spreading the basic notions of descriptive geometry into secondary schools during the last decades of the nineteenth century when the first treatises on descriptive geometry appeared written in Greek.

Keywords Descriptive geometry · Perspective · *École polytechnique* · Ioannis Carandinos · Ionian Academy · Military School · Konstantine Negris · George Bouris · University of Athens · Polytechnic School · Ioannis Papadakis · Louis-Benjamin Francœur · Jean-Pierre Thénot · Technical education

1 Introduction

It might be emphasized that even though Greece was occupied, it was not an intellectual desert. Before the proclamation of the War of Independence in 1821, and more specifically between 1770 and 1821, many mathematical textbooks were

---

C. Phili (✉)
Department of Mathematics, Greece National Technical University, Athens, Greece
e-mail: xfili@math.ntua.gr

published in Greek in Europe (Karas 1992). The main aim of all these books was the education of Greek readers, an indispensable condition for their revolt against the Ottoman Empire.

However, it would be an omission here not to mention the edition of various journals and periodicals published in Vienna (Phili 2010), which had the objective of functioning as a channel that would popularize scientific and literal knowledge. Among them, *Hermes the Scholar* (Vienna 1811–1821) played a decisive role as a bridge diffusing Western ideas in the East (Karas 2003). In this journal, Greek intellectuals read for the first time the modern editions of books devoted to descriptive geometry. Thus, in the issue of 15 June 1817, the Greeks learned that in Brunswick a treatise of descriptive geometry was published, *Traité de Géométrie descriptive á l'usage des éléves de l'institut des voyes de communication, par M. Potrér avec grav.*[1] Vol. gr in 8° á Brunswick, Pluchart.[2]

The next year, in this same journal in the issue of 1 August 1818, p. 422, an anonymous Greek writer from the Greek diaspora informed his compatriots about and presented his comments on new mathematical publications. Thus, the Greeks learned that the following books had recently been published: *Traité de la Géométrie descriptive par M. Potrér, Géométrie descriptive par Monge (sénateur) avec les deux suppléments de la Géométrie descriptive par Hachette*.[3]

We consider that this information regarding Monge's new geometry probably constitutes the very first reference to descriptive geometry in the Greek diaspora, as well as for those who lived under Ottoman rule. We must take into consideration that the journal Hermes the Scholar was quite well diffused in more than 40 Greek and European cities through its numerous subscribers. Therefore, it became an important channel for transferring the ideas of the French Enlightenment in Greece.

2 The Flourishing Epoch of the Ionian Academy Under Carandinos' Dominance

Thanks to the numerous and persistent efforts of Frederick North,[4] the 5th Earl of Guilford as well as his influence on the British Government, the Ionian Academy

---

[1]Despite the orthographical error, it concerns the book by Charles Marie Potier, *Géométrie descriptive* Paris 1817 (96 pages in 8°).

[2]*Hermes the Scholar*, News regarding French books. Vienna, 15 June 1817, p. 284.

[3]From this notification results that it concerns Hachette's *Supplément á la Géométrie descriptive de Gaspard Monge*, Paris 1811 (published according to the 3rd edition of Monge's Descriptive Geometry), as well as, *Second supplément de la Géométrie descriptive ... suivi de l'Analyse géométrique* de M. John Leslie, Paris, 1818. *Hermes the Scholar*, 1 August 1818, p. 422.

[4]Frederick North Guilford (1766–1827), the younger son of Prime Minister Frederick North, 2nd Earl of Guilford, showed early evidence of his capacities by promoting and developing the status of education during his mandate (1798–1805), as the governor of Ceylon. Later, he traveled across Europe in order to enrich his knowledge in the leading universities. His relationship with the former

Fig. 8.1 A gravure representing the inaugural ceremony of the Ionian Academy (Gazzetta Ufficiale degli Stati Uniti delle Isole Ionie n. 335, 17/29, May 1824)

(Fox 2012), i.e. the first university, was established by the Legislative Assembly (*Gazzetta Ufficiale degli Stati Uniti delle Isole Ionie* 284, 26 May–7 June 1823) in Corfu in 1823 with Greek as its official language (Idem 339, 17 May–29 May 1824). However the Ionian Academy opened in 1824 (see Fig. 8.1).

Professor Ioannis Carandinos the Ephorus (Έφορος) or Rector was entirely responsible for mathematical education, as well as administrative matters.[5]

Ioannis Carandinos (1784–1834), a penniless child from the island of Cephalonia studied at the first public school created in Corfu under the regime of the Septinsular Republic (Phili 2006). Later during the second French occupation of the Ionian Islands, he had the opportunity to study mathematics privately (analysis and mechanics) under Charles Dupin's guidance. After Dupin's departure from Corfu (Bradley 2012, pp. 73–75), Carandinos returned as the teacher at this public school and was in the position "to teach the young pupils concerning the Lacroix's and Laplace's systems and other contemporary French [Scientists]".[6] The meeting of Carandinos with Lord Guilford was decisive in his career. The early education

---

Minister of Foreign Affairs, Lord Bathrust, facilitated his task organizing the Ionian Academy. In 1819, he was named Chancellor (Άρχων) of that institution.

[5]Guilford proposed Carandinos for this position: "I will take the liberty of proposing, for that office, our well deserving senior professor John Carandinos". 17th May 1824. Plans submitted to the government, for the establishment and regulation of the Ionian Academy, Corfou Reading Society. Guilford's Archives. file V5.

[6]See Proselantis' letter to the Review, *Hermes the Scholar*, 1812, p. 190.

of Carandinos facilitated Guilford's options. Through Guilford's scholarship, his young *protégé* attended lectures at the *École polytechnique* in 1821 as free auditor (auditeur libre) (Phili 1996, p. 307).

After the inaugural ceremony, Carandinos started his lectures. Fortunately, we found his weekly program in his book, which can be found now in the Gennadius Library in Athens. Thus, Carandinos declared in his autographed notes that:

> I give three lectures per week...complement of Algebra [of] Lacroix as well as to the primary class from 1 November 1825 in the first class, which comprises of 11 students, I did 10 hours per week, and I presented the following authors...text [of] Monge descriptive geometry and the above mentioned introduction (Carandinos 1826).

After Carandinos' departure in 1832 due to health reasons, a new epoch started for the Ionian Academy that was never again able to reach the previous high level of mathematical education except during the presence of Ottaviano Fabrizio Mossotti (1836–1840). It might be stressed that the preference for applied mathematics of the new High Commissioner Lord Howard Douglas (1776–1861) and Mossotti's preference for the same topic were the main reasons for the reform of 1837, which created the Faculty of Civil Engineers. Mossotti, who was associated with its establishment, proposed that the candidates for this new faculty followed preparative lectures on analytic and descriptive geometry, on optical instruments as well as on elements of surveying (Phili 2012a). Nevertheless, as the students were not adequately qualified and as the Professor of Mathematics Ioannis Kontouris resigned in order to attend the lectures at the *École polytechnique* as an auditeur libre, the faculty of civil engineers never opened.

During the academic year 1837–1838, lectures in the Faculty of Philosophy on pure mathematics were attended by several students, although in this same year the Othonian University was officially established. Professor Kontouris taught stereometry,[7] elements of algebra and trigonometry. For these courses, Kontouris mainly utilized Adrien-Marie Legendre's book translated by Carandinos.

After Mossotti's departure in 1840, the curriculum of the Faculty of Philosophy was modified, and descriptive geometry was no longer included in the curriculum.

3 The Military School of Evelpides

When Ioannis Kapodistrias arrived in Greece on 6 January 1828, he found a country without determined borders, devastated by the War of Independence, as well as by internal conflicts. In this atmosphere of disorder and misery, Kapodistrias undertook the first measures in order to establish a well-organized state worthy of ensuring a successful outcome of this disastrous war and the recovery of the Greek people.

After the War of Independence, primary schools developed a quite well-balanced curriculum between classical studies and sciences. But in the Central School

---

[7]We cannot be certain if the lectures on stereometry comprised elements of descriptive geometry.

(Κεντρικὸν Σχολεῖον), established by Kapodistrias in Aegina in 1829, pupils had the opportunity to have the most modern manuals of that time: Carandinos' translations on arithmetic and algebra. Specifically about 600 books of Caradinos' translations of Legendre's Elements of Geometry were distributed in 1830–1832, an impressive number for that period.

However, the re-organization of the army remained one of the main aims of the governor. So, on 1 July 1828, he established the Company of Evelpides[8] in Nafplion, the first capital of Greece from 1829 to 1834. However, due to the poor condition of the building, the Military School was housed in an orphanage in Aegina from 1834 to 1837.

We owe a special mention to the subsequent translation of Francœur's book, *Dessin Linéaire et arpentage* ...[9] (Linear Drawing and land surveying...) (Francoeur 1819, 1827). During the first years of his mandate, Ioannis Kapodistrias ordered Konstantinos Kokkinakis (1781–1831), the co-editor of the journal Hermes the Scholar, to translate Louis-Benjamin Francœur's book. Finally, this book was published (Fragkirou 1831) posthumously in 1831, as its translator had died and the final revision was undertaken by Ioannis Kokkonis (1795–1864), a member of the educational commission, a general inspector of the schools in the Peloponnese peninsular and an ardent partisan of the mutual teaching method due to his studies in Paris (1824–1829) with Louis-Charles Sarazin. We must stress that Kapodistrias invited two distinguished architects to teach linear drawing in the School of Aegina: Stamatis Kleanthis (1802–1862) and Eduard Schaubert (1804–1860).

This book largely contributed to the diffusion of linear drawing, as well as some elementary methods of projection, leveling and rules of perspective (see especially the "Descriptive Geometry in Nineteenth Century Spain: From Monge to Cirodde" chapter on the rules of dioptics, i. e. Greek translation of the word perspective) and became an indispensable tool for artisans, carpenters, etc. Thus, we could consider that this manual became the preliminary tool for the teaching of descriptive geometry in several secondary schools. However, we must take into consideration that according to the educational planning of the Regency (3/15 July 1833) (Greece adopted the Gregorian calendar only in 1923), the theory of shadows (σκιαγραφία) was included in the curriculum of secondary schools, as well as "Euclidean geometry and the geometry of Diesterweg (1828)" [sic] Project of the committee regarding the public education. Nafplion 3/15 July 1833 in (Antoniou 1992, p. 109). It might be stressed that after Kapodistrias' assassination, the teaching of design was established by the royal decree of 6 February 1834, art. 1 (*Journal of the State* no 11, 3 March 1834), while the teaching of painting became obligatory

---

[8]In Greek, this adjective (here it is used as noun) means hopeful, promising, and this name was given to the very young students of the Ionian Academy. We consider that this nomination, which is used until now, was given also to these young students, who represented the hope for the Greek nation.

[9]The complete title is: *Dessin Linéaire et arpentage pour toutes les écoles primaires quel que soit le mode d'instruction qu'on y sait*, 1e éd. 1819, 2e éd. 1827, Paris. Francœur dedicated this book to the Duc of Gazes.

by the royal decree of 31 December 1836 art. 7 (*Journal of the State* no 87, 31 December 1836). Unfortunately, we have not been able to find any official list regarding their syllabi.

On 2 December 1828, Kapodistrias accepted Jean-Pierre-Augustin Pauzié's proposal for the founding of a military school (Kastanis 2003, p. 125), inspired by the model of the *École polytechnique* in order to supply the country with qualified officers, who would provide for the administration of the state and contribute to its growth. Thus, the Military School was officially established on 28 December 1828. Kapodistrias entrusted the direction of the school, the first in Greece, to Pauzié (a former student at the *École polytechnique* in 1812).

In the General State Archives, we found an important manuscript (General State Archives. Secretary of Military Affairs doc. 54 January 1829, f. 102) (see Fig. 8.2) in French, probably dictated by Pauzié, in which he revealed, among other things, the regulation of the curriculum.

From this important French manuscript, which in fact constitutes a relevant document for an historian of mathematics, we will focus on the 64th article of the regulation regarding studies. This article ordered that the Military School should deliver the following indispensable tools to each student, such as a drawing board, a box of pencils, two rulers, an inkpot, an elastic gum, a French-Greek dictionary,

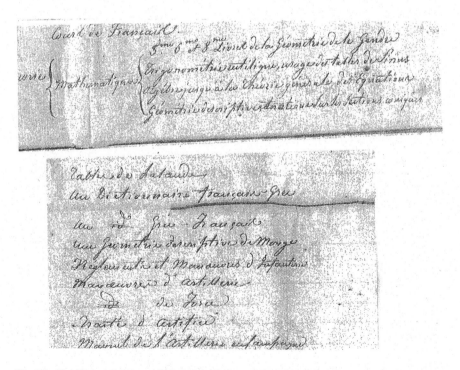

Fig. 8.2 The 64th article regarding the regulation of the curriculum (General State Archives. Secretary of Military Affairs doc 54 January 1829. f. 102)

books on algebra and arithmetics by Pierre-Louis-Marie Bourdon, Legendre's geometry and "une géométrie descriptive de Monge" (sic), i.e. one descriptive geometry by Monge. According to this list, we consider that as Carandinos' translation was never edited, Kapodistrias was able to provide several copies of Monge's treatise via his supporters abroad to make up for the lack of that manual in Greek.

After King Otto's crowning and his decrees establishing primary and secondary education, a new era began in education. The secondary schools that were named Hellenic Schools were similar to the German *Lateinschulen*. The pupils learnt mainly ancient Greek and Latin and less mathematics, physics, zoology, etc. The mathematics curriculum comprised of, firstly, arithmetic, based mainly on Bourdon's book, elements of arithmetic and, secondly, geometry, based on Legendre's, Elements of Geometry. Christos Vafas, a former student in the Ionian Academy, translated in 1837 Louis Lefébure de Fourcy's book, *Éléments d'Algèbre*, which he taught in the first Gymnasium in Athens. However, under Otto's reign, we can see a "shift" from the French mathematical ideal, which was inaugurated in Greece with Carandinos' translations, to the German one. In 1842, George Gerakis, Professor of the Gymnasium, translated "Elementary Geometry and Trigonometry" by Friedrich Snell (Snell 1799, 1819) and a few years later presented Karl Koppe's *Arithmetic und Algebra* (Koppe 1836a) as well as "Plane Geometry (1836b) and Stereometry" (Koppe 1836b) in Greek (Kopp 1855, 1857, 1858).

In 1834, the Regency re-organized the Military School, whose first 4 years were mainly dedicated to a preparatory curriculum. Henceforth, the studies lasted 8 years in order to become the highest institution in the country's educational system. The Bavarians modified the curriculum and, apart from German, introduced differential and integral calculus, spherical trigonometry, geodesy, mechanics and fluid mechanics and hydraulics, etc. Of course, descriptive geometry maintained an important role within the curriculum. Especially in the fifth year when students attended lectures regarding elements of differential, integral calculus and descriptive geometry, while in the sixth year the mathematical curriculum included the continuation of lectures on differential and integral calculus and descriptive geometry and a calculus of variations, the last being an extremely innovative course for that epoch in Greece.

There was a significant lack of didactic books in 1840. Thus, on 6/18 June 1840 the new commander of the Military School Colonel Spyridon Spyromilios (1800–1880), who replaced the Prussian Colonel Eduard von Rheineck (1796–1854), emphasized the lack of handbooks. In his report (General State Archives. Othonian period Ministry of Defense f. 372, no 323) to the Ministry of Defense, he remarked that this lack meant professors were obliged to translate or to compile European treatises as students otherwise tried with difficulty to note or to partially copy the lectures. This situation constitutes one of the main factors that prevented their progress. Thus, the commander proposed that the Military School should offer lithograph handbooks in Greek to its students. However, despite our research, we could not find any mathematical manuals from that period.

From the list of professors (General State Archives. Ministry of Defense. Othonian period M/B f. 372) in the academic year 1841, we can ascertain that Major Dimitrios Stavridis (1803–1866) was appointed to teach architecture, descriptive geometry, leveling and surveying of buildings and machines for 20 h per week, while Dimitrios Despotopoulos, a former student of Carandinos, taught mathematics for 18 h per week.

The report of the Council of Studies on 17 March 1842 revealed the course material. Regarding descriptive geometry, the course material of the 5th grade comprised of intercepts and applications of projectivity to the theory of shadows and scenography[10] (sic) for $1\frac{1}{2}$ h per week and the surveying of buildings and machines, which also included the presentation of woodcutting and the construction of the five capitals of columns for 2 h per week.

In the 4th grade, after the course of projective geometry, the lessons in descriptive geometry included leveling were taught $3\frac{1}{2}$ h per week. It might be stressed that the coefficient of 8 (General State Archives. Ministry of Defense. Othonian period M/B f. 421) for this course was quite high.

A letter (General State Archives. Ministry of Defense M/B f. 421) (written in French) on 11/23 June 1843 by Adolph Hast, a bookseller in Athens, to the Royal Ministry of War regarding the books, which would be distributed by the King as prizes to the diligent students, revealed that Monge's treatises were abandoned, although Monge's *Géométrie descriptive*—after its 6th edition in Paris 1838— was later edited for a 7th edition in Paris and in Brussels in 1854 (Taton 1951, p. 383). The bookseller's list, among others (General State Archives Ministry of Defense. M/B f. 421), contains the recently edited book by the Professor of the Polytechnic School, Charles-François-Antoine Leroy (Leroy 1837).[11] This permits us to conclude that the teaching of descriptive geometry was modified slightly and that Monge's classic treatise was relinquished (on Leroy's textbook, see Barbin, Chap. 2, this volume).

It might be stressed that King Otto showed evidence of his sincere interest in the Military School, as during his reign he attended exams and several times visited the school in Piraeus in order to attend lectures and regularly received reports regarding the progress of students. Therefore, Otto decided that the studies should last 7 years and the preparatory year was abolished in 1842. Descriptive geometry was taught in the fourth year for 16 h per week, and the students, along with the well known chapters regarding the surfaces' intersections, evolute and evolutionary, were initiated to study the applications of projectivity: theory of shadows and perspective. It is quite impressive that almost at the same time these chapters were also included

---

[10] It is interesting that the erudite Greeks named the science of perspective as scene painting based on Geminus' classification of the mathematical sciences (σκηνογραφικὴ in Greek).

[11] Charles-François-Antoine Leroy published his treatise on descriptive geometry in Bruxelles in 1837 (Leroy 1837). See *Traité de Géométrie Descriptive avec une collection d'épures, composée de 60 planches*.

in the curriculum of the Polytechnic School, but, unfortunately, for the purpose of this chapter we ignore the course material.

However on 16 September 1858, Lieutenant Dimitrios Antonopoulos (1821–1885), a former student of the *École polytechnique* and the professor of descriptive geometry and the theory of shadows and scenography (General State Archives. Ministry of Defence M/B f. 407), was replaced by officer Vassilios Romas. On 2 September 1859, Dimitrios Tournakis, a lieutenant of artillery was appointed to teach descriptive geometry.

Spyromilios' mandate was also characterized by an important innovation as in 1840 he introduced written exams. He considered that the oral ones could not offer an accurate idea of students' background as the examiner could intervene in order to facilitate the level of the questions. Thus, thanks to this modification, several copies are preserved in the General State Archives. The copies regarding descriptive geometry reveal the topics of the exams, as well as the name of the examiner.

On 14 October 1854, Professor V. Romas asked his students to determine the figure of the shadow and the perspective of a triangular pyramid (General State Archives. Ministry of Defense f. 427) (see Fig. 8.3). On 6 November 1857, Professor D. Antonopoulos demanded the intercept of a girder by a vertical plane, the shadow of the girder as well as its perspective (Idem, f. 424). On 13 September 1858, the same professor asked his students to describe a sphere in a triangular pyramid (Idem f. 429). On the 16 September 1859, the problem for the students was the intersection of a rectilinear cone by a cylinder (Idem).

After Otto's expulsion in 1863, the Military School was reformed. Thus, the new King, George I, a former student of a Danish naval school, introduced in 1864 the entry exams and decided that the studies should last 4 years for students who desired to study infantry and cavalry weapons and 6 years for those who opted for technical weapons: artillery and engineering. The first 3 years were preparatory and common for all the Evelpides, and, of course, descriptive geometry had a prominent role in the first 2 years. During the third year, the students learned about the intersections of surfaces, while during the fourth year those who were to become officers of the artillery and engineering corps were introduced to the applications of descriptive geometry: shadows, perspective, wood cutting.

Nevertheless, during this period, Greece was attempting to adopt modern technology, and military staff were one of the main supporters of its administration, and so a new reform of the Military School was announced on 31 October 1866 in order to speed up studies, which were reduced. From then on, studies lasted 5 years, three of which were preparatory as before. Descriptive geometry, having ten as a coefficient, remained an important part of the curriculum. However, this reform lasted only 1 year. In July 1867, the authorities closed the Military School and its re-opening in January 1868 was marked by the re-adoption of the 1864 program.

According to the new reform of 1870, lessons in the Military School lasted 7 years. During the first 5 years, students attended several courses, including mathematics. Descriptive geometry remained an important part of the curriculum as it retained the highest coefficient of ten; it was taught in the third year in its basic form. In the fourth year, its applications, gnomonics and wooden frameworks were

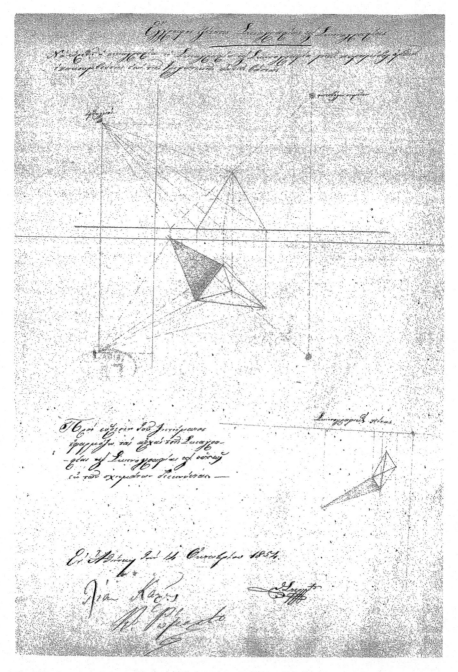

Fig. 8.3 A copy of a student's work from the Military School of a figure's determination of a shadow and the perspective of a triangular pyramid (14 October 1854) (General State Archives. Ministry of Defence f. 427)

introduced, while the fifth year was more focused on applications like the theory of shadow, perspective and enumerated plans (Poulos 1988, p. 137).[12]

From 1880 to 1886, the Professor of Descriptive Geometry was a major in the engineering corps, Miltiadis Kanellopoulos, who wrote one of the first treatises on descriptive geometry. Indeed his treatise,[13] "Lectures on descriptive geometry", based on his lessons given during the academic year 1882–1883 in the Military School appeared in 1883 (Kanellopoulou 1883). In this same year, Timoleon Moschopoulos' book, "Elementary Descriptive geometry" (Moschopoulos 1883), based on his lessons given in the school to non-commissioned officers, also appeared (Poulos 1988, p. 133). During the same period, Theodore Libritis, a captain of the engineering corps and Professor of Descriptive Geometry at the Military School, presented his books on the topic (Libritis 1881, 1886, 1888).

We must take into consideration that the demanding curriculum of the Military School became an extremely rigorous filter for young students. Thus, between 1831 and 1860, only 138 students managed to graduate (Stasinopoulos 1933). The majority of them completed their studies in the *Grandes Écoles of France* (General State Archives. Ministry of Defence. Othonian period f. 444).

Konstantine Chatzis notes (Chatzis 2018) that from 1830 to 1860 these well-educated officers started to translate a multitude of terms in their manuals and in their articles in military reviews into Greek. However, most mathematical terms had already been translated into Greek during the eighteenth century. Therefore, Ioannis Carandinos introduced the teaching of descriptive geometry by literally translating the adjective "descriptive" into Greek as "περιγραφική" (Phili 2012b).[14] Moreover, Dimitrios Stavridis, a graduate of the Polytechnic School of Vienna, contributed to the translation of some terms in these disciplines in his lectures in the Military School on descriptive geometry (Efimeris tou Stratou 1860), leveling and surveying of buildings (*Journal of Army*, Veteran no 21, 30 September 1860, p. 336). Nevertheless, the introduction of scientific terms in Greek during the nineteenth century demands special and meticulous research.

4 The University of Athens

By the royal decrees of 14/26 April 1837 and of 22 April/4 May 1837, King Otto established the first University, the Othonian University, a unique high institution in the Balkans and the Near East.

---

[12] In 1887 in Piraeus (where the Military School was located since 1837), an anonymous translation of Charles-François-Antoine Leroy treatise entitled "On Enumerated Plans" was published (Stratiotiko Scholeio ton Evelpidon 1887). In his book, Greek Mathematical Bibliography, Andreas Poulos considers that this translation followed the 4th edition of the book (Leroy 1855).

[13] His book on the theory of shadows was published in 1884.

[14] See, for example, Carandinos' notification on 22 Mai 1827 or his letter to Fourier on September 1828.

The teaching of mathematics was ensured by Konstantine Negris (1804–1880), a former student of the *École polytechnique* in Paris. In his autographed note on 21 of July 1836 (General State Archives. Othonian period f. 32) addressed to the Secretary of Education, he proposed, among other things, to teach descriptive (διαγραφική from the verb διαγράφω = to trace) geometry, dioptics (perspective) and the theory of shadows. However, Negris' ambitious curriculum was only partially realized.

Thus, for the first academic year 1837–1838, the curriculum consisted of the last five "Books" of Legendre's Elements of geometry, Legendre's rectilinear trigonometry, the general properties of numbers, algebra and Hachette's descriptive geometry[15] (Phili 2001, p. 84). These lectures were given from 4 to 5 p.m. every Monday, Wednesday and Friday.

It might be stressed that in the very first years, special care was taken to teach students the principles of practical geometry such as land surveying, leveling and of course, the use of geometrical instruments. These last instructions were indispensable not only for those following a career in civil engineering, but also for those who following a military career. The students conceived that mathematics in its applied form was the basis of astronomy, mechanics, architecture, fortification and navigation, etc. (Phili 2001, p. 85).

In 1840, according to the program, Konstantine Negris lectured on differential and integral calculus and continued lectures on descriptive geometry, focusing on the intersections of second degree surfaces and three-dimensional analytic geometry. He also taught rectilinear trigonometry and algebra considering known the Newton's binomial. During his teaching, the former student in the *École polytechnique* tried to educate his very few[16] students on the basics of mathematics.

By the royal decree of 19 May/31 May 1842 regarding exams, students of the mathematical department in order to obtain their graduation were obliged to be examined in the following subjects: "high pure mathematics and applied mathematics, i.e. the analysis of finite quantities, differential and integral calculus, research regarding various curves, descriptive and practical geometry, mechanics and astronomy" (Phili 2001, p. 86). So, this royal decree confirmed that descriptive geometry through Negris' teaching remained in the curriculum.

However, we must stress that after Negris' dismissal in 1845, due to the election of the university deputy, lectures in descriptive geometry were taught again in the Othonian University only for a while (1886–1887) with Cyparissos Stephanos (1857–1917).

Nevertheless, in the "Instruction to the Students..." (Odigiai pros tous fititas ekastis scholis peri allilouchias ton diaforon epistimon 1838,1853, p. 27) edited by

---

[15] It is not clear which book was followed in these lessons. According to this source, we could suppose that it refers to J.N.P. Hachette (Hachette 1811).

[16] The fees probably made the universitarian studies imperative. However, the majority of students opted for the faculties of law and medicine. It might be stressed that during Negris' short mandate, no more than six students attended his lectures. However, we must note that only five students graduated from the mathematical department from 1837 until 1866.

the university in 1853, we can read that students had to attend six semesters of the following lessons:

1th semester: rectilinear trigonometry, algebra and algebraic application to two-dimensional geometry
2nd semester: continuation of algebra's application to two-dimensional geometry, statics and the beginning of descriptive geometry
3rd semester: spherical trigonometry, algebraic application to three-dimensional geometry and continuation of descriptive geometry
4th semester: end of descriptive geometry and differential calculus
5th semester: integral calculus and mechanics
6th semester: integral calculus and mechanics.

But this curriculum, along with that of 1838, once more contradicts the official program of 1853 as it is recorded that Professor George Bouris (1802–1860), a graduate from Vienna University and Negris' substitute, was to teach[17] a simplified syllabus based mainly on geometry and stereometry, every Tuesday, Thursday and Saturday from 5 p.m. to 6 p.m. Unfortunately, there is no mention of whether these lectures comprised elements regarding descriptive geometry.

Therefore, after the departure of Professor Negris, Professor Cyparissos Stephanos during the academic year 1886–1887 taught descriptive geometry every Tuesday and Saturday from 11 a.m. to 12 a.m. However, we can consider that the spirit and the method of Monge were abandoned at the university, and that his method was mainly used and developed in the Military School and in the Polytechnic School, instead.

5 The Polytechnic School of Athens

The construction of the Royal Palace in Athens according to the plan of the Bavarian architect Friedrich von Gärtner (1792–1847) revealed a lack of qualified Greek builders (craftsmen, stonemasons, bricklayers, etc.). Moreover, the needs for new techniques in order to build the new capital exceeded the actual technical abilities of Greek artisans. Thus, at the end of 1836, a noble Bavarian officer Friedrich von Zentner (1777–1847), as he revealed in his book "The Kingdom of Greece...", conceived the idea of creating a technical school based on the model of the Royal School of Building (*Königliche Baugewerkeschule*) in Munich (established in 1826) as well as the technical school in Lyon, *La Martinière*, (Zentner 1844, pp. 11–13).

The first lectures were elementary mathematics, architecture and drawing. The Professor of Drawing (liberal and linear) was Christian Hansen,[18] a distinguished

---

[17] Since 1850 Ioannis Papadakis ensured the teaching of astronomy and analysis.
[18] In 1839, his brother Theofile Hansen (1813–1891), whose buildings constitute even today the architectural ornaments of Athens, was appointed to give lessons in drawing, too.

Danish architect. The lessons in plastic constructions were given by the French architect Charles Laurent and his assistant, a Bavarian sculptor called Karl Heller. A 2nd lieutenant of the engineering corps, Theodore Komninos (1807–1883), taught mathematics, i.e. practical arithmetic and elementary geometry, until 1854 when Komninos was appointed to give lessons in mathematics, mechanics, geometry and architecture (Biris 1952, pp. 486–487).

As the demand for educated craftsmen, surveyors and technicians increased, the school was transformed into a daily one in 1840, and consequently its curriculum was enlarged. Thus, this newly established daily school functioned side by side with the Sunday school. Nonetheless, mathematics remained an important part of the curriculum in both schools.

The next year, Zentner proposed that the French architect Charles Laurent teach mathematics and descriptive geometry side by side with Komninos. The mechanics lectures were covered by other professors at the university. Indeed, George Bouris was invited to teach physics and elements of mechanics.

After the revolution on 3 September 1843, all the foreign professors like Friedrich Zentner, Charles Laurent, Christian Hansen and Theofile Hansen were expelled, and, henceforth, only native Greeks were appointed to the administration. The School of Arts[19] was re-organized and divided into three distinct schools: the Sundays School, the Every Day School and the Higher School, exclusively dedicated to the instruction of Fine Arts (*Official Gazette* no 38, 9 November 1843. Decree regarding the organization of the School of Arts).

The curriculum of the Sundays School comprised elementary algebra, principles of practical geometry, arithmetic but also courses on drawing, construction of objects, courses which from the didactic point of view should follow the Greek translation of Francœur's book, Linear drawing, while their teaching was covered by Michael Georgiades, an architect, who taught the construction of objects 11 h per week, as well as 11 h per week of drawing.

The curriculum of the Every Day School contained among other elements, construction of objects, elements of algebra and geometry, applications to the arts, elements which undoubtedly permit us to suppose that the applications included at least some elements of perspective. M. Georgiades was appointed to teach the construction of objects and drawing 3 h per week, respectively.

In 1844, Lyssandros Kavtanzoglou (1811–1885), a distinguished architect, who had graduated from the Fine Arts Academy of Rome, was appointed to succeed Zentner, as director of the Polytechnic School, in which he remained until Otto's dethronement. However, according to Konstantine Biris' book, in the first year of his mandate the students could not attend the lessons of descriptive geometry, as well as those of building and architecture (Biris 1952, p. 70).

---

[19]This institution is known under several names: Polytechnic School, School of Constructing Arts and Professions, School of Craftsmen, Royal School of Arts, School of Industrial Arts. We usually utilize the name of Polytechnic School here.

However, Kavtanzoglou, taking into consideration the complicated situation regarding the studies in the Polytechnic School and in his report of 5th May 1851 addressed to the Ministry of Internal Affairs tried to elucidate it. In this interesting document (General State Archives, Secretariat of the Ministry of Internal Affairs (Otto's reign) f. 2185), which we found in the General State Archives, among others, he exposed a deficiency regarding mathematical education, as well as the lack of scenography (i.e. perspective) lessons which, as a main application of descriptive geometry, was an indispensable course, which should have been taught practically like its other applications. Therefore, a priori, Kavtanzoglou opted for practical teaching, probably based on design. In this same framework, we could include his remarks regarding the affinity of this Greek institution to the similar French Schools: *École des Beaux Arts*, (School of Fine Arts), *École des Arts et Métiers*, (School of Arts and Crafts), *École préparatoire du dessin*, (Preparatory School of Design) and his proposition for free access.

In 1853, according to the former tradition, Ioannis Papadakis (1820 or 1825–1876), professor of the University of Athens was invited to teach descriptive geometry and perspective. Although he had previously studied in Munich, he attended the courses of the "physicomathematical department"[20] of Athens University and graduated from it. As a distinguished student (Proceedings of the Philosophical Department, session of 28 December 1840 (M.S.), in Greek) and thanks to the favourable opinion of Konstantine Negris he obtained a stipend from the Greek government to study in Munich and since 1842 in Paris at the *École polytechnique* and at the *École des Mines* (1844) (Assimakopoulou et al. 2009, p. 40).

Papadakis was quite adequate in ensuring the teaching of descriptive geometry and its applications to the Polytechnic School, as he was trained in the French *Grandes Écoles*. Therefore, he started to teach descriptive geometry in the Polytechnic School twice a week, firstly from 1853 until 1856. "The lessons of perspective, as application of descriptive geometry were firstly introduced by I. Papadakis and were taught with zeal and success. These courses were mostly indispensable for the progress of any art" (Journal *Helios* of the 12th October 1855. [in Greek]). We must take into consideration that these lessons regarding perspective were mainly considered as a substitute for elementary architecture. The above statement makes clear that the teaching of descriptive geometry actually began between 1853 and 1855.

However, an ordinary event changed this apparently calm situation. In 1856, the lesson of stenography was introduced in order to re-compensate the lack of manuals. Joseph Mindler (1808–1868), a Bavarian officer at the Royal court and stenographer of the Parliament was appointed as professor of stenography. As his salary was superior to that of Papadakis, the professor of descriptive geometry resigned as a mark of protest.

---

[20]The university of Athens was established according to the German model of ordus philosophicus. The physicomathematical department gained its autonomy only in 1904.

As a temporary solution, the Director of the School of Arts, Lyssandros Kavtanzoglou proposed that the Ministry of Internal Affairs authorize Sotirios Pilotos (Salvatore Pilotto) (Assimakopoulou et al. 2009, p. 40), who was born in Corfu and studied at the *École Centrale des Arts et Manufactures* (sic) in Paris graduating in 1855 as an "ingénieur métallurgiste" (sic), to give lessons in mechanics and, especially, in descriptive geometry and its applications, as well as in geodesy, gratis. In the ministerial report was also stressed Pilotos anterior experience as "until then he had successfully given lessons to 14 regular auditors" (General State Archives. King George's reign unclassified). Thus, the ministerial report ended by entreating the King to appoint Sotirios Pilotos (or Pilottos) as full professor. This demand was fulfilled, and Pilotos was appointed as professor of descriptive geometry and its applications, as well as geodesy in the 1860s.

Meanwhile in 1856 in Hermoupolis (Syros) (an important town of the Cyclades), the French treatise of Jean-Pierre Thénot (Thénot 1834) was translated[21] into Greek by the Secretary of the Prefecture of the Cyclades, Panos Pleskas. The translation was probably requested by A. Kriezis, who, along with the publisher M. Petridis, covered the costs of the edition (see Fig. 8.4). Andreas Kriezis (1813–1878), who

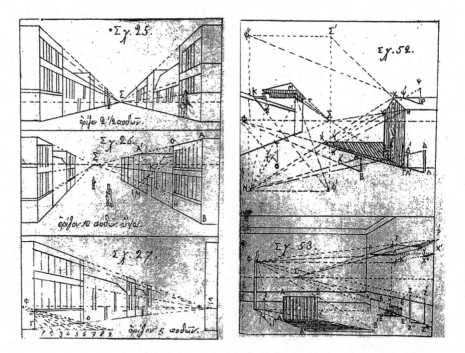

Fig. 8.4 Exercises (Tenetou 1856, p. 28), Hermoupolis (Syros) 1856

---

[21]The word perspective was translated in Greek as dioptics and the same word was used in Francœur's book, too.

after his studies in painting in Paris, was appointed to teach linear design and painting in the gymnasium in Syros (the Greek Liverpool of the nineteenth century). We must stress that Kriezis' designed the 66 figures (Mykoniatis 1995, p. 347) of the treatise. It is possible that this book covered the needs for the teaching of perspective in the Polytechnic School.

Nevertheless, at the end of the same year, on 15 December 1860, the new minister of Internal Affairs repeated the request concerning the teaching of descriptive geometry. In his report to the King, he stressed the existing gap after Ioannis Papadakis' resignation, as his courses were indispensable for the training of surveyors. Ending his report, the Minister of Internal Affairs proposed that Ioannis Papadakis be appointed in order to teach rectilinear trigonometry, descriptive geometry and its applications, as well as elements of statics. This proposal was accepted in 1863 and Papadakis was appointed again by a royal decree and continued to teach until his death in 1876.

In October 1862, after Otto's dethronement, a new era began for the School and the decree on 26 August 1863 marked the re-organization of its aims. Thus, henceforth the institution would ensure: "Craftsmen's theoretical and practical education, as well as to the owners of the manufacturing, in the most necessary arts, i.e. building construction, smothery, joinery sculpture, painting, ceramics, tanning and soap-making" (*Gazette of the State* no 33. Decree regarding the new organization and direction of the School of Arts).

Dimitrios Skalistiris (1815–1883) who from 1859 was engaged by a royal decree to teach physicomechanics (sic.), became the new director of the Polytechnic School (1864–1873). His first task was to improve studies, in line with the standard in France. Thus, in his letter on 12 October 1864 addressed to the Ministry of the Internal Affairs, he noted the existence of three technical schools in France, the *Écoles des Arts and Métiers* (more precisely he noted that one of them is situated in Aix, near Marseille), and considered that all three constituted an appropriate model to follow. Moreover, in his letter, he emphasized that France "owed a lot to these schools regarding the diffusion and the perfection of the arts" (Biris 1952, p. 181).

He openly stated that after the revolution of 1843, the Polytechnic School had not reached the targets that were cultivated in the French institutions. Thus, Skalistiris revealed that his very first intention was to re-organize the Greek Polytechnic School based on the French model.

The new director D. Skalistiris was a captain of the engineering corps and a graduate from the Military School, who attended lectures at the *École polytechnique* and later the *École des Ponts et Chaussées* (Assimakopoulou et al. 2009, p. 40). Returning to Greece, he was appointed professor of bridge construction in the Military School in 1846, as well as professor of physicomechanics in the Polytechnic School, as we have already mentioned.

At the end of 1876, Papadakis died and, thus, in January 1877, Dimitrios Tournakis (1820–1902) a lieutenant of artillery started to teach descriptive geometry in the Polytechnic School. It might be stressed that Tournakis had significant experience as since 1859 he had taught descriptive geometry in the Military School. However, Tournakis' career in the Polytechnic School was brief. In February 1878, he was removed and replaced by an officer of the engineering corps Nikolaos Solomos, a professor of the Military School, whose teaching career also lasted just a year (he was removed in December 1878). N. Solomos (1840–?) side by side with his educational duties devoted his life to writing manuals for Greek artisans and workers (Chatzis 2003, pp. 83–86).

Solomos' successor in the Polytechnic School was Andreas Zinopoulos (1842–1890), a graduate of the *École Centrale des Arts et Manufactures* in Paris (Assimakopoulou et al. 2009, p. 40) and a civil engineer, who started to teach descriptive geometry from 1878 until 1882 and was then appointed to the direction of public works.

In the General State Archives, a certificate of 1882 regarding the training of a civil engineer translated into French presents the curriculum of that time, and shows the high level of teaching, which permitted the young graduate to continue his studies in Ghent, Belgium.

Solomos' resignation matched the plan of the new director,[22] Gerasimos Mavrogiannis (1828–1905), a former consul in Marseille and Trieste and an erudite man specializing in history, who opted for the demilitarization of the school (Biris 1952, p. 287). Of course the military staff were against his decision, which reversed its long-standing tradition. After Mavrogiannis' dismissal, the new director of the Polytechnic School was Anastasios Theofilas (1827–1901) (Assimakopoulou et al. 2009, p. 348) a graduate from the Military School, who later completed his studies at the *École de Saint-Cyr* (Assimakopoulou et al. 2009, p. 41). Theofilas started the reform of 1887, which transformed the Polytechnic School into a university institution, although the military spirit and austere discipline were maintained.

In October 1882, the Council of Instruction at the Polytechnic School ordered several books, scientific journals as well as some drawings regarding descriptive geometry, in order to enrich its library "because the non-experienced students had an absolute need to understand this course" (Biris 1952, p. 287). In the General State Archives, we found several receipts from an international bookshop in Athens during Theofilas' direction (1878–1901). Among the books, the Polytechnic School bought four copies of Ernest Lebon's book on descriptive geometry,[23] while in the same year, Theofilas approved the expenses for 300 copies, a most impressive number for 1886, a lithographic leaflet of 168 pages on descriptive geometry. (General State Archives. Ministry of Internal Affairs. King George's reign, unclassified, f. 32, no 89).

Immediately after A. Zinopoulos' dismissal in November 1882, Apostolos Apostolou (1840–1918), an officer of the engineering corps, taught descriptive geometry until 1905. It might be stressed that Apostolou presented one of the first treatises on descriptive geometry (Apostolou 1883). Thus, his lectures during the academic

---

[22] His mandate was very short, from 1876 to 1879.
[23] Maybe it concerns the book of Ernest Lebon (Lebon 1881).

year 1883–1884 were edited in 1883 in Athens, thanks to Thomaidis' will.[24] This treatise was re-edited in 1890 and adding a second part (Apostolou 1890) devoted to the theory of shadows (Poulos 1988, p. 140).

In 1887, French became an obligatory course. So, in December of this year, Joseph Cellar was appointed to teach a French course. In this same year, a new decree modified the status of the school, whose name now became the "School of Industrial Arts", which comprised of mainly two faculties: the faculty (School in Greek) of civil engineers and that of mechanical engineers. Henceforth, the institution would provide for the scientific education of engineers, who were then ready to face the challenges of the great technical projects in the country (railways, road constructions, as well as the construction of the Isthmus of Corinthos).

In January 1888, for the first time, an open selection was announced for the appointment of an assistant in topography. Nikolaos Karakatsanidis (1852–1920), a graduate of the Polytechnic School, was appointed the position and subsequently became professor of descriptive geometry in 1905 and taught this course until his death in 1920. In 1917, he published his own treatise of descriptive geometry, "Lectures on descriptive geometry" (Karakatsanidis 1917), which constitutes a complete manual as it covers the complete curriculum.

However, we must state that September 1890 constitutes a turning point in the history of the Polytechnic School as its subordination was modified; henceforth, it was under the direction of the public works of the Ministry of Interior Affairs.

At the end of the nineteenth century, descriptive geometry was recognized as an indispensable tool for the studies of civil and mechanical engineering. Thus, the proclamation of 12 July 1891 regarding the admission of forthcoming students is quite impressive, as among other disciplines, the elements of descriptive geometry (Biris 1952, p. 322) were also included.

The French dominance in descriptive geometry in respect of the professorship and literature lasted until 1897. After the marriage of the crown prince to the Kaiser's sister, subsequent staff sought their training in Germany (Munich, Berlin, Dresden, Karlsruhe, etc.) (Biris 1952, p. 365). They introduced the German model of engineering into Greece, which largely contributed to promoting the industrialization of the country as well as its technical progress.

6 Conclusion

The introduction, the adoption and the teaching of descriptive geometry into higher technical education in Greece during the nineteenth century had French origins. Every manifestation in descriptive geometry revealed its French affinity.

---

[24]Dimitrios Thomaidis (?–1878) originated from Metsovo (Epirus) became a wealthy merchant in Constantinople. Established since 1873 in Athens, he decided to use his legacy, valid until today, to support the studies (edition of books, scholarships) at Polytechnic School.

The diffusion of this branch of geometry in Greek educational institutions was impressionable thanks to some former students of the French *Grandes Écoles*, who, after returning to Greece, spread the discipline by teaching and for holding important posts in the administration. For example, regarding the administration of the Polytechnic School, we must take into consideration that among the six directors from 1837 until 1901, four were officers of the engineering corps: Friedrich von Zentner (1837–1843), Dimitrios Skalistiris (1864–1873), Dimitrios Antonopoulos (1873–1876), (Nikolaidis 2000) and Anastassios Theofilas (1879–1901) (Chatzis 2003, pp. 81–83).

During the nineteenth century, technical education was monopolized by the Military School and the Polytechnic School, which was founded a few years later. Until 1880, the profession of engineer was exclusively bestowed to the officers of the engineering corps who had graduated from the Military School. A great number of the didactic books of the Military and Polytechnic Schools were written by officers of the engineering corps.

The teaching of the Mongean method of projections became a quite important tool not only in developing geometric knowledge, but also in familiarizing the Greek students with graphic procedures, which were useful for engineers and builders. Via this training, the graduates of the Military School and the Polytechnic School were able to participate in the modernization of the Greek state due to their contribution to urban planning, buildings, roads and railway construction, etc.

References

Sources

General State Archives (Γενικὰ Ἀρχεῖα τοῦ Κράτους) Secretary of Military Affairs (Γραμματεία Στρατιωτικῶν), Ministry of Internal Affairs (Ὑπουργεῖο Ἐσωτερικῶν)

- Othonian period (Ὀθωνικὴ περίοδος)
- King George's reign (Βασιλεία Γεωργίου Α')

Ministry of Defence (Ὑπουργεῖο Πολέμου)

- Othonian period (Ὀθωνικὴ περίοδος)
- King George's reign (Βασιλεία Γεωργίου Α')

Proceedings of the Sessions of the Philosophical Department (Πρακτικὰ Συνεδριάσεων Φιλοσοφικῆς Σχολῆς)
Gazzetta Ufficiale degli Stati Uniti delle Isole Ionie.
Journal of the State.
Gazette of the State.
Corfou Reading Society (Ἀναγνωστικὴ Ἑταιρεία Κερκύρας) Guilford's Archives.

Publications

Antoniou, David. 1992. *The Origins of the Educational Planning in the Neo-Hellenic State: The Plan of the Commission of 1833* (in Greek). Athens: Pataki.

Assimakopoulou, Fotini, Konstantinos Chatzis, and Anna Mahera. 2009. Éléve en France, enseignant en Grèce. Les enseignants de l'École Polytechnique d'Athènes (1837–1912) formés dans les Écoles d'ingénieurs en France. *Jogos de Identicade Professional: Os engnheiros entre a formação e acçāa*. 2009. Ana Cardoso de Matos, Maria Paulo Digo, Irina Gouzévitch, André Grelon (eds) Lisboa, Edicoĕs Colibri /DIDHUS-UE/CIUHCT, pp. 25–41.

Assimakopoulou, Fotini, Konstantinos Chatzis, and Georgia Mavrogonatou. 2009. Implanter les Ponts et Chaussées européens en Grèce: le rôle des ingénieurs du corps du Génie 1830–1880. *Quaderns d'Historia de l'Enginyera* X: 331–350.

Biris, Konstantine. 1952. *History of the Polytechnic School* (in Greek). Athens.

Bradley, Margaret. 2012. *Charles Dupin and His Influence on France*. New York: Cambria Press.

Carandinos, Ioannis. 1826. *On Some Theorems on Polygonometry* (in Greek). Corfu in Gennadius Library, MGL 277.

Chatzis, Konstantinos. 2003. Des ingénieurs militaires au service des civiles: les officiers du génie en Grèce au XIXe siècle. In *Science, Technology and the 19th Century Army*, 69–87. Conference Proceedings edited by Konstantinos Chatzis, Konstantinos and Efthymios Nikolaidis. National Hellenic Research Foundation, Laboratoire Techniques, Territoires et Sociétés/C.N.R.S., Athens.

———. 2018. Grec ancien et modernité: l'officier militaire – traducteur et la constitution de l'État hellénique». In *La Révolution française*, 13. http:journals.openedition.orglIrpl1878

Diesterweg, Adolph. 1828. *Raumlehre, oder Geometrie nach jetzigen Anforderung der Pädagogik für Lehrende und Lernende*. Bonn.

Fox, Robert. 2012. *The Savant and the State: Science and Cultural politics in Nineteenth Century, France*. Baltimore: John Hopkins University Press.

Francoeur, Louis-Benjamin. 1819, 1827. *Dessin Linéaire et arpentage pour toutes les écoles primaires quel que soit le mode d'instruction qu'on y sait*. Paris.

Gazzetta Ufficiale degli Stati Uniti delle Isole Ionie n. 335 17 May-29 May 1824

Hachette, Jean-Nicolas-Pierre. 1811. *Supplément á la Géométrie descriptive de Gaspard Monge*. Paris.

Karas, Yiannis. 1992. *Science during the Ottoman Rule* (in Greek). vols. 1–2. Athens.

———. 2003. Journals edited before the war of Independence. In *History and Philosophy of Science in Greek Regions (17th–19th Centuries)*, 685–690. Athens Metaixmio.

Kastanis, Andreas. 2003. The teaching of mathematics in the Greek military academy during the first years of its foundation (1828–1834). *Historia Mathematica* 30: 123–139.

Koppe, Karl. 1836a. *Die Arithmetik und Algebra und allgemeine Groessenlehre fuer Schulenunterricht*. Bädeker Essen.

———. 1836b. *Die Planimetrie und Stereometrie fuer Schulenunterricht*. Bädeker Essen.

Lebon, Ernest. 1881. *Traité de géométrie descriptive pour l'enseignement secondaire*. Paris, Fréres Delalain.

Leroy, Charles-François-Antoine. 1837. *Traité de Géométrie Descriptive avec une collection d'épures, composée de 60 planches*. Bruxelles.

Leroy, Charles-François-Antoine. 1855. *Traité de Géométrie descriptive: suivi de la méthode de plans cotés et de la théorie des engranages cylindriques et coniques, avec une collection d'épures, composée de 71 planches*. Tome 1, Paris: P. Mallet-Bachelier.

Monge, Gaspard. 1811. *Géométrie descriptive*. Nouvelle édition avec un supplément par M. Hachette. Paris: P. Klostermann.

Mykoniatis, Ilias. 1995. The first publishing Neo-Hellenic treatise on perspective. The European art and its reception in Greece of the 19th century (in Greek). *Ellinica* 45: 341–352.

Nikolaidis, Efthymios. 2000. Les élèves grecs de l'Ecole Polytechnique: 1820–1921. In *La Dispora Hellénique en France* ed. G. Grivaud Athénes Ecole Française d'Athénes, 55–65.

Phili, Christine. 1996. La reconstruction des mathématiques en Grèce: l'apport de Ioannis Carandinos (1784–1834). In *L'Europe Mathématique*, ed. Catherine Goldstein, Jeremy Gray, and Jim Ritter, 305–319. Paris Édition de la Maison des Sciences de l'Homme, Paris.
——. 2001. Mathematics and mathematical education in the university of Athens from its foundation to the beginning of the XXth century. In *Archives Internationales d'Histoire des Sciences* 51: 74–98.
——. 2006. Ioannis Carandinos (1784–1834): l'initiateur des mathématiques françaises en Grèce. *Archives Internationales d'Histoire des Sciences* 56: 156–157 (Juin-Décembre 2006, pp. 79–124).
——. 2010. Greek mathematical publications in Vienna in the 18th–19th centuries in Mathematics in the Austrian-Hungarian Europe. *Proceedings of a symposium held in Budapest on August 1, 2009 During the XIII ICHST*. Prague 2010, 137–147.
——. 2012a. L'Académie Ionienne et le Risorgimento. In *Europa Matematica e Risorgimento Italiano*, a cura di Luigi Pepe CLUEB Bologna, 67–80.
——. 2012b. An unpublished letter of Ioannis Carandinos to Fourier.... (in Greek). *Proceedings of the Academy of Athens* 87: 27–32.
Poulos, Andreas. 1988. *Greek Mathematical Bibliography (1500–1900)* (in Greek). Athens: Greek Mathematical Society.
Stasinopoulos, Epameinondas. 1933. *History of the School of Evelpides. Epoch of Kapodistria and Otto* (in Greek). Athens.
Taton, René. 1951. *L'Œuvre Scientifique de Monge*. P.U.F. Paris.
Thénot, Jean-Pierre. 1834. *Traité de perspective pratique pour dessiner d' après nature*. Paris.
Zentner, von Friedrich. 1844. *Das Königreich Griechenlands in Hinsicht auf Industrie und Agrikultur*. Augsburg: J. Wirth.

Greek References

Fragkirou, Ioannis. 1831. *Didaskalia tis diagrafikis i grammikis ichnografias*. En Aiginei.
Odigiai pros tous fititas ekastis scholis peri allilouchias ton diaforon epistimon kai peri tis tiriteas methodou kai taxeos kata tas akadimaïkas spoudas. 1838,1853. En Athinais.
Tenetou Dioptiki (perspective). 1856. metafrasthisa men ek tou Gallikou ypo Panou N. Pleska Grammateos tis Nomarchias Kykladon ektdothissa de dapani M. P. Pieridou. En Hermoupoli ek tis typografias M. P. Pieridou.
Efimeris tou Stratou, *Apomachos* no 21, 30 Septembriou 1860.
Libritis, Theodoros. 1882. *Pinakes Perigrafikis Geometrias*. Ek tou Lithografiou tou Stratiotikou Scholeiou. En Peiraiei (containing only exercises).
Apostolou, Apostolou. 1883. *Mathimata Perigrafikis Geometrias*, paradidomena kata to scholiko etos 1883–1884. Lithografithenta dia tou pros ekdosin sygrammaton orizomenou posou en ti diathiki tou aoidimou D. Thomaidou. En Athinais.
Kanellopoulou, Miltiadou. 1883. *Mathimata Perigrafikis Geometrias*, paradothenta kata to scholikon etos 1882–1883. Stratiotikon Scholeion ton Evelpidon. En Peiraiei.
Kopp, Karolos. 1855. *Arithimitiki Kai Algevra*. En Athinais
Kopp, Karolos. 1857. *Epipedometria*. En Patrais
Kopp, Karolos. 1858. *Stereometria*. En Patrais
Moschopoulos, Timoleon. 1883. *Stoicheiodis Perigrafiki Geometria*. Paradothisa en to Scholeio ton ypaxiomatikon. En Athinais.
Libritis, Theodoros. 1881. *Pinakes Perigrafikis Geometrias lithografo*. En Athinais
Libritis, Theodoros. 1886. *Mathimata Perigrafikis Geometrias*. Paradothenta kata to scholiko etos 1885–1886. Lithografo – [Meros A]. En Peiraiei.
Libritis, Theodoros. 1888. *Mathimata Perigrafikis Geometrias*. Lithografo – [Meros B]. En Peiraiei.

Stratiotiko Scholeio ton Evelpidon. 1887. *Peri ton Irithimenon Epipedon*. Metafrasis anonimos ek tou sygrammatos tou C. F. A. Leroy (Traité de Géométrie descriptive...) En Peiraiei.

Apostolou, Apostolou. 1890–1891. *Mathimata Perigrafikis Geometrias*. To B' Meros periechei Mathimata Skiagrafias (the second part contains lectures on perspective). En Athinais.

Karakatsanidis, Nikolaos. 1917. *Mathimata Parastatikis Geometrias (lithografimena) Ethniko Metsovio Polytechneio*. En Athinais.

Part II
Installation of Descriptive Geometry in Europe

Chapter 9
A German Interpreting of Descriptive Geometry and Polytechnic

Nadine Benstein

Abstract 30 years after the publication of Gaspard Monge's *Géométrie descriptive* and the foundation of the tightly connected *École polytechnique* in France, the impact of these events could be fully noticed in Germany. The ideas radiating from France were taken up, but applied in a way that led to decisive differences from the French role models. These differences, referring to descriptive geometry as a mathematical discipline and to inherent educational institutions, will be dealt with in this contribution. In Germany, the establishment of descriptive geometry as a mathematical discipline heavily depended on the standing of the relevant technical, secondary and professional education institutions, which mainly differ from the *École polytechnique*. The same holds for descriptive geometry itself, which at first was transmitted in a "Mongean way", but was rapidly put into the context of other projection methods. Indeed, despite crucial differences, the French developments finally led to an intensive treatment of geometrical representation methods and the emancipation of technical subjects as scientific branches in the nineteenth century in Germany in general.

Keywords Descriptive geometry · Geometry teaching · *Technische Hochschule* · Polytechnic school · Secondary school · Technical sciences · Professional education · Gaspard Monge · Guido Schreiber · Bernhard Gugler · Friedrich Weinbrenner · Karl Pohlke

1 A German Definition of Descriptive Geometry?

In France, the connection between the denotation "*géométrie descriptive*" and the work of Gaspard Monge is relatively clear. The meaning of its German translation

---

N. Benstein (✉)
Bergische Universität Wuppertal, Faculty of Mathematics and Natural Sciences, Didactics and History of Mathematics, Wuppertal, Germany
e-mail: benstein@uni-wuppertal.de

© Springer Nature Switzerland AG 2019
É. Barbin et al. (eds.), *Descriptive Geometry, The Spread of a Polytechnic Art*, International Studies in the History of Mathematics and its Teaching,
https://doi.org/10.1007/978-3-030-14808-9_9

is not as definite. At first, in Germany,[1] there did never exist a coherent denotation, meaning that the terms *darstellende, beschreibende* and *descriptive Geometrie* were often used synonymously for descriptive geometry (cf. Papperitz 1909, p. 520).[2] Secondly, the general understanding of descriptive geometry in Germany is and was wider by integrating into this mathematical discipline several different projection methods, the methods of plane and elevation as systematised by Monge being one of them. In general, descriptive geometry was (and is) seen as the mathematical discipline that treats planar depictions of three-dimensional objects (e.g. cf. Wiener 1884, p. 1; Papperitz 1909, p. 521; Stäckel 1914, p. 165), which includes, next to orthogonal parallel projection, oblique parallel and central projection (perspective). This is illustrated, for example, by the tables of contents of later school and textbooks on descriptive geometry or, for example, by Christian Wiener's[3] history of descriptive geometry, integrated in his book on the matter (1884), in which he dealt with the developments of all the different projection, or representation methods, respectively.

2 The Transmission of Descriptive Geometry: German Textbooks

In Germany, the literature on and the teaching of descriptive geometry was tightly connected to the French works (cf. Stäckel 1915, p. 133) until novel treatments came up through Bernhard Gugler, Karl Pohlke or later Wilhelm Fiedler,[4] who connected descriptive with projective geometry. How tight the connection to the French sources was will be explored by analysing the differences between the first profound German textbook by Guido Schreiber with the first translation of Monge's work by Haussner (Monge 1900). Before Schreiber's, two other books on descriptive geometry were published by Friedrich Weinbrenner[5] and Michael

---

[1] For "Germany", it is generally difficult to make statements, i.e. about (secondary or tertiary) education during the considered time, since it consisted of 39 loosely connected states (kingdoms, princedoms, duchies, etc.) in 1815 (Deutscher Bund) and 25 of these in 1871 when the German Empire was established (cf. Schubring 2014, p. 242). For the sake of simplicity, "Germany" will be used for the different stages of its history. Territorial restrictions and differences within Germany will be commented on in the relevant contexts.

[2] In the following, these three terms will all be translated with "descriptive geometry".

[3] Ludwig Christian Wiener (1826–1896), professor at the polytechnic school and later *Technische Hochschule* in Karlsruhe (1852–1896), had studied building at the university in Gießen and worked as a teacher at the trading school in Darmstadt. Later, he did his doctorate in mathematics at the university of Gießen (cf. Wiener (1879)).

[4] Wilhelm Fiedler and the synthesis of descriptive and projective geometry will be dealt with in (Volkert, Chap. 10, this volume).

[5] Friedrich Weinbrenner (1766–1826), architect and head of the construction department in Karlsruhe, also taught architecture at the construction school in Karlsruhe (cf. Katzenstein 1896).

Creizenach,[6] which will be presented at first. In the end, further developments of descriptive geometry in a wider sense as treated by several German authors in the nineteenth century will be discussed.

2.1 First Attempts in 1810 and 1821

The first German textbook introducing Monge's method was published by Weinbrenner in 1810 (*Architektonisches Lehrbuch*) and appeared in four parts. It was addressed to students of architecture at a construction school in Karlsruhe (German State of Baden). Weinbrenner distinguished between geometric and perspective drawing manners and denoted the first as *Géométrie descriptive*, which at the same time provided the content of his first book (39 pages, six plates). At first, Weinbrenner introduced 20 paragraphs with basic notions (*Allgemeine Lehrsätze*), which deal with characteristics of plane and elevation and the orthogonal projections of angles, straight lines, planes and solids, all reduced to three cases: parallel, rectangular or oblique to the drawing plane. Five chapters follow dealing with straight lines, planes (and different combinations of these) and, finally, solids. In the end of his book, Weinbrenner dissociated from the work of Monge, who (among others) had integrated "the most difficult" mathematical tasks into the teaching of geometric drawing, which contradicted the requirements of artists (cf. Weinbrenner 1810, p. 39). Weinbrenner reduced his work to the depiction of straight lines, planes and solids in plane and elevation and excluded any curved lines and surfaces, which results in quite a basic level of his work. Wiener called Weinbrenner's book *Anfangsgründe* (basic/elementary treatise) of descriptive geometry, which he called theory of projections (*Projektionslehre*) (cf. Wiener 1884, p. 35).

The work *Anfangsgründe der darstellenden Geometrie oder der Projektionslehre für Schulen* (1821), published in Mainz (German State of Hessen), by Creizenach was a book for "schools", which were not specified further by the author neither in his title nor in his preface. The author aimed at providing an elementary work on descriptive geometry and the theory of projections, which Germany lacked at this point (cf. Creizenach 1821, p. IV), also to prepare students for further studies in descriptive geometry (cf. ibid., pp. VI-VIII). The book is divided into four chapters: 1. "On points, straight lines and planes", 2. "On curved surfaces", 3. "On sections of curved surfaces" and 4. "On the perspective depiction". Creizenach himself admitted that he had to restrict the contents of the book so that it was impossible to cover the complete science (cf. ibid., p. VII). The understanding of descriptive geometry was denoted by the author (similar to Monge) as the means to depict geometric objects in space on the plane and to draw from these depictions geometric properties and mutual relations of the respective objects (cf. ibid., p. 1). In contrast, perspective depictions provided a vivid representation, an "illusion"

---

[6]Michael Creizenach (1789–1842) was a teacher for mathematics (cf. Brüll 1903).

(cf. ibid., pp. 91–92). The author subsumed these two projection methods as both belonging to the theory of projections (cf. ibid., p. III). In fact, the former comprises 89 of the total 108 pages of the book. Moreover, Creizenach acknowledged the work of "the French" in his preface because they had created a new science, at least regarding the method, and had achieved to reduce the "essential theorems" to a conveniently small number (cf. ibid., p. VI). He claimed to have used the works of Monge, Hachette, Lacroix and Potier, while his book cannot be seen as a translation of neither the contents nor their structure (cf. ibid., p. VI). In general, also considering the targeted readership, the book remains on a relatively low level: indeed, Creizenach covers all the topics that Monge included, but only went so far as to treat intersections of two spheres (which provides the last task in the chapter about orthogonal parallel projections).

2.2 Guido Schreiber's Textbook After Monge's Géométrie Descriptive from 1828/1829

Even though Creizenach preceded with his *Anfangsgründe*, Schreiber's[7] *Lehrbuch der Darstellenden Geometrie nach Monge's Géométrie descriptive* (cf. Fig. 9.1), which he published during his professorship in Karlsruhe, was considered the first profound textbook in the German language about descriptive geometry (cf. Wiener 1884, p. 36; Obenrauch 1897, p. 81) or even perceived as the first work of the topic in general (cf. Stäckel 1915, p. 133). This probably results from the fact that Schreiber treated the topic on a scientific level and relatively completely referring to the work of Monge in contrast to his predecessors. However, Schreiber himself did not only acknowledge Creizenach's book as the first German work on descriptive geometry, but also lists it among the works he consulted for his own (cf. Schreiber 1828, p. IX). In his preface, Schreiber claims to have oriented himself primarily towards the edition of Monge from 1811 including the first supplement of Hachette, while refraining from the distinction the latter made into three-dimensional and descriptive geometry to avoid that the latter was reduced to a "naked and dry lesson of projections" (cf. 1828, p. VII). Moreover, he claimed that he tried to stay as closely as possible to the progression and structure of the original, but that a lot of its unity and roundness had been lost by interweaving Hachette's work (cf. ibid., pp. VI–VII).

Schreiber's work is divided into five chapters. The first three chapters provide the first part of Schreiber's work published in 1828, the fourth and fifth chapters are integrated in the second part published in 1829. The structure of Schreiber's

---

[7]Guido Schreiber (1799–1871) was the first professor for descriptive and practical geometry (1826–1852) at the first German polytechnic school in Karlsruhe. He never received an academic education himself but, "talented in teaching", had already taught at the artillery school during his military service (1813–1827) (cf. Renteln 2000, p. 296).

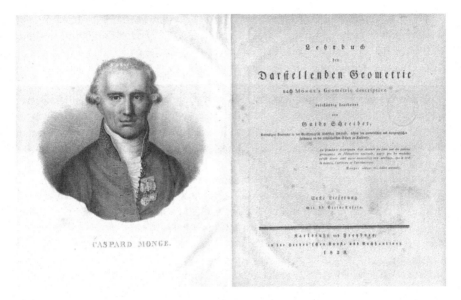

Fig. 9.1 Title page of the first part of Guido Schreiber's *Lehrbuch der darstellenden Geometrie* (1828)

work is similar to Monge's; the headings of the chapters differ sometimes, but the global order of contents was mostly maintained. In general, Schreiber seems to have taken Monge's work, interwove tasks from Hachette's supplement, added further explanations in some places and rearranged some contents. Schreiber's first chapter exemplifies this manner.

The first chapter deals with straight lines and planes and mostly resembles or even equals Monge's first part "Tasks and methods of descriptive geometry". In this chapter, the contents and their order are mostly alike. This includes, for example, the introduction of the notion of *rabattement*, which he explained with the rotation of the vertical plane around its intersection line with the horizontal plane like it happens with a hinge (cf. ibid., p. 10). Yet, Schreiber did not explain that the point of the observing eye (*Augpunkt*) has to be rotated into the plane as well. In between, Schreiber integrated a few paragraphs, e.g. on the (German) terminology or explanations of the graphic execution. He went so far as to dictate the way in which a sheet of paper should be prepared before any drawing could be made: it should be divided into four equal parts, each part providing space for one drawing (cf. ibid., p. 11). Furthermore, Schreiber suggested to use different kinds of lines depending on what they represent ("full" lines for the actual depicted object, dotted lines for auxiliary lines, etc.) (cf. ibid., p. 17). Interestingly, the author moreover explained within these additional paragraphs the orthogonal projection method with the help of the image of positioning the observing eye at infinity (cf. ibid., p. 16). So, he directly located Monge's method within the context of central projection and already anticipated later developments of descriptive geometry in Germany, namely the connection to projective geometry. A further alteration is that Schreiber did not include curved surfaces into the first chapter as Monge did, but instead into the following (cf. ibid., p. 15). In the end of the chapter, Schreiber integrated 14 tasks (instead of Monge's nine) of which he took eight from Monge and four from Hachette.

Schreiber proceeded in this manner. In the second chapter with the title "Curved surfaces" (while Monge's second part is called "Normals and tangential planes of curved surfaces"), Schreiber introduced the formation of curved surfaces through the motion of a curved line just as Monge did in his first part with the help of the examples of cylindrical, conical and rotational surfaces (cf. ibid., pp. 34–41), but more extensively. The first few following explanations and tasks about tangent planes and developable surfaces are similar regarding order and content. In general, Schreiber covered everything that Monge treated in his second part, but included 30 pages dealing with considerations and tasks on surfaces of second order and their rotations, which Monge shortly addressed in his third part. Interestingly, in the third chapter with the title "Surface section" (in Monge "Sections of curved surfaces"), Schreiber did not include the method of Roberval. In Schreiber's work, this method is found in the appendix and the author commented that there existed "plenty of other methods", which can be applied in more cases than the method of concern (cf. Schreiber 1829, p. 298). Furthermore, Schreiber included oblique parallel and perspective projection at the end of this chapter. The chapters four and five are called "Different tasks" (in Monge "Application of the method provided for the construction of the sections of curved surfaces for the solution of different tasks") and "Theory of curved lines and curved surfaces" ("Curvature of twice curved curves und curved surface"). So, the global structure of Schreiber's book resembles Monge's and within the different chapters, he rearranged some parts.

Despite some structural differences and several additional paragraphs (Schreiber's book counts over 300 pages without plates, Haussner's translation 176 including illustrations) Schreiber's work resembles the German translation of Monge's work often even word by word, which indicates that Schreiber closely worked with Monge's book when he was writing his own. Paul Stäckel even called it a reproduction of Monge's (cf. Stäckel 1915, p. 133). However, Schreiber's book already contained the five types of surfaces of second order, skew surfaces and especially those of second order (cf. Wiener 1884, p. 36), which constitutes a significant difference from Monge's original. So, Schreiber's work can neither be seen as a translation of Monge's work nor as an original work. Indeed, Schreiber was the first author who transmitted descriptive geometry in the spirit of Monge to Germany.

2.3 Further Developments After 1839

In later German textbooks on descriptive geometry, new aspects referring to a wider understanding of descriptive geometry were integrated and developed. For example, also projective geometry, which had been emerging in Germany during the 1830s, was dealt with in this context, but these two disciplines remained widely separated until the first actual conflation was provided by Fiedler (cf. ibid., p. 38). Indeed, the connection between projective and descriptive geometry was used or mentioned by other German authors before.

Opposing the French tradition, Schreiber was also the first geometer to try to make descriptive geometry independent from analytic geometry and instead to build it upon projective geometry, which he treated in *Geometrisches Port-Folio, Curs der darstellenden Geometrie in ihren Anwendungen* (1839 and 1843) (cf. Obenrauch 1897, p. 82). In this work, he conducted tasks on sections, distances and angles of points, straight lines and planes, and the shadow constructions of cylinders, cones and several rotational surfaces in the perspective depiction and developed the theorems about ranges of points (*Punktreihen*) and pencils of rays (*Strahlenbüschel*), the projective and focal point features of conic sections, including especially their polarity, and derived from them properties of surfaces of second order, mainly of rotational surfaces (cf. Wiener 1884, p. 37). The integration of projective into descriptive geometry meant to put parallel projection in relation to central projection (the former being a special case of the latter). Among several other works connecting these two projection methods were, for example, *Die darstellende Geometrie im Sinne der neueren Geometrie* (1870) by Joseph Schlesinger, Wiener's *Lehrbuch der darstellenden Geometrie* part I and II (1884 and 1887) or Fiedler's *Die darstellende Geometrie in organischer Verbindung mit der Geometrie der Lage* (1871). The connection of descriptive and projective geometry was supposed to lead to a more careful investigation of the projection traits, to a simplification of constructions and especially to a geometric exploration of lines and surfaces of second degree (cf. ibid., p. 36). In general, also central projection was scientifically enlarged upon in the relevant works on a purely geometric basis (cf. Obenrauch 1897, p. 88).

For Gugler's[8] *Lehrbuch der descriptiven Geometrie* (1841), it was claimed that it followed the ideas of Monge (cf. Papperitz 1909, p. 567). Nevertheless, descriptive geometry was built up independently (cf. Wiener 1884, p. 37), in particular independent from analytic geometry (cf. Obenrauch 1897, p. 83). The author himself explained in his preface that he included analytic proofs in the appendix which were to be seen as an addition, because they did not essentially belong to descriptive geometry since no constructions were based on them (cf. Gugler 1841, p. III). Gugler denoted the French works as the "best models", but decided to deviate from the originals in basically two aspects (cf. ibid., p. V). Firstly, the author claimed to have arranged the contents in a clearer and more scientific way by uniting related contents and segmenting complex tasks into several individual ones (cf. ibid., pp. V–VI). Secondly, Gugler shifted the main focus from curved surfaces to tasks on plane

---

[8]Bernhard Gugler (1812–1880), former student of Karl G. C. von Staudt, was professor for descriptive geometry at the polytechnic schools in Nürnberg and Stuttgart. He visited polytechnic schools as well as the *Gymnasium*, studied at universities and technical universities, and attained teaching qualifications for agricultural and trading schools as well as for *Gymnasien* (cf. Böttcher et al. 2008, pp. 63–64).

and straight line, which in the French works were rather used as a means to an end limited to fundamental tasks and those needed for the later tasks on curved surfaces; in his work, Gugler decided for a more extensive treatment of tasks on straight lines and planes since they made the learner perceive the nature of descriptive geometry best and hence fostered their inner perception most effectively (cf. ibid., p. VI). Next to these deviations from the "French models", Gugler also integrated parts of projective geometry, e.g. polarity or affine relations, but did not use it as a foundation for descriptive geometry (cf. Wiener 1884, p. 37). It is claimed that Gugler's work was the first German book to unite the entire knowledge about descriptive geometry at this time (cf. Böttcher et al. 2008, p. 63). In the presented German works, the need was felt to make descriptive geometry independent from analytic geometry and to bring it onto a purely synthetic level. Therefore, projective geometry was used, which at the same time led to an intensification of the works on central projection.

Likewise, Pohlke[9] used theorems of so-called new ("*neuere*") geometry in his *Darstellende Geometrie* (1860), but without coining its character onto his investigations of orthogonal projection, axonometry and perspective, which basically provide the topics of his book (Obenrauch 1897, p. 83). Furthermore, Pohlke himself is supposed to have developed theorems of "new" geometry (cf. Wiener 1884, p. 38). However, the main contribution Pohlke made to descriptive geometry in a wider sense concerns the field of axonometry. In Germany, not only the connection between projective and descriptive geometry was a matter of innovation, but also the (further) development of axonometric methods. These methods developed out of the wish to be able to draw information from depictions as well as to visualise the depicted object (cf. Papperitz 1909, p. 573). Found in 1853 (cf. ibid., p. 573), Pohlke published his theorem in his book from 1860. The Theorem of Pohlke states that any three rays in the plane, which start in the same point, can be perceived as the parallel projections of three rays which are orthogonal to each other in one point of space (cf. Küpper 1888, p. 448). Consequently, the laws of parallel projection can be used in any coordinate system.

Next to the connection of descriptive with projective geometry and the further development of axonometry, German geometers dedicated themselves to shadow constructions (cf. Wiener 1884, 36). This aspect was taken up by Johann Hönig,[10] in his *Anleitung zum Studium der darstellenden Geometrie* (1845), and by Ludwig Burmester[11] in his *Theorie und Darstellung der Beleuchtung gesetzmäßig gestalteter Flächen* (1875). In general, the methods of Monge were embedded and applied

---

[9]Karl Wilhelm Pohlke (1810–1876) studied art in Berlin and worked as a free artist before he became a teacher and later (1860) professor for descriptive geometry in Berlin (cf. Institut für Mathematik und Informatik der Universität Greifswald 2011).

[10]Johann Hönig (1810–1886) was professor for descriptive geometry at the polytechnic institute in Vienna where he had studied himself (cf. Stachel, Chap. 9, this volume).

[11]Ludwig Burmester (1840–1927), who did not only accomplish a mechanical apprenticeship and gained practical experience in industry and craft, but also studied at different universities, later became professor for descriptive geometry at the polytechnic institute in Dresden (cf. Löbell 1957, p. 55).

in other contexts of the theory of projections or geometry. The publication of Monge's work at the end of the eighteenth century led to an intensive treatment of descriptive geometry in a wider sense, as it was perceived in Germany, during the nineteenth century.

3 Polytechnic Schools in Germany

To understand the development and the position of polytechnic schools (or other possible equivalents of the French *École polytechnique*) in Germany during the nineteenth century, it is important to consider their origin or predecessor institutions, which were mainly professional training schools (different kinds of *Fachschulen*, also including *Gewerbeschulen*, etc.) of which some had already existed in the eighteenth century. These different kinds of schools were mainly in charge of the training of tradesmen (*Gewerbetreibende*), craftsmen or sometimes technical civil servants. The structure of the technical education system became quite complex in the nineteenth century, since new institutions (as polytechnic schools) developed to adapt technical education to the new conditions evolved in the course of the beginning industrialisation. In general, polytechnic schools in Germany represent an important stage in the nineteenth century within the development of *Technische Hochschulen*[12] out of professional training schools, which at the same time represents the process of the raise of technical education to a scientific discipline in the German Reich and Empire. However, German polytechnic schools reveal characteristics that differ from the French original.

3.1 The Foundation of Polytechnic Schools and Gerwerbeinstitute from the 1820s on

In the aftermath, it is often claimed for *Technische Hochschulen* in general (cf. Manegold 1970, p. 22) and for single higher technical institutions when they were founded in Germany that they were instituted after the model of the *École polytechnique* in Paris. Yet, right from the beginning, the relevant institutions in Germany differed decisively from the French original.[13] At first, their internal structure was different, sometimes through the integration of not only preparatory

---

[12]*Technische Hochschulen* (TH) represent an interim stage within the institutional development of higher technical education institutions out of schools for professional education into technical universities. An important step for this development was the attainment of the right to award doctorates for the TH Berlin in 1899. A possible, but still not adequate translation could be "technical college".

[13]In Chap. 22 of this book, the role model function of the *École polytechnique* will be discussed.

mathematical courses but also *Fachschulen* (cf. Klein and Schimmack 1907, p. 177; Stäckel 1914, p. 151; Stäckel 1915, p. 2) and sometimes because they resembled secondary schools rather than institutions of higher education (cf. Schubring 1989, p. 174). Secondly, "[...] their resources and general level of instruction were comparably poorer" (ibid., p. 179) at least in the beginning. Nevertheless, the model function of the *École polytechnique* contributed to the establishment of independent technical schools of higher level in general (cf. Stäckel 1914, p. 151), also by "[...] drew[ing] much of their legitimation from the fame of the *École polytechnique* [...]" (cf. Schubring 1989, p. 179).

Polytechnic schools, or similar institutions as trading institutes,[14] developed at the beginning of the nineteenth century to provide higher, in the sense of scientific, professional education in the technical disciplines. The education at the existing *Fachschulen* did not meet the newly emerged technical requirements and, in the rise of industrialisation, more engineers were needed (cf. Lexis 1904, pp. 4–5). The new technical institutions were supposed to exceed the aim of practical training at *Fachschulen* by targeting to educate scientifically, technically and economically thinking graduates for leadership in economy (cf. Stäckel 1915, p. 8).

In Prussia, however, the original idea of the *École polytechnique*, to provide students with preparatory education in mathematics and natural sciences for further special studies, was not realised; here, the aforementioned trading institutes were established to provide school education for tradesmen (cf. Jeismann and Lundgreen 1987, p. 295). In other parts of Germany, polytechnic schools were founded, which, in contrast to the French original, integrated next to preparatory mathematical studies several different *Fachschulen*. These institutions incorporated the entire technical education sector in Prussia (or France) under a single roof (cf. ibid., p. 295). Despite their initial differences, trading institutes as well as polytechnic schools are considered in the context of higher technical education (or possible equivalents of the *École polytechnique*) since they have in common their further development into *Technische Hochschulen*. Not all of these institutions underwent this transformation; some stayed at quite a low level. Furthermore, territorial restrictions have to be made. If the entire territory of the German speaking countries was considered, the first equivalent institutes to the French *École polytechnique* were established in Prague (1806) and Vienna (1815).[15] They distinguished from other "German" institutions, because they both claimed the status of *Hochschule* from the beginning (cf. Manegold 1970, p. 34), also because of their "casual character" resulting from the fact that they addressed not only future civil servants but numerous other visitors (cf. Schoedler 1847, p. 89). In the following, only those

---

[14]Trading institutes (*Gewerbeinstitute*) may not be confused with trading schools (*Gewerbeschulen*). The former had a higher status or level than the latter.

[15]For information on the developments in Austria see Chaps. 11 and 12 in this volume.

institutions that were opened in parts of Germany that belong to it until today will be considered.[16]

In the different states of Germany, the developments of the relevant institutions proceeded quite differently. These in total eight German institutions for higher technical education developed throughout the first half of the nineteenth century out of several different schools. The first institution to provide higher education for tradesmen was the trading institute in Berlin (opened in 1821), where general education and technical knowledge for professional practice were taught at the same time (cf. Scharlau 1990, p. 16).[17] The trading schools in Stuttgart (1829), Hannover (1831) and Darmstadt (1836) became polytechnic schools in 1840, 1847 and 1868, respectively. The trading school in Stuttgart was even attached to a *Realschule* (another type of secondary school) in the beginning (cf. ibid., p. 245). The predecessor institution of the polytechnic school in Dresden (renamed in 1851) was a technical school founded in 1828 (cf. ibid., p. 84), and the polytechnic school in Braunschweig (1862) developed out of the technical department of the *Collegium Carolinum*, an institution which originally provided preparation for university studies (cf. ibid., p. 57). The schools in Karlsruhe (1825) and Munich (1827) were directly instituted as polytechnic schools, whereas in Munich actually two different polytechnic schools existed (a second one was founded in 1833), which underwent several transformations until in 1868 one final polytechnic school was reopened. Then, it was comprised of five different *Fachschulen* (cf. ibid., p. 216) just as the first polytechnic school (referring to the name) in Karlsruhe. There had been more similar institutions in Germany as, for example, the polytechnic schools in Augsburg and Nürnberg (opened together with the school in Munich in 1833). All three schools were originally designed as specific *Fachschulen*, but only the one in Munich developed further into a school for higher technical education, in the same manner as the other seven mentioned institutions (cf. Lexis 1904, p. 225). The polytechnic schools in Augsburg and Nürnberg are not considered and mostly not even mentioned in the literature. This also concerns the institutions in Kassel or Chemnitz. The polytechnic school in Augsburg, for example, was coequal to the *Gymnasium* (grammar school), so, to a secondary school (cf. Schoedler 1847, p. 34), whereas polytechnic schools in Germany were usually located in professional education and later achieved a university-like status.

Because the polytechnic school in Karlsruhe was not only the first of its kind but also illustrates significant differences from its proclaimed model, the *École polytechnique*, it will be covered in more detail here. The polytechnic school in

---

[16]This restriction also excludes the *Eidgenössische Polytechnische Schule* in Zürich, which will be dealt with in Chap. 11 and which, moreover, also differed decisively from the other German institutions. Opened in 1855, the school revealed university characteristics in some respects from the beginning, too, e.g. by integrating humanistic and political subjects (cf. Manegold 1970, p. 55) and providing training for teachers (cf. Stäckel 1915, p. 9).

[17]At the same time, a so-called *Bauakademie* (construction academy), opened in 1799, existed in Berlin. The trading institute (*Gewerbeakademie*) and this *Bauakademie* would merge to become the *Technische Hochschule* in Berlin in 1879.

Karlsruhe was the result of the merging of a professional school for construction (run by Weinbrenner) and an engineering school. The school was supposed to be designed after the model of the *École polytechnique* (cf. Scharlau 1990, p. 175), but concerning its internal organisation, it resembled a secondary school rather than an institution for higher education, on the one hand, because of the age of admission (here, 15 years of age) and, on the other hand, because of their strict division into classes (cf. Schubring 1990, p. 273), which additionally opposed the structural organisation of universities in Germany, where *Studienfreiheit* was obtained. Seven years after its foundation, the polytechnic school had integrated a preschool, mathematical courses and five *Fachschulen* (cf. Renteln 2000, p. 5). The preschool offered two preparatory courses focusing on mathematics, drawing and physics, the general basics of technical knowledge (cf. Stäckel 1915, p. 7). The integration of five *Fachschulen* (engineering, construction, forestry, industry and commerce) (cf. ibid., p. 7) constituted a significant difference from the proclaimed model. Moreover, the mathematical studies were not mandatory for every student (cf. ibid., p. 8). The polytechnic school in Karlsruhe rather was a unification of different *Fachschulen*, a trend common at that time also to guarantee their survival (cf. Stäckel 1914, p. 151; Stäckel 1915, p. 8), since the majority of these institutions dedicated to the preparation for a specific profession were closed again (cf. Stäckel 1910, p. III). In the case of Karlsruhe and Germany in general at that time, the term "polytechnic" rather referred to the meaning of the integration of several technical disciplines than to the French institution. In 1885, the polytechnic school changed its name to *Technische Hochschule*. This upgrade was experienced by all of the eight considered polytechnic institutions in Germany during the nineteenth century (the dates can be found in Table 9.1).

3.2 The Upgrade to Technische Hochschulen in the 1870s

The original purpose of the technical institutions which were founded at the beginning of the nineteenth century was to provide higher (in the sense of theoretical) education, nevertheless, for practitioners in the trading or industrial sector. The orientation towards professional practice contradicted the contemporary conception of universities in Germany, which aimed at the transmission of theoretical or scientific knowledge. This difference provided the first reason for the comparatively low standing of higher technical institutions and their allocation within the professional training sector. The second reason often was the standing of technical sciences within the concept of humanism, which was especially predominant in the Southern parts of the German Empire, in general, namely not being a relevant science at all. The structural differences (the course system, the young age of admission, etc.), a third reason for the deviation from universities, soon faded due to the organisation into faculties (cf. Jeismann and Lundgreen 1987, p. 286). Nevertheless, despite structural similarities between *Technische Hochschulen* and universities, not until the twentieth century did the former achieve an equal status including the right

Table 9.1 Overview of the mandatory weekly ("Lecture" and "Practice") and total (referring to the entire academic year) lessons in descriptive geometry at German technical universities for the academic year 1913/1914

Technical university	Course	Term	Lecture	Practice	Total
Berlin (1821/1879)	Darstellende geometrie I	Winter	4	4	240
	Darstellende geometrie II	Summer	4	4	(120)
Karlsruhe (1825/1885)	Darstellende geometrie I	Winter	4	4	240
	Darstellende geometrie II	Summer	4	4	(120)
Munich (1827/1877)	Darstellende geometrie I	Winter	4	4	240
	Darstellende geometrie II	Summer	4	4	(120)
Dresden (1828/1890)	Darstellende geometrie I	Winter	3	4	210
	Darstellende geometrie II	Summer	3	4	(120)
Stuttgart (1829/1890)	Darstellende geometrie I	Winter	3	4	246
	Darstellende geometrie II	Summer	4	6	(144)
Hannover (1831/1879)	Darstellende geometrie	Winter	3	6	270
		Summer	3	6	(180)
Braunschweig (1835/1877)	Darstellende geometrie	Winter	6	4	300
		Summer	6	4	(120)
Darmstadt (1836/1877)	Darstellende geometrie I	Winter	4	6	300
		Summer	4	6	(180)

Under the total amount of lessons the share of practice lessons is found in brackets (cf. Stäckel 1915, p. 136)

to award doctorates (except for Berlin in 1899). The primary fostering factor for the emancipation of *Technische Hochschulen* was the education of civil servants.

Originally "modest" education institutions for the trading sector (cf. Lexis 1904, p. V), it is claimed for *Technische Hochschulen*, or for specific professional training schools out of which they developed, respectively, that they grew out of the needs of industrialisation (cf. Stäckel 1914, p. 153). Yet, the actual upgrade to an institution of higher (tertiary) education did not result from growing demands initiated by industrialisation and economy, an argument that was then used by advocates of this development (cf. Jeismann and Lundgreen 1987, p. 298). Finally, it was the conformation of the education of technicians to the education of technical civil servants that fostered the academisation of the engineering sciences (cf. ibid., p. 297). The orientation at civil service served as a legitimisation of technical education due to the

> [...] functional relationship between the educational system and the employment system: the types of careers accessible to graduates by virtue of the training they receive in academic institutions largely mold the institutionalization of the related disciplines. The autonomy of an academic discipline depends upon the existence of specialized professional careers for the discipline's graduates (Schubring 1989, p. 174).

3.3 Abstract vs. Applied Mathematics

The institutionalisation of the technical sciences was mainly achieved by the means of mathematics, namely by grounding technical education on mathematics (cf. Jeismann and Lundgreen 1987, p. 298). At first, the function of mathematics within the technical education sector was perceived as providing fundamental basic knowledge for further special studies (cf. Lexis 1904, p. 49) and later as taking the part of an auxiliary science (cf. Papperitz 1899, p. 9; Stäckel 1910, p. IX; Scharlau 1990, p. 151). However, as basic or auxiliary science, the main focus was put on the applicability of knowledge: mathematics in general was not taught for its own sake but rather embedded within the purposes and tasks of future engineers. "That the type of institution in which mathematicians teach their subject determines to a certain degree the style and substance of their mathematics would appear to be undeniable" (Schubring 1989, p. 173). So, the "applied" branches of mathematics fell within the remit of *Technische Hochschulen* (or higher technical institutions). In the course of the emancipation of the technical sciences and its inherent education institutions in the last third of the nineteenth century, mathematics assumed the new role of providing a scientific foundation. In order to achieve a more scientific mathematical level, scientific qualification, among other consequences, was demanded from teachers at higher technical institutions, so that mathematics graduates from universities were recruited as professors (cf. Schubring 1990, p. 273), who did neither have experience in actual professional practice nor were able to apply mathematics in the relevant technical disciplines due to the predominant status of pure, or abstract, mathematics[18] at universities. In the early stages of higher technical institutions, their teachers had had some connections to the technical disciplines, either through their own practical experience or apprenticeship in a relevant profession, through the visit of one of the technical predecessor institutions or the teaching at one of these. This, among other but connected reasons, led to the "anti-mathematics" movement among German engineers in the 1870s (cf. Schubring 1989, p. 181). The academisation of the technical education sector through the inclusion of abstract mathematics was being criticised for its decreasing fit to the needs of trade and industry (cf. Jeismann and Lundgreen 1987, p. 298) and led to a dissociation from professional practice as needed by engineers (cf. Manegold 1970, pp. 144–146).

---

[18]Indeed, the distinction between pure and applied mathematics, as we know it today, was not familiar at the beginning of the nineteenth century (cf. Wußing 1975, p. 243). The term "abstract" mathematics here denotes the contrast to "applied" mathematics as in "applied in other disciplines, subjects or in professional practice". For example, mechanics and geodetics were included under this notion (cf. Lexis 1904, p. 52).

4 Descriptive Geometry at Polytechnic Schools and *Technische Hochschulen*

Indeed, descriptive geometry was regarded as belonging to pure mathematics at the beginning of the twentieth century (cf. Ott 1913, p. 2; Lexis 1904, p. 52), but, before, due to the focus on its applicability in the technical disciplines and in professional practice, and especially due to its general association with the technical education sector during the nineteenth century, it was generally referred to as applied mathematics. Within the internal structure of German polytechnic schools (or their equivalents), mathematics in general, and with it descriptive geometry, was embedded in the so-called general departments (*Allgemeine Abteilungen* or the like)[19] or, earlier, in the preparatory (mathematical) courses. For the latter, again, the polytechnic school in Karlsruhe will serve as an example: from 1832 to 1843, there existed two preschool courses (age of admission 13 or 14, respectively) and two mathematical courses (15 or 16, respectively), from 1843 to 1863, the preparatory and three mathematical courses had to be taken before entrance into one of the seven *Fachschulen* and in 1863 the preschool and the first mathematical course were transferred to a secondary school (cf. Renteln 2000).[20] Descriptive geometry was taught in these first years, being attributed the role of fundamental basic knowledge. Before their upgrade to *Technische Hochschulen*, mainly teachers or external docents taught the different fields of mathematics (cf. Scharlau 1990). Only from the 1850s, chairs for geometry and sometimes especially for descriptive geometry were established. An exception provided the polytechnic school in Karlsruhe where the first professor for geometry (descriptive and practical geometry), Guido Schreiber, held the chair from 1826 to 1852, followed by Christian Wiener from 1852 to 1896 (cf. Renteln 2000, p. 11).

4.1 Teacher Education

In the aforementioned general departments, "basic knowledge" relevant for all further studies in the different *Fachschulen* was taught. This includes natural sciences and mathematics as well as subjects for general knowledge as, for example,

---

[19]Lexis claims at the beginning of the twentieth century that 60–80 years before, general departments had existed (cf. 1904, p. 50), but he probably referred to mathematical and/or preparatory courses, since the official denotation and integration occurred later in the nineteenth century. In Munich, a general department was integrated after the refoundation in 1868 (cf. Hashagen 2003, p. 40). In Stuttgart, a *Fachschule* for general education was established next to a one for mathematics and natural sciences in 1870 (cf. Böttcher et al. 2008, pp. 14–15).

[20]The mentioned outsourcing of preparatory studies into secondary education was part of the development into *Technische Hochschulen* and will be dealt with again later in the context of secondary education.

economics, modern languages, history, literature or philosophy (cf. Lexis 1904, p. 51). At the general assembly of the Association of German Engineers in 1864, the presumed task of *Technische Hochschulen* was defined as follows: they are supposed to aim at the scientific education for those technical jobs of civil service and private business which are based on mathematics, natural sciences and drawing arts and also at the education of teachers for the technical sciences represented at school and their auxiliary sciences (cf. Lexis 1904, p. 60). The last task, the education of teachers for mathematics and natural sciences, was assigned to the general departments. In some states, the training of teachers was integrated into the general departments of *Technische Hochschulen* (or their predecessor institutions) earlier than in Prussia. At the TH Munich and Dresden, prospective teachers for *Gymnasien* (humanistic schools) and for *Realschulen* ("realistic" or "real" schools)[21] could accomplish their full education in natural sciences and mathematics since the 1860s (cf. Lorey 1916, p. 153). At other institutions (Stuttgart, Dresden, Karlsruhe and Darmstadt), a partial completion of the scientific teacher training in mathematics, physics or chemistry was possible (cf. Lexis 1904, p. 59). In Prussia, only with the installation of the exam regulations for teachers in 1898, which determined applied mathematics including descriptive geometry as a possible examination subject, prospective teachers could complete parts of their training (up to three out of six semesters) at *Technische Hochschulen* (cf. Scharlau 1990, p. 33). Before, universities had fully been in charge of the scientific teacher training, where "applied" mathematics was not cultivated, so that *Technische Hochschulen* were supposed to make up for this vault. So, applied mathematics, including descriptive geometry, at *Technische Hochschulen*, just like abstract mathematics at German universities had experienced before, gained some of its legitimisation due to the educational function for the scientific teacher training (cf. Schubring 1990, p. 264).

4.2 The Role of Descriptive Geometry for the Education of Engineers

According to Erwin Papperitz, descriptive geometry had become indispensable at *Technische Hochschulen* by the end of the nineteenth century (cf. Papperitz 1899, p. 44). It had been established at German polytechnic institutions by the middle of the nineteenth century at the latest, in individual cases much earlier (cf. Jeismann and Lundgreen 1987, p. 298). Descriptive geometry was the only mathematical discipline that had been treated more intensively at *Technische Hochschulen* than at universities (cf. Papperitz 1899, p. 44). By that time, lectures on descriptive geometry were occasionally given at some German universities, as, for example, at the university in Gießen by L. Heffter, first professor of applied mathematics

---

[21] The different types of secondary education institutions will be explained in the next section of this chapter.

(1891–1897) there (cf. Scharlau 1990, p. 113) or in Leipzig by Felix Klein (1881–1882) and Walther von Dyck (1883) (cf. Hashagen 2003, p. 123, 129), but these occasions are hardly worth mentioning. At the beginning of the twentieth century, the teaching of descriptive geometry was still focused on practical application (cf. Stäckel 1915, p. 140). In the curricula of the by then technical "universities", the teaching of descriptive geometry in a wider sense was determined as exemplified by the curricula of the institution Darmstadt: oblique as well as orthogonal parallel projections, the method of plane and elevation, shadow constructions, perspective, axonometry, elements of projective geometry, etc. (cf. ibid., pp. 137–138). Table 9.1 reveals the amount of mandatory lessons for students of engineering at the eight German institutions of concern during the academic year 1913/1914. For Berlin and Munich, the number of lessons for students of architecture was higher and for Stuttgart lower than for engineers; for the others, the numbers hold for both. For the academic year 1903/1904, the mandatory lessons in descriptive geometry had been even higher (cf. Lexis 1904, pp. 86–87). In comparison, within the entire course of studies for students of construction engineering, descriptive geometry sometimes had (roughly) the same share as the subject construction (*Baukonstruktionslehre*) (in Berlin, Braunschweig and Darmstadt) and in Dresden even a higher share (cf. ibid., pp. 86–89). Furthermore, in states as Württemberg, for example, where descriptive geometry was taught intensively at realistic secondary schools,[22] only those students that had visited a *Gymnasium* before had to study *"Darstellende Geometrie I"* at the TH Stuttgart (cf. Stäckel 1915, p. 137). In general, the lessons in descriptive geometry were to be taken in the first semesters (cf. Lexis 1904, p. 81), so as a preparation for further, more specialised studies.

5 Descriptive Geometry in Secondary Education

At the beginning of the nineteenth century, the secondary and the professional education system in the different German states existed relatively parallel to each other: on the one hand, *Gymnasien* and universities presumably provided "higher" education, and, on the other hand, professional and technical schools provided education for the trade and commerce sector. Due to its association with the technical sciences, the teaching of descriptive geometry, or "applied" mathematics in general, fell within the remit of the latter. Parts of both systems would soon merge in the course of the attempt to integrate realistic education in the German secondary as well as tertiary education system and later to provide prior education for the technical education sector. The development of the German education system did not only differ decisively in different states, but is also complex within itself,

---

[22]This notion includes next to the aforementioned *Realschulen* so-called *Realgymnasien* and *Oberrealschulen*, "higher" secondary schools integrating realistic subjects. "Realistic" in this context means the opposition of humanistic studies (Latin, Greek etc.).

since numerous different school types with numerous different purposes existed, structurally changed and sometimes even intersected throughout their developments (e.g. some schools changed from 2/3 years courses to 6 year courses, etc.). Restrictions and simplifications are necessary for the sake of comprehensibility: the following explanations provide a selection relevant for the considered context, namely for the role of descriptive geometry in secondary education and for the schools' function as preparatory institutions for higher technical studies. In the context of this chapter, it is impossible to elaborate on all occurring differences in the different states.[23] This section should merely provide an overview of existing school types, so the institutional context for the teaching of descriptive geometry.

5.1 The Establishment of "Realanstalten"

The classical school types at the beginning of the nineteenth century were the *Volksschule*, visited by the vast majority of the population (cf. Klein and Schimmack 1907, p. 10), for lower and the *Gymnasium*, which developed out of *Gelehrtenschulen* or *Lateinschulen* (older grammar schools or Latin schools) in the sixteenth century (cf. Lexis 1904a, p. 68), for higher education. At this time, the *Gymnasium* was the only existing higher secondary school, which also stayed predominant throughout the nineteenth century (cf. ibid., p. 217), primarily because of its function to grant entrance qualification for (any kind of) higher studies and for occupations in civil service. Mainly within the first decades of the nineteenth century (in some states earlier as, e.g. in Bavaria and Württemberg at the end of the eighteenth century), new schools developed, which will be summarised by the term *Realanstalten* (*Realschulen, Realinstitute, Gewerbeschulen, Bürgerschulen*, etc.).[24] Even though these institutions assumed different functions within either secondary or professional education in different German states, in general, they searched to fill a gap in classical secondary education: the contemporary disagreement about the purpose and contents of (general) education as should have been provided by secondary education.*Gymnasien* put their focus on humanistic studies, i.e. in Latin and Greek and were supposed to prepare their students for further university studies. However, the persuasion arose that professional practice should also be preceded by some kind of general education, but appropriate for the purposes of practical life (cf. Stäckel 1910, p. III). Supposedly developed out of the spirit of enlightenment and the needs of the cities (cf. Jeismann and Lundgreen 1987, p. 153), different kinds of *Realanstalten* were expected to provide education for the preparation of

---

[23] For the single states, the German subcommittee of the *Internationale Mathematische Unterrichtskommission* (IMUK) published reports on the situation of mathematics education and the development of the respective education system, which provide detailed information about the developments of single *Realanstalten*.

[24] For the sake of simplicity, the term *Realanstalten* will denote all relevant school types and their various names, etc. in the different states.

those professions that did not need classical languages (cf. Klein and Schimmack 1907, p. 75) and that were tightly connected to practical life (mainly tradesmen and technicians) (cf. Lexis 1904a, p. 71). These *Realanstalten* integrated or put the focus on "realistic" subjects (including mathematics, natural sciences and modern languages). However, these "new" schools, i.e. *Realschulen*, did not have a legal foundation (cf. Klein and Schimmack 1907, p. 75) or a clearly defined form yet and were to be located in between a professional school and a school for general education overloaded with technical subjects (cf. Lexis 1904a, p. 72). The situation of the different *Realanstalten* was not only inconsistent within single states, but also "[...] the situation and level of this school type was not comparable at all among the various German states" (Schubring 2012, 530). In the beginning, each single *Realanstalt* was a "special case".

However, throughout the nineteenth century, a share of these institutions (sometimes with the interim stage of *Realschulen I. Ordnung* and *II. Ordnung*, which means of I. and II. order) evolved into *Realgymnasien* and *Oberrealschulen*, both school types offering a 9-year course then and to be located in the secondary education sector. The latter provided an education excluding Greek as well as Latin and put the main focus on natural sciences, mathematics and modern languages; the former for their part integrated natural sciences as well as Latin into their curricula. In Prussia, for example, the graduation from *Realschulen* of I. order, or *Realgymnasien*, respectively, originally qualified for professions that did not need any university studies (e.g. construction, postal services and mining) (cf. Klein and Schimmack 1907, p. 86). *Oberrealschulen* were allegedly founded, e.g. in Prussia and Württemberg, in order to prepare students for polytechnic schools or *Technische Hochschulen*, respectively (cf. Jeismann and Lundgreen 1987, p. 303; Geck 1910, p. 46). However, in Prussia, for example, from 1892 on, university studies in mathematics and natural sciences and the admission to civil service in the field of construction were permitted for absolvents of *Oberrealschulen* (cf. Lexis 1904a, p. 82). Here, even earlier, from 1870, admission to university studies in mathematics, natural sciences and modern languages was granted for former students of *Realschulen* of I. order, or *Realgymnasien*, respectively (cf. Stäckel 1915, p. 78). Finally, both *Realgymnasien* and *Oberrealschulen* throughout the entire German Reich did achieve the same legal status as *Gymnasien* in 1900 by being qualified to a grant full university entrance qualification (cf. Lexis 1904a, 87), namely the *Abitur*.

5.2 Preparatory Institutions for Higher Technical Studies

At the beginning of the twentieth century, the functions and educational aims of the different secondary schools in Germany were finally relatively clear. Graduation from one of the three higher secondary schools (*Gymnasium*, *Realgymnasium* and *Oberrealschule*), which could all grant the *Abitur* by then, permitted entrance into all kinds of universities. The demand for equality to universities by *Technische*

Hochschulen, namely the entitlement to provide scientific education as well, on the one hand, led to increased demand of entry criteria on the other hand: "In 1899 they were officially recognized as institutions of higher education for which the *Abitur* [...] became obligatory for admission" (Schubring 1989, p. 180). On the part of the engineers, this was seen as a necessary sacrifice connected with heavy concerns for the sake of an appreciation of their branch (cf. Stäckel 1915, p. 81). However, throughout the nineteenth century, the area of responsibility between secondary schools and tertiary education had not always been clearly distributed.

Polytechnic schools in Germany were mostly founded out of professional training institutions that originally required prior knowledge accomplished at *Volksschulen*, sometimes a few classes of the *Gymnasium* (Lexis 1904b, pp. 41–43), and/or practical experience in the relevant craft. However, when some technical schools developed further into institutions of higher technical education, they lacked preparatory institutions, so they instituted them themselves (Schoedler 1847, p. 117) either internally as preparatory (mathematical) courses or externally in *Gewerbeschulen* or similar institutions. As mentioned before, descriptive geometry was, for example, taught in these preparatory courses. When the mathematical preparatory courses were dissolved throughout the process of the emancipation of technical education institutions, the imbalance had to be equilibrated also by secondary education. The boundaries between mathematics in tertiary and secondary education had to be redefined, since the *Technische Hochschulen* passed a decisive part of mathematical basics on to schools of general education and extended the curricula in the direction of the technical disciplines (cf. Schubring 1990, pp. 274–275). The school types that were supposed to make up for this vault were "middle schools", namely the different kinds of *Realanstalten* (cf. Lexis 1904, p. 5). So, logically, descriptive geometry should have been outsourced into these institutions as well. However, the *Realanstalten* could not completely fulfil their presumed task due to the parallel growing admission conditions of higher technical institutions. For the qualification for civil service the *Abitur* was needed, which could only by granted by *Gymnasien* until 1900. So, the original aim to educate ("higher") technicians, who needed a special education different from that offered at *Gymnasien* (which actually was a motivating factor for the foundation of technically oriented or "realistic" schools) was in conflict with the entry criteria for technical civil services. The original preparatory classes and *Gewerbeschulen* were fitted to the needs of future technical professions (cf. Stäckel 1914, p. 153) by focussing on mathematics and drawing (cf. Stäckel 1915, p. 80), while the education at *Gymnasien* had other purposes. So, the student body at higher technical institutions always included students that accomplished their secondary education at *Gymnasien*, which led to a wide range of prior knowledge in the relevant subjects among the students.

5.3 Prior Knowledge in Descriptive Geometry

After 1900, "[...] the mathematical curricula in all the various secondary and tertiary institutions would have [had] to be redefined in order to make the free transition from each school type to both higher education types [possible]" (Schubring 1989, p. 185). Before, the prior education in mathematics was often insufficient. At first, many students proceeded to *Technische Hochschulen* after having accomplished their general education at a *Gymnasium*. The discrepancy, referring to prior knowledge, in contrast to the absolvents of *Realgymnasien* and *Oberrealschulen*, was especially perceivable in descriptive geometry (cf. Papperitz 1899, p. 31), which had to be compensated for by the introduction of basic notions at the *Technische Hochschulen* themselves (cf. Müller 1910, p. 19). Students from *Gymnasien* lacked the skills in drawing necessary for a scientific treatment of descriptive geometry especially in the northern parts of Germany (cf. Stäckel 1915, p. 114). Indeed, in Prussia, even students from realistic secondary schools lacked basic knowledge and skills (cf. Müller 1910, p. 19). Here, all secondary school types were short of trained teachers (cf. Zühlke 1911, p. 10) until 1898 when descriptive geometry finally had been integrated (as subsidiary subject) into the curricula for prospective teachers at *Gymnasien* and *Realanstalten*. In other states, at least a distinction was made between realistic and humanistic teacher education as, for example, in Baden and Bavaria (cf. ibid., p. 30). However, before 1898, descriptive geometry as a school subject was taught throughout the entire German Empire mostly by drawing teachers (cf. Zühlke 1910, p. 54). In particular, graduates from *Gymnasien* often had to take special courses in descriptive geometry at *Technische Hochschulen*, especially in states (mostly Southern Germany), where an intensive treatment of descriptive geometry happened at realistic secondary institutions. At *Technische Hochschulen*, descriptive geometry had the status of an auxiliary science that should have been taught with a main focus on applications in technics (cf. Müller 1910, p. 19), which was not possible if an adequate prior knowledge had not been transmitted during secondary education.

5.4 Descriptive Geometry as a School Subject

For a general overview of the dissemination of descriptive geometry as a school subject, a selection of curricula as well as a selection of schoolbooks will be analysed in order to get an impression of the position of descriptive geometry within the secondary school system and which understanding of this mathematical discipline was transmitted in secondary education. Thereby, not only the curricula for mathematics, but also those for drawing (i.e. linear drawing) will be considered. In 1911, Paul Zühlke published an elaborate survey about descriptive geometry and linear drawing as school subjects at *Realanstalten* in the German Empire including a tabular overview for several German states from which the following information is taken (cf. Zühlke 1911, p. 13). Throughout the different German states, several different titles for the subjects, if an independent subject existed, concerned with the teaching of descriptive geometry were used: linear drawing in Prussia and Saxony, geometric drawing in Hessen and Württemberg or technical drawing in Bavaria. The different denotations mirror the wider understanding of descriptive geometry (as the accumulation of mathematical drawing methods). However, descriptive geometry or related subjects were not always obligatory and the general distribution of lessons is quite low (1–2 weekly lessons, mainly in the last 3–4 school years). In general, the subject of linear drawing only started to develop independently from "artistic" drawing in the middle of the nineteenth century (cf. Zühlke 1910, p. 51).

As a guideline, the curricula for Prussia will be treated exemplarily even though, for the case of descriptive geometry, their degree of representativity in comparison to other German states is quite low.[25] For the Southern parts of Germany, it was claimed that descriptive geometry as a school subject had been "appreciated" since the 1840s (cf. Zühlke 1911, p. 10). A careful study of each state would be necessary for a complete overview, which would exceed the scope of this chapter. In this context, a few regional differences will be commented on.[26]

The first curricula in Prussia determining the contents for the education at *Realschulen* of I. and II. order were published in 1859. For *Realschulen* of I. order, descriptive geometry was integrated into the mathematics lessons and a connection to the drawing instructions was scheduled; after the stereometry lessons, descriptive ("*beschreibende*") geometry, shadow constructions and perspective were supposed to be taught. However, this seems to have been seldom realised (cf. Brennecke 1869, p. III), also due to a lack of trained teachers. Looking at schoolbooks from this time, it is obvious that a coherent understanding of relevant contents of descriptive geometry for secondary education did not exist. According to the author, Brennecke's schoolbook on descriptive geometry from 1869 deals with those contents of descriptive geometry that are required from aspirants of French institutions as the *École polytechnique*, military colleges, the navy and the forestry academy and only treats plane and elevation. In Scherling's book (1870), the author deals with projective geometry before addressing orthogonal and central projections to achieve a more profound understanding. Butz integrated into his work (1870) descriptive geometry in a "broader sense" as he claimed by dealing with descriptive geometry in a "narrower sense" (after Monge), axonometry, linear perspective and shadow constructions—in general, mathematical representation methods. In

---

[25] Prussia comprised at times approximately two thirds of the Deutscher Bund (around the 1860s) excluding the Austrian Kingdom and at least for Northern states the developments in education were often similar (cf. Klein and Schimmack 1907, pp. 78–79). In Hessen, for example, the need was felt to adapt to Prussian education policy in the case of *Realschulen* due to political and economic dependence (cf. Schnell 1910, p. 7). Moreover, the Prussian developments concerning *Oberrealschulen*, for example, also had an influence on these schools in Baden in Southern Germany (cf. Cramer 1910, p. 17). Moreover, 60–65% of *Gymnasien, Realgymnasien* and *Oberrealschulen* and 45% of *Realschulen* belonged to Prussia in 1902 (cf. Lexis 1904a, p. 217).

[26] For the role of mathematics in general in education in different German states, consult (Schubring 2012).

the appendix of his book, the author added three different depictions ("with the method of descriptive geometry", axonometric and perspective) of a house which are compared. This instance provides the only "external" application within the three schoolbooks: the different projection methods are only applied for geometric objects, hence, within geometry itself. So, these schoolbooks (all published in Prussia) have in common that descriptive geometry (by whatever means) was to be taught for its own sake within the context of pure mathematics. Furthermore, these three books are "preparatory" or "introductory" works (*Anfangsgründe, Vorschule, Einführung*) for *Realschulen*, at a stage, when these schools were supposed to prepare their students for later technical or industrial studies or professions.

With the official installation of *Realgymnasien* and *Oberrealschulen* in Prussia, curricula for these schools were published in 1882. The contents were similar for both schools; there were just more lessons assessed for several subjects as, e.g. mathematics and drawing for the *Oberrealschule*. For the mathematics lessons the "preparation of perspective drawing" was designated in the context of stereometry, and students could choose "elements of descriptive geometry" as an optional course within the drawing lesson. In the curricula of 1892, no significant changes were made in the relevant subjects. In 1901, new curricula determined the treatment of the "perspective drawing of spatial objects" in the 10th and "basic conceptions of descriptive geometry" in the 12th and 13th grade. Linear drawing ("depiction of different objects from different perspectives" and "introduction to descriptive geometry, shadow constructions and perspective") could be studied as an optional subject from 9th–13th grade.[27] In Bork's mathematics textbooks *Mathematische Hauptsätze* (Bork and Max Nath 1903, 1904), the notions of different projection methods are explained and perpendicular projections onto one plane are treated. Yet, in addition, a book just about descriptive geometry, claiming to deal with its "basics", was published in 1903 by Wilhelm Gercken 1903.[28] It deals with oblique and orthogonal parallel projections, central projections and shadow construction in this order which, according to the author, results from the increasing difficulty of the different projection methods. In his preface, the author defined descriptive geometry as the "depiction of solid objects with geometric rules through projections" and distinguished between visual methods (central projection) and projection methods for technical purposes (parallel projection). Another book on the theory of projections 1903 by Müller and Presler deals with oblique parallel projections (part I) and orthogonal parallel projections including central projections (part II). This schoolbook is coined by an intensive treatment of applications in other subjects (physics, astronomy, geography, etc.). Because in Southern Germany, Austrian books were widely used, too (cf. Zühlke 1911, p. 4), they can be considered

---

[27] At *Oberrealschulen* in Baden, Bavaria and Hessen, for example, descriptive geometry was taught as an independent subject, while at *Realgymnasien* it was integrated within the mathematics curricula. In Württemberg, descriptive geometry was an independent subject in both school types.

[28] In his survey on the contents of mathematics education in Northern Germany based on schoolbooks, Lietzmann considered this book and the following (Müller and Presler) in the context of descriptive geometry (Lietzmann 1909).

in this context. In a book by Suppantschitsch (1910), the order of projection methods is arranged by their vividness. At first, plane and elevation are introduced and "disdained" for the lack of presentiveness since the task of descriptive geometry was supposed to enable students to read and understand representations like these. So, oblique parallel projections including shadow constructions are treated before "fundamentals of central projections" being denoted the most visual part of descriptive geometry. From this small sample of schoolbooks for *Realgymnasien* and *Oberrealschulen* in Northern and Southern states, a certain tendency of competence orientation is revealed, either by arranging the contents in a way that the students' visual perception is trained or by making it applicable in other subjects.

Even though, traditional *Gymnasien* were actually not considered adequate preparatory schools for further technical institutions, there were times in the nineteenth century when higher technical institutions asked prior education at a *Gymnasium* as entrance qualification, in the beginning, if a student wanted to proceed in civil service, later because of the increased status of *Technische Hochschulen*. Not surprisingly, no mentioning of any contents connected with descriptive geometry can be found in any of the curricula from 1837, 1856 and 1882, neither for mathematics nor for drawing. Consequently, no contents of descriptive geometry are found in the editions of (Mehler 1859/1872/1885). Interestingly, in the preface of a geometry book (Gruber 1854) for *Gymnasien* in Baden, it is stated that figures from the stereometry part should be drawn in plane and elevation, while the topic itself is not treated. Another geometry book (Schlömilch 1855) from Sachsen made use of descriptive geometry to visualise and facilitate analytic geometry by introducing coordinates as the projections of points onto three planes. The curricula of 1901 constituted a landmark for descriptive geometry at Prussian *Gymnasien*: in the last 2 years, "instructions in perspective drawing of spatial object" was integrated within the mathematics lesson. In a geometry book (Henrici and Treutlein 1901), parallel and central projections are introduced as affine or perspective projections, respectively, and a chapter on the "depiction of spatial objects in the plane" is added in the appendix. Another Austrian geometry book (Suppantschitsch 1910a) contains plane and elevation as "support" at the beginning of the stereometry chapter and deals with oblique and orthogonal parallel projections. In these two examples, as well as in the earlier geometry books from 1854 and 1855 mentioned before, descriptive geometry was used as a "(drawing) tool" within the field of geometry as a visualisation of or support for either analytic geometry or stereometry, or mappings. Seemingly, at *Gymnasien*, too, descriptive geometry assumed the role of an auxiliary science.

6 Unfortunate Conditions for Descriptive Geometry

From the beginning, descriptive geometry was perceived as the science that provides two-dimensional representations of three-dimensional objects in "Germany". In this wider sense, it had been cultivated and innovated at technical institutions, where it continuously presumed an important role for the education of engineers. Mathematical innovations concerned different projection methods, including especially central projection and axonometric representations. This wider German understanding of descriptive geometry also arrived in secondary education, where different projection methods seem to have been taught—at the beginning of the twentieth century even at *Gymnasien*. Through the wider perception, as the science that deals with planar representations of solid objects, consequently its area of application finally also became wider. Despite its original association with the technical sciences, descriptive geometry had not merely been perceived as "mathematically valid drawing techniques" for technical purposes, but rather as a general means of representation. Within secondary education, for example, its function was more global, e.g. for the training of the visual perception or the application in other subjects. However, the actual problem was that the possible establishment of descriptive geometry as a mathematical discipline in Germany fell into a time when technical subjects or sciences were not acknowledged within education, be it secondary or tertiary, and when the relevant technical or vocational institutions struggled for acknowledgement and emancipation. As in France, descriptive geometry developed together with technical education institutions, i.e. polytechnic schools (cf. Wiener 1884, p. 35). However, it took the entire nineteenth century for the latter to emancipate in Germany.

References

Böttcher, Karl-Heinz, Bertram Maurer, and Klaus Wendel. 2008. *Stuttgarter Mathematiker: Geschichte der Mathematik an der Universität Stuttgart von 1829 bis 1945 in Biographien.* Stuttgart: Universitätsarchiv.

Bork, Heinrich, and Max Nath. 1903. *Mathematische Hauptsätze: Ausgabe für Realgymansien und Oberrelaschulen, 1. Teil: Prensum der Unterstufe (bis zur Untersekunda einschl.)* (4th ed.). Leipzig: Verlag der Dürr'schen Buchhandlung.

———. 1904. *Mathematische Hauptsätze: Ausgabe für Realgymansien und Oberrelaschulen, 2. Teil: Pensum der Oberstufe (bis zur Reifeprüfung)* (3rd ed.). Leipzig: Verlag der Dürr'schen Buchhandlung.

Brennecke, Wilhelm H. 1869. *Einführung in das Studium der darstellenden Geometrie (Ergänzung zu jedem Lehrbuch der elementaren Stereometrie).* Berlin: Verlag von Th. Chr. Fr. Enslin.

Brüll, Adolf. 1903. Creizenach, Michael. In *Allgemeine Deutsche Biographie* 47: 546–549. https://www.deutsche-biographie.de/gnd11775160X.html#adbcontent. Accessed 20 Mar 2017.

Butz, Wilhelm. 1870. *Anfangsgründe der darstellenden Geometrie, der Axonometrie, der Linear-Perspective und der Schattenconstruction für den Schul- und Selbst-Unterricht.* Essen: Bädeker.

Cramer, Hans. 1910. Der mathematische Unterricht an den höheren Schulen nach Organisation, Lehrstoff und Lehrverfahren und die Ausbildung der Lehramtskandidaten im Grossherzogtum Baden. In *Abhandlungen über den mathematischen Unterricht in Deutschland veranlasst durch die Internationale Mathematische Unterrichtskommission Band II, Heft 4*, ed. Felix Klein. Leipzig/Berlin: B.G. Teubner.

Creizenach, Michael. 1821. *Anfangsgründe der darstellenden Geometrie oder der Projektionslehre für Schulen.* Mainz: Florian Kupferberg.

Geck, Erwin. 1910. Der mathematische Unterricht an den höheren Schulen nach Organisation, Lehrstoff und Lehrverfahren und die Ausbildung der Lehramtskandidaten im Königreich Württemberg. In *Abhandlungen über den mathematischen Unterricht in Deutschland veranlasst durch die Internationale Mathematische Unterrichtskommission Band II, Heft 3*, ed. Felix Klein. Leipzig/Berlin: B.G. Teubner.

Gercken, Wilhelm. 1903. *Grundzüge der darstellenden Geometrie für die oberen Klassen höherer Lehranstalten*. Leipzig: Verlag der Dürr'schen Buchhandlung.

Gruber, Karl. 1854. *Der Unterricht in der Planimetrie, Stereometrie und ebenen Trigonometrie zum Gebrauche an Gymnasien und höheren Bürgerschulen*. Karlsruhe: Hofbuchhandlung Braun.

Gugler, Bernhard. 1841. *Lehrbuch der descriptiven Geometrie*. Nürnberg: J. L. Schrag.

Hashagen, Ulf. 2003. *Walther von Dyck (1856–1934): Mathematik, Technik und Wissenschaftsorganisation an der TH München*. Stuttgart: Franz Steiner Verlag.

Henrici, Julius, and Peter Treutlein. 1901. *Lehrbuch der Elementar-Geometrie, 3. Teil: Die Gebilde des körperlichen Raumes. Abbildung von einer Ebene auf eine zweite. (Kegelschnitte.)* (2nd ed.). Leipzig: B.G. Teubner.

Institut für Mathematik und Informatik der Universität Greifswald. 2011. Mathematik und Kunst: Karl Wilhelm Pohlke. http://stubber.math-inf.uni-greifswald.de/mathematik+kunst/kuenstler_pohlke.html. Accessed 10 June 2016.

Jeismann, Karl-Ernst, and Peter Lundgreen, eds. 1987. *Handbuch der deutschen Bildungsgeschichte, Band III 1800–1870: Von der Neuordnung Deutschlands bis zur Gründung des Deutschen Reiches*. München: C.H. Beck.

Katzenstein, Louis. 1896. Weinbrenner, Friedrich. In *Allgemeine Deutsche Biographie* 41: 500–502. http://daten.digitale-sammlungen.de/~db/bsb00008399/images/index.html?id=00008399&groesser=&fip=eayaxsqrsxdsydenxdsydeayayztseayafsdryzts&no=23&seite=503. Accessed 20 Mar 2017.

Klein, Felix, and Rudolf Schimmack. 1907. *Vorträge über den mathematischen Unterricht an den höheren Schulen, Teil 1: Von der Organisation des mathematischen Unterrichts*. Leipzig: B.G. Teubner.

Küpper, C. 1888. Der Satz von Pohlke. In *Mathematische Annalen* 33: 474–5. https://www.deutsche-digitale-bibliothek.de/item/F5TNIEX4TG7XYQ27EVE3VME3TN7ITZC3. Accessed 10 June 2016.

Lexis, Wilhelm, ed. 1904. *Das Unterrichtswesen im Deutschen Reich*, IV. Band: Das Technische Unterrichtswesen, 1. Teil: Die Technischen Hochschulen. Berlin: Asher.

———. 1904a. *Das Unterrichtswesen im Deutschen Reich* IV, II. Band: Die höheren Lehranstalten und das Mädchenschulwesen. Berlin: Asher.

———. 1904b. *Das Unterrichtswesen im Deutschen Reich*, IV. Band: Das Technische Unterrichtswesen, 3. Teil: Der mittlere und niedere Fachunterricht. Berlin: Asher.

Lietzmann, Walther. 1909. Stoff und Methode im mathematischen Unterricht der norddeutschen höheren Schulen auf Grund der vorhandenen Lehrbücher. In *Abhandlungen über den mathematischen Unterricht in Deutschland veranlasst durch die Internationale Mathematische Unterrichtskommission Band I, Heft 1*, ed. Felix Klein. Leipzig/Berlin: B.G. Teubner.

Löbell, Frank. 1957. Burmester, Ludwig Ernst Hans. In *Neue Deutsche Biographie* 3: 55. http://www.deutsche-biographie.de/pnd117174742.html. Accessed 10 June 2016.

Lorey, Wilhelm. 1916. Das Studium der Mathematik an den deutschen Universitäten seit Anfang des 19. Jahrhunderts. In *Abhandlungen über den mathematischen Unterricht in Deutschland veranlasst durch die Internationale Mathematische Unterrichtskommission Band III, Heft 9*, ed. Felix Klein. Leipzig/Berlin: B.G. Teubner.

Manegold, Karl-Heinz. 1970. *Universität, Technische Hochschule und Industrie*. Berlin: Duncker & Humblot.

Mehler, Ferdinand. 1859/1872/1885. *Hauptsätze der Elementar-Mathematik zum Gebrauche an Gymnasien und Realschulen* (1st/6th/13th ed.). Berlin: Georg Reimer.

Monge, Gaspard. 1900. *Darstellende Geometrie*. Trans. R. Haussner. Leipzig: Wilhelm Engelmann.

Müller, Carl H., and Otto Presler. 1903. *Leitfaden der Projektions-Lehre, Ausgabe A: Vorzugsweise für Realgymnasien und Oberrealschulen*. Leipzig/Berlin: B.G. Teubner.

Müller, Emil. 1910. Anregungen zur Ausgestaltung des darstellend-geometrischen Unterrichts an den technischen Hochschulen und Universitäten. In *Jahresbericht der Deutschen Mathematiker-Vereinigung* 19: 19–24.

Obenrauch, Ferdinand J. 1897. *Geschichte der darstellenden und projektiven Geometrie mit besonderer Berücksichtigung ihrer Begründung in Frankreich und Deutschland und ihrer wissenschaftlichen Pflege in Österreich*. Brünn: Carl Winiker.

Ott, Karl. 1913. Die angewandte Mathematik an den deutschen mittleren Fachschulen der Maschinenindustrie. In *Abhandlungen über den mathematischen Unterricht in Deutschland veranlasst durch die Internationale Mathematische Unterrichtskommission Band IV, Heft 2*, ed. Felix Klein. Leipzig/Berlin: B.G. Teubner.

Papperitz, Erwin. 1899. *Die Mathematik an den Deutschen Technischen Hochschulen: Beitrag zur Beurteilung einer schwebenden Frage des höheren Unterrichtswesens*. Leipzig: Veit & Comp.

———. 1909. Darstellende Geometrie. In *Encyklopädie der mathematischen Wissenschaften mit Einschluss ihrer Anwendungen, 3. Band: Geometrie, 1. Teil, 1. Hälfte*, ed. Fr. Meyer and H. Morhmann, 520–601. Leipzig: B.G. Teubner.

Pohlke, Karl W. 1860. *Darstellende Geometrie, zunächst für den Gebrauch bei den Vorträgen an der Königlichen Bau-Akademie und dem Königlichen Gewerbe-Institut zu Berlin*. Berlin: Gaertner.

Renteln, Michael von. 2000. *Die Mathematiker an der Technischen Hochschule Karlsruhe (1825–1945)*. Karlsruhe: Ernst Grässer.

Scharlau, Winfried, ed. 1990. *Mathematische Institute in Deutschland 1800–1945*. Braunschweig/Wiesbaden: Vieweg.

Scherling, Christian. 1870. *Vorschule und Anfangsgründe der descriptiven Geometrie. Ein Cursus für die Secunda einer Realschule erster Ordnung*. Hannover: Hahn'sche Hofbuchhandlung.

Schlömilch, Oskar. 1855. *Lehrbuch der analytischen Geometrie, 2. Teil: Analytische Geometrie des Raumes*. Leipzig: B.G. Teubner.

Schnell, Heinrich. 1910. Der mathematische Unterricht an den höheren Schulen nach Organisation, Lehrstoff und Lehrverfahren und die Ausbildung der Lehramtskandidaten im Grossherzogtum Hessen. In *Abhandlungen über den mathematischen Unterricht in Deutschland veranlasst durch die Internationale Mathematische Unterrichtskommission Band II, Heft 5*, ed. Felix Klein. Leipzig/Berlin: B.G. Teubner.

Schoedler, Friedrich. 1847. *Die höheren technischen Schulen nach ihrer Idee und Bedeutung*. Braunschweig: Vieweg.

Schreiber, Guido. 1828. *Lehrbuch der Darstellenden Geometrie nach Monge's Géométrie descriptive, Erster Theil*. Karlsruhe/Freiburg: Herder'sche Kunst- und Buchhandlung.

———. 1829. *Lehrbuch der Darstellenden Geometrie nach Monge's Géométrie descriptive, Zweiter Theil*. Karlsruhe/Freiburg: Herder'sche Kunst- und Buchhandlung.

Schubring, Gert. 1989. Pure and applied mathematics in divergent institutional settings in Germany: The role and impact of Felix Klein. In *The History of Modern Mathematics*, ed. David E. Rowe, and John McCleary, 171–220. Boston: Academic.

———. 1990. Zur strukturellen Entwicklung der Mathematik an den deutschen Hochschulen 1800–1945. In *Mathematische Institute in Deutschland 1800–1945*, ed. Winfried Scharlau, 264–276. Braunschweig/Wiesbaden: Vieweg.

———. 2012. Antagonims between German states regarding the status of mathematics teaching during the 19th century: processes of reconciling them. In *ZDM Mathematics Education* 44: 525–535.

———. 2014. Mathematics Education in Germany (Modern Times). In *Handbook on the history of mathematics education*, ed. Alexander Karp and Gert Schubring, 241–255. New York: Springer.

Stäckel, Paul. 1910. Einführung. In *Abhandlungen über den mathematischen Unterricht in Deutschland veranlasst durch die Internationale Mathematische Unterrichtskommission Band IV, Heft 2*, ed. Felix Klein. Leipzig/Berlin: B.G. Teubner.

———. 1914. Die mathematische Ausbildung der Ingenieure in den verschiedenen Ländern: Gesamtbericht der Internationalen Mathematischen Unterrichtskommission erstattet auf der Zusammenkunft in Paris am 3. April 1914. *Zeitschrift des Vereins Deutscher Ingenieure* 58: 149–169.

———. 1915. Die mathematische Ausbildung der Architekten, Chemiker und Ingenieure an den deutschen Technischen Hochschulen. In *Abhandlungen über den mathematischen Unterricht in Deutschland veranlasst durch die Internationale Mathematische Unterrichtskommission Band IV, Heft 9*, ed. Felix Klein. Leipzig/Berlin: B.G. Teubner.

Suppantschitsch, Richard. 1910. *Leitfaden der darstellenden Geometrie für die V. und VI. Klasse der Realgymnasien*. Wien: Tempsky.

———. 1910a. *Lehrbuch der Geometrie für Gymnasien und Realgymnasien Mittelstufe*. Wien: Tempsky.

Weinbrenner, Friedrich. 1810. *Architektonisches Lehrbuch*, 1. Theil. Tübingen: Joh. Georg Cottaische Buchhandlung.

Wiener, Christian. 1884. *Lehrbuch der darstellenden Geometrie*, 1. Band. Leipzig: B.G. Teubner.

Wiener, Hermann. 1879. Wiener, Christian. *Allgemeine Deutsche Biographie* 42: 790–792. http://www.deutsche-biographie.de/sfz31203.html. Accessed 10 June 2016.

Wußing, Hans. 1975. Die Mathematik der Aufklärungszeit: Überblick. In *Biographien bedeutender Mathematiker: Eine Sammlung von Biographien*, ed. Hans Wußing, and Wolfgang Arnold, 242–247. Berlin: Volk und Wissen.

Zühlke, Paul. 1910. Mathematiker und Zeichenlehrer im Linearzeichenunterricht der preußischen Realanstalten. In *Berichte und Mitteilungen durch die Internationale Mathematische Unterrichtskommission IV*, 51–54. Leipzig/Berlin: B.G. Teubner.

———. 1911. Der Unterricht im Linearzeichnen und in der darstellenden Geometrie an den deutschen Realanstalten. In *Abhandlungen über den mathematischen Unterricht in Deutschland veranlasst durch die Internationale Mathematische Unterrichtskommission Band III, Heft 3*, ed. Felix Klein. Leipzig/Berlin: B.G. Teubner.

Chapter 10
Otto Wilhelm Fiedler and the Synthesis of Projective and Descriptive Geometry

Klaus Volkert

Abstract We study the contributions by Wilhelm Fiedler (1832–1907) to the development of descriptive geometry in the German speaking countries, in particular his idea to provide a synthesis of descriptive and projective geometry.

Keywords Wilhelm Fiedler · Descriptive geometry · Projective geometry · Polytechnic at Zürich

1 Introduction

Around 1870, several German authors expressed the need to integrate descriptive geometry into "new geometry" (or "geometry of position", as it was often called in the tradition of von Staudt, that is, projective geometry from our modern point of view) (cf. Schlesinger 1870 and Scherling 1870).

This synthesis of descriptive and projective geometry was later considered as a genuine German way (cf. Wiener 1884, p. 36).[1] Of course, this is not completely true; in particular, Luigi Cremona (1830–1903) in Italy had similar ideas (Menghini, Chap. 4, this volume). The most prominent author in the German speaking countries, who expressed this idea, was Otto Wilhelm Fiedler. In the following, we will discuss some of his ideas and give some information on him.

Before doing this, let us make an important remark. As we have seen above in Chap. 9, descriptive geometry in the Monge'an style was the brand mark of technical education in Germany. Technical education as a whole was considered for a long time as a minor part of German education. So, it had to fight for its

---
[1] See also the citation by Paul Stäckel at the end of this paper.

K. Volkert (✉)
University of Wuppertal, Wuppertal, Germany
e-mail: klaus.volkert@math.uni-wuppertal.de

© Springer Nature Switzerland AG 2019
É. Barbin et al. (eds.), *Descriptive Geometry, The Spread of a Polytechnic Art*,
International Studies in the History of Mathematics and its Teaching,
https://doi.org/10.1007/978-3-030-14808-9_10

emancipation. One way of upgrading its status was to link it to highly estimated fields of traditional science. I think that this was an important motivation for the proposition to link descriptive to projective geometry: the latter had a high position in mathematics during the second half of the nineteenth century. It was then a very active field of research with several prominent researchers. Other reasons, in particular of pedagogical nature, will be discussed below.

2 Fiedler's Life and Opus

Let me start with some facts about Fiedler's life: Otto Wilhelm Fiedler was born in Chemnitz (Saxony) in 1832; his father was a shoemaker there. Chemnitz was one of the centres of early industrialization in Germany; it was sometimes called "Manchester of the East". Young Fiedler attended the *höhere Gewerbeschule* (higher trading school) in Chemnitz and then the *Bergakademie* in Freiberg; there, he was influenced by Julius Weisbach (1806–1872), an engineer and mathematician who taught technical sciences.[2] So, Fiedler never had the opportunity to enter regular higher education institutions (*Gymnasium*, university); he was completely educated in the parallel world of technical education—that meant, among others, no Latin, no Greek, but descriptive geometry/technical drawing and mechanics. In 1852, he became teacher of mathematics at the *Werkmeisterschule* in Freiberg, and 1853 at the *Gewerbeschule* in Chemnitz. In 1859, Fiedler got a Ph.D. by the university of Leipzig with his dissertation "*Die Zentralprojektion als geometrische Wissenschaft*" (central projection as a geometric science). Möbius' report[3] on this paper is not very enthusiastic—in particular, he criticized that it could be shortened to the half without a real loss of content. But he attested a broad knowledge of the methods of new and descriptive geometry to the author (Fig. 10.1).[4]

From 1859 on, Fiedler was occupied by producing German editions of some of the works of George Salmon (1819–1904). This is the reason why Fiedler is today remembered mainly by the combination "Fiedler-Salmon".[5] We do not consider this

---

[2]Weisbach is well known for his work in geodesy.

[3]August Ferdinand Möbius was the leading mathematician of the university at Leipzig at that period. It was quite natural for Fiedler to present his dissertation at Leipzig because he worked and lived in Saxony and Leipzig was the only university in the kingdom of Saxony. As far as I know there was no relation before the dissertation between Fiedler and Mübius. But afterwards, Mübius sent all his papers to Fiedler—as is reported proudly by the latter (cf. Fiedler 1905, p. 494).

[4]The report is in the university archives at Leipzig (UAL Phil. Fak. Prau. 370 Bl. α). I thank Mrs. Letzel at Leipzig for providing me a copy of Möbius' report. As is shown by a letter to the Dean of the Faculty of Philosophy the graduation of Fiedler was a delicate question because of his lack of a classic education (UAL Phil. Fak. Prau. 370, Bl. 1). Fiedler's dissertation was published in 1860 as a "*Schulprogramm*" by the *Gewerbeschule* in Chemnitz.

[5]There were four of them: *Analytische Geometrie der Kegelschnitte* (1860), *Die Elemente der neueren Geometrie und die Algebra der binären Formen* (1862), *Vorlesungen zur Einführung in*

Fig. 10.1 The printed version of Fiedler's dissertation

important aspect of his work here. I only want to state that this was another aspect of Fiedler's great synthesis of geometry: the integration of algebraic methods into it. Towards the end of his life, Fiedler became also interested in Salmon's ideas on theology. By the way, Fiedler wrote a paper on mythology (under the pseudonym

---

die Algebra der linearen Transformationen (1863) and *Analytische Geometrie des Raumes* (in two volumes: 1863, 1865). Many of those books went through several editions.

Dr. H.F. Willers); it was published like his dissertation by the *Gewerbeschule* in Chemnitz in 1860. Fiedler was also very gifted in artistic drawing (Fig. 10.2).

In 1864, Fiedler was called to the (German and Czech speaking) Polytechnic at Prague as a professor for descriptive geometry; in 1867, he moved to the *ETH* at Zürich as a professor for descriptive and geometry of position (projective geometry in modern terms). This denomination was quite unusual at that time; it is well known that Carl Culmann (1821–1881), a leading engineer and an influential professor at the *ETH*, promoted Fiedler's nomination because he wanted to introduce projective geometry into the teaching program for engineers and

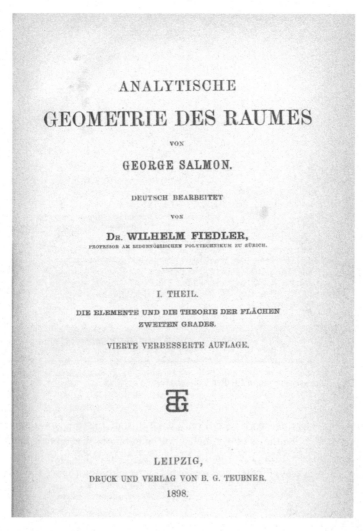

Fig. 10.2 Title page of the German edition of Salmon's "Analytic geometry of three dimensions". One of the books which made Fiedler well known

Fig. 10.3 Otto Wilhelm Fiedler (Grossmann 1913, p. 615)

architects.⁶ After Elwin Bruno Christoffel had left Zürich, Fiedler became head of the VI. *Abteilung* (department) of the *ETH* (1868–1881), dedicated to the general education of future engineers and to the training of future teachers of sciences and mathematics and of the *Mathematische Seminar*. In the German speaking countries, this was a rather unusual situation because the training of future teacher was usually the responsibility of the universities during that period.⁷

Besides Fiedler, there were always one or two chairs in pure mathematics at the ETH; they were members of the VI. department. During the time Fiedler was in service, there were among others Elwin Christoffel (1862–1869), Hermann A. Schwarz (1869–1875), Heinrich Weber (1869–1875), Ferdinand G. Frobenius (1875–1892), Friedrich Schottky (1882–1892), Adolf Hurwitz (1892–1919) and Hermann Minkowski (1896–1902). Fiedler retired in 1907 and died in Zürich in 1912. Students of Fiedler were Giuseppe Veronese (until 1876, he was due to a grant member of the Allgemeine Abteilung), Marcel Grossmann and Emil Weyr; Hendrik de Vries and Grossmann were assistants to Fiedler; Grossmann became his successor. The fact that there were two assistants to the chair of descriptive geometry marks also a difference to other mathematicians who—at that period—had no assistants at all (Fig. 10.3).⁸

---

⁶Culmann is well known for the invention of graphic statics (cf. Scholz 1989 or Maurer 1998); he himself gave lecture courses on projective geometry before the arrival of Fiedler. Th. Reye was a *Privatdozent* at the *ETH* lecturing on projective geometry when Fielder was called to it. It seems that Reye was disappointed by Fiedler's nomination. By the way, Einstein attended lectures by Fiedler on descriptive and on projective geometry. Thanks to T. Sauer (Mainz) who informed me about this.

⁷There were two exceptions: Dresden (1862) and Munich (1868).

⁸Seemingly, Felix Klein was the first mathematician who got an assistant (Walther von Dyck at Leipzig [1880]).

The germ of Fiedler's idea to integrate descriptive geometry into projective geometry can be found as early as in his dissertation, finished in 1858. Fiedler himself reports that he disliked the teaching of descriptive geometry at the *Gewerbeschule* he had encountered, because it was so "empiric" (cf. Fiedler 1905, p. 493).[9] Fiedler began to teach descriptive geometry in 1857 because he had to replace a colleague who was fallen ill. On several occasions, Fiedler expressed his feeling that the Monge'an heritage was conserved in France in a way which was too conservative: a reason why France lost his leading position in mathematics around the middle of the nineteenth century. In Fiedler's eyes it was an important task to enhance descriptive geometry and not to petrify it.[10]

The central goal of Fiedler dissertation was to "provide a systematic exposition of the method of perspective drawing" (cf. Fiedler 1860, p. 1); following Fiedler, this method is to be considered as genuine geometric and not only as a tool for artists. Because "descriptive geometry has to convey the general and complete methods, which serve to treat spatial objects in a graphical way" (cf. ibid., p. 1), perspective should be taught within the subject of descriptive geometry Fiedler criticized the tradition to locate the perspective in an appendix and the emphasis being laid onto the method of parallel projection. He refers to the traditional method of constructing a perspective image by using ground plan and elevation (the method of intersection described by Leon Battista Alberti (1404–1472) and others). At the end of his paper, Fiedler gives a short conclusion in stating that he is confident for having given "a complete treatment of parallel and central projection at the same time" (ibid., p. 39). He underlines "that the method of central projection is much more prolific from the point of geometry than that of parallel projection" (cf. ibid., p. 39). The latter is a special case of the further. This is true if one looks at the situation from the point of view of projective geometry: here, one may say that a parallel projection is nothing but a central projection with its centre at infinity. This idea goes back to Johann Heinrich Lambert (1728–1777) and his *Freye Perspective* (1759, § 252), a work often mentioned by Fiedler. Lambert is besides Monge the second father of descriptive geometry *á la* Fiedler.

Some years later, Fiedler formulated his basic ideas in his paper "On the system in descriptive geometry" (Fiedler 1863).[11] Note the term "system", which is one of Fiedler's favourites—together with "organic", "fundamental" and "natural". These terms reflect the influence of Jakob Steiner (1796–1863), in particular of his first great book (1832). We will come back to this.

---

[9] In difference to that, Weisbach's teaching at Freiberg was more inspiring: here, Fiedler learned about axonometry (cf. ibid.).

[10] Cf. Fiedler (1876, p. 65). It is not the purpose of this contribution to discuss whether Fiedler's view was right or wrong.

[11] The journal "Zeitschrift für Mathematik und Physik" in which Fiedler published his article was edited by Oscar Schlömilch (1823–1901) an important person in the academic world of Saxony and professor at the Polytechnic at Dresden (1849). He also worked for the ministry of education of Saxony.

Fiedler stated:

> The system of descriptive geometry must include the constructive methods by central collineation or by solid homographic transformation. It is sure that it will be able to take important advantages of them in all its parts. This is well known to certain persons since a long time, and it cannot be avoided that the systematic introduction of the cited theories of the new geometry[12] into descriptive geometry will be realized because they belong to the latter in a natural way. (cf. Fiedler 1863, p. 445)

In this short paper, which is cited by Josef Schlesinger in the preface of his book, we find some indications to certain ideas which were elaborated in detail by Fiedler later. Fiedler himself called this paper "the program of my activities as a professor which began in 1864" (Fiedler 1905, p. 495). The program was detailed in a long paper presented to the Academy of Sciences in Vienna in 1867[13] under the title "The methodology of descriptive geometry as an introduction to the geometry of position". The detailed elaboration of these ideas was reserved to Fiedler's book "Descriptive geometry in an organic connection with the geometry of position" (1871). Fiedler described his intentions concerning the reform of the teaching of descriptive geometry also in an article (1877) which was also translated into Italian; a retrospective commentary on those questions is provided in Fiedler (1905).

Fiedler positioned himself in the broad movement proposing a new way to teach geometry—that is a profound reform. In his eyes, the most important deficit of the traditional way to teach geometry was that the students do not get "the impression of a well-ordered whole" (cf. Fiedler 1877, p. 82). They learn a lot of details, which are easily forgotten, because they stay isolated; their cultural value (*Bildungswert*) is minor. His conclusion was

> All geometry must become descriptive, it must proceed by projecting, in order to become projective .... (cf. ibid., p. 92)[14]

In a long historical excursion, Fiedler explained that in his eyes the arrival of descriptive geometry was decisive for the development of projective geometry, in particular for "Poncelet's renewal of the general method of perspective" (cf. ibid., p. 86).

Following Fiedler, projective geometry provides the foundation of descriptive geometry; it explains all that is done in the latter in providing the true reasons. In a letter to Fiedler (9.2.1864), Alfred Clebsch (1833–1872) stated:

---

[12] From our modern point of view this is more or less the same as projective geometry. This term became only popular in the 1870s (cf. Voelke 2010, pp. 239–260). Another term was "geometry of position" (*Geometrie der Lage*) used by Christian von Staudt and Theodor Reye. Strangely enough, the German term for "new geometry" was "neuere Geometrie". It was used by Moritz Pasch in the title of his famous book (1882).

[13] This paper was written during Fiedler's stay at Prague. So, it was natural to present it to the Academy at Vienna because Prague and Bohemia were parts of the k. and k. monarchy.

[14] The original is: "...die ganze Geometrie muss darstellend werden, muss projicirend verfahren, um projectivisch zu sein...". In this paper, Fiedler often used the term "projective" (projectivisch).

Your engagement in descriptive geometry will surely be very fruitful. The mistake, which is always made, of exclusively treating specific projections and not giving the true origin, out of which everything coalesces, has to revenge itself delicately in many respects—(cf. ibid., p. 92)[15] (translated by the author).

In providing such a framework, the mere practice is transformed into a part of theory. This is in full coherence with Fiedler's idea how to teach. On several occasions he criticized the dogmatic style—proposing nothing but facts without motivation and justification—and voting for a teaching really explaining the facts ("genetic teaching").

Moreover, descriptive geometry was, in Fiedler's eyes, the best way to enter geometry, in particular projective geometry, because its starting point—central projection—is close to our visual experiences. And, of course, it is very useful.

Let me just summarize some of Fiedler's sources of inspiration which become rather obvious in this paper:

1. Very important was Steiner's idea of geometry as an organic whole. At its basis, there are some fundamental entities and some fundamental principles which allow to get new entities from the old ones, and so on. It is very important to understand that this program is conceived as an alternative to traditional axiomatics.
2. Fiedler was influenced by Möbius' idea of a hierarchical ordered system of transformations[16] which structures geometry.
3. Projective geometry is conceived in the traditional way, that is, as the extension of Euclidean space by elements at infinity. It has a metric structure. Fiedler was not at all interested in von Staudt's idea to get an autonomous projective (and non-metrical) geometry.
4. Plane and solid geometry should not be separated.

Let me briefly indicate how Fiedler proceeded in his book on descriptive geometry (1871). His starting point was the problem of projection. Descriptive geometry is the practice of projecting from a centre (which, of course, may be at infinity):

– A plane onto a plane
– The space (or a part of it) onto a plane
– A part of space onto another part of space.[17]

---

[15]The original quotation is: "Ihre Beschäftigung mit der darstellenden Geometrie wird gewiss sehr fruchtbar werden. Der Fehler, der immer gemacht wird, indem man spezielle Projektionsarten ausschließlich behandelt und die wahre Quelle nicht angibt, aus der alles zusammenfließt, muss sich in vielem empfindlich rächen" (cited in Scholz 1989, p. 303). Scholz also cites a letter by Cremona to Fiedler who is very enthusiastic concerning Fiedler's way.

[16]Möbius' term was *Verwandtschaften*; examples are (in modern terms) congruence, similarity, affinity, etc. Of course, this can be seen as an anticipation of the *"Erlangen Programme"*.

[17]In this context, Fiedler spoke of a "model"; he refers to the practice of producing a scenery in theatre. Another term he uses is "Reliefperspective"; this term was also used by J. V. Poncelet.

So, in short, descriptive geometry is considered as the part of geometry studying projections. Following Fiedler, it is quite natural to start with central projection, because it is the way in which seeing is modelled.

From the scratch, Fiedler used the terminology of projective geometry in speaking of straight lines as ranges of points and describing the projection by pencils of lines and pencils of planes. These are the fundamental entities of first rank, the plane as a field of points, bundles of rays and bundles of planes are those of second rank and the space is of third rank. The fundamental operations are cutting (forming sections) and projecting. All this is taken from Steiner. So, Fiedler's readers—and in particular his students—have to learn a rather complicated terminology to describe rather simple situations. One can easily imagine that his students—often they were future engineers—were frustrated because they wanted to learn how to draw machines and not to express themselves in such an esoteric way.

A very important tool is the cross ratio. Because of its invariance, it is a characteristic of all kinds of projections.

I illustrate Fiedler's style with an example. It is taken from the beginning of his book where he is studying the central projection of a plane onto a plane. This is done by transporting everything into one plane (by a so-called *Umklappung* (rabattement)). The situation is that of Fig. 10.4.

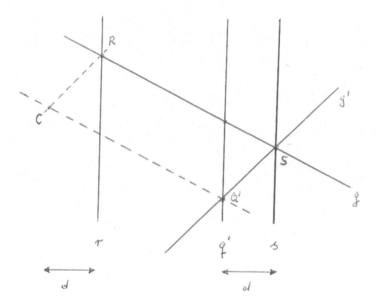

Fig. 10.4 C is the centre of the projection, s is the ground line, q' is the horizon, r is the line of vanishing points. Points, lines, etc., with an apostrophe belong to the image. The image g' of the line g is constructed in this picture

This is a central collineation.[18] Fiedler is now interested in the question: if there is given a segment of a certain length on g, is it possible to find a segment on g' with the same length? (cf. Fiedler 1875, pp. 36–38). His solution—the answer is "yes, we can"—is the result of a lengthy calculation.

The intersection of g with r is denoted by R, the intersection with s by S. S is also a point on g'. We determine the image Q' of the point at infinity Q of the straight line g. This is the point of intersection of the parallel to g through C with q'. The points C, Q', S and R are the vertices of a parallelogram (by their construction).

Now, we look at two points A and B on g and their images are A' and B' on g'. We want to relate the length of the segment AB to that of the segment $A'B'$.

The triangles ARC and $CQ'A'$ are similar. Therefore, we have: $AR : RC = CQ' : Q'A'$ or $AR\,Q'A' = RC\,CQ' = SQ'\,RS$ (because of the parallelogram).

By analogy (for B and B'):

$$BR \times Q'B' = RC\,CQ' = SQ'\,RS.$$

The quantity $SQ'\,RS$ is independent of the special choice of A and B; it is the same for all points on g (or g'). Let $SQ'\,RS = k^2$.

Then we have

$$A'B' = Q'B' - Q'A' = k^2/BR - k^2/AR = k^2(1/BR - 1/AR)$$
$$= (k^2 \times AB)/(AR \times BR) \qquad (10.1)$$

and by analogy:

$$AB = (k^2\,A'B')/(A'Q'\,B'Q') \qquad (10.2)$$

We now use the fact that the cross ratio is an invariant of central collineations (and—of course—of central projections). Therefore (by (10.1)), the length of $A'B'$ depends only on the distances of the points A and B to the point R on r and the length of AB depends only (by (10.2)) on the distances of A' and B' to Q' on q? Therefore, it is easy to find segments with equal lengths.

This is a typical example for Fiedler's style. It can be characterized by the term problem solving. That was Monge's style, too. Fiedler's book proceeds from one problem to the next, a lot of work is delegated to the problem of sections. His style is narrative, that is there is no arrangement in theorems, demonstrations, lemmata, definitions and so on. As a genuine problem solver, Fiedler did not pay much attention to the methods used, that is, he was not interested in reducing the number of

---

[18] Remember that it was Möbius, who introduced the notion of collineation.

used methods or any sorts of methodological purity. In particular, he did not refrain from using metric and algebraic techniques. The result justifies the methods.[19]

Of course, there is an enormous number of drawings in the book, and there are a lot of commentaries on historical backgrounds, sources and so on (in the appendix).[20]

The way of introducing the idea of duality was also important to Fielder because it was at the origin of his ideas about descriptive and projective geometry (cf. Fiedler 1905, pp. 494–495): He understood how to introduce duality in the framework of central projection (cf. Fiedler 1875, pp. 76–80).[21]

We illustrate this again by the simplest situation, that is, the central projection of a plane onto a plane. Let P be a point of the plane under consideration. How can we find its dual, that is a straight line in the given plane? The answer is: Join P to the centre C of the projection. In C, take the plane which is orthogonal to PC. This plane will cut the given plane into a straight line, this is the dual of P. If we start with a straight line g in the given plane and if we continue like above but in reversed order, we get the dual point to g. This procedure is called "construction of the orthogonal system" by Fiedler.[22]

The results on the simple case that we have discussed here shortly are proudly summarized by Fiedler:

> In this way, the natural system of geometry is constructed out of the basic ideas [*Grundanschauungen*] and the methods of descriptive geometry In this way, the difference between plane geometry and solid geometry is suspended [*aufgehoben*]. (cf. Fiedler 1875, p. 76)

Just to give an overview on the rich content of Fiedler's book let me summarize its table of content (cf. ibid., XXVIII–XLI)[i]:

I. Part: The doctrine of methods developed through the investigation of elementary geometric figures and their combinations

  A. Central projection developed as a method of representation and by its general law
  B. Constructive theory of conic sections as projections of the circle
  C. Central collineations of spatial systems as the theory of the methods of modelling
  D. Basic principles of orthogonal parallel projection, their transformations and axonometry.

---

[19] So, Fiedler was in complete opposition to mathematicians like von Staudt (1798–1867) or Pasch (1843–1930)—perhaps even to the famous *Zeitgeist*?!

[20] In his paper of 1877 cited above, Fiedler declared himself as a follower of the *genetic method*: teaching should follow grosso modo the historical development of its subject.

[21] Of course, there are points at infinity. So, Fiedler worked—always—in the projective plane or space.

[22] The idea is the same as has been used for a long time in spherical geometry to construct the polar of a given point or the pole of a given straight line (that is a great circle).

II. Part: Constructive theory of curved lines and surfaces

    A. About curves and developable surfaces
    B. About curved surfaces in general and surfaces of second degree in particular
    C. About skew ruled surfaces
    D. About surfaces of revolution.

III. Part: Geometry of position and projective coordinates

    A. Basic facts and coordinates
    B. The parameters of figures and projectivities; products of projective forms of first degree
    C. Forms of second and third degree and the results of their combination.

Note that the reader has to wait until section D of the first part to learn something on descriptive geometry in the Monge'an style (pp. 154–210). The second part of the book begins on page 211, the third on page 495; it ends on page 728. The list of figures occupies 13 pages.

There are some hints to the fact that Fiedler has had no great success with his teaching for future engineers—as indicated above, this is not hard to understand (cf. Voss 1913, p. 100 and pp. 103–104).[23] He had some followers at polytechnic schools (for example, Guido Hauck [1845–1905] at Berlin, and Luigi Cremona in Italy) and also at German universities where the introduction of descriptive geometry was discussed for a certain period (e.g., Klein introduced courses of descriptive geometry at Leipzig). So, we may conclude that in sum Fiedler's way was a dead end. His dream of the great synthesis was overthrown by the strong tendency in modern mathematics to separate fields and to look for autonomous structures with unique methods. But this conclusion may be overhasty. To end this story, I cite Paul Stäckel in his report on the technical teaching in Germany:

> In the first half of the nineteenth century, in Germany, descriptive geometry was in the wake of France, where Monge had established this science [...]. Later, productive achievements occurred in Germany, too. Peculiar about the German perception was the endeavour to organically interweave so-called new geometry with descriptive geometry und to see in this connection its completion. [...] It is the merit of Fiedler, to have realised the importance of geometry of position for descriptive geometry at first in its full extent (cf. Stäckel 1915, pp. 133–134).[24]

---

[23] An interesting question, which I cannot answer yet, is the one about his teaching for future teachers. It is obvious that they had (and still have) different needs than future engineers.

[24] The original quotation is as follows: "In der ersten Hälfte des 19. Jahrhunderts lief in Deutschland die darstellende Geometrie im Schlepptau Frankreichs, wo Monge diese Wissenschaft begründet hatte, [...] Später kam es auch in Deutschland zu schöpferischen Leistungen. Der deutschen Auffassung eigentümlich ist das Bestreben, die sogenannte neuere Geometrie organisch in die darstellende Geometrie einzuarbeiten und in dieser Zusammenfassung deren Vollendung zu sehen. [...] Es ist das Verdienst Fiedlers, zuerst in vollem Umfange die Wichtigkeit der Geometrie der Lage für die darstellende Geometrie erkannt zu haben". See also (Wiener 1884, p. 36).

So, in short, Stäckel states that Fiedler's way was an important step in the emancipation of descriptive geometry from the origins: from a mere technique to a complete theory. But history decided to restore the original state: in the twentieth century, descriptive geometry returned to its origins as a mere tool. Consequently, it was often called technical drawing.

Endnotes

[i]Original text:

I. Theil: Die Methodenlehre, entwickelt an der Untersuchung der geometrischen Elementarformen und ihrer einfachen Verbindungen

   A. Die Centralprojection als Darstellungsmethode und nach ihren allgemeinen Gesetzen
   B. Die constructive Theorie der Kegelschnitte als Kreisprojectionen
   C. Die centrische Collineation räumlicher Systeme als Theorie der Modellierungs-Methoden
   D. Die Grundgesetze der orthogonalen Parallelprojection, ihre Transformationen und die Axonometrie

II. Theil: Die constructive Theorie der krummen Linien und Flächen

   A. Von den Curven und den developpablen Flächen
   B. Von den krummen Flächen im Allgemeinen und den Flächen zweiten Grades insbesondere
   C. Von den windschiefen Regelflächen
   D. Von den Rotationsflächen

III. Theil: Die Geometrie der Lage und die projectivischen Coordinaten

   A. Grundlagen und Coordinaten
   B. Die Parameter der Gebilde und die Projectivität; Erzeugnisse der projectivischen Gebilde erster Stufe
   C. Die projectivischen Gebilde zweiter und dritter Stufe und die Erzeugnisse ihrer Verbindung

References

Fiedler, O. Wilhelm. 1860. Die Centralprojektion als geometrische Wissenschaft. In *Programm zu der am 29., 30. Und 31. März 1860 zu haltenden Prüfung der Schüler der Königlichen Gewerbeschule, Baugewerkenschule und mechanischen Baugewerken- und Werkmeisterschule zu Chemnitz*. Leipzig: Brockhaus.

———. 1863. Ueber das System in der darstellenden Geometrie. *Zeitschrift für Mathematik und Physik* 8: 444–447.

———. 1871. *Die darstellende Geometrie in organischer Verbindung mit der Geometrie der Lage. Für Vorlesungen an technischen Hochschulen und zum Selbststudium*. Leipzig: Teubner [Second edition extended 1875; third edition in three volumes 1883–1888, fourth edition of volume 1 in 1904. Translation into Italian *"Trattato di geometria descrittiva"* (1874) by Ernesto Padova (Pisa) and Antonio Sayno (Milano)].

———. 1876. Ueber die Symmetrie. *Vierteljahresschrift der Naturforschenden Gesellschaft in Zürich* 21: 55–66.

———. 1877. Zur Reform des geometrischen Unterrichts. *Vierteljahresschrift der Naturforschenden Gesellschaft in Zürich* 22: 82–97.26

———. 1905. Meine Mitarbeit an der Reform der darstellenden Geometrie in neuerer Zeit. *Jahresbericht der Deutschen Mathematiker-Vereinigung* 14: 493–503.

Grossmann, Marcel. 1913. Prof. Dr. Otto Wilhelm Fiedler (1832–1912). *Vierteljahresschrift der Naturforschenden Gesellschaft in Zürich* 57: 614–618.

Maurer, Bertram. 1998. *Karl Culmann und die Graphische Statik*. Berlin/Diepholz/Stuttgart: VGN.

Scherling, Christian. 1870. *Vorschule und Anfangsgründe der Descriptiven Geometrie. Ein Cursus für die Secunda einer Realschule Erster Ordnung*. Hannover: Hahn'sche Hofbuchhandlung.

Schlesinger, Josef. 1870. *Die Darstellende Geometrie im Sinne der Neueren Geometrie für Schulen Technischer Richtung*. Wien: Carl Gerold's Sohn.

Scholz, Erhard. 1989. *Symmetrie. Gruppe. Dualität*. Basel/Boston/Berlin: Birkhäuser.

Stäckel, Paul. 1915. Die mathematische Ausbildung der Architekten, Chemiker und Ingenieure an den deutschen Technischen Hochschulen. In *Abhandlungen über den Mathematischen Unterricht in Deutschland Veranlasst durch die Internationale Mathematische Unterrichtskommission Band IV, Heft 9*, ed. Felix Klein. Leipzig/Berlin: B.G. Teubner.

Voelke, Jean Daniel. 2010. Le développement historique du concept d'espace projectif. In *Éléments d'une Biographie de L'espace Projectif*, ed. L. Bioesmat-Martagon, 207–286. Nancy: Presses Universitaires de Nancy.

Voss, Aurel. 1913. Wilhelm Fiedler. In *Jahresbericht der Deutschen Mathematiker-Vereinigung* 22: 97–113.

Wiener, Christian. 1884. *Lehrbuch der darstellenden Geometrie*, 1. Band. Leipzig: B.G. Teubner.

Chapter 11
The Evolution of Descriptive Geometry in Austria

Hellmuth Stachel

Abstract In comparison with France, the development of descriptive geometry in Austria started with a delay of approximately 40 years and reached a first culmination in education and research in the era of Emil Müller, during the first decades of the twentieth century. With respect to education, emphasis was mostly placed on the practicability of descriptive geometry methods, and 'learning by doing' was seen as an important methodological principle. At some schools and in variable degrees, the syllabus of descriptive geometry was extended by closely related geometric subjects like kinematics, photogrammetry, nomography, or elementary differential geometry.

In view of research, during the nineteenth century, the synthetic method of reasoning dominated; descriptive geometry was seen as a counterpart to analytic geometry. Later this puristic point of view became obsolete. Descriptive geometry found its justification as a method to study three-dimensional geometry through two-dimensional views, thus providing insight into structure and metrical properties of spatial objects, processes, and principles. This is independent of the tools and still valid when computers take over computational and drawing labour.

Keywords Austro-Hungarian empire · Cyclography · Descriptive geometry · Differential geometry · Kinematics · Projective geometry · Relief perspective · Johann Hönig · Josef Krames · Erwin Kruppa · Emil Müller · Gustav Peschka · Rudolf Staudigl · Walter Wunderlich

---

H. Stachel (✉)
Vienna University of Technology, Wien, Austria
e-mail: stachel@dmg.tuwien.ac.at

© Springer Nature Switzerland AG 2019
É. Barbin et al. (eds.), *Descriptive Geometry, The Spread of a Polytechnic Art*, International Studies in the History of Mathematics and its Teaching,
https://doi.org/10.1007/978-3-030-14808-9_11

1 Introduction

It was Gaspard Monge's revolutionary merit

- to extract the geometric methods from their various applications in fields like architecture, the art of painting, stone cutting, civil and mechanical engineering,
- to put them onto a common scientific basis, and
- to combine all in a separate discipline under inclusion of related mathematical topics.

Immediately after its foundation, the new science "descriptive geometry" was implemented in the curricula of French polytechnic schools. In Austria[1] it took a couple of years until this new discipline found acceptance in the curricula of the new Austrian polytechnic schools, which were founded around 1810 in Prague, Vienna, Graz, Brünn (today Brno/Czech Republic), and Lemberg (today Lviv/Ukraine).

The delayed acceptance by these new Austrian schools was mainly caused by their founders' belief that not science but experience in practice should be the ultimate goal for their students. Consequently, during the first decades of the nineteenth century, the geometric methods necessary for engineering and architecture were still taught within traditional courses like engineering drawing or architectural drawing (Fig. 11.1). However, soon it became obvious that there is no progress in technical practice without scientific achievements. Shortcomings were also observed in education: teaching mere practice without any scientific background was not satisfying. In this sense, the original mission statement of the polytechnic schools was recognized as being too narrow.

In 1834, Johann Hönig (1810–1886) became the first to teach an optional course on descriptive geometry in Vienna, and in 1843 he became the first professor of descriptive geometry at a new chair (German: *Lehrkanzel*) that had been founded 1842 at the Polytechnicum in Vienna. In Prague first optional lectures had already been held 1830; a new chair was established in 1853. This was the beginning of a flourishing era for descriptive geometry in Austria, in education as well as in research.

Below we provide an overview on how the contents of descriptive geometry education and research have changed in Austria over the course of time. Further details can be found in Benstein (Chap. 9, this volume) and in the references therein, or in Loria (1908). For more details concerning Vienna, the reader is referred to Binder (Chap. 12, this volume).

2 Descriptive Geometry Education at University Level

In the nineteenth century, all professors in Austria were totally free in the selection of topics to teach in their lectures. Therefore, the development of descriptive geometry can be studied only on the basis of related textbooks. Below we concentrate

---

[1] In this context, *'Austria'* stands until 1918 for the Austro-Hungarian Empire, mainly for the German-speaking part, and afterwards for the country with its today's extension.

11 The Evolution of Descriptive Geometry in Austria

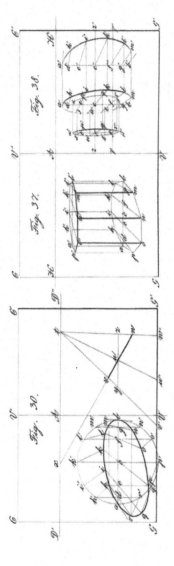

Fig. 11.1 Drawings before Monge (Rittinger 1839, supplement, Table 3)

mainly on books of authors who worked at the descriptive geometry chair in Vienna, which was more or less the leading institution in Austria. It is worth to notice that about 50% of the authors originated from Bohemia, which today is part of the Czech Republic.

There is a visible difference between the drawings displayed in the literature before and after Monge (compare Figs. 11.1 and 11.2). Figure 11.1, selected from Table 3 in the book by Rittinger (1839), shows the strategies to produce simple axonometries or perspectives. It is a purely planar process without any attempt to reveal the included geometric relations, though perspective affine and projective transformations clearly play a role in these routines. The labeling of points corresponds to the planar construction and has no relation to the spatial situation. Before Monge, the main purpose of drawings was to produce pictures which came close to a real impression of the depicted object. Therefore, even in mechanical engineering, there was a priority of axonometric views and perspectives.

Monge defined representing and analysing three-dimensional objects as the two main objectives of the science of descriptive geometry.[2] Drawings reduce spatial geometry problems to planar problems. Hence, in the period after Monge, drawings served as a tool to determine metric properties, but also, to create objects that satisfy given conditions. The latter can be traced to the usage of graphical methods in the design of military fortifications, on account of which it is reported that for some time descriptive geometry methods were even handled as a military secret. By virtue of Monge's theories, priority was given to the principal views: top view, front view, and side view.

2.1 Hönig and Staudigl

The first Austrian textbook on descriptive geometry was published by Johann Hönig (Hönig 1845). It shows a consequent use of spatial coordinates (x, y, z) and a consistent labeling of points and lines (cf. Fig. 11.2): one prime for the top view, two primes for the front view, and three primes for the side view. This tradition is still valid in Austria and Germany. The axis between the image planes, the so-called "hinge line" (German: *Rissachse*, denoted by x in Fig. 11.2), played an important role. Various constructions were based on the traces of planes and the trace points of lines (see Barbin, Chap. 2, this volume). It is surprising that Emil Müller already

---

[2]G. Monge (Monge 1811, p. 1): *"La Géométrie descriptive a deux objets:*

Le premier, de donner les méthodes pour représenter sur une feuille de dessin qui n'a que deux dimensions, savoir, longueur et largeur, tous les corps de la nature qui en ont trois, longueur, largeur et profondeur, pourvu néanmoins que ces corps puissent être définis rigoureusement.

Le second objet est de donner la maniére de reconnaître, d'aprés une description exacte, les formes des corps, et d'en déduire toutes les vérités qui résultent et de leur forme et de leurs positions respectives."

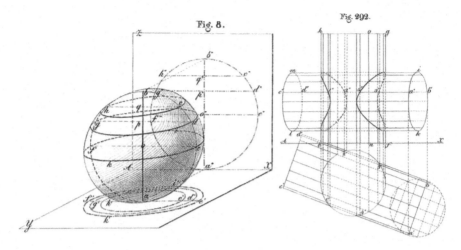

Fig. 11.2 Drawings after Monge (Hönig 1845, supplement, left: Table 1, Fig. 8, right: Table 17, Fig. 292)

recommended (Müller 1911, p. 62) to omit the hinge line for several reasons: This axis is not visible in technical drawings, and constructions based on trace points of lines or traces of planes often fail since the trace elements lie far beyond the limits of the drawing board. Moreover, a translation of a single image plane does not change the corresponding view. This meets Felix Klein's general definition, according to which geometry has to study invariants. It turned out that, from Müller onward, the hinge line was more or less banned at university level but not in high schools. This may be due to a didactical reason: This axis supports the pupils' imaginations as something that is fixed in space and pupils can rely upon.

Let us return to Hönig's book on descriptive geometry (Hönig 1845): the figures in the 26 tables, which are added as a supplement, also show various curves, such as trochoids, evolutes, and involutes, as well as surfaces of revolution and helical surfaces, defined as trajectories of points or lines under particular movements. Differential-geometric aspects are not addressed: there are no tangents drawn to the depicted curves. Moreover, the contours of displayed surfaces are often missing, and no contour points have been constructed. Ellipses are depicted without indicating their axes of symmetry. Of course, it was not until 1845 that the famous construction of vertices from given conjugate diameters was found by the Swiss mathematician David Rytz von Brugg.

This changed soon. Twenty years later, Rudolf Staudigl recommended in his textbook (Staudigl 1875) to construct not only points but also tangents to the displayed curves, in order to obtain a higher precision. In Fig. 64 of Staudigl's book already the Rytz construction is presented, but no mention is made of the name Rytz. Staudigl's textbook also contains a detailed description of how to find the contours of surfaces of revolution and even possible cusps. Since, for these surfaces, the shade lines, i.e., the boundaries of shades, are also constructed, we already recognize something which is characteristic of descriptive geometry: a clear distinction between the 'true' contour on the surface in space and its image, the 'visual' contour in the image plane. The same distinction exists between shade lines and shadow lines, terms which already date back to Monge. In his textbook, Staudigl also shows the construction of tangents to the visual contour, while tangents to the true contour are missing, though in an earlier paper (Staudigl 1843) the author already presented a pertinent result. However, he did not mention that, due to Charles Dupin (Dupin 1813, p. 48), these tangents are conjugate to the lines of sight.

2.2 Descriptive Geometry and Projective Geometry

Staudigl was the first Viennese descriptive geometer who focussed also on projective geometry. In his textbook (Staudigl 1870), he called it *'Neuere Geometrie'*, and according to the book's preface, he considered it as "geometry of position". However, his book is actually a comprehensive introduction to projective geometry, treated in a "synthetic" way, i.e., without any computation. The book starts with perspectivities in the plane and ends with projective properties of spatial cubics.

The incorporation of projective geometry into descriptive geometry was perfectly done by Wilhelm Fiedler (1832–1907), who was professor at the Polytechnicum in Prague and later in Zürich. The presentation in his textbook (Fiedler 1871) deviated in several respects from the other ones: He approached the topic in a deductive way. Beginning with central projections and projective transformations, the usual mappings, transformations, and constructive methods of descriptive geometry are developed step by step, in a top-down approach, by successive specialization. Fiedler also broke with another tradition. He was the first to include analytic representations. More details about Fiedler can be found in Volkert (Chap. 10, this volume).

In Austria, exactly in Brünn and later in Vienna, descriptive geometry and projective geometry have been bound together by Gustav A.V. Peschka (1830–1903). His comprehensive textbook (Peschka 1883–1885) consists of four volumes, which amount to a total of 2553 pages and additional tables with 1140 figures. Each single volume is dedicated to emperor Franz Joseph's son, Kronprinz Rudolph, and, as proudly stated in the book's subtitle, "in keeping with the latest scientific developments". Volume 1 presents the traditional topics of descriptive geometry and the basics of projective geometry, often structured as a series of more or less 'academic' exercises, for example: Find graphically the position of a balloon which is seen from three given points on earth under given slope angles. In Austria, it was for the first time that descriptive geometry was presented without direct engineering applications.

Volume 2 continues with a synthetic treatment of algebraic curves and surfaces of arbitrary degree. This is pure "geometry of position". Volume 3 focusses on surfaces of 2nd degree and presents their projective, metric, and differential-geometric properties. Finally, volume 4 treats algebraic ruled surfaces, surfaces of revolution and helical surfaces, and on approximately 200 pages, shades, and shadows. The last exercise deals with the curves of constant illumination on a Dupin cyclide (however, without identifying this class of surfaces).

2.3 Further Development

With Emil Müller (1861–1927), the successor of Peschka on the chair in Vienna, the focus of descriptive geometry education in Vienna returned again to engineering applications. Müller's most successful era will be presented in the fourth section.

After one century of descriptive geometry education, another characteristic of descriptive geometry became visible: Its algorithms never remain restricted to generic elements, but they always focus on particular cases like contour points or singularities of curves or surfaces, too. Till today, such a point of view is advantageous for the development of computational algorithms, because it forces to look carefully for cases, where general algorithms fail.

In the nineteenth century, additional topics were implemented into descriptive geometry courses. Contour maps of geometric elements, in particular of surfaces, combined with marked altitudes, were the basis of a topic, which in German is called *Kotierte Projektion* (topographic mapping). Here the students of civil engineering learned graphical methods to solve geometrical and topographical problems, e.g., for the design of roads. It needs to be noted that Monge was already familiar with this method, not least because of its military importance. The latter is confirmed by the fact that the first textbook on this topic (Noizet 1823) was written by a *Captaine du Génie* (captain of the corps of engineers).

On the other hand, at some universities, in particular in Brünn, the students of mechanical engineering learned some basics of kinematics, i.e., about point trajectories and instantaneous poles or axes of planar or spatial motions. In Josef Krames' textbook on descriptive geometry (Krames 1947), the word 'kinematics' even appears in the title.

3 Scientific Progress in and Around Descriptive Geometry

It is quite natural that some topics of scientific research during the early days of descriptive geometry were later included into descriptive geometry courses for engineers. This holds, for example, for the theory of "shadows and shading", and more general, for geometric lightning models, the *illumination* of surfaces including isophotic lines, i.e., curves of constant illumination (Fig. 11.3), or for the detection of brightest points.

This happened hand in hand with research on "differential geometry". Based on Dupin's results on the curvature of surfaces (Dupin 1813), the asymptotic lines and the curvature lines of particular surfaces inclusive torsal and non-torsal ruled surfaces were studied. However, in differential-geometric research, synthetic methods soon reached their limits, and with Carl Friedrich Gauß and Bernhard Riemann, the intrinsic differential geometry of surfaces came into the focus of interest. A late example of a textbook with constructive applications of differential geometry was published by Kruppa (1957). He used Dupin indicatrices, for instance, to determine tangents at singularities of curves of intersection.

Another topic intimately connected with descriptive geometry was the "relief perspective" (in Italian: *prospettiva solida*), used, e.g., on stages. Here, a perspective collineation maps a half space, which includes the depicted scene, onto a layer bounded by two parallel vertical planes (Fig. 11.3). The construction of a relief perspective of any given object is based on two theorems, which are attributed to Gournerie (1859) and Staudigl (1868), respectively. The first addresses the front view of the wanted relief, the other the top view. Both coincide with particular perspective views of the given object.

A new field of research in descriptive geometry started with *Pohlke's theorem*, which states that each axonometry is the composition of a parallel projection and a scaling. For more details about Karl Pohlke see Benstein (Chap. 9, this volume). Subsequently, new proofs of this theorem were given (cf. Müller and Kruppa 1923), and the underlying problem, i.e., the decomposition of any transformation into a product of simpler ones, could, of course, lead to various generalizations.

A topic that has its origin in descriptive and projective geometry became famous under the name "geometry of position" (German: *Geometrie der Lage*). It was developed as a counterpart to analytic geometry and focussed on geometric theorems which are independent of any metric. However, it exceeded the borders of projective geometry toward algebraic geometry, since the question of constructability with ruler and compass had no importance. The proofs of

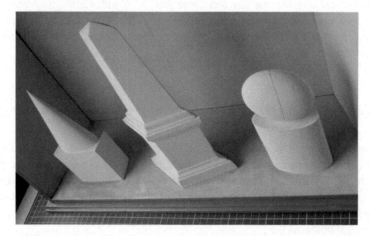

Fig. 11.3 Relief perspective: model no. 1, designed by L. Burmester (photo by the author)

algebraic statements were mainly based on results of the French mathematician Michel Chasles concerning algebraic (m, n)-correspondences. Prominent German-speaking representatives were K. G. Christian von Staudt, Fiedler, and Theodor Reye. Among Austrian's descriptive geometers, only Peschka was involved. Von Staudt's occupation with questions of algebraic geometry was probably also the origin for his work on imaginary elements. He demonstrated that even pairs of complex conjugate elements are accessible for graphic constructions, which even became standard in Müller's teacher training.

Another topic which evolved from descriptive geometry and separated soon was "photogrammetry" (today also known under the name "remote sensing"). It started with the question of how to recover metrical data from perspectives and reached high actuality with the invention of photography. Two fundamental theorems are attributed to Sebastian Finsterwalder, a mathematician and surveyor in Munich. Soon the economical importance of this field was recognized, also in view of military applications, and much effort was made in the design of mechanical devices for transforming aerial photographs into maps. In Vienna, Eduard Doležal was a pioneer in photogrammetry.

In the field of *kinematics*, we owe remarkable progress to descriptive geometers. For example, even in the present day, the contributions of Ludwig Burmester in Dresden (the same, who designed the relief perspective shown in Fig. 11.3) and Martin Disteli in Karlsruhe are well known in the scientific community. Prominent kinematicians originating from the Austrian school of descriptive geometry include Wilhelm Blaschke, Josef Krames, Hans Robert Müller, and Walter Wunderlich.

In "cyclography", descriptive geometers studied a new type of mapping, where points were no longer sent to points but to oriented circles (called "cycles") in the plane. By virtue of the fundamental theorem, points belonging to the same line in space with an inclination of 45° correspond to circles with oriented contact. This mapping, which is attributed to Fiedler (Fiedler 1882), was the beginning of the geometry of circles and spheres, and further on of conformal differential geometry. It also opened a door to pseudo-Euclidean geometry (or classical Minkowski geometry), which gives spacetime, i.e., the geometric standard model of Albert Einstein's special relativity, in four dimensions.

Finally, it must be mentioned that, at the end of the nineteenth century, German companies like M. Schilling and B.G. Teubner started to produce mathematical models of curves and surfaces, for instance, from gypsum or brass and strings (cf. von Dyck 1892).[3] The intention behind these collections was to support the student's spatial ability and intuition as well as to demonstrate the beauty of mathematics. Many models visualize results of descriptive geometry; Fig. 11.3 shows one example out of these collections.

---

[3] A collection of mathematical models is, e.g., provided at http://www.geometrie.tuwien.ac.at/modelle/.

4 The High Standard of Descriptive Geometry in Emil Müller's Era

Emil Müller was an engaged and inspiring teacher, famous also for his ingenious drawings on the blackboard. His academic career started rather late, after a 10-year career as a teacher at the *Baugewerkschule* in Königsberg i. Pr./Germany. In 1902, he was appointed professor for descriptive geometry at the *Technische Hochschule* Vienna.

In a couple of papers, he presented his ideas on education in descriptive geometry (e.g., Müller 1910b or Müller 1911). He emphasized its importance in civil and mechanical engineering, and he was convinced that spatial ability could only be trained with objects of our physical world. Therefore, in his eyes, projective geometry was of less importance. And he avoided too much abstraction or even a flavour of an axiomatic treatment.

Müller gradually published a collection of applied descriptive geometry exercises based on his ideas (Müller 1910–1926), which consists of six volumes with 60 exercises.[4] Some of the examples included in the collection were incredibly rich

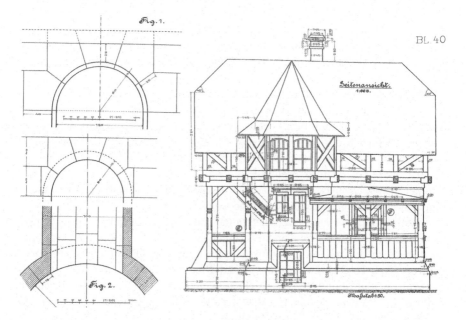

Fig. 11.4 Exercises out of E. Müller's printed collection (Müller 1910–1926, issue IV, left: sheet 34, right: sheet 40)

---

[4]Later editions, co-edited by Erwin Kruppa, were still available in the 1950s of the last century.

in detail (note Fig. 11.4 or Fig. 10.2). Certain solutions, produced by professionals, have even been printed in large format and made available as pieces of fine art (Wildt 1895, 1902, note Fig. 11.5). Parallel to this collection of examples, mainly for civil engineering and architecture, Müller's colleague in Vienna, Theodor Schmid, edited a collection of 25 examples for mechanical engineering (Schmid 1911, note Fig. 10.4).

Müller's textbook on descriptive geometry (Müller 1908, 1916) became a standard reference work in Austria, where it continued to be used as such until the sixties of the twentieth century. From the 4th edition onward, it was edited by Kruppa, who would only publish one volume; the last edition appeared in 1961. Later it was replaced with Wunderlich's pocket-books (Wunderlich 1966, 1967) and Fritz Hohenberg's textbook (Hohenberg 1956). The first one was outstanding because of its precise formulations and elegant reasoning, the latter because of the high-quality figures and its focus on various applications of geometry recovered in almost all branches of engineering.

Müller was the first one to create a particular program for high-school teachers in descriptive geometry. While in former time this training consisted only of standard lectures and exercises for civil and mechanical engineers and occasional courses on projective geometry, Müller gave lectures on the 'geometry of mappings', on 'cyclography', 'ruled surfaces', and 'constructive treatment of helical and translational surfaces'. Three of Müller's special courses were later published as lecture notes, which confirmed the successful evolution of descriptive geometry from a mere technique to a science.

I. The first volume was co-edited by Müller and Kruppa (1923) and presents the theory of linear mappings in a visual and constructive way, quite contrary to today's treatment in linear algebra. Besides, the mapping of lines onto their trace points in given planes is considered from a general synthetic point of view. A comprehensive synthetic treatment of the "kinematic mapping" is also included. It was discovered in 1911 by Josef Grünwald and Blaschke, independently of each other. Due to this intuitively introduced mapping, points in space are in one-to-one correspondence to planar displacements. This was the forerunner of a method which continues to be of great significance in robotics: curves in a 7-dimensional space correspond to one-parameter movements of the end effector. Volume 1 of Müller's lecture notes concludes with a survey on Sophus Lie's "line-sphere-transformation".

II. The courses of type two were later elaborated and edited by Krames (Müller and Krames 1929) under the title *Die Zyklographie* (Cyclography). This is an intuitive introduction into the Möbius-, Laguerre-, and Lie-geometry of oriented circles and spheres, but also into pseudo-Euclidean geometry with its indefinite metric, here under the name "C-geometry". This volume also provided extensive information on the cyclographic images of curves and surfaces and hence, e.g., caustics.

III. The third volume, again elaborated and edited by Krames, treats ruled surfaces (Müller and Krames 1931). It provides a synthetic differential geometry of

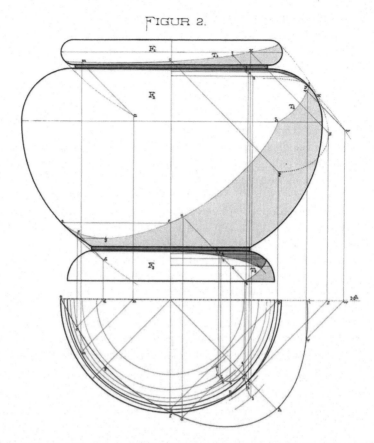

Fig. 11.5 Elaborated solution of one of E. Müller's exercises, as published in (Wildt 1895, 1902, 2nd release, sheet no. 12)

ruled surfaces, including striction curves, as well as bendings of ruled surfaces. Besides, this volume presents a unique graphics-oriented analysis of algebraic ruled surfaces of degrees 3 and 4. It demonstrates Müller's and Krames' mastery in synthetic reasoning; for Krames, descriptive geometry was "die Hohe Schule des räumlichen Denkens und der bildhaften Wiedergabe" (the high art of spatial reasoning and its graphic representation) (Krames 1947, p. 1). This book is still a storehouse for experts on computer graphics who like to produce colourful realistic pictures of spectacular ruled surfaces.

While in education Müller avoided excessive abstraction and preferred the synthetic method, his scientific publications demonstrate a mastery of analytic reasoning. This is why Kruppa in his obituary (Kruppa 1931, p. 50) characterized Müller as a geometer in the middle between the purely synthetic treatment and the analytic method.

Müller's scientific œuvre reveals him to be an expert in Grassmann's theories on multi-dimensional geometry. He authored an article (Müller 1910a) on coordinate systems in the prestigious *Enzyklopädie der Mathematischen Wissenschaften*, which outlined the state of the art of mathematics at the beginning of the twentieth century. Another subject, where Müller's scientific achievements have not lost their significance, is "relative differential geometry", where, instead of the unit sphere, an appropriate surface is used to define the normalization.

5 Conclusion

The scientific foundation of descriptive geometry by G. Monge had a tremendous impact on the education in Austrian schools and polytechnical institutes. From the mid to the nineteenth century until the end of the twentieth century descriptive geometry was, beside mathematics, mechanics, and physics, one of the basic sciences, which were taught in the first semesters of almost all technical studies. Furthermore, till today descriptive geometry is a topic in vocational high schools and selected gymnasia, and about 50% of Austrian pupils in the age of 13 or 14 years become acquainted with a light version of descriptive geometry in a subject called *Geometrisches Zeichnen* (geometric drawing).

However, a worldwide scan at the begin of the twenty-first century reveals that outside of France the name of Gaspard Monge is almost forgotten and the topic descriptive geometry is more or less unknown. There are only a few exceptions: it was in Ukraine that the national Association of Applied Geometry devoted its 1995 annual meeting to the 200th anniversary of Monge's *Géométrie descriptive*. Moreover, the Serbian Society for Geometry and Graphics continues to use the sophisticated name *moNGeometrija* for their biannual international scientific conferences.[5]

With the rise of computers, manual constructions have been replaced by CAD software. Instead of sheets with drawings, we use 3D-databases, and with 3D printers, we can produce 3D models of virtual shapes of any complexity. Nevertheless, only people with a profound knowledge of descriptive geometry are able to make extended use of CAD programs since the interface is usually based on 2D images only. The more powerful a modeling software, the higher the required geometric knowledge. Although the name "descriptive geometry" is gradually vanishing, the science is still in use (cf. Cocchiarella 2015), and parts of it are included in different fields like engineering drawing, architectural drawing, computer graphics, computer vision, virtual reality, or computer-aided design. Moreover, in a graphics-oriented world, a specific training of spatial ability is inevitable for many professions.

---

[5] Note http://www.mongeometrija.com/konferencije/mongeometrija-2016.

References

Cocchiarella, Luigi (ed.). 2015. *The Visual Language of Technique*, vol. 1: *History and Epistemology*, vol. 2: *Heritage and Expectations in Research*, vol. 3: *Heritage and Expectations in Education*. Basel: Springer International Publisher.

Dupin, Charles. 1813. *Développements de Gèomètrie*. Paris: V. Courcier.

Fiedler, O. Wilhelm. 1871. *Die darstellende Geometrie in organischer Verbindung mit der Geometrie der Lage*. B.G. Teubner, 754 pages. 2nd ed. 1875, from 3rd ed. on in 3 volumes 1883–1888, 4th ed. 1904.

———. 1882. *Cyklographie oder Construction der Aufgaben über Kreise und Kugeln, und Elementare Geometrie der Kreis- und Kugel-Systeme*. Leipzig: Teubner.

Gournerie, Jules M. de la. 1859. *Traité de la perspective linéaire*. Paris: Dalmont et Dunod.

Hohenberg, Fritz. 1956. *Konstruktive Geometrie in der Technik*. 2nd ed. 1961, 3rd ed. 1966. Wien: Springer.

Hönig, Johann. 1845. *Anleitung zum Studium der Darstellenden Geometrie*, with 26 copper plates, Wien: Carl Gerold.

Krames, Josef L. 1947. *Darstellende und Kinematische Geometrie für Maschinenbauer*, 2nd ed. 1952. Wien: Franz Deuticke.

Kruppa, Erwin. 1931. Emil Müller. *Jahresbericht der Deutschen Mathematiker-Vereinigung* 41: 50–58.

———. 1957. *Analytische und Konstruktive Differentialgeometrie*, 191. Wien: Springer.

Loria, Gino. 1908. Perspektive und Darstellende Geometrie. In *Vorlesungen über Geschichte der Mathematik*, ed Cantor, Moritz, vol. 4, 579–637.

Monge, Gaspard. 1811. *Géométrie descriptive par G. Monge avec un supplément par M. Hachette*. Paris: Klostermann fils.

Müller, Emil. 1908, 1916. *Lehrbuch der Darstellenden Geometrie für technische Hochschulen*, I, II. B.G. Teubner, Leipzig und Berlin. 2nd ed. 1918, 1919; 3rd ed. 1920, 1923; 4th ed., elaborated by E. Kruppa, 1936; 5th ed. 1948; 6th ed. 1961, Wien: Springer.

———. 1910a. Die verschiedenen Koordinatensysteme. In *Encyklopädie der math. Wiss.* Band III, 1. Teil, 1. Hälfte, no. AB 7, 596–770. Leipzig 1910: B.G. Teubner.

———. 1910b. Anregungen zur Ausgestaltung des darstellend-geometrischen Unterrichts an technischen Hochschulen und Universitäten. *Jahresbericht der Deutschen Mathematiker-Vereinigung* 19: 19–24.

———. 1910–1926. *Technische Übungsaufgaben für Darstellende Geometrie*. 6 issues (60 sheets in total). later eds. co-edited by E. Kruppa. Leipzig und Wien: Franz Deuticke.

———. 1911. Der Unterricht in der Darstellenden Geometrie an den Technischen Hochschulen. In *Berichte über den mathematischen Unterricht in Österreich*, Heft 9, 37–124. Wien: Alfred Hölder.

Müller, Emil, and Erwin Kruppa. 1923. *Vorlesungen über darstellende Geometrie, I. Die linearen Abbildungen*. Leipzig and Wien: Franz Deuticke.

Müller, Emil, and Josef Leopold Krames. 1929. *Vorlesungen über Darstellende Geometrie, II. Die Zyklographie*. Leipzig and Wien: Franz Deuticke.

———. 1931. *Vorlesungen über darstellende Geometrie, III. Konstruktive Behandlung der Regelflächen*. Leipzig and Wien: Franz Deuticke.

Noizet, F. 1823. *Mémoire sur la Géométrie appliquée au dessin de la fortification*, vol. 6, 5–224. Paris: Mémorial de l'Officier du Genie.

Peschka, Gustav A.V. 1883–1885. *Darstellende und Projektive Geometrie*. vol. 1: 1883, vols. 2,3: 1884, vol. 4: 1885. Wien: Carl Gerold's Sohn.

Rittinger, Peter. 1839. *Anfangsgründe der freien Perspektivzeichnung zum Selbstunterrichte*. Wien: Carl Gerold.

Schmid, Theodor. 1911. *Maschinenbauliche Beispiele für Konstruktionsübungen zur Darstellenden Geometrie*. 2nd ed. 1925, 25 sheets. Leipzig and Wien: Franz Deuticke.

Staudigl, Rudolf. 1843. Bestimmung von Tangenten an die Selbstschattengrenze von Rotationsflächen. *Sitzungsber. Abth. II der kais. Akad. d. Wissensch., Math.-Naturw. Cl.*, 68, 1–7.
———. 1868. *Grundzüge der Reliefperspektive*. Wien: L. W. Seidel & Sohn.
———. 1870. *Lehrbuch der Neueren Geometrie*. Wien: L. W. Seidel & Sohn.
———. 1875. *Die Axonometrische und Schiefe Projektion*. Wien: L. W. Seidel & Sohn.
von Dyck, Walter F.A. 1892. *Katalog Mathematischer und Mathematisch-Physikalischer Modelle, Apparate und Instrumente*. München: C. Wolf & Sohn.
Wildt, Josef (ed.). 1895, 1902. *Praktische Beispiele aus der Darstellenden Geometrie*, 1st and 2nd issue. Vienna: A. Pichler's Witwe & Sohn.
Wunderlich, Walter. 1966, 1967. *Darstellende Geometrie I, II*. BI-Hochschultaschenbücher Bd. 96, 133. Mannheim: Bibliographisches Institut.

Chapter 12
The Vienna School of Descriptive Geometry

Christa Binder

Abstract The *Vienna School of Descriptive Geometry* played a leading role in the development of all branches in the field, including freehand drawing, the construction of machines, projective geometry, and, of course, theoretical descriptive geometry, as well. Extensive teaching combined with thorough drawing training as well as high level research characterizes the institution for more than 100 years. Emphasis was always given to practical applications and to geometric imagination. We shall describe the early years, beginning with the founding of the Vienna *Polytechnicum* in 1815, the school's slow start, the first chair, and the first institute in 1843. We shall also describe—by introducing the chief characters and their role—the growing importance of descriptive geometry in the subsequent years due to industrial development and the introduction of studies for future teachers at *Realschulen*. Because of the increasing number of students the institute was divided into two parts in 1870, and a second institute was established in 1896 (see also Stachel, Chap. 11, and Moravcová, Chap. 16, this volume).

Keywords Descriptive geometry · Nineteenth century · Twentieth century · Vienna · Vienna University of Technology · Johann Hönig · Rudolf Staudigl · Rudolf Niemtschik · Gustav Peschka · Emil Müller · Theodor Schmid · Erwin Kruppa · Ludwig Eckhart · Josef Krames · Walter Wunderlich

1 First Years

The needs of the growing industries and trades led to the founding of various special schools and proposals for new organizations (see, for example, Lechner 1940) until the *Polytechnicum* of Vienna was founded in Vienna in 1815 modeled on the *École polytechnique* in Paris. From the beginning, geometric drawing

C. Binder (✉)
Institute for Analysis and Scientific Computing, Vienna University of Technology, Wien, Austria
e-mail: christa.binder@tuwien.ac.at

was an important topic since it was needed for most studies, including building, engineering, architecture, and so on. But contrary to the Paris *École polytechnique*, there was no chair devoted to it. The necessary courses—"machine drawing" and "preliminary technical drawing"—were held as part of the normal curricula. Soon this situation was considered as unsatisfactory, and in 1834 Johann Hönig,[1] an assistant at the Faculty for Mechanical Engineering (*Maschinenlehre*), voluntarily offered to give courses on descriptive geometry and drawing. In 1839, he left Vienna to become a professor for *Darstellende Geometrie und Zivilbaukunst* at the *Berg- und Forstakademie* in Schemnitz[2] where he stayed until 1843. From 1843 to 1870, he was a professor for descriptive geometry at the *Polytechnicum* in Vienna and a rector from 1868 to 1869 (Ottowitz 1992, p. 499, Obenrauch 1897).

The situation in Vienna was considered to be very bad, and the board of professors repeatedly (1827, 1835, and 1839) demanded the installation of a chair for descriptive geometry, and—with Hönig away—the situation became even worse. In 1841, another application was made "auf das Dringlichste" (most urgent) with reference to Germany where similar chairs had already been installed and were very successful (see Benstein, Chap. 9, this volume).

In April 1842, a *Konkurs* was announced, and Hönig—having already successfully passed such an exam in Schemnitz—was declared to be the best candidate, and he became a professor of Descriptive Geometry in 1843.

In order to show the importance of this new chair, three assistants and nine rooms for drawing were assigned to it.

Hönig had to teach a course of 3 h (5 h from 1850) and 10 h of construction practice per week.

In addition, he gave a 2-h popular course on Sundays. For many years, his book *Anleitung zum Studium der Darstellenden Geometrie* (Hönig 1845) was also used at the other *Polytechnica* of the Austro-Hungarian Empire (Graz, Prague, Brno, and Budapest—all of them German-speaking *Polytechnica*, also in Lemberg (Lvuv) where also some courses were taught in German). Another influential book was written by a professor of the Military Academy (Stampfl 1845).

A big step forward happened in the middle of the nineteenth century caused by the growing importance of the *Realschulen*. In the middle of the nineteenth century, more than 60 such schools were founded and also many special schools for the military, mining, arts, and trade, which all needed descriptive geometry.

Their curricula contained a great deal of geometrical drawing and ornaments and descriptive geometry (much more than in the corresponding schools in the other German-speaking countries (see Benstein, Chap. 9, this volume)). The teachers needed for this were educated at the *Polytechnicum* (for more details about the teaching of descriptive geometry in Austria, see Stachel, Chap. 11, this volume).

---

[1] Johann Hönig (March 9, 1810, Karlsbrunn–October 26, 1886, Pressbaum).

[2] The Academy of Mining and Forestry in Schemnitz (Germany), later Selmecb'anya (Hungary), and now Bansk'a Štiavnica in Slovenia. The academy moved to Sopron in 1919 (Hungary).

2 Second Half of the Nineteenth Century

Hönig was a professor for 27 years until 1870. He was supported by Rudolf Niemtschik[3] as his assistant from 1857 to 1861 and then by Rudolf Staudigl as *Privatdozent*. Rudolf Niemtschik worked in the building industry and studied at the Polytechnicum Vienna from 1852 to 1856. He was also an assistant at the Polytechnicum Vienna from 1857 to 1861 and was a professor for descriptive geometry at the Joanneum in Graz from 1861 to 1870. He was a professor and chair of the First Institute for Descriptive Geometry at the TH (formerly *Polytechnicum*) Vienna until 1877. Rudolf Staudigl[4] studied in Vienna with Hönig and was his assistant from 1861 to 1867; in 1865, he obtained a teacher's degree for descriptive geometry, mechanics, and theory of machines at the *Oberrealschulen*. In 1866, he received the *Habilitation* for technical and freehand drawing, and in 1868, he was promoted in absentia at the University of Rostock (the Vienna Polytechnicum had not yet the right to promote), and he obtained the *Habilitation* for ornamentics and newer geometry in Vienna in 1869. From 1870 to 1877, he was *extraordinarius* and a professor and chair at the Institute for Descriptive Geometry at the TH Vienna from 1877 to 1891 (Ottowitz 1992, pp. 499–500).

In 1870, the chair in Vienna was divided into two, and Niemtschik followed Hönig as chair of the institute until his early death in 1877. He became a professor, and Staudigl was *extraordinarius ad personam*.

Niemtschik took over the courses for mechanical engineering students and Staudigl for the various building students, and both taught courses for future teachers of descriptive geometry. Both were also very much engaged in scientific publishing, often in competition as they considered the same kind of problems. While Staudigl used projective geometry for his solutions, Niemtschik used more original methods in the field of elementary descriptive geometry (Niemtschik 1866a,b).

Teachers for descriptive geometry spread all over the Austro-Hungarian Empire and most were also very interested in the field of research. They could publish their results in various journals, for example, in the so-called *Schulprogramme*, which were edited yearly by the schools. The Austrian Academy of Sciences (founded in 1848) also provided the possibility to publish and promote the field of descriptive geometry—each mentioned professor was a member of the academy.

After Niemtschik's death, Staudigl took over all his courses and became chair of the institute.

In 1872, the *Polytechnicum* became the Technical High School of Vienna (*Technische Hochschule*, abbreviated as TH), and it was granted the right to promote students in 1903 (Neurath 1915).

---

[3] Rudolf Niemtschik (Němčik) (April 28, 1831, Frýdek–March ?, 1877, Vienna).
[4] Rudolf Staudigl (November 11, 1838, Vienna–February 2, 1891, Vienna).

The Staudigl's courses had a very tight structure and combined various methods, and he always had practical applications as his goal. It was Staudigl, for sure, who laid the foundation for what later became the famous Vienna School of Descriptive Geometry. For his students, he distributed the first collections of reproduced examples. From 1884 on, the training of teachers took 6 years (instead of 4 years as before), and Staudigl taught courses on modern geometry and selected fields of modern geometry (Staudigl 1868, 1870, 1875).

Staudigl was responsible for both parts of the chair and the following list of his courses in 1890/1891 is a good example of the extent and the contents:

> Descriptive geometry: orthogonal, axonometric, skew and perspective projection of points, lines and planes; exercises on the relations between these elementary fundamental elements, first by orthogonal projection and then by all other methods of projection; drawings of simple technical objects, bounded by planes including shadows; pyramids and prisms, curves and curved surfaces, mainly cones and cylinders; rotational surfaces, regulated, skew and envelopes surfaces. 4 hours per week, 4 hours for a seminar, and 10 hours for construction drawings (Vorlesungsverzeichnis der TH Wien, translated by the author).
>
> Newer Geometry: basic figures (*Grundgebilde*) of the first degree (*Stufe*), their projective relationship, confocal and involutory basic figures of the first degree, products (*Erzeugnisse*) of these figures, curves of the second degree, *collinear* and *reciprocal* figures of the second degree, their relationship and products, surfaces of the second degree, and space curves of the third degree, *collinear* and *reciprocal* systems in space. For engineers, mechanical engineers and building engineers: 2 semesters, 4 hours per week, and 10 hours for construction drawings (Vorlesungsverzeichnis der TH Wien, translated by the author).

After the death of Staudigl in 1891 a second chair was installed and Gustav Peschka became his successor. Peschka studied at the *Polytechnicum* and the University of Prague, and worked as a constructor in a factory for machines. Then, he had positions at the *Polytechnicum* in Prague (from 1852 to 1857, he was "Adjunct" for mechanics, drawings of machines, and physics), at the TH Lemberg (from 1857 to 1863, he was a professor for mechanics, mechanical engineering and drawings of machines, and descriptive geometry), and at the German TH Brno (from 1863 to 1891, he was a professor for mechanics, mechanical engineering, and constructive drawing), Peschka (1877, 1882) he was dean from 1880 to 1882. From 1891 to 1901, he was chair for descriptive geometry at the TH in Vienna. Gustav Peschka[5] was the successor of Staudigl though he was only ranked second place by the committee. He is said to have been a "favorite of the court" (and the ministry). He was, in fact, not a good choice since Peschka had no interest in the practical applications, and he gave no courses for the teacher students. His main scientific contributions were in steam machines and their construction and safety (Einhorn 1985, pp. 565–571) (Lechner 1940, pp. 154–155).

In Vienna, Peschka taught courses for mechanical engineers, and it was said that he—contrary to his colleagues before and after—graded the student's works only by their size and number, and not by their content. It was also said that he was a master in drawing on the blackboard. He did most of his scientific work during his

---

[5]Gustav Peschka (August 30,1830, Joachimsthal–August 29, 1903, Vienna).

stay in Brno, including the four volumes of *Darstellende und projektive Geometrie* (Peschka 1883–1885, 1899). This book had an important influence, and his book (Peschka 1868, 1882) was famous for the excellent typography and its 336 wood carvings.

The second chair was first held by Franz Ruth (1850–1905) from 1891 until 1895 and then from 1897 until 1899 by Jan Sobotka (1862–1931)—both as extraordinarius, and both later went to the Czech countries: Ruth to Prague and Sobotka to Brno.

3 The Era Emil Müller–Theodor Schmid

At the turn of the century, both chairs were vacant, and excellent mathematicians were found for both: Emil Müller[6] and Theodor Schmid. Both were students of Staudigl and followed his tradition, thereby making Vienna the center of descriptive geometry in the German-speaking countries. Both were interested in various fields of mathematics, in applications for all engineering studies, and in the education of student teachers. They were dominating the field for many years and solidified the reputations of the *Wiener Schule der Darstellenden Geometrie* (Vienna School of Descriptive Geometry). Müllers interests were widespread. He made many contributions to Grassmannian methods—he was considered to be one of the main experts in this field (*Grassmannsche Ausdehnungslehre*)—and he applied these methods to different aspects of geometry. He also developed the theory itself, and introduced the calculation of "Faltprodukte" (inner products), stressing the advantage of his methods for the theory of invariants.

During his long career at the TH in Vienna, he continued to work in different fields, but, of course, his main interest became descriptive geometry. He taught the introductory courses for building engineers and architects and enlarged the curriculum for the student teachers. Up till then the student teachers had to take the general introductory courses for the engineers and some specialized courses. Müller changed this situation and introduced a special seminar and a four semester cycle of courses: methods of projections in descriptive geometry, cyclographic and stereographic projection, constructive treatment of regulated surfaces, and

---

[6]Müller, Emil Adalbert (April 22, 1861, Landskron–September 1, 1927, Vienna). He studied mathematics and descriptive geometry at the TH and the University of Vienna and obtained teacher's degree in 1885. He was assistant to Staudigl, from 1890 to 1892 "Supplent" (a substitute) at the *Technologisches Gewerbemuseum*, and from 1892 to 1902 at the Baugewerksschule in Königsberg (where he was promoted and obtained "Habilitation" in 1898–1899 on Grassmannian methods). From 1902 to 1927, he was a professor and the first chair for descriptive geometry at the TH Vienna and in 1912/1913, he was a rector. He received many honors, and was a founding member of the Vienna Society of Mathematics (Einhorn 1985, pp. 572–587, Kruppa 1928, Schmid 1928).

constructive treatment of *Schraub- und Schiebflächen* (helicoidal and translation surfaces) Müller 1908, 1918, 1920 and Müller 1916, 1919, 1923.

His goal was not only to produce good teachers—with great success because teachers of descriptive geometry from the TH Vienna spread all over the whole Austro-Hungarian Empire—but also to further the scientific treatment of the field. His goal was to establish descriptive geometry as a part of geometry, in combination with projective geometry, non-Euclidean geometry, group theory, and so on. Müller was a very inspiring teacher with many students and followers. He also expressed his ideas in lectures given at the *Versammlung der Naturforscher* (the predecessor of the German Mathematical Society) in Germany (Müller 1910), where he also organized an exposition of his students' drawings. These construction exercises were always very important and much time was devoted to them. In order to provide examples from which the students could choose, he edited collections of sample drawings (Müller 1910, 1911, 1920, 1926).

Figures 12.1 and 12.2 show the level of complexity. The first one uses trimetric projection, a parallel projection with three different scale factors on the three orthogonal axes. The second one uses perspective drawing. You also should keep in mind that they had to be done with Indian ink; thus, one wrong line or a single splash would ruin it, and you would have to do it all over again.

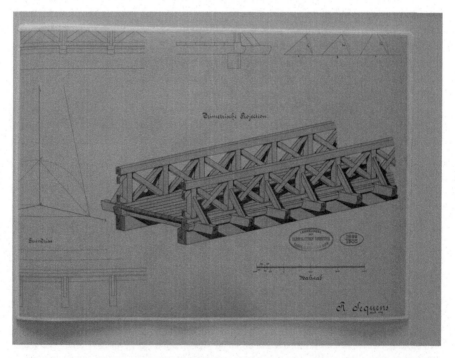

Fig. 12.1 Bridge, by H. Sequenz, student of architecture in his first year under Peschka, in 1899/1900[7]

---

[7] All figures in the present chapter are in the possession of the Institute of Algebra and Geometry, University of Technology Vienna.

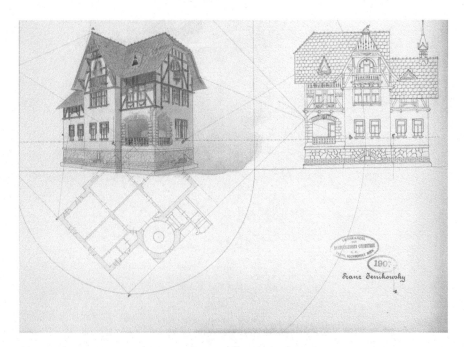

Fig. 12.2 House, by J. Jenikowski, student of architecture 1900/1901

Theodor Schmid[8] held the Second Chair for Descriptive Geometry for 29 years. He regularly delivered the introductory courses in descriptive geometry for future mechanical engineers and on projective geometry for the student teachers. His examples of machine drawings (Schmid 1911, 1925) were very highly regarded in many places. Figures 12.3 and 12.4 show drawings by his students.

Schmid worked on transmission devices (*Getriebe*)—very important for machines—on the construction of various curves and on the building of geometrical models. He was also interested in the shape of the earth and in photogrammetry. His goal was to combine descriptive geometry with projective geometry (Schmid 1912, 1919, 1922 and 1921, 1923).

He often complained about having too many students, so he was not able to write down his ideas on projective geometry. He was a very inspiring teacher, but he always remained a bit in the shadow of his charismatic colleague Emil Müller.

---

[8]Theodor Schmid (December 6, 1859, Erlau–October 30, 1937, Vienna). He studied in Vienna at the TH and the university and got his teacher's degree for mathematics and descriptive geometry in 1882 and for physics in 1886. After some difficult years as teacher at various schools, he got the Second Chair of Descriptive Geometry at the TH Vienna, from 1900 to 1906 as extraordinarius and from 1906 to 1929 as ordinarius (Einhorn 1985, pp. 633–643; Sequenz 1965, pp. 137–138; Dolezal 1937/1938, pp. 85–94).

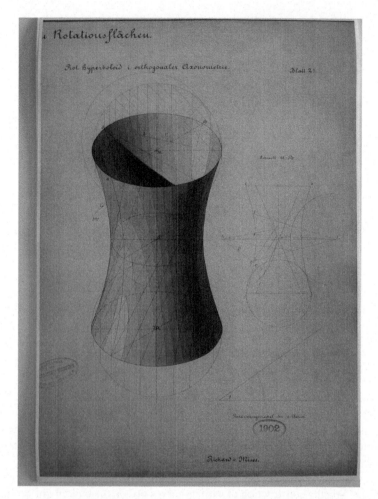

Fig. 12.3 Surfaces of revolution: the hyperboloid in axonometric projection. The drawing is from 1902 and by Richard von Mises (1883–1953), who was studying mechanical engineering in his first year

Nevertheless, he received many honors, including that of dean. He was also elected to be a rector of the TH but could not accept this honor because of health problems.

4 The Era Erwin Kruppa–Ludwig Eckhart–Josef Krames

After Müller's death in 1927 and Schmid's retirement in 1929, both chairs were vacant. Müller's assistant Josef Krames substituted the chair of the First Institute

12 The Vienna School of Descriptive Geometry

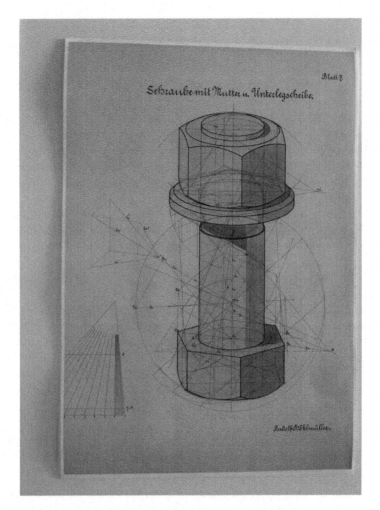

Fig. 12.4 Reproduction of a screw by R. Höhlmüller, a student of Schmid

for 2 years until Erwin Kruppa[9] became successor of Müller. Kruppa succeeded in maintaining the international reputation of the Vienna School of Descriptive Geometry.

---

[9]Erwin Kruppa (August 11, 1885, Biala, Galicia–January 26, 1967, Vienna). He studied mathematics and descriptive geometry in Graz and Vienna; in 1911, he was promoted in Graz and obtained his *Habilitation* in Czernowitz. From 1911 to 1918, he was a *Privatdozent* in Czernowitz and a teacher at a German-speaking *Realschulen*. During his war service from 1914 to 1918, he was wounded and then became a teacher in the Infanterie-Kadettenschule in Wien-Breitenlee. From 1918 to 1921 he was a *Privatdozent* at the TH Graz, then from 1921 to 1922 extraordinarius in Graz, from 1922 to 1929 a professor for mathematics at the TH Vienna, from 1929 to 1957 a professor and chair of the First Institute of Descriptive Geometry at the TH Vienna, and in 1953/1954 a rector (Einhorn 1985, pp. 588–603, Krames 1968).

His education and former occupation as a professor of mathematics and his knowledge of modern analysis influenced his treatment of descriptive geometry (Kruppa 1932, 1933, 1936). He introduced exact limits instead of an intuitive treatment of neighboring elements, he also treated non-Euclidean geometry and geometry in higher dimensions. He edited and enlarged Müller's book (Müller 1923), which consisted of three volumes, had at least four editions, and was very popular (Kruppa 1936, 1948, 1961).

In his later years, he also worked on differential geometry (Kruppa 1957).

Ludwig Eckhart[10] became chair of the Second Institute of Descriptive Geometry in 1929. He had already found a new *Schrägrissverfahren* (a particular form of axonometry), which soon found its way into many text books. He introduced cinematics as a basis for *Getriebelehre* into the courses for mechanical engineers and *Ausdehnung des Einschneideprinzips* in die *schiefe Axonometrie* (extension of the method of sections in skew axonometry). He was dismissed from office in 1937, a decision he could not understand and that hit him very hard. It seems that the reasons for this dismissal were political since he was active in the *Vaterländische Front* (Fatherland Front),[11] and one of his assistants was Jewish. He committed suicide in 1938.

Krames[12] was always very much engaged in teaching student teachers and he played a major role in introducing courses on selected topics in descriptive and projective geometry in Brno. In Graz, he was also responsible for the final examinations of teachers in descriptive geometry. But, in spite of his many duties, he

---

[10] Ludwig Eckhart (March 28, 1890, Selletitz bei Znaim–October 5, 1938, Vienna). He studied building engineering, mathematics, and descriptive geometry at the University and the TH in Vienna and he was in war service from 1914 to 1917 (Reseveoffizier des Infanterieregiments Nr. 99, Znaim, getting many honors as: silberne Tapferkeitsmedaille Signum laudis mit den Schwertern, Karl-Truppenkreuz, and Verwundetenmedaille). He was teacher at the military academy, and at various schools. He was promoted in 1918 and got his *Habilitation* in 1924 at the TH Vienna in descriptive geometry; from 1924 to 1929, he was a *Privatdozent*; and from 1929 to 1937 a professor and chair of the Second Institute of Descriptive Geometry, TH Vienna, and the dean from 1935 to 1937.

[11] The Fatherland Front (*Vaterländische Front*) was the ruling political organization of "Austrofascism". It aimed to unite all the people of Austria, overcoming political and social divisions. Established on 20 May, 1933, by the Christian Social Chancellor Engelbert Dollfuss, advocating a one-party system along the lines of Italian Fascism, it advocated Austrian nationalism and independence from Germany on the basis of protecting Austria's Catholic religious identity from what they considered a Protestant-dominated German state. The Fatherland Front was immediately banned after the "Anschluss" (annexation) of Austria to Germany in 1938.

[12] Josef Krames (October 7, 1897, Vienna–August 30, 1986, Salzburg). He studied, at the TH Vienna, mathematics and descriptive geometry to became a teacher. He was promoted with Müller and in 1923 he obtained the "Habilitation" at the TH Vienna. From 1924 to 1929, he was an assistant at the TH (during that time he edited (Müller 1929) and (Müller 1931)), and from 1929 to 1932 extraordinarius at the German Technical High School of Brno, from 1932 to 1939 a professor at the Technical Highschool Graz, from 1939 to 1945 and from 1957 to 1969 a professor of descriptive geometry at the TH Vienna, and he was a rector in 1961/1962. From 1948 to 1956, he worked in the *Bundesamt für Eich- und Vermessungswesen* (Einhorn 1985, pp. 604–622; Wunderlich 1987).

continued to work scientifically. He published a series of articles on *symmetrische Schrotungen* (a type of symmetric motions in space), which were much valued by his French colleagues. From 1937 on, he studied the so-called *gefährliche Flächen* (dangerous surfaces), a problem in photogrammetry, and solved it using purely geometrical methods.

Descriptive geometry lost its importance after 1938: The hours required in engineering studies were reduced, and descriptive geometry was dropped in schools. Krames protested against this. In 1939, after the death of Eckhart, he became chair of the Second Institute of Descriptive Geometry at the TH Vienna. His main duties were the courses for machine engineering (Krames 1947, 1952). After the war in 1945, he was dismissed for political reasons and for the next years he worked at the *Bundesamt für Eich-und Vermessungswesen* (Federal Office of Metrology and Surveying), mostly in photogrammetry and improving instruments.

In 1946, the chair of the second institute was given to Walter Wunderlich[13] as extraordinarius. He was a pupil of Kruppa and had been working as an assistant and a *Privatdozent* during the war. In 1955, he became a professor.

He had many interests in all areas of geometry, especially in kinematics, *höhere Radlinien* (higher cycloidal curves), wobbly structures, and much more generally solving them in a very original way using intuitive geometric arguments.

In 1957, Krames was back at the TH Vienna succeeding Kruppa as chair in the First Institute of Descriptive Geometry. In the same year, both institutes changed their names to the First and the Second Institute of Geometry and became open to other fields of geometry such as differential geometry and computer aided design.

Together with his colleague Wunderlich, Krames reorganized the studies for teachers. Alternatingly they gave courses on projective geometry I and II. They also introduced two new seminars; Krames gave a 2-year course on *Konstruktive Abbildungsmethoden* (constructive methods of representation) and *Konstruktive Strahlgeometrie* (constructive line geometry), whereas Wunderlich taught *Konstruktive Differentialgeometrie* (constructive differential geometry) and *Nichteuklidische und mehrdimensionale Geometrie* (non-Euclidean and higher dimensional geometry) (Wunderlich 1966, 1967).

With Krames' retirement in 1969 and Wunderlich's in 1980, we come to the end of the classical period of descriptive geometry in Vienna. In 1975, the TH became the University of Technology of Vienna (Technische Universität Wien, TU).

---

[13]Walter Wunderlich (March 6, 1910, Vienna–November 3, 1998, Vienna). He first studied building engineering at the TH Vienna, then in 1933 he also obtained the teacher's degree of mathematics and descriptive geometry, and since there was no possibility to get a teacher's position, he also obtained the teacher's degree for stenography in 1935, which helped a bit. In 1934 he graduated in descriptive geometry. In 1938 he was an assistant, and in 1940 he got the *Habilitation*; during the war he served at the *Physikalische Versuchsanstalt* of the marine. In 1946, back in Vienna, he got the chair of the Second Institute of Descriptive Geometry as extraordinarius and since 1955 till 1980 as a professor, he was a rector in 1964/1965 (Sequenz 1965, pp. 139–140; Binder 2016, p. 119, Stachel 1999).

References

Binder, Christa. 2016. Mathematik an der TH/TU Wien, 1965–2014. In *Die Fakultät für Mathematik und Geoinformation, The Faculty of Mathematics and Geoinformation*, ed. Michael Drmota, 111–126. Vienna and Cologne: Böhlau.
Dolezal, Eduard. 1937/1938. Hofrat Professor Theodor Schmid. In: *Bericht Über das Studienjahr*, 85–94, TH Wien.
Einhorn, Rudolf. 1985. *Vertreter der Mathematik und Geometrie an den Wiener Hochschulen, 1900–1940*. Dissertation, TU Wien: Verein wissenschaftlicher Gesellschaften.
Hönig, Johann. 1845. *Anleitung zum Studium der Darstellenden Geometrie*. Wien.
Krames, Josef. 1947, 1952. *Darstellende und Kinematische Geometrie für Maschinenbau*. Wien: F. Deuticke.
———. 1968. Nachruf (Erwin Kruppa). *Almanach der Österreichischen Akademie der Wissenschaften* 117, 246–258.
Kruppa, Erwin. 1928. Emil Müllers Leben und Wirken. *Monatshefte für Mathematik und Physik* 35: 197–218.
———. 1932, 1933, 1936. *Technische Übungsaufgaben für Darstellende Geometrie I, II, II*. Leipzig-Wien: F. Deuticke.
———. 1936, 1948, 1961. *Lehrbuch der Darstellenden Geometrie (vollständige Neubearbeitung des Buches von Müller)*, 4th ed. B. G. Teubner, Leipzig and Berlin, 316 pages. 5th ed., Springer, Wien, 413 pages. 6th ed., Springer, Wien, 404 pages.
———. 1957. *Analytische und konstruktive Differentialgeometrie*. Wien: Springer.
Lechner, Alfred. 1940. *Die Geschichte der TH in Wien 1815–1940*. ed. TH Wien.
Müller, Emil. 1908, 1918, 1920. *Lehrbuch der Darstellenden Geometrie für Technische Hochschulen*, Band I, Leipzig u. Berlin, 384 pages.
———. 1916, 1919, 1923. *Lehrbuch der Darstellenden Geometrie für Technische Hochschulen*, Band II, Leipzig u. Berlin, 372 pages.
———. 1923. *Vorlesungen Über Darstellende Geometrie, Band I: Die linearen Abbildungen*, bearbeitet von E. Kruppa, Leipzig u. Wien, 303 pages.
———. 1929. *Vorlesungen über Darstellende Geometrie, Band II: Die Zyklographie*, aus dem Nachlaß herausgeg. von J. Krames, Leipzig u. Wien, 485 pages.
———. 1931. *Vorlesungen über Darstellende Geometrie, Band III: Konstruktive Behandlung der Regelflächen*, bearbeitet von J. Krames, Leipzig u. Wien, 311 pages.
———. 1910. *Anregungen zur Ausgestaltung des darstellend-geometrischen Unterrichts an technischen Hoschschulen und Universitäten*, Jahresber. DMV, 19–24.
———. 1910, 1911, 1920, 1926. *Technische Übungsaufgaben für Darstellende Geometrie*, Hefte I, II, III (1910), Heft IV (1911), Heft V (1920), Heft VI (1926). Wien und Leipzig: Deuticke.
Neurath, J. (ed). 1915. *Die Technische Hochschule in Wien 1815–1915*. Wien.
Niemtschik, Rudolf. 1866a. *Neue Constructionen der auf ebenen und krummen Flächen erscheinenden Reflexe und hierauf bezügliche Theoreme*. Wien: Gerold.
———. 1866b. *Directe Constructionen der Contouren von Rotationsflächen in orthogonalen und perspectivischen Darstellungen*. Wien: Gerold.
Obenrauch, Ferdinand Josef. 1897. *Geschichte der darstellenden und projectiven Geometrie mit besonderer Berücksichtigung ihrer Begründung in Frankreich und Deutschland und ihrer wissenschaftlichen Pflege in Österreich*. Brünn: Carl Winiker.
Ottowitz, Nikolaus. 1992. *Der Mathematikunterricht an der Technischen Hochschule in Wien 1815–1918*. Dissertation. TU Wien: VWG (Verein wissenschaftlicher Gesellschaften).
Peschka, Gustav. 1868, 1882. *Freie Perspektive in ihrer Begründung und Anwendung*. Hannover, Leipzig: E. Koutny.
———. 1877, 1882. *Kotierte Ebenen und deren Anwendung; Kotierte Projektions-methoden und deren Anwendung*, 2nd ed., Brno.
———. 1883–1885, 1899. *Darstellende und projektive Geometrie*, 4 volumes, Wien, 2553 pages, second edition, first volume, Wien-Leipzig, 719 pages.

Schmid, Theodor. 1911, 1925. *Maschinenbauliche Beispiele für Konstruktionsübungen zur Darstellenden Geometrie*, 2nd ed. Leipzig, Wien.
———. 1912, 1919, 1922. *Darstellende Geometrie*, I. Band. W. de Gruyter. Berlin-Leipzig. 283 pages and 170 figures.
———. 1921, 1923. *Darstellende Geometrie*, II. Band, W. de Gruyter, Berlin-Leipzig, 340 pages and 163 figures.
———. 1928. Nachruf auf Emil Müller. *Almanach der Österreichischen Akademie der Wissenschaften* 78: 183–188.
Sequenz, Heinrich. 1965. *150 Jahre Techn. Hochschule Wien*, Bd. 2, Wien.
Stachel, Hellmuth. 1999. *Walter Wunderlich (1910–1998)*. http://www.geometrie.tuwien.ac.at/stachel/nachruf_wunderlich.pdf
Stampfl, Josef. 1845. *Lehrbuch der darstellenden Geometrie und ihrer Anwendungen auf die Schattenbestimmung, Perspectivlehre und den Steinschnitt*, Wien, 638 pages, 395 figures
Staudigl, Rudolf. 1868. *Grundzüge der Reliefperspektive*. Wien.
———. 1870. *Lehrbuch der neueren Geometrie*. Wien: Seidel & Sohn
———. 1875. *Die Axonometrische und schiefe Projektion*. Wien.
Wunderlich, Walter. 1966. *Darstellende Geometrie I*, Bibl. Inst., 187 pages.
———. 1967. *Darstellende Geometrie II*, Bibl. Inst., 234 pages.
———. 1987. Nachruf auf Josef Krames. *Almanach der Österreichischen Akademie der Wissenschaften* 137: 285–295.

Chapter 13
At the Crossroads of Two Engineering Cultures, or an Unedited Story of the French Polytechnician Charles Potier's Descriptive Geometry Books in Russia

Dmitri Gouzevitch, Irina Gouzevitch, and Nikolaj Eliseev

Abstract In the early nineteenth century, the ideas of Gaspard Monge's descriptive geometry spread to Russia. As in other countries, the transfer of knowledge in this field was connected to the French *École polytechnique*. Its disseminators were both the first Russian students at the polytechnique and its French graduates, who were invited to Russia by Alexandre I in 1810. Two of the latter, Alexandre Fabre and Charles Michel Potier, initiated the teaching of descriptive geometry according to Monge's principles at the newly created Institute of the Corps of Engineers of Ways of Communication (1810–1811). Appointed professor of descriptive geometry (1815–1818), Potier completed and published, for the first time in Russia, a series of original works on descriptive geometry and its applications, which became essential for the teaching, the dispersion and the spreading development of this science in Russia for the decades to come. However, according to standard Russian historiography, it was not Potier but his Russian translator, former student and disciple Jakov Sevastianov, who was seen as the founding father of descriptive geometry in Russia. Our paper examines the career in Russia and the mathematical works of this French engineer who, despite his pioneering books preserved in major Russian and French libraries, including that of the *École polytechnique*, remains, paradoxically, an "illustrious unknown" in both countries.

D. Gouzevitch
Centre d'études des mondes russe, caucasien et centre européen, École des hautes études en sciences sociales, Paris, France
e-mail: dmitri.gouzevitch@ehess.fr

I. Gouzevitch (✉)
Centre Maurice Halbwachs, École des hautes études en sciences sociales, Paris, France
e-mail: irina.gouzevitch@ens.fr

N. Eliseev
State University of Ways of Communication of Saint-Petersburg, Saint-Petersburg, Russia

Keywords Gaspard Monge · Descriptive geometry · Its applications · Charles Potier · Âkov Sevast'ânov · Russia · *École polytechnique* · Institute of the Corps of Ways of Communication · Scientific school

1 Introduction

In the early nineteenth century, the ideas of Gaspard Monge's descriptive geometry spread to Russia. As in other countries, the transfer of knowledge in this field was connected to the French *École polytechnique*. Its disseminators were both the first Russian students at this school, such as Petr Rahmanov and Aleksej Majurov, and its French graduates, who were invited to the Russian Empire under the reign of Alexandre I in 1810. Two of the latter, Alexandre Fabre (1782–1844) and Charles Michel Potier (1785–1855), initiated the teaching of descriptive geometry according to Monge's principles at the newly created Institute of the Corps of Engineers of Ways of Communication (1810–1811). Appointed professor of the discipline, which he taught from 1815 to 1818, Potier completed and published, for the first time in Russia, a series of original works on descriptive geometry and its applications, which became essential for the teaching, the dispersion and the development of this science in Russia for many decades to come. However, according to standard Russian historiography, it was not Potier but his Russian translator, former student and disciple Jakov Sevast'janov (Âkov Sevast'ânov[1]), who was seen as the founding father of descriptive geometry in Russia. Our paper examines the career in Russia and the mathematical works of this French polytechnician who, despite his pioneering books preserved in major Russian and French libraries, including that of the *École polytechnique*, remains, paradoxically, an "illustrious unknown" in both countries.

2 From the *École Polytechnique* to the Service of the Russian Crown

Charles Michel Potier was born in Paris on 16 November 1785. The death of his father Jean Charles Potier in 1792, and the family's subsequent financial difficulties were factors that led to the late conclusion of his secondary education.[2] In the

---

[1] This version of the name transliterated according to the current bibliographical norms, the system ISO (International Organization for Standardization, for the transliteration from Cyrillic to Latin—NF ISO 9), will be used for the bibliographical references in order to ensure they are found by non-Russian readers.

[2] His mother Marie Anne Geneviève Vallée, "épicière, rue de la Tournelle no. 8" died in 1809. See AÉP, registre de matricules des élèves, vol. 3, ff. 59, 67; ANP, acte de décès, V3E/D 1444, 1809, no. 30.

1804/1805 academic year, he entered the Lyceum Napoleon where his teacher Charles Louis Félix Dinet prepared him in 1 year for the entrance exam to the *École polytechnique*. Potier entered Polytechnique on 20 November 1805 at the age of 20, 22nd on the list of more than 400 candidates. He had among his classmates Augustin Louis Cauchy, and among his teachers Gaspard Monge and Jean-Nicolas-Pierre Hachette for geometry, Jean-Baptiste Labey, André-Marie Ampère and Gaspard Prony for analysis and its application to mechanics, Sylvestre-François Lacroix and Siméon-Denis Poisson for mechanics, Jean-Henri Hassenfratz for physics, Joseph-Mathieu Sganzin for public works and François-Marie Neveu for drawing (Fourcy 1828, pp. 376–379/1, pp. 143–193/2; Callot 1982, pp. 475–478).

After the *École polytechnique*, Potier completed his engineering education at the *École des ponts et chaussées*, where he enrolled on 1 November 1807, no. 3 on the list of candidates (Cauchy was no. 1) and from which he graduated in 1810 (Brunot and Coquand 1982, p. 66).[3] In the same year, he was invited to Russia together with three other polytechnicians and engineers of *ponts et chaussées*, Alexandre Fabre, Pierre-Dominique Bazaine and Maurice Destrem.

We have analysed in previous works the circumstances of their invitation to Russia within the context of the brief political alliance between France and Russia in 1807–1812 (Guzevič and Guzevič 1995; Guzevič and Guzevič 2015; Gouzévitch and Gouzévitch 1993; Gouzévitch and Gouzévitch 1996), particularly so in the case of Potier in the biography that was published in 1814 (Gouzévitch et al. 2013). Suffice to say here that although their invitation to Russia was considered as a temporary mission and that they were simultaneously in the service of both countries, their plans were altered by historical circumstances and the four engineers stayed on in Russia after the end of the war with Napoleon and the Restoration in France until the end of their careers in the cases of Fabre and Bazaine and until their deaths in the cases of Destrem and Potier.

The four polytechnicians arrived in Saint-Petersburg in June 1810. In Russia, engineers of the Corps of Waterways and Land Communications (since 1811—the Corps of Engineers of Ways of Communication, CEWC) were subjected to a parallel system of and military ranks. Thus, in July 1810, Fabre and Bazaine received the civil rank of director-conductor of works and the military rank of lieutenant-colonel, whereas Potier and Destrem became engineers of the 2d class and captains (promoted to majors in June 1811). Bazaine and Destrem were assigned to Southern Russia, whereas Fabre and Potier were appointed to the newly created "Institute of the Corps of Engineers of Ways of Communication" (ICEWC) as professors of applied mathematics.[4] The teaching staff of the ICEWC consisted of five professors, with Augustin Betancourt as its Chief Inspector. In October 1810, Fabre and Potier took part in the examination of the applicants for the first admission of the institute.

During the first academic year 1810/1811, Potier taught arithmetic, algebra, geometry and trigonometry, whereas Fabre taught, for the first time in Russia, a

---

[3] AÉP, registre de matricules des élèves, vol. 3, ff. 59, 67.
[4] RGIA, F. 159, op. 1, d. 520, ff. 1–2, 12.

course on descriptive geometry. Unfortunately, there is little information about this course. Petr Rahmanov (Guzevič and Guzevič 2003), the first Russian student at the *École polytechnique*, while describing the public exams at the ICEWC in April 1811, briefly mentioned:

> [...] descriptive geometry with all its applications [...] is taught in Russia at the only Institute of Ways of Communications where the famous General Betancourt introduced it quite recently (Izvestiâ 1811, pp. 79–80, All translations are by the authors).

Potier took over the next year 1811/12, adding to his course a once-a-week lecture on fortification, in which he examined sources on the subject (Istoričeskie svedeniâ... 1842, p. 22). It is sufficient to mention here that descriptive geometry had been created by Gaspard Monge as a tool for the solution of theoretical problems in fortification. In 1826 Jakov Sevastianov, describing the textbook that served as a basis for the teaching, had Monge's book on descriptive geometry clearly in mind:

> One of the published guides in this science, or at least the order and methods adopted in it [were to be found] in the manuscript, which both allowed the acquisition of theoretical knowledge and entertained the students with its beautiful style and drawings, and was from the beginning adopted by the direction of the Institute of Engineers; and this continued almost until 1816 (Sevast'ânov 1829, p. 228; Sevast'ânov 1826, no. 2, pp. 30–31).

In the other words, during the first years, descriptive geometry was taught using a manuscript copy of Monge's book illustrated with beautiful drawings by hand.

During the campaign against Napoleon from the summer of 1812 until the spring of 1815, the four polytechnicians lived in exile in Siberia (Irkutsk). On their return to Saint-Petersburg, Potier was appointed professor of the course on descriptive geometry. He occupied this chair for three and a half years, until the spring of 1818, and he managed to extend the course, which he taught in French, by adding such sections as *La théorie des projections* (1815), *La coupe des pierres* and *La perspective aérienne* (Stone cutting and Aerial perspective, 1818) (*Programme pour l'Examen...* 1816, pp. 72–76; *Programme pour l'Examen...* 1818, pp. 64–69). During this period, Potier wrote three handbooks on descriptive geometry and its applications. These works were immediately translated into Russian by his pupil and assistant Sevastianov (graduated from the Institute of Ways of Communication in 1814) and published both in Russian or in a bilingual French–Russian version.

3 Potier's Textbooks of Descriptive Geometry

The first textbook to be published in autumn 1816 was the course on descriptive geometry, *Traité de Géométrie descriptive à l'usage des élèves de l'Institut des voies de communication* (Potier 1816), followed by the Russian version entitled *Osnovaniâ načertatel'noj geometrii dlâ upotrebleniâ vospitannikami Instituta korpusa inženerov putej soobŝeniâ* (Pot'e 1816; Pot'e corresponds to the Russian transliteration of Potier), which appeared at the end of the same year.

The second textbook, published in 1817, dealt with the *Application de la Géométrie Descriptive à l'art du dessin; à l'usage des élèves de l'Institut des Ingénieurs des voies de Communication* (Potier 1817c). Its Russian translation, *Priloženie načertatel'noj geometrii k risovaniû dlâ upotrebleniâ Vospitannikami Instituta korpusa inženerov putej soobŝeniâ*, came the following year (Pot'e 1818).

As for the *Traité de la coupe de pierres, à l'usage des élèves de l'Institut des voies de communication* = *Načal'nye osnovaniâ razrezki kamnej: Dlâ upotrebleniâ vospitannikami Instituta Korpusa inženerov putej soobŝeniâ* (Potier 1818), the book and its translation were published under the same cover in 1818 as a French–Russian bilingual edition.

The fourth work, which completes this collection, is the *Cours de constructions = Kurs postroeniâ* (Pot'e 1818, p. 6, line 131), which was then considered as an application of descriptive geometry. Co-authored by Charles Potier and Pierre-Dominique Bazaine in 1816, this work was completed and re-edited in 1819. In 1827, it was once more completed and revised by André Guillaume Henry (Henry 1827, 1828). Sevastianov, who replaced Potier as professor, wrote his own textbooks: *Priloženie načertatel'noj geometrii k risovaniû; Teoriâ tenej, linejnaâ perspektiva, optičeskie izobraženiâ* (Application of descriptive geometry to drawing; Theory of shadows, linear perspective and optical representations) (Sevast'ânov 1830), *Osnovaniâ načertatel'noj geometrii* (Foundations of descriptive geometry) (Sevast'ânov 1834) and *Kurs plotničnago Iskusstva* (Course of carpentry) (Sevast'ânov 1840) et al. Nevertheless, Potier's works continued to be published in Russia and abroad for the following three decades. In 1817, his *Traité de géométrie descriptive* was issued in France by the *École polytechnique*, which might be explained by the lack of textbooks on the subject after a long series of wars. However, the fact that the *Société d'encouragement pour l'instruction élémentaire* in Liège republished Potier's textbook in 1842 (Potier 1842), when literature dealing with descriptive geometry was readily available, testifies to its great didactic value.

3.1 The Traité de Géométrie Descriptive: *An Original Didactic Approach*

In the introduction to the Russian version of his *Traité*, Potier wrote:

Descriptive geometry brings the greatest benefit to the art of engineering. Its countless applications permit the most precise determination of places where work is to be carried out, the representation and definition of lines and surfaces which form various parts of a project and, thus, serve to draw and measure them. Descriptive geometry is needed when making various types of maps; using it one can proceed with confidence to distribute fortification works and very often even the astronomical activities can be helpfully conducted thereby with success. Its benefits are also seen in the circumstances, although not as important but no less useful, when an engineer has to pass on their concepts to people having little or no practice in the art as, only descriptive geometry in its applications for drawing and perspective offers a means to do it.

Then, the author offered a short historical overview of the methods of representation and stressed that he had based his work on that of Monge, "in which the foundations of the science, as well as all the works of this famous geometer, are written clearly, simply, understandably, and brought to perfection" (Pot'e 1816, pp. II–III).

Potier's textbook was conceived in the spirit of the "Elements" of Euclid. At the same time, it was not a simple reproduction or exposition of the work of the "famous geometer" Monge. Its originality was specifically highlighted by the French editors of the *Traité*, version "*École polytechnique*":

> The plan for this Work makes it totally different from anything that has appeared to date on Descriptive Geometry and even led to theorems of geometry, new by their generality. Deprived of all kinds of help and guides, the Author has endeavored, as far as it was possible, to be concise, exact and clear (Potier 1817b, p. 5).

Potier's work was, indeed, extremely concise, as it contained only a few necessary illustrations (2 sheets with 15 drawings) and was presented according to the axiomatic principle. Simply structured, it included three parts: definitions, preliminary theorems and applications. The author examined the most frequent positions necessary to solve practical problems. This approach avoided any complex constructions, which would have overcomplicated the drawing (*épure*), as the reasoning could be applied to any other example. To simplify the construction and the reading of the drawings, Potier used dotted lines to mark the plans and the methods of analysis to explain the graphic constructions. The limited number of drawings in the book aimed to stimulate students towards independent work.

Potier explained:

> I followed the rules and methods of Monge and tried [...] to divide the most practically used surfaces into classes [...], and to make those as simple as possible; various proposals are given so that one can clearly see their relationship and avoid duplication; the figures are only concerned with the initial theorems, as all the other theorems are their derivations; presenting initial theorems latter under the name of applications, I could explain them more briefly and with greater accuracy; I separated them from the demonstrations so that the exposition of the first did not impede the clear understanding of the latter (Pot'e 1816, p. VIII).

He enriched the theory of the formation and the construction of surfaces by the addition of new types of surfaces: the left cylinder and the left plan (or, in modern terms—the Catalan rules of surfaces: cylindroids and conoids). These were applicable not only to projects on ships and the working elements of machines but also to the works of fortification and of ways of communication and the cylinder and the cone of revolution (envelopes) useful for the construction of shadows and perspectives.

Exposing the essence of an orthogonal projection on two mutually perpendicular planes—having showed how to set points, lines and planes for the drawing—he paid particular attention to problems such as the intersection of lines and planes with different surfaces, the intersection of surfaces or the carrying out of tangents and normals to the surfaces. In his later works, all these ideas were successfully applied to the construction of shadows and perspective projection (Potier 1817c), to the practice of stone cutting (Potier 1818) and to the solution of metric problems.

In the preface to the Russian edition of this book, his translator Sevastianov underlined "its critical nature, its clear and concise demonstrations" that put it among the best in the genre, while "its orderliness made the translation especially pleasant" (Pot'e 1816, p. I).

In recognition of his books Potier was promoted to the rank of colonel. The next year, 1817, the book was republished in France (Potier 1817b).[5]

3.2 Application de la Géométrie Descriptive à l'art du dessin...

Published in French (Potier 1817c) and then in Russian (Pot'e 1818), this second textbook by Potier addressed the fundamental problems of descriptive geometry—its application to the construction of true and falling (from the sun and the point of light) shadows, black and light points, linear and aerial perspective which studied the rules for the representation of the gradual increase and decrease of light and shade depending on the degree of density or transparency of the air between the represented object and the pictorial plane. The author also paid attention to the rules for the application of paint in the colouring of the drawings.

Regarding the graphics in two projections of a work of art, the scale was scrupulously respected; one determined the true and falling shadows that made the picture more visible and easier to measure, replacing the third projection of the object. When engineers performed a drawing projection, shadows replaced the second projection. Such a design, arrived by taking into account the known direction of light rays, offered a fairly complex idea of the object.

Potier's book was very timely since students at the ICEWC learned from the first year how to produce accurate drawings, mostly using ink with washing and, thus, familiarized themselves with the symbolic representation of the used material. When studying the basic theory of machines, they executed drawings in accordance with the theory of shadows and colouring that was followed by an explanation of their operating principle. As for graduate students, they had to analyse the graphic models of architectural structures and machinery while carrying out the projects.

Potier paid great attention to enveloping surfaces, very important for finding shadows: the enveloping cylinders for the object lit by the sun (parallel to the beam) and the enveloping cones—from the light point. A true shadow was constructed as a curved tangent line of the enveloping surfaces with the studied object[6] and falling shadows were constructed as an intersection line of the enveloping surfaces with a given plane. The perspective was constructed as a line of intersection of the enveloping cone with the vertex in the point of view of the painting surface.

---

[5]The book was published in Paris not later than in September 1817 because, it was already announced in the *Livres nouveaux* of the *Journal des savans* in October (Livres nouveaux 1817).

[6]The lines which separate the shadows from light.

By constructing an aerial perspective of an object and eliminating the true shadows of the object, Potier solved two main problems: how to represent the contour of a shadow and how to find nuances in any shadowed part of the object. The book had the same structure and was organized in the same axiomatic form as the *Traité de la géométrie descriptive*. In the sections propounding the theory and application of shadows and perspective, he used only two drawings, thereby giving students the opportunity to learn the other principles during practical lessons.

According to Sevastianov, Potier, who had attended Monge's lectures, applied the ideas of his teacher to the aerial perspective (Potier 1817c; Pot'e 1818) much earlier than Vallée (1821) and Hachette (1822) (Sevast'ânov 1826, no. 3, pp. 9–10) (on Hachette and Vallée's textbooks, see Barbin, Chaps. 1 and 2, this volume).

It is relevant to quote the view of a user of Potier's textbook, the engineer A. Nordštejn, who graduated from the ICEWC in 1829:

> Potier wrote an excellent little book: descriptive geometry, theory of shadows and perspective; we never had before anything as clear and lucide (Nordštejn 1905, p. 244).

3.3 Other Developments

The third of Potier's textbooks, *Traité de la coupe de pierres* (Potier 1818), dedicated to the application of descriptive geometry to the problems of masonry stone arcs, domes and spiral staircases, was a substantial aid in the teaching of engineering. It was a large (4000 mm × 5000 mm), richly bound bilingual edition with an elegant gold-lettered cover and well produced drawings; the original text was on the right side, its Russian translation by Sevastianov[7] on the left. Potier's work proved to be a classic example of the application of the theory of descriptive geometry to the solution of construction problems: the cutting of stones, the determination of their limiting surfaces and their graphic representation.

The problems dealing with the intersection of surfaces and the drawing of tangent planes, which were profoundly explored in the first theoretical treatise, were successfully used by the author in the theory of arcs in conjunction of various surfaces (Tarasov 1995, pp. 42–43).

This overview of Potier's works would be incomplete without mentioning his *Cours de constructions* (Potier and Bazaine 1816) which was seen as a demonstration of one of the applications of descriptive geometry.

In these works, Potier outlined the main directions of the new science descriptive geometry that was to be later developed by his Russian disciples and followers: Sevastianov, Aleksandr Reder, Nikolaj Durov, Nikolaj Makarov, Valerian Kurdjumov, Nikolaj Rynin and others. Indeed, scientists at the ICEWC paid great attention to

---

[7]In June 1818 for his translations of Potier's two last books Sevast'janov was granted the Order of Sainte Anne 3d class.

the application of the methods of descriptive geometry to the solution of various practical engineering problems.

One of the earliest developments was the work of Sevastianov, *Osnovaniâ načertatel'noj geometrii* (Foundations of descriptive geometry) published in 1821 (Sevast'ânov 1821). In his introduction, the author clearly stressed that:

> The proposed [...] foundations are written according to the model of the *Traité de Géométrie descriptive* [...] by Colonel Potier (Sevast'ânov 1821, p. II).

In 1834, Sevastianov re-edited a new, corrected and complete version of his course (Sevast'ânov 1834). He extended the theoretical part and created Russian terminology, which has remained virtually unchanged up to the present time. Nevertheless, he once again recognized in his introduction that his work was modelled on Potier's treatise.

In 1830, Sevastianov published his own *Priloženie načertatel'noj geometrii k risovaniû* (Application of descriptive geometry to drawing) (Sevast'ânov 1830), which was not only a natural continuation of the theory of descriptive geometry and of its applications to the construction of perspective projections and shadows but also a wonderful guide for studying the basis of drawing. In 1831, his new work *Priloženie načertatel'noj geometrii k vozdušnoj perspektive, k proekcii kart i k gnomonike* (Application of descriptive geometry to aerial perspective, projection of maps and gnomonic) was awarded the Dmidov Prize of the Academy of Science of Saint-Petersburg (Sevast'ânov 1831; Tarasov 1995, pp. 145–163). Continuing his work on the application of descriptive geometry to practice, in 1840, he finished the *Kurs plotničnago iskusstva* (Course of carpentry) (Sevast'ânov 1840).

However, in spite of these new developments, decades after Potier's courses had been firstly published, their application and didactic value remained so great that they were repeatedly re-issued. His *Traité de géométrie descriptive...* of 1816 was re-issued in 1834: it was, as we might say today, a "stereotypical" edition (Pot'e 1834). In 1849, his other course was lithographed under the title *Osnovaniâ načertatel'noj geometrii. Priloženie načertatel'noj geometrii k risovaniû* (Foundations of descriptive geometry. Application of descriptive geometry to drawing) (Pot'e 1849). Here, compared with the earlier editions, two sections were excluded, "The aerial perspective" and "About colours". Judging by the fact that in 1916 Rynin referred to this work (Rynin 1916, p. 251), it remained known in the early twentieth century before being completely forgotten thereafter.

That, Potier's textbooks were still pre-eminent and rivalled but not replaced by Sevastianov's works in the 1830s, is confirmed by Valerian Kiprijanov who studied at the ICEWC between 1832 and 1839. When listing the lithographically printed textbooks and courses they had used during this period, he mentioned regarding "descriptive geometry and its applications" the works of both "General-Major Potier and Colonel Ja. Sevastianov" (Kipriânov 1882, pp. 183, 193).

The same fact was mentioned at the meeting of the Conference of the ICEWC on 6 October 1836 when Potier resigned as director of the institute. The conference examined the Sevastianov's reports, which clearly stipulated that Potier's books were used for teaching descriptive geometry in the fourth class and stone cutting in the third class. Aerial perspective in the second class was taught using both Potier's and Sevastianov's courses. The projection of maps and gnomonic were the only subjects taught entirely using Sevastianov's textbook. By the autumn of 1836, all these books were to be found in sufficient number in the ICEWC's library (*Žurnaly konferencii...* 1836, ff. 79–79v).[8] However, the first edition of Potier's book is absent, perhaps because it had fallen to pieces from constant reading.

4 The Influence of Potier's Works in Russia

In Russia, the influence of Potier's works went far beyond the ICEWC, and Sevastianov's translations greatly contributed to their popularity. Contemporaries were well aware of this fact. Thus the teaching of descriptive geometry was introduced into the Main Engineering School (MES; *Glavnoe inženernoe učilišê*) and at the Cadet Corps of Mines (CCM; *Gornyj kadetskij korpus*) in 1819. In both cases, the course was taught using the Russian versions of the three Potier's treatises of 1816 and 1818 (Pot'e 1816; Potier 1818; Pot'e 1818; Sevast'ânov 1821, p. 3; Sevast'ânov 1819, p. 1; Sevast'ânov 1829, p. 229; Sevast'ânov 1826, no. 2, p. 32; Fabricius 1903, p. 176).

According to Evgenij M. Zablockij (Zablockij, *Gornoe učilišе...*), the teaching of descriptive geometry at the CCM started in 1816, which is doubtful but, in principle, possible for the 1816/1817 academic year although that of 1817/1818 seems more likely. Until 1839, it was taught by a graduate of the Pedagogical Institute (1815), Kondratij Shelejkovskij, who from 1820 also taught the applications of algebra to geometry and descriptive geometry. Potier's books are still preserved in the library of this institution (*Leningradski... gornyj institut...* 1973, p. 31).[9]

The introduction of descriptive geometry into the MES was attributable to the Count Egor Sivers, a member of the Board of Ways of Communication and, from 1820 simultaneously, the head of this school. Sivers insisted on the thorough mathematical training of military engineers. It is a matter of note that Sevastianov thanked him as well as Evgraf Mechnikov, who was director of the CCM from 1817 to 1824, in the preface of his 1821 treatise, "with the ardent feelings of a son of the fatherland" (Sevast'ânov 1821, pp. 6–7). Descriptive geometry in the MES was studied 4 h a week in both elementary and higher officer classes given by Colonel Valuev and Captain Lebedev[10], respectively (Maksimovskij 1869, pp. 33–48).

---

[8]In 1828–1829, Potier's printed course of "Stone cutting" (Potier 1818) was in sale at the institute and costed 10 rubles (CGIA SPb., F.381, op. 13, d. 139, f. 30). This book was still in use for teaching.

[9]The library of the State Mine University of Saint-Petersburg still possesses (Pot'e 1818)—code B.967, and (Potier 1818)—code B28601.

[10]Most probably, it was the captain of the CEWC Vassilij Lebedev, who graduated from the Institute of Ways of Communication in 1816 (Čarukovskij 1883, pp. 3–5/2, 74/2, 19/3), and then, Potier's disciple.

Descriptive geometry was also taught at the Technological Institute (*Tehnologicheskij Institut*) and at the Institute of Civil Engineers (*Institut Grazhdanskih Inzhenerov*). However, they opened much later (the first in 1828 and the second in 1832), and it is not surprising that in the libraries of these schools one can find Sevastianov's adaptations of Potier's books rather than those of Potier himself (Sevast'ânov 1830, 1831, 1834).[11] Especially, as Sevastianov personally taught at the MES from June 1832.

By 1834, descriptive geometry was introduced into the Artillery School (*Artillerijskoe uchilishche*), the Cadet Corp of Navy (*Morskoj kadetskij korpus*) and the School of Sea officers (*Uchebnyj Morskoj Èkipazh*) (Sevast'ânov 1834, pp. I–II).[12]

During the same years, descriptive geometry was progressively introduced into the civil educational institutions.

At the University of Saint-Petersburg, it was taught first by the above mentioned Shelejhovskij (1818–1824), then by Nikolaj Shcheglov, Ivan, Konovalov (1831) and Fedor Chizhov (1832–1840).[13] At the University of Kazan, the professor was Nikolaj Lobachevskij (from 1822). In the 1830s, descriptive geometry was introduced into the University of Kiev. In 1843, a graduate of the officers' classes of the Navy Cadet School, Aleksandr Zelenoj, prepared for his students a "Brief guide of descriptive geometry with annexes" (*Kratkoe rukovodstvo po načertatel'noj geometrii s priloženiâmi*), which was awarded the Demidov Prize in 1845. New school regulations adopted in 1828 specified that the elements of descriptive geometry were to be introduced into secondary schools (Mezenin 1987, pp. 15, 191).

But the most striking story relates to the Polish-speaking regions of the Russian empire, where Potier's *Traité* was published in a Polish version: *Wykład geometryi rysunkowey dla użycia uczniow Instytutu Drog Kommunikacyynych*—in Vilnius and Warsaw as early as in 1817 (Potier 1817a). The translator was Gregor Hreczyna, a Polish mathematician and pedagogue and Sevastianov's coeval, who had recently graduated from the University of Vilnius and was only 21 years old when he translated the book. Like Sevastianov, he had to invent Polish terminology for descriptive geometry for this work. Thereby, Potier's work proved ground-breaking not only for Russian but also for Polish mathematical literature. Potier's treatise in Polish was published by the University of Vilnius. Most likely, this treatise served as a basis for the lectures on descriptive geometry at the university that were given by the magister Hipolit Rumbovich from 1823. In 1829, he published his own

---

[11] In the library of the State Technological Institute in Saint-Petersburg, their codes were: II.7232, II.7298, III.1774/T-A; in the library of the Institute of Civil Engineers (today State University of Architecture and Building)—15.984, 15.985.

[12] RGIA, F. 207, op. 16, d. 104, f. 115–116.

[13] In the library of Saint-Petersburg State University, there is an important collection of Potier's and Sevast'ânov's books (Pot'e 1816; Potier 1818; Pot'e 1818; Idem 1834; Sevast'ânov 1821; Idem 1830; Idem 1831; Idem 1834) (codes: g.II.168, g.II.169, g.II.7943, g.III.123, g.III.124, g.III.795, g.III.796, g.IV.11). However, it does not necessarily mean that they were used for the teaching because the library accepted a significant collection of books from the Censorial Committee, and one needs to study the stamps on each exemplar.

textbook on descriptive geometry in Polish, which was printed in Vilnius at the same university Joseph Zavadsky's printing house as Potier's treatise (Rumbowicz 1829; Voronkov 1977).

As for Hreczyna, he became a professor of mathematics at the *Liceum of Volynsk*, and in 1834–1839, he also taught mathematics and descriptive geometry at the University of Kiev and later in 1839–1840, at the University of Kharkov. However, his name does not appear in reference books on mathematical literature although sometimes he is mentioned in encyclopaedias (in *Brockhaus and Efron*, and thereafter—in some others).

The most important article concerning him appeared in the supplementary volume of the Russian Biographical Dictionary (*Russkij biografičeskij slovar'*), which was published only in 1997 (Grečina 1997). All these editions usually mention the Polish version of Potier's treatise by Hreczyna but do not point out the role he played in the elaboration of the Polish terminology of descriptive geometry nor in the teaching of this discipline. In the same way, that the *École polytechnique* became the centre from which descriptive geometry spread worldwide, the Institute of the Corps of Engineers of Ways of Communication was responsible for its spreading throughout of the Russian Empire.

5 Influence of Potier's Books Outside Russia

In this section, we will examine two groups of Potier's publications: those printed in Russia and those printed abroad. Varied information about Potier's books published in Russia (Potier 1816, 1817c, 1818), as well as their analysis and even mention of their Russian versions by Sevastianov (Pot'e 1816; Potier 1818; Pot'e 1818) can be found in several publications from the 1820s, including the *Bulletin des sciences mathématiques* and the *Bulletin des sciences technologiques* by André E. J. P. J. F. d'Audebard, baron de Férussac (Du[leau] 1828; *Sur les travaux des ingénieurs des voies de communication* 1828). Globally, they all go back to Sevastianov's article about the theoretical works of the CEWC, which was published in the *Žurnal Putej Soobšeniâ/Journal des voies de communication*, in 1826 and in the *Journal du Génie Civil* in Paris, in 1829 (Sevast'ânov 1829; Idem 1826). This article was widely known.

Joseph Marie Quérard provides a description of one French and two Russian publications together with a short biographical note on Potier himself (Quérard 1835, pp. 295–296). Quérard's description includes some strange information that can be explained by a simple error or by the fact that there are some more reprints we don't know about. He, thus, asserts that the *Traité de Géométrie descriptive* was published in Petersburg and in Brunswick (Braunschweig), that they had the one and the same editor—Pluchart, and that the book was issued in 1816 and 1817, respectively. He also differentiates between these two publications from the one that appeared in Paris. A similar situation arises with the book on the application of descriptive geometry to drawing (Quérard 1835, p. 295).[14] However, as yet, no trace of these supposed publications has been found.

It should be noted that all this information concerns almost exclusively mentions and/or reviews of Potier's books. As for the Russian editions, they are very little known in Europe. One can find his applications of descriptive geometry to drawing (Potier 1817c) at the Royal Military Academy in Breda (Holland) and at the *École des ponts et chaussées* in Paris (*Catalogus van de Bibliotheek* ... 1869, p. 189; Catalogue des bibliothèques de l'ENPC). In this context, Germany proves an exception. In 1828 in Leipzig, a military engineer Baron Hungern-Sternberg, a native of the Baltic States and captain in the Russian service, published a book entitled *Projectionslehre* (Treatise of geometrical projections), which was structured according to the plan proposed by Potier in his *Traité* of 1816. The author himself acknowledged this, and it was also pointed out in a review of his book published in the *Revue encyclopédique* (Ungern-Sternberg 1828, C. 7; Livres étrangers 1828). The German title reflects the author's search for a term that would be comprehensible to German readers, but it did not catch on since Poncelet had already created projective geometry.

The situation with the French publication of Potier's *Traité* (1817b) is quite different. First of all, it was immediately included in the list of the ten basic monographs on descriptive geometry cited by Hachette in the foreword to his fundamental *Traité de géométrie descriptive* (Hachette 1822, p. xx), which summed up the achievements in this science before 1822 and was re-issued in 1828. Even in 1916, Rynin referred to it as one of the basic foreign works (Rynin 1916, pp. 224, 252). It was regularly included in libraries' and book-sellers' catalogues (*Catalogue général des livres français* ... 1857, p. 162; *Catalogue général et raisonné de livres français* ... 1828, p. 263) and is mentioned in a history of descriptive geometry (Obenrauch 1897, p. 76).

One can find the Parisian edition of Potier's *Traité* in the libraries of the most important engineering and military schools all over the world: at the *École polytechnique*, *École des ponts et chaussées*, *West-Point Military Academy* and in the libraries of various academies (*Catalogue of the Library*... 1853, pp. 171, 172; *Overzigt van de boeken*... 1860, p. XXIX; *Catalogue des bibliothèques de l'ENPC*). It would seem to indicate that this book was used for teaching purposes. The *Cours industriels*, which opened in the late 1820s at the Royal Academy of Metz, simply took Potier's treatise, together with the course of Monge, as a basic teaching guide (*Mémoires de l'Académie royale de Metz* 1830, p. 113).

In Belgium, the situation was even more striking. In various Belgian and German catalogues—of libraries, book-sellers and recommendatory (*Catalogue d'une nombreuse* ... 1848, p. 18; *70e Catalogue d'une belle collection* ... 1857, p. 32; *Catalogue de la Bibliothèque populaire* ... 1869, p. 40)—there are references to the fact that in Liége in 1842 the *Société d'encouragement de l'instruction élémentaire* republished the *Traité de géométrie descriptive* by Potier (1842). It is

---

[14] Place of publication: Brunswick, publisher: Pluchart.

difficult to think of a better indicator of its didactic importance than that 25 years after its Parisian publication and although other literature dealing with the discipline was subsequently published that Potier's treatise was still being read.

This brief overview allows us to glimpse the significant role that Potier's books, and especially the Parisian publication of his first *Traité*, played in the development and dissemination of descriptive geometry and of its teaching in Europe.

6 The Rise of the Russian School of Descriptive Geometry

Unfortunately, Russian historiography presents endless terminological confusion in its attempts to describe the key positions occupied by Potier and Sevastianov in the history of descriptive geometry in Russia. Some of them are quoted below.

Lidia Pugina: "Ja. A. Sevastianov is the founder of the school of descriptive geometry in our country" (Pugina 1992, p. 75). Vladimir Shulzhevich: "The founder of the Russian school of engineering graphic Ja. A. Sevastianov", "the school of descriptive geometry laid down by Ja. A. Sevastianov" (Šul'ževič 1960, p. 13). This position deprives the Russian school of the right to be considered as a direct branch of the school of Monge. Boris Tarasov's description provokes fewer objections for he tries to avoid use of the word "school": "founder of national descriptive geometry" (Tarasov 1995, pp. 122, 140). But this is also a strained interpretation. Some authors try to go from Sevastianov back to Monge, making a direct link between Monge's treatise of 1799 and Sevastianov's book of 1821 and ignoring what happened in between (Bibikov 1984).

However, in order to evaluate a phenomenon, it is necessary to begin by defining it in order to avoid terminological problems. The most general and operative definition of the word "school" seems to be the following:

> School is a community of people which emerged during their joint activity and is composed of at least two generations that elaborated an epistemological system possessing some specific features, and ensure its inheritance (Guzevič 2003, p. 76).

If we try to identify the individuals who could form the first generation of the Russian school of descriptive geometry, we propose the following names:

- French polytechnicians: Fabre, Bazaine, Destrem and Potier but also Joseph Marie Anne Jean Antoine Auguste Gleize, who taught at the ICEWC in 1813–1814;
- Russian auditors at the *École polytechnique* (Petr Rahmanov, Mihail and Andrej Golicyn, Aleksej Majurov) (Guzevič and Guzevič 2003); a graduate of the *École du génie de Mézières*, Etienne-François de Sénovert (Gouzévitch and Gouzévitch 1995);
- a trainee of the *École des ponts et chaussées* in Paris: Augustin Betancourt.

All of them, except Betancourt, were direct pupils of Monge himself or of his disciple Hachette (for example, the Golicyn brothers, who studied in 1810). It must be emphasized that the frequent assertion that Betancourt was a pupil of Monge is not sustainable: he was simply well aware of Monge's work. However, once enunciated, such an assertion takes on a life of its own contrary to the facts. One can read, for example, that Betancourt followed Monge's lectures in Paris in 1784 (Tarasov 1995, p. 16). Another version: "Betancourt was at that time in Paris where he often met Monge" (Dem'ânov 1986, p. 131). However, there is still no evidence of their acquaintance, and Betancourt can only be considered as a follower of Monge and not as his disciple, which would imply personal contact. In any case, all these individuals who had acquired their knowledge in France and applied it in Russia naturally acted as agents of intellectual transfer.

Two other conditions required for a school to be established are the formation of a specific corpus of knowledge and skills (epistemological system), which has its own distinctive features and provides its inheritance.

Of all the people who have been mentioned, only a few were directly (or hypothetically) related to the teaching of descriptive geometry: Fabre, Potier, to some degree Betancourt and possibly Senovert. Until 1816, teaching was based upon a manuscript copy of Monge's lectures. Thus, during this period, teachers only ensured the transfer of knowledge imported from France, and did not create a new course of their own. In 1815, Potier was in charge of the course. Any one of the four engineers could have taken his place. It would seem that Potier's previous experience of teaching the course in 1811/1812 proved a decisive factor. The decision taken, most probably by Betancourt, turned out to be the right one. Potier, using his previous teaching experience, succeeded in creating new areas of knowledge related to descriptive geometry and its applications which, compared to the Monge's course, had original methodological and didactic features that Potier was able to incorporate in a series of textbooks.

Can we consider them as Russian textbooks, given that they were written in French, or simply as books in a foreign language published in Russia?

Two points seem to justify a positive answer to the first question. Firstly, the French language was one of the main languages in Russian culture, which for nearly two centuries was based on the Russian–French diglossia. Secondly, Potier's works were immediately translated into Russian, and one of them was even published as a bilingual textbook (Potier 1818). To deny Potier's works a place within Russian culture on the basis of language would be similar to denying the Russianness of the ICEWC during the first 25 years of its existence when the language of instruction was French or of the Academy of Sciences in Saint-Petersburg where the working languages at different times were Latin, German or French.

Thus, the first generation of individuals who had studied descriptive geometry under Monge in France, and precisely with Monge, and had returned later to Russia, counted in their number someone whose activity may be considered as founding the Russian school of descriptive geometry, and taught students who became followers in the second generation.

The example of Fedor Rerberg and Sergej Stroganov who graduated from the ICEWC in 1813 is illustrative. In the 1812/1813 academic year Rerberg engaged in teaching, and the next year after, he himself taught descriptive geometry at the institute before being sent to the south of Russia. In 1825, Stroganov established a school of drawing for arts and trade and assumed the role of organizer (since 1860: Stroganov School of Technical Drawing; today: S. Stroganov Moscow State University of Arts and Industries, or, in popular language—"Stroganovka"). In 1825–1827, Rerberg became its first director and taught descriptive geometry (Tarasov 1995, p. 31).

The second generation also includes Aleksandr Devjatnin (Devâtnin) and Vsevolod Denisov. In 1815, they both received the rank of ensign although Devjatnin graduated from the ICEWC in 1817 and Denisov in 1819. Devjatnin became a tutor of the course of construction, which formed part of the applications of descriptive geometry. But, firstly, the course of construction rapidly outgrew its original brief, thanks to Devjatnin who completed and republished its guide-notes in 1819 (Potier et al. 1819). However he was dismissed from the ICEWC in 1823. Denisov also remained at the institute where he initially taught drawing, architecture and rudiments of topographical surveying. And although he was still part of the ICEWC's teaching staff in 1834, he had to abandon teaching the course in drawing because of his deteriorating eyesight, and he left no published work (Larionov 1910, pp. 59, 62, 74, 81).

The most significant of Potier's disciples was, however, Jakov Sevastianov. At the institute, he firstly followed Potier's lectures, then those of Rerberg, and he dedicated his life to the teaching and development of descriptive geometry, first as a tutor, then as a "chief lecturer" (Tarasov 1995, p. 43) and finally as a professor. He started as the translator of Potier's works and became an independent author and the creator of the Russian terminology for descriptive geometry and thereby, doing his utmost to promote it as a background of higher technical education in Russia.

Were Sevastianov works directly linked to those of the first generation?

As the translator of Potier's texts and as a tutor attached to Professor Potier, he logically followed Potier's methodology of teaching and presentation. Having prepared his first independent book and subsequent then its re-editions, he always stressed his link with the works of Potier and their pioneering character (Sevast'ânov 1821, p. II; Sevast'ânov 1834). Sevastianov retained in his own books the axiomatic principle of presentation that was typical of Potier's works. Throughout his professional career and even after, he had resigned, Potier's works were edited and used alongside his own. Finally, Potier and Sevastianov worked at one and the same institution—the ICEWC which served as the institutional frame for the consolidation of the school, and Sevastianov followed Potier as "chief lecturer" and professor of descriptive geometry.

7 Conclusion

The French polytechnician Charles Potier was the founder and first leader of the Russian school of descriptive geometry and engineering drawing (in 1815–1818). He also contributed to the transfer of the discipline to Russia, ensuring, thereby, a genetic link between the Russian school and that of Gaspard Monge. From this point of view, the Russian school can be considered as deriving from Monge's. Sevastianov was a disciple of Potier and the second leader of the school (in 1818–1843), who ensured not only the consolidation and development of the knowledge transferred from France by his teacher but also its extremely rapid and widespread dissemination in Russia. The years 1815–1818 were for Sevastianov a period of apprenticeship under Potier as the head of the scientific school.

Their "areas of influence" were also very different. Potier acted as the agent of the transfer of knowledge from France to Russia, and his books had an influence on the teaching of descriptive geometry in Europe and America. Sevastianov worked within Russia, and his national fame was much broader than that of Potier as the first ten books in the field of descriptive geometry and its applications were translated or written in Russian by him. Outside Russia, however, he is practically unknown as a result of not employing the international languages of science of the nineteenth century: French and, later, German.

This brief survey allows us to access Charles Potier's significant role in the dissemination of descriptive geometry and its applications in Europe. A role all the more significant in that Potier never saw himself as a mathematician but only as a practical engineer, who, having left the chair of descriptive geometry in the ICEWC, never touched mathematical problems again. He was, in short, a polytechnician, the pupil of great teachers and a conscientious man. That is why, having been entrusted with the creation of a school course to teach a new science to future Russian engineers, he pitched it at the highest scientific and didactic level. *La noblesse polytechnicienne oblige!*

Sources

Archives and Manuscripts

Archives de l'École polytechnique, Palaiseau (France) [AÉP]: Registre de Matricules des élèves. Vol. 3: 1803 à 1809. F. 59, 67.
Archives [numérisées] de Paris, État civil de Paris (France) [ANP]: Acte de décès. V3E/D 1444, 1809, no. 30: Vallée, Marie Anne Geneviève.
Central'nyj gosudarstvennyj istoričeskij arhiv S.-Peterburga (Russia) [CGIA SPb]: F. 381. Op. 13. D. 139. Rossijskij gosudarstvennyj istoričeskij arhiv (Russia) [RGIA]. F.159. Op. 1. D.520, č. I; F. 207. Op. 16. D. 104. L. 110–122.
Istoričeskie svedeniâ ob osnovanii instituta (après 1842). [SPb.]. (Manusript. Peterburgskij gosudarstvennyj universitet putey soobŝeniâ, Library [PGUPS]. KP.IV.49).
Žurnaly konferencii Instituta korpusa putej soobŝeniâ: 1836. 1836. [SPb.]. (Manusrit. PGUPS. KP.IV.42).

The Works of Charles Potier and Those of Âkov Sevast'ânov Linked with the Firsts[15]

Printed Works[16]

Potier, Charles. 1816. *Traité de Géométrie descriptive à l'usage des élèves de l'Institut des voies de communication*. St.-Pétersbourg: Chez Pluchart. [2], XII, 129, [1] p., 2 f.

Pot'e, Karl Ivanovič. 1816. *Osnovaniâ načertatel'noj geometrii dlâ upotrebleniâ Vospitannikami Instituta Korpusa Inženerov Putej Soobŝeniâ / Sočinenie* G. Pot'e, Korpusa Inženerov Putej Soobŝeniâ Podpolkovnika i ordena sv. Vladimira 4-j stepeni Kavalera; Perevod [s fr. i predisl.] Inžener-Poručika Sevast'ânova. SPb.: Pri Imp. Akademii Nauk. [6], X, 119, [13] p., 2 f. de pl. (Censorial autorisation 16.6.1816).

Potier, Charles. 1817a. *Wykład geometryi rysunkowey dla użycia uczniow Instytutu Drog Kommunikacyynych* / Przez Potier; gprzetł. G. A. Hreczyna. Wilno; Warszawa: Cnakładem i drukiem Józefa Zawadzkiego, typografa Uniwersytetu Wileńskiego. [12], 95, [1] s., [2] k. tabl. złoż. – N.V.

Potier, Charles. 1817b. *Traité de Géométrie discriptive* / Par M. Potier, élève de l'École Polytechnique. Paris: Chez Firmin Didot. 96 p.

Potier, Charles. 1817c. *Application de la Géométrie descriptive à l'art du dessin; à l'usage des élèves de l'Institut des Ingénieurs des voies de Communication* / Par M. Potier, Élève de l'École Polytechnique, Colonel au Corps des Ingénieurs des voies de Communication, Chevalier de St.-Wladimir 4e classe. SPb.: Chez A. Pluchart. 96 p.

Potier, Charles. 1818. *Traité de la coupe de pierres, à l'usage des élèves de l'Institut des voies de communication* / Par Mr Potier, élève de l'École polytechnique, Colonel au Corps des voyes de communication, chevalier de St.-Wladimir 4e classe = Pot'e K.I. Načal'nye osnovaniâ razrezki kamnej: Dlâ upotrebleniâ vospitannikami Instituta Korpusa Inženerov Putej Soobŝeniâ / Sočinenie Gna. Pot'e, vospitannika Politehničeskogo učiliŝa, Korpusa Inženerov Putej Soobŝeniâ Polkovnika i ordena Sv. Vladimira 4-j stepeni Kavalera; Perevod [s fr.] Inžener-Kapitana Sevast'ânova. SPb.: De l'imprimerie de P.P. Alexandre Pluchart = SPb.: V tip. A. Plûŝara, MDCCCXVIII. 80 p., 8 f. de pl. (Censorial autorisation - 1.5.1818).

Pot'e, Karl Ivanovič. 1818. *Priloženie načertatel'noj geometrii k risovaniû dlâ upotrebleniâ vospitannikami Instituta Korpusa Inženerov Putej Soobŝeniâ / Sočinenie G. Pot'e, Korpusa Inženerov Putej Soobŝeniâ Polkovnika i ordena sv. Vladimira 4-j stepeni Kavalera. Perevod [s fr. i predisl.] Inžener-Kapitana Sevast'ânova*. SPb.: Pri Imp. Akademii Nauk. VI, [2], 90, [7] p., 1 f. de pl. (Censorial autorisation - 30.1.1818).

Pot'e, Karl Ivanovič. 1834. *Osnovaniâ načertatel'noj geometrii, dlâ upotrebleniâ vospitannikami Instituta Korpusa Inženerov Putej Soobŝeniâ / Sočinenie G. Pot'e, Korpusa Inženerov Putej Soobŝeniâ Podpolkovnika i ordena sv. Vladimira 4-j stepeni Kavalera. Perevod [s fr.] Inžener-Poručika Sevast'ânova*. 2-e izd. s 1-go, otpečatannogo v 1816 g. SPb.: Tip. Glavn. Upr. Putej soobŝ. i publičnyh zdanij. [8], VI, 105, XI p., 2 f. de pl.

Potier, Charles. 1842. *Traité de Géométrie descriptive. Nouvelle édition augmentée et publiée par la Société d'encouragement pour l'instruction élémentaire à Liège*. Liège: Dessain. 96 p., 4 f. de pl. – N.V.

---

[15] For Potier's works on descriptive geometry and for the associated works of Sevast'ânov, complete bibliographic descriptions are given because of their low dissemination and difficult access.

[16] In the bibliographical database OCLC WorldCat, the second version of the title is orthographied differently: "Wykład geometrii rysunkowej dla użycia uczniów Instytutu Dróg Komunikacyjnych".

Lithographed Courses

Potier, Charles Michel, Bazaine Pierre Dominique. 1816. *Cours de construction.* SPb., (1816/17). (Lithogr.). – N.V.
Potier, Charles Michel; Bazaine Pierre-Dominique, Déviatnine Alexandre. 1819. *Cours de construction.* SPb. (Lithogr.). – N.V.
Henry, André Guillaume, [Potier Ch. M., Bazaine P. D., Déviatnine A.] 1827. *Cours de construction à l'usage de l'institut des Ingénieurs des voies de communication.* St. Pétersbourg: De l'imprimerie Lithographique de Lange. 277, [7] p., XXXIV f. de pl. (Lithogr.; Ex. in: Rossijskaâ Nacional'naâ biblioteka [RNB]. 15a.51.1.4, 15a.58.4.73, B.7/16.0; PGUPS. 6053).
Pot'e Karl Ivanovič. 1849. *Osnovaniâ načertatel'noj geometrii: Priloženiâ načertatel'noj geometrii k risovaniû* / Soč. Pot'e; Institut Korpusa Inženerov Putej Soobŝeniâ. SPb.: I.K.I.P.S. 32 f. (IKIPS, no. 10). (Lithorg.; Ex. in: RGIA. F.1609. Op.1. D.395. 37 f.).

The Works of Â. Sevast'ânov Going Back to Potier's Works and/or Being Their Developments

Sevast'ânov, Âkov A. 1821. *Osnovaniâ načertatel'noj geometrii*/ Izdannye Â.A. Sevast'ânovym, Korpusa Inženerov Putej Soobŝeniâ Majorom i Kavalerom. SPb. [7], XII, 186, [1] p., XII f. de pl.
Sevast'ânov, Âkov A. 1830. *Priloženie načertatel'noj geometrii k risovaniû; Teoriâ tenej, linejnaâ perspektiva, optičeskie izobraženiâ* / Izdannoe Â.A. Sevast'ânovym, Korpusa Inženerov Putej Soobŝeniâ Polkovnikom i Kavalerom; Čerteži v osobennoj knižke, sostoâŝej iz dvadcati listov. [2 t.]. SPb. [T.1: Tekst]. [4], IV, XI, 152, [1] p.; [T.2]: Sobranie čertežej, sostavlennyh Korpusa Inženerov Putej Soobŝeniâ Poričikom Demidovym. [2] p., 20 f. de pl.
Sevast'ânov, Âkov A. 1831. *Priloženie načertatel'noj geometrii k vozdušnoj perspektive, k proèkcii kart i k gnomonike; Čerteži na devâtnadcati listah* / Izdannoe Â.A. Sevast'ânovym, Korpusa Inženerov Putej Soobŝeniâ Polkovnikom. [2 t.]. SPb. T.[1]: [Tekst]. [8], III, XII, [1], 156 p.; T.[2]: Sobranie čertežej, sostavlennyh Korpusa Inženerov Putej Soobŝeniâ Poručikom Demidovym. XIX f. de pl.
Sevast'ânov, Âkov A. 1834. *Osnovaniâ načertatel'noj geometrii* / Â. A. Sevast'ânova, Korpusa Inženerov Putej Soobŝeniâ Polkovnika, i ordenov: Sv. Vladimira 3 stepeni i sv. Anny 2 st. s almaz. ukraš.kavalera. Izd-e vtoroe, ispr. I dop.; čerteži na četyrnadcati listah. SPb.: Tip. Gl. upr. Putej soobŝeniâ i publičnyh zdanij. [4], III, [5], XIV, 204 p., 14 f. de pl.
Sevast'ânov, Âkov A. 1840. *Kurs plotničnago iskusstva: S 8û Čertežami.* SPb. 33 f., [8] f. de pl. (Subhead at the 1st page of the text: Načal'nyâ Osnovaniâ Plotničnago Iskustva / Polkovnika Sevast'ânova). (Lithogr.; Ex. in: PGUPS. E 1140).

References

70e Catalogue d'une belle collection de livres…20 octobre 1857. 1857. Gand: Salle de ventes publique de livres.
Bibikov, E. 1984. Student, sozdavšij novuû nauku. *Tehnika – molodeži* 11: 11.
Brunot, André, and Roger Coquand. 1982. *Le corps des ponts et chaussées.* Paris: CNRS.
Callot, Jean Pierre. 1982. *Histoire de l'Ecole polytechnique.* Paris-Limoges: Charles Lavauzelle.
Čarukovskij, Aleksej. 1883. Spisok lic, okončivšhih kurs nauk v Institute inženerov putej soobŝeniâ imperatora Aleksandra I s 1811 po 1882 g. SPb.: Tip. A. Benke.
Catalogue d'une nombreuse et riche collection de bons livres de …Lundi 22 Mai 1848. 1848. Gand: Salle de ventes publique de livres.

Catalogue de la Bibliothèque populaire communale (Ville de Liège). 1869. Liège: J. Ledoux.
Catalogue des bibliothèques de l'ENPC. http://lib.enpc.fr/loris_internet/jsp/system/win_main.jsp.
Catalogue général des livres français, italiens, espagnols, etc., tant anciens que modernes, qui se trouvent chez Barthés et Lowell. 1857. Londres; Paris: Barthés et Lowell.
Catalogue général et raisonné de livres français, italiens, espagnols, partugais, latins, etc. 1828. London: A. Dulau & Co.
Catalogue of the Library on the U.S. Military Academy, West Point N.Y. 1853. New York: Jons F. Trow.
Catalogus van de Bibliotheek der Koninklijke militaire akademie. 1869. Breda: Brosse.
Dem'ânov, Vladimir P. 1986. *Geometriâ i marsel'eza.* Moskva: Znanie.
Du[leau], Alphonse J. Cl. 1828. Sur les travaux des ingénieurs des voies de communication, en Russie. (Journal des voies de communication, publié à Saint-Pétersbourg, pour 1826, no. 2, p.26). *Bulletin des sciences technologiques* 10: 170.
Fabricius, Ivan. 1903. *Voenno-inženernoe vedomstvo v carstvovanie imperatora Aleksandra I (Očerk I).* Paris: SPb.
Fourcy, Ambroise. 1828. *Histoire de l'École Polytechnique.* Paris: L'École Polytechnique.
Gouzévitch, Dmitri. 2001. *Mise en place et développement des sciences techniques dans les années 1820–1870.* Paris: SPb.
Gouzévitch, Dmitri, and Irina Gouzévitch. 1993. Les contacts franco-russes dans le monde de l'enseignement supérieur technique et de l'art de l'ingénieur. *Cahiers du Monde russe et soviétique* 34 (3): 345–368.
Gouzévitch, Dmitri, and Irina Gouzévitch. 1995. Etienne-François de Sénovert, traducteur en français des oeuvres de J. Steuart: Trois volets d'une vie: 1753–1831. In *James Steuart: Colloque International (14-15-16 septembre 1995, Château de Vizille),* t.2: 1–22. [Grenoble].
Gouzévitch, Dmitri, and Irina Gouzévitch. 1996. Note de l'ingénieur-colonel Raucourt de Charleville concernant des voies de communication en Russie (Avec la particip. de W.Bérélowitch). *Cahiers du monde russe* 37 (4): 479–504.
Gouzévitch, Dmitri, Irina Gouzévitch, and Nikolaj Eliseev. 2013. *Karl Ivanovič (Charles Michel) Potier (1785–1855): Troisième directeur de l'Institut du Corps des ingénieurs des voies de communication.* SPb.: PGUPS.
Grečina, Grigorij Vasil'vič. 1997. *Russkij biografičeskij slovar',* t.: Gogol' – Gûne, 449–450. Moskva: Aspekt-press.
Guzevič, Dmitrij. 2003. Naučnaâ škola kak forma deâtel'nosti. *Voprosy istorii estestvoznaniâ i tehniki* 1: 64–93.
Guzevič (Gouzévitch), Dmitrij, and Irina Guzevič (Gouzévitch). 1995. *Petr Petrovič Bazen:* 1786–1838. SPb.: Nauka.
Guzevič, Dmitrij, and Irina Guzevič. 2003. Pervyj russkij učenik école polytechnique (Petr Rahmanov). *Istoriko-matematičeskie issledovaniâ,* II ser., 8 (43): 186–208.
Guzevič, Dmitrij, and Irina Guzevič. 2015. *Gabrièl' Lame v Rossii, ili odin iz likov Ânusa.* SPb.: Poltorak.
Hachette, Jean Nicolas Pierre. 1822. *Traité de Géométrie discriptive, comprenant les application de cette géométrie aux ombres, à la perspective et à la stéréotomie.* Paris: Cirby by Guillaume et Cie.
Henry, Guillaume. 1828. Recherches sur le plan d'un traité théorique et pratique des constructions, ou d'un cours complet de construction, à l'usage des ingénieurs des voies de communication. *Journal des voies de communication* 12: 61–80.
Izvestiâ. 1811. *Voennyy žurnal* 15: 79–80.
Kipriânov, Valerian A. 1882. K vospominaniâm o Korpuse inženerov putej soobšeniâ i ego Institute. In K.V.A. *Očerki: Iz zapisok V.A.K.* Vyp.1, 163–235. Moskva: È. Lissner i Û. Roman.
Larionov, Aleksej M. 1910. *Istoriâ Instituta inženerov putej soobšenia imperatora Aleksandra I za pervoe stoletie ego sušestvovaniâ: 1810–1910.* SPb.: Tip. Û. N. Èrlih.
Leningradskij ordena Lenina i ordena trudovogo krasnogo [znameni gornyj institut imeni G.V. Plehanova: 1773–1973]. 1973. Leningrad: LGI.

Livres étrangers. 1828. Projectionslehre. – Traité des projections géométriques, par le baron d'Ungern-Sternberg, capitaine du génie russe. Leipzig, 1828...*Revue encyclopédique ou analyse raisonnée des productions les plus remarquables*... 39 (juil.): 671.

Livres nouveaux. 1817. *Journal des savans*. Oct.: 637.

Maksimovskij, Mihail S. 1869. *Istoričeskij očerk razvitiâ Glavnogo inženernogo učiliŝa*: 1819–1869. SPb.: Tip. AN.

Mémoires de l'Académie royale de Metz: Lettres, sciences, arts, agriculture. 1830. XIe année: 1829–1830. Metz: Mme Thiel; Paris: Bachelier.

Mezenin, Nikolaj A. 1987. *Laureaty Demidovskih premij Peterburgskoj Akademii nauk*. Leningrad: Nauka.

Nordštejn, Aleksandr. 1905. Vypiski iz tetradej inženera Nordštejna. *Russkij arhiv* 3 (10): 233–270.

Obenrauch, Ferdinand Josef. 1897. *Geschichte der darstellenden und projectiven Geometrie: mit besonderer Berücksichtigung ihrer Begründung in Frankreich und Deutschland und ihrer wissenschaftlichen Pflege in Österreich.*.., S.76. Brünn: C. Winiker.

Overzigt van de boeken, kaarten, penningen enz., ingekomen bij de Koninklijke Akademie van wetenschappen, te Amsterdam, van junij 1857 tot april 1860. 1860. Amsterdam: C.G. van der Post.

Programme pour l'Examen des élèves de l'Institut des ingénieurs des voyes de communication. 1816. SPb.

Programme pour l'Examen des élèves de l'Institut des ingénieurs des voyes de communication. 1818. SPb.

Pugina, Lidiâ V. 1992. *Stanovlenie Peterburgskoj matematičeskoj školy*: Diss. ...kand. fiz.-mat. nauk. Moskva: IIET RAN.

Quérard, Joseph Marie. 1835. *La France littéraire, ou Dictionnaire bibliographique*..., t.7. Paris: Firmin Didot frères.

Rumbowicz, Hipolit. 1829. *Geometrya wykreślna, czyli wykład rzutowych i obrazowych wykreśleń z dodatkiem prawidełoznaczania cieni i stopniowania światła, tak w rzutach jako też i w perspektywie, napisana dla użytku uczniów uniwersyteckich przez Hipolita Rumbowicza*. Wilno: Nakładem autora, drukiem Zawadzkiego.

Rynin, Nikolaj A. 1916. *Načertatel'naâ geometriâ: Metody izobraženiâ*. Petrograd: Tip. A.È. Kollins.

Sevast'ânov, Âkov A. 1819. *Načal'nye osnovaniâ analitičeskoj geometrii*. SPb.: Pri Imp. AN.

Sevast'ânov, Âkov A. 1826. Des travaux des officiers du Corps des ingénieurs des voies de communication: Partie théorique (Trad. du russe). *Journal des voies de communication* 2: 26–32; 3: 1–15; 5: 1–18.

Sevast'ânov, Âkov A. 1829. Notice sur les travaux des officiers du Corps des ingénieurs des voies de communication de l'Empire de Russie. *Journal du Génie civil* 5: 225–249.

Šul'ževič, Vladimir A. 1960. *K 150-letiû pervoj kafedry načertatel'noj geometrii v Rossii: Ûčebnoe posobie po teme «Istoriâ načertatel'noj geometrii i ee priloženij»*. Leningrad: LII'T.

Sur les travaux des ingénieurs des voies de communication. 1828. (Journ. des voies de communication; 1826, nr 2, p. 26; nr 3, p. 1; nr 5, p. 1). *Bulletin des sciences mathématiques, astronomiques, physiques et chimiques*, 9(mai): 149.

Tarasov, Boris F. 1995. *Âkov Aleksandrovič Sevast'ânov: 1796–1849*. SPb.: Nauka.

Ungern-Sternberg, Eduard, baron, Kais. Russ. Ingenieur-Capitain. 1828. *Projectionslehre: Géométrie descriptive*. Leipzig: F.A. Brockhaus. (Lith.).

Vallée Louis-Léger. 1821. *Traité de la science du dessin ; contenant la théorie générale des ombres, la perspective linéaire, la théorie générale des images d'optique, et la perspective aérienne appliquée au lavis: pour faire suite à la Géométrie descriptive*, vol. 2. Paris: Mme Ve Courcier.

Voronkov, Boris A. 1977. Matematičeskoe obrazovanie v Vil'nûsskom universitete v 1803–1831 gg. In *Voprosy istorii nauki i tehniki Pribaltiki: Tezisy dokladov XI Pribaltijskoj konf. po istorii nauki i tehniki*, 56–61. Tartu.

Zablockij, Evgenij M. Gornoe učiliŝe – Gornyj institut imp. Ekateriny II: administraciâ, prepodavateli, činy gornogo vedomstva // Gornoe professional'noe soobŝestvo dorevolûcionnoj Rossii. (http://russmin.narod.ru/bioGi.html).

Chapter 14
Engineering Studies and Secondary Education: Descriptive Geometry in the Netherlands (1820–1960)

Jenneke Krüger

Abstract Descriptive geometry was mentioned in a plan for lessons for the first time in the Netherlands in 1819, at the School for Artillery and Military Engineers in Delft. The teacher was Isaac Schmidt, who in 1821 published a translation of *Essais de géométrie* by Lacroix. In 1828 the school was moved to Breda and at the same time became the Royal Military Academy, with an updated and more demanding curriculum, for military and some civilian (non-military) engineers. In 1842 the Royal Academy, for civilian engineers only, was established in Delft. In 1840–1841 Hendrik Strootman, teacher at the Royal Military Academy, published the first original Dutch textbook on descriptive geometry; his colleague Jacob Badon Ghyben published a textbook in 1858. The books by these two authors were widely used until the twentieth century, at the Royal Military Academy, at the Royal Academy and also in some secondary schools.

The lack of reliable secondary education remained a problem for both Academies until the law on secondary education of 1863. This law defined a new type of secondary school, the *Hogere Burger School* (Higher School for Citizens), and also a Polytechnic School, as successor to the Royal Academy. The HBS (*Hogere Burger School*) provided a good preparation for this Polytechnic School. Descriptive geometry was a subject in the final exams of the HBS until 1958. The Polytechnic School developed into the Delft University of Technology; descriptive geometry was mentioned in the formal program until 1969.

Keywords Descriptive geometry · Projective geometry · Isometric perspective · Methods of projection · Royal Military Academy · Isaac Schmidt · Hendrik Strootman · Jacob Badon Ghijben · Royal Academy · Polytechnic School · Secondary school · HBS · Institute for Technology

---

J. Krüger (✉)
Freudenthal Institute, Utrecht University, Utrecht, Netherlands
e-mail: j.h.j.kruger@uu.nl

© Springer Nature Switzerland AG 2019
É. Barbin et al. (eds.), *Descriptive Geometry, The Spread of a Polytechnic Art*,
International Studies in the History of Mathematics and its Teaching,
https://doi.org/10.1007/978-3-030-14808-9_14

1 Politics and Education in the Netherlands

In the United Provinces, the name used for the Dutch Republic from the sixteenth century until the Batavian Revolution in 1795, mathematical sciences were appreciated especially for their usefulness in practical applications. The renowned *Duytsche Mathematique* (1600–1679) set an example for the teaching of mathematical subjects to (future) practitioners.[1] In the eighteenth century mathematics instruction took place in small private schools and institutions such as the Foundation of Renswoude, in which talented orphans learned mathematical sciences within the framework of an education for technical professions (Krüger 2012, Roberts 2012). Later in the century, drawing schools were founded in some larger towns to enable the craftsmen to learn drawing, architecture and mathematics, without having to pay the high fees demanded by the guilds (Lottman 1985).

Throughout the eighteenth century the lack of well-trained engineers was seen as one of the causes of many problems in the country; in particular as a major cause of the continuous lack of military success of the Dutch Republic. Private initiatives, such as the *Fundatie van Renswoude* (Foundation of Renswoude), attempted to improve the training of some engineers by providing a theoretical mathematics foundation in combination with practical training to able students (Krüger 2014), but that was only scratching the surface. In 1789, after the Patriotic revolution of 1785–1787, the first three national schools for artillery were established, in the Hague, Breda and Zutphen, all with the same curriculum (Janssen 1989).

During the period 1795–1815 there were many political changes, with a much more centralized government as one of the outcomes. This government took some responsibility for education, which would result in a law on primary education (1806), followed by the creation of institutes for the formation of engineers and finally in 1863 in a law on secondary education, including a type of school in which sciences, mathematics and modern languages formed a large part of the curriculum. To the Dutch, at that time water management seemed more important than secondary education; already by 1803 water management had developed from a loose collection of regional organizations into a fairly large and bureaucratic national institute, *Waterstaat* (Water board). This is relevant to the history of descriptive geometry in the Netherlands, as water management traditionally had close ties with military engineering (Lintsen 2009).

After the defeat of Napoleon the economic situation of the new Dutch monarchy was rather dismal. Improving industry and commerce were priorities of the new king and his ministers; but development of a reliable army and thus military education were equally important. In 1814 a new military school, the School for Artillery and Military Engineers, was established in Delft. This School had as its major aim the formation of military engineers; however, it was also possible to study, as a civilian,

---

[1]The *Duytsche Mathematique* (Dutch Mathematics) was the first Dutch course for engineers, established at Leiden University on request of Maurits van Nassau, with a teaching plan by Simon Stevin (Krüger 2010).

water management or shipbuilding (Janssen 1989; Lintsen 2009). From around 1820 descriptive geometry was taught at this school. In 1828 the school moved to Breda, under a new name: the Royal Military Academy. As such it is still situated in Breda.

The combination of military and civilian engineering proved unsatisfactory; in 1842 a Royal Academy, for the formation of civilian engineers only, was founded, again in Delft. At this Royal Academy one could study engineering (water management, shipbuilding and mining), chemistry, mechanics and calibration; there were also courses for civil servant in the East-Indies (at present Indonesia). Descriptive geometry was part of the curriculum.

On a lower level technical training of craftsmen traditionally happened mainly through the guilds, however, these were abolished in 1798. Initially several towns established one or more drawing schools, in which geometry, perspective and drawing were taught in the evenings to (future) craftsmen (Krüger 2014). In 1829 the king issued a decree on a national curriculum for all 400 drawing schools. During the first 2 years there would be a communal program for the artistic and technical stream, during which drawing techniques and theoretical and practical geometry were taught. After the second year descriptive geometry, architecture, drawing and constructions were major subjects in the technical stream (Lintsen 2009). So from 1829 some descriptive geometry was taught in drawing schools to craftsmen.

A problem common to the Royal Military Academy and the Royal Academy was the lack of a reliable system of secondary education, with a qualified curriculum for all pupils. Consequently the first-year students at the Royal Military Academy and at the Royal Academy were often not well prepared, with a lack of mathematical knowledge and insufficient command of languages. This situation improved somewhat through the introduction of admission exams for both the Academies, but overall the level of education after primary school was rather low. Legislation was largely dependent on the good-will of the king, general secondary education did not have priority. However in 1848 a new constitution, proposed by the liberal statesman Johann Rudolf Thorbecke (1798–1872), was accepted by the king, with a shift in power from the monarch to the parliament.

In 1863 Parliament accepted for the first time a proposal for a law on secondary education, by minister Thorbecke. Two important features of this law were the introduction of a new type of secondary school, the *Hogere Burger School* or HBS (Higher School for Citizens) and the introduction of a Polytechnic School, the replacement for the Royal Academy. At the HBS mathematics, sciences and modern languages formed an important part of the curriculum and descriptive geometry was part of the mathematics curriculum. This school served also as preparation for the Polytechnic School (Krüger and van Maanen 2014). Similar to the Royal Academy, the Polytechnic School was not part of the structure of higher education, formally it was secondary education. This rather strange situation lasted until 1905, when the Polytechnic School became the *Technische Hogeschool* (Institute of Technology) which was then considered higher education; eventually it became the Delft University of Technology.

At the universities descriptive geometry was taught from 1877, following the law on higher education from 1876, which allowed a more prominent position for mathematics. The increasing relevance attached to descriptive geometry is also visible in the occurrence of this topic in the program of the science department of some *gymnasia*[2] between 1838 and 1863, for instance, in Deventer, Maastricht and Leiden (Smid 1997).

2 The School for Artillery and the Royal Military Academy

By 1814 descriptive geometry had been a feature of the formation of craftsmen and engineers in France for at least 20 years (Chap. 2, this volume). However, it was as far as is known, not a part of the curriculum of the School for Artillery and Military Engineers in Delft during its first years. The first professor in advanced mathematical sciences and physics was Jacob de Gelder (1765–1848); his assistant was Isaac Schmidt (1782–1826). As had been the case in the *Duytsche Mathematique* in the seventeenth century and in the Foundations of Renswoude in the eighteenth century, theory and practice were combined in the curriculum, which took 4 years to complete (Janssen 1989). De Gelder was a highly praised mathematician and teacher, author of several textbooks and articles. As far as is known he did not write about nor teach descriptive geometry; he was more interested in analysis, in the spirit of Lagrange (Beckers 1999). It is not known whether his assistant Schmidt taught some descriptive geometry before 1819. In that year, after disagreement with the director of the school on the amount and the style of mathematical theory which was to be taught, De Gelder was appointed professor at the University of Leiden and Schmidt took care of the mathematics lessons at the School for Artillery (Janssen 1989). Schmidt was trained as a painter before his appointment as assistant in mathematics, so he was skilled in drawing, he knew about perspective and how to represent objects on the plane. He published several textbooks on mathematics, mostly translations. In 1821 he published a translation of *Essais de Géométrie sur les plans et les surfaces courbes (Elémens de géométrie descriptive)* by Lacroix (1808), with the title *Grondbeginselen der beschrijvende meetkunst*. This was the first textbook on descriptive geometry in Dutch language. So at least from 1820, perhaps 1819, descriptive geometry was taught to the cadets of the School for Artillery and Engineers, based on a translation of the textbook by Lacroix. An outline for a curriculum, dating from 1819, mentions in year one technical drawing and in year two descriptive geometry in combination with 'perspective' (NL-HaNa 2.13.22, inv. 261).

---

[2]From 1838 until 1863 a gymnasium consisted of two departments: a Latin school, with emphasis on Latin and Greek, and a Science department, with modern languages and sciences. The Latin school prepared for admission to the university. From 1864 the Science departments were closed and the students were transferred to a HBS.

In 1829, during the first year of the Royal Military Academy, a committee to review the mathematics curriculum proposed a number of changes to 'the course of Mr Schmidt', as the program was still called (NL-HaNa 2.13.22, inv. 164). The resulting plan covered a 4-year course. In the first year the cadets were taught plane and solid geometry and geometric constructions; in the second year descriptive geometry, trigonometry, algebra, applications of algebra to geometry and physics. Differential and integral calculus were taught in the third and fourth year. Until the end of the century there were frequent adaptations of the curriculum and of the admission exams of the Royal Military Academy, but descriptive geometry remained part of the curriculum.

The successors of Schmidt were Hendrik Strootman (1799–1851) and Jacob Badon Ghijben (1798–1870). Both favoured a thorough theoretical mathematics instruction before the cadets started on the military subjects. In 1828 Isaac Delprat (1793–1880) became head of mathematics education; he had studied at the *École des Ponts et Chaussées* during 1811–1813 (NNBW, vol. 6). Delprat himself published many books and articles during his long and successful career; he encouraged his teachers to write new textbooks for the courses at the Military Academy. The first textbook by a Dutch author, which was not a straightforward translation, was published by Strootman in 1840, followed in 1841 by an expanded version, *Gronden der beschrijvende meetkunst* (Elements of descriptive geometry). This last version had a second edition in 1847, when Strootman was teaching at the Royal Academy in Delft, where he was reader in mathematics from 1844 to 1851. His former colleague, Badon Ghijben, published *Gronden der beschrijvende meetkunst* (Elements of descriptive geometry) in 1858, some years after Strootman had died. Strootman and also Badon Ghijben followed more or less Schmidt/Lacroix in their approach, mainly the same topics and the same order of treatment, divided over two sections. In all three textbooks the applications treated are perspective and light and shadow. The book by Badon Ghijben is the most elaborate, regarding text and figures. These and other books by Strootman and Badon Ghijben were also used at some *gymnasia* and the HBS (Krüger 2014). On comparing the book by Schmidt/Lacroix with the books by Strootman and Badon Ghijben, differences become apparent as well. We will discuss some differences.

The introduction and first paragraphs are illustrative. Schmidt-Lacroix started straight away with the position of a point, determined by the intersection of two lines.

We shall start by elaborately explaining how to represent the various parts of space on some proposed planes. The place of a point on any plane is determined by the position of two lines passing through this point (Schmidt 1821, p. 1).[3]

Strootman and Badon Ghijben both gave a definition or rather description of descriptive geometry (see below). Strootman was reminiscent of Monge (Chap. 1, this volume) in the reasons given for the importance of descriptive geometry and

---

[3] All translations of quotations are by Jenneke Krüger.

the professions for which it was useful: mechanical engineers, architects, artists, craftsmen, military engineers. He also stated that

> Descriptive geometry facilitates the learning of analytic geometry, it trains and strengthens the intellect and imagination and it provides those who practice it with an attractive and serious pastime (Strootman 1841, p. 2).

Towards the end of the century and even more so in the twentieth century the argument that descriptive geometry 'trains and strengthens the intellect and imagination' would be used regularly to defend its position in the curriculum of secondary schools.

Strootman then discussed the position of a point as the centre of a sphere, described several ways of determining the position of a point in space and concluded that there was an easier way: the method of projections. Again, this was similar to what Monge (1811) wrote. Badon Ghijben started straight away with the easiest way to determine the position of a point in space. They both emphasized constructions as practical exercises.

Badon Ghijben explained that when solid figures are depicted on a plane through perspective, the size and angles change. His text is not very different from the text by Strootman, though slightly more accurate and he also refers to Monge.

> Descriptive geometry thus teaches: how it is possible to represent solids in an exact way by means of a drawing; reversely how to construct from such a representation the true size and shape of the solids and their parts and lastly, by performing constructions in these drawings, the ways to find all properties one wishes to know of the figure represented (Badon Ghijben 1858, p. 2).

The mathematical treatment differs slightly. In the book by Schmidt-Lacroix there is more emphasis on propositions, corollaries, consequences of the proven propositions and general mathematical methods. In the books by Strootman and Badon Ghijben the emphasis is more on constructions, of projections and of solids; nevertheless there are also proofs of some statements, usually by the same methods as used by Schmidt.

Strootman discussed right at the start the projection of a curve, and mentioned the cylinder (Fig. 14.1).

Badon Ghijben started with points and lines and explained more thoroughly (Fig. 14.2).

A difference between the French authors, including Schmidt, and the Dutch authors is in the choice of the applications. Lacroix took perspective as an application of descriptive geometry, he also made some remarks on shadows and on sundials. As Schmidt produced a faithful translation, these applications are in his textbook as well. In Hachette (1822) one finds in Chap. 2 the applications locus, shadow and perspective, anamorphosis and stereographic projection and in Chap. 3, stereotomy and the construction of architectural structures. Strootman and later Badon Ghijben discussed shadow and perspective, but Strootman also discussed isometric perspective. Badon Ghijben did the same. Isometric perspective was used in Great Britain as a form of graphical representation (Chap. 18, this volume).

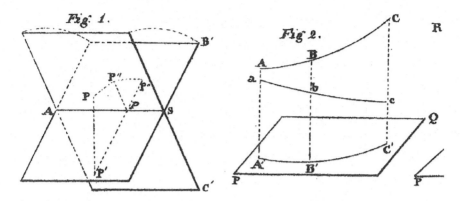

Fig. 14.1 First figures in Strootman (1841, 1)

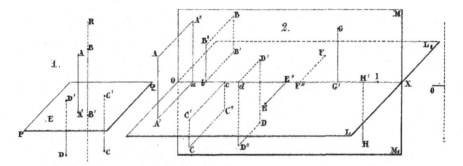

Fig. 14.2 First figures in Badon Ghijben (1858, I)

On this type of representation Strootman wrote (p. 174): "Now we shall say something about the isometric perspective of professor Farish.[4]" This was followed by an example: the cube in isometric perspective, followed by a more general treatment (Fig. 14.3), curved lines and some solids.

He finished this part with

> The isometric perspective [...] is often preferable above the usual perspective; the rules are more practical and simple and thus give rise to more accurate constructions. That parallel lines of equal length remain parallel and of equal length in isometric perspective is a great advantage [...]. This type of perspective is especially suited to represent all kind of implements and physics instruments, when it is important to show clearly the relation between the parts and their functioning; it is often used for those aims, especially in England (Strootman 1841, p. 178).

---

[4]Professor William Farish (1759–1837) of Cambridge University provided detailed rules for isometric drawing. He published his ideas in an article "On Isometrical Perspective", in *Cambridge Philosophical Transactions*. 1 (1822), In this paper he discussed the "need for accurate technical working drawings free of optical distortion".

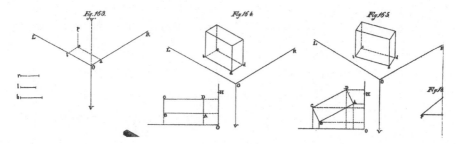

Fig. 14.3 Isometric perspective, a parallelepiped in two projections (Strootman 1841, IX)

Fig. 14.4 Grinding-engine model, Farish (Strootman 1841, IX)

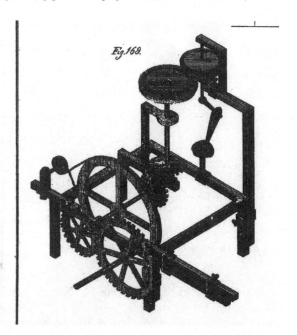

Strootman evidently considered isometric perspective a useful technique for his engineering students, whether they were military or civilian. As a practical example of the use of isometric perspective, he took the drawing of the optical-grinding engine model from Farish' publication (Fig. 14.4). See for more information about Farish in (Lawrence, Chap. 18, this volume). We may conclude that Strootman and probably the teachers at the Royal Military Academy looked further than the Polytechnic School in Paris and the French literature. At the time Strootman wrote his textbook, dissatisfaction with the limits of descriptive geometry to solve practical problems, for instance, with the construction of arches in the network of railways in Great Britain, had become apparent (Sakarovitch 1993).

Strootman passed away in 1852. Badon Ghijben followed Strootman in the treatment of isometric perspective, without mentioning Farish. He also did not include an illustration of an instrument.

3 The Royal Academy–The Polytechnic School–The Institute for Technology

3.1 The Royal Academy

The combination of the formation of military engineers together with civilian engineers proved to be problematic. In 1842 the Royal Academy was founded in Delft, with the aim to offer a thorough scientific education to future industrialists and engineers. Lintsen (2009) points out that the Polytechnic School in Karlsruhe probably served as an example for the Royal Academy. The programs of the Polytechnic School in Karlsruhe (Benstein, Chap. 9, this volume) were indeed requested by the Ministry of Inland Affairs (NL-HaNA 3.12.08.01, inv. 39). At the Academy one could study engineering,[5] but it was also possible to follow a course for civil servant in the colonies in the East Indies (at present Indonesia).

After a few years some problems became apparent. The courses for civil servant in the East Indies proved to be more attractive than engineering; the number of engineers graduating was disappointing. Moreover, there were financial struggles, because the Royal Academy had to provide its own income. In 1842 it had received ƒ 10,000, books and some instruments to enable a start, thereafter it received each year a small sum for the East-Indian studies, but that was all the support the government provided. The income had to come from student's fees, which was insufficient for the expenditure.

The first director was Antoine Lipkens (1782–1846), who for many years had promoted the establishment of a Polytechnic Institute. Lipkens was an engineer and an inventor, who studied at the *École polytechnique*, followed by a career in France, before returning to the Netherlands. He had been employed at the ministry of Internal Affairs since 1828.

Rehuel Lobatto (1797–1866) was appointed professor of mathematics, an indication of the importance of mathematics as a central subject for the Academy. Lobatto was a respected mathematician, who had frequent contact with Adolphe Quetelet, wrote many publications, published in *Crelle's Journal* and, like Lipkens, worked for the government (Krüger 2014; Stamhuis 1989). He did not publish on descriptive geometry.

Descriptive geometry was part of the curriculum for engineers during the first 3 years of the 4-year course (NL-HaNA 3.12.08.01, inv. 32). The first teacher in mathematics and physics, apart from professor Lobatto, was Willem Lodewijk Overduyn (1816–1868), who received his doctorate at Leiden University. Strootman was his mathematics colleague from 1844 until 1851, when health problems prevented him to continue teaching. The successor of Strootman was Lewis Cohen Stuart (1827–1871), who also taught descriptive geometry. Cohen Stuart was a

---

[5]Baudet (1992) mentions as graduates: civil engineers, shipbuilding engineers, mining engineers, surveyors of weights and measures, civil servants for the East Indies.

student at the Academy from 1844 to 1848 and in 1864 would become the first rector of the newly established Polytechnic School. From 1854 the former student, Ryklof van Goens,[6] taught descriptive geometry. Van Goens was an admirer of Monge; according to Cardinaal his lectures were much appreciated by the students (Cardinaal 1906). Probably Strootman taught descriptive geometry to Cohen Stuart and to van Goens.

In 1843 the program for descriptive geometry consisted of the theory as in the first volume of the book by Strootman (straight line and plane, the trihedral angle, polyhedron and sphere), in combination with drawing of constructions. In the second year the timetable for engineering students contained on each day 1 h theory of descriptive geometry, followed by 2 h technical drawing. About 20 years later, in 1860–1861, the mathematics program of engineers was as follows:

Year 1: algebra, elementary geometry, trigonometry, start of analytic and of descriptive geometry.
Year 2: advanced algebra and differential calculus, continuation of analytic and descriptive geometry.
Year 3: descriptive geometry, including construction of several implements, integral calculus and statics.
Year 4: integral calculus and statics, hydrostatics, dynamics and hydrodynamics, theory of steam engines (NL-HaNA 3.12.08.01, inv. 32).

By then the program for descriptive geometry had expanded to include curves, surfaces and the construction of representations of implements. Evidently drawing remained an important part of the lessons. Dissatisfaction with the number of graduates in engineering, with student behaviour and with the quality of the Royal Academy in general resulted in its closure in 1864. The Academy was replaced by a Polytechnic School, for which the government took financial responsibility. All the teachers at the Royal Academy were dismissed; some of them were appointed as professors at the Polytechnic School.

3.2 The Polytechnic School

The law on secondary education of 1863 stated in the chapter on the Polytechnic School that this institute was meant for the formation of

1. future industrialists or technologists, who desire more theoretical and technical knowledge than is provided by a *hogere burger school* [HBS]
2. those who wish to qualify for civil engineer, architect, naval engineer, mechanical engineer or mining engineer.

---

[6] Van Goens graduated in 1851.

The Polytechnic School had a 2-year communal program for all engineers (the B-program), followed by a specialization of 2 years in one of the five different engineering disciplines mentioned in the law on secondary education (the C-program). It combined thus preparatory years and specialization for civilian engineers. Mathematics and sciences were the main subjects. As the new type of secondary school, the HBS, provided a preparatory program, equal to the propaedeutic exam (A-exam) of the Polytechnic School, the starting point for the teaching could be at a more advanced level than previously.

Mathematics was the basis for all students. At the start there were five professors of mathematics. Three of them came from the Royal Academy: Lobatto, honorary professor, Cohen Stuart, director of the Polytechnic School, and van Goens. Newly appointed were Franciscus van den Berg (1833–1897), a water management engineer and former student of Lobatto, and Lewis Cohen and George Baehr (1822–1898). Baehr received his first mathematical education while in the army, studied mathematics at the university of Groningen and was a mathematics teacher at a gymnasium, before he was appointed professor at the Polytechnic School. Both Baehr and van den Berg favoured French textbooks; indeed next to a textbook by Lobatto, books by Sturm, Briot and Bouquet, Frenet and Lefébure de Fourcy were used in analysis and a book by Legendre was used for spherical trigonometry (Krüger 2014).

Van Goens and van den Berg both taught descriptive geometry. The textbooks used were the book by Badon Ghijben, which had several new editions throughout the years, and *Géométrie descriptive* by Leroy, teacher of descriptive geometry at the *École polytechnique* until 1849. Van den Berg also had in his possession *Traité de géométrie descriptive* by Lefébure de Fourcy and three books by Strootman (Cardinaal 1906; Mannoury 1907). The professors apparently preferred descriptive geometry as taught at the *École polytechnique* at the time of Leroy to the method of changes as promoted by Olivier and followers (Chap. 2, this volume). Leroy's textbook is quite extensive, mainly concerned with the theory. Leroy himself had no practical experience in engineering (Sakarovitch 1993).

Mathematics for the B-exams contained descriptive and analytic geometry, spherical trigonometry and analysis (algebra, differentiation, integration). The lectures in descriptive geometry started with the sphere, continuing with perspective, shadows, curves and surfaces, curves of intersection, tangent planes, development of surfaces and shadows of solids with curved surfaces. In 1870–1871 central, parallel and axonometric perspective were added to the program, possibly influenced by developments in German countries (Chap. 8, this volume) and in France, where de la Gournerie introduced lessons in axonometry into the courses of descriptive geometry (Sakarovitch 1993). Monge omitted the mention of axonometric (or isometric) perspective.

At the Polytechnic School in Delft the course of descriptive geometry consisted of 1–2 h theory and 2–4 h drawing practice each week. Though the description of the program remained the same until 1905 (*Programma der lessen*), the actual content changed gradually. An exam of 1870 consisted of three problems, respectively on central perspective, the construction of a cone enveloping an ellipsoid and the construction of the shadow of a screw with square thread (Onderwijsverslagen 1864–1877).

An example of a problem in the B-exam of 1899 is on axonometric projection.

$\angle ZO'Y = 135°$, $\angle ZO'X = 120°$. A cylinder of revolution (radius=4 cm, height=15 cm.) has its base on XOY and touches the planes ZOX, ZOY (O is not specified in the text). The cylinder is illuminated from a point L on the axis OX; $OL = 18$ cm. Determine the axonometric projection of the cylinder, its shade and its cast shadow on the plane ZOY (Well 1899).

Axonometric projections were part of the exams from at least 1880 (van de Well 1899).

Jacob Cardinaal (1848–1922), professor of mathematics at the Polytechnic School from 1894, recounted in 1906 that, during the first years after the start of the Polytechnic School, mathematical subjects and engineering subjects seemed to grow somewhat apart, the role of mathematics as basis for engineering studies and the ideal of mathematics as a pure science, eminently suitable to form the logical mind, proved a difficult combination. Gradually the German work in geometry became more influential, especially that of Fiedler (see Menghini, Chap. 4 and Volkert, Chap. 10, this volume) and Wiener.

> In descriptive geometry the comparative study of methods of projection was introduced, with the result that perspective and axonometry no longer seemed arbitrary chapters, but were seen as belonging to a coherent whole; the treatment of curves and ruled surfaces also became connected to the general newer methods. That is how the monotony of the older French textbooks was broken and the work was more in the spirit of W. Fiedler and Chr. Wiener. The big step to incorporate projective geometry in the course was not taken...(Cardinaal 1906, pp. 143–148)

The addition in 1870 of axonometric and parallel projection to the program may have been an attempt to diminish the perceived gap between theory and practice, mentioned above.

From 1902 to 1906 Hendrik de Vries (1867–1954), former assistant to Fiedler, was professor in Delft, teaching descriptive geometry, including different methods of projection.[7] He published articles and textbooks on descriptive and projective geometry (see below) and on differential and integral calculus.

The position of the Polytechnic School as part of secondary education was too restrictive to enable keeping up with new developments in technology and science. From 1862, when the law on secondary education was proposed to Parliament, members of Parliament and other parties had argued that the Polytechnic School ought to be part of higher education, comparable to a university. However, at that time the very influential minister Thorbecke insisted that the main criterion to distinguish between higher and secondary education should be whether or not Latin was the teaching language and not the age of the students (Krüger 2014). Finally, in 1905, the Polytechnic School became the *Technische Hogeschool* (Institute of

---

[7]De Vries was assistant to Fiedler at the ETH in Zürich from 1890 to 1894, with descriptive and projective geometry as his areas of work.

Technology), part of the structure of higher education, with a similar organization and rights as the four universities and usually indicated as TH. Since 1986 it is the Delft University of Technology.

At the Polytechnic School there had been relatively small changes in the teaching of descriptive geometry, nevertheless gradually aspects of a more modern program were introduced.

3.3 The Institute of Technology (TH)

The faculty for General Science of the TH took care of the propaedeutic program, which took 2 years. During the first years after 1905 all mathematics was taught in General Science, which meant that mathematics was all-important for the propaedeutic exam (Alberts 1998). The subjects were higher algebra, principles of differentiation and integration, analytic geometry, descriptive geometry and spherical trigonometry. Descriptive geometry was considered by many mathematics professors in General Science as the most important subject for engineering students (Bijl 1966). Cardinaal wrote in 1909 in his report on the mathematics curriculum:

> It is of the utmost importance to develop the imagination of the future engineer. The lessons and exercises in descriptive geometry are most suited to this aim. [...] In all mathematical topics the teaching of descriptive geometry is central (in Bijl 1966, p. 8).

During the first half of the twentieth century the yearly published 'program of lessons' for descriptive geometry mentioned methods of projection, curves and surfaces, just as during the nineteenth century. That does not necessarily mean that the mathematics was the same. In 1905 professor de Vries and professor Cardinaal taught descriptive geometry. In 1908 de Vries, who by then was professor at the University of Amsterdam, published a textbook on descriptive geometry, *Leerboek der beschrijvende meetkunde* (Textbook of descriptive geometry). De Vries wrote in his foreword that the publisher, Waltman, had asked him to write a textbook in between a small elementary book and the large scientific French and German volumes, in which descriptive and projective geometry were fully treated. In his introduction de Vries mentioned the aim of descriptive geometry.

> The aim is to perform geometrical constructions on figures which are not situated in one plane; usually these figures are not present, but they are determined by some characteristics. So before constructions can be performed these figures in space have to be represented in one plane, following simple set rules: projections (Vries 1908, pp. 1–2).

The author stated that the book was written for readers who knew something of descriptive geometry, which would be the case for students who came from the HBS.

The first chapter was on central projection, starting with a pencil of rays, reminiscent of Fiedler's *Die darstellende Geometrie* (1871); the chapter finished with the theorem of Pascal and the theorem of Brianchon. The second chapter was on perspective, with among others the theorem of Dandelin and finishing with stereographic projection. The third chapter treated orthogonal parallel projection, with a referral to Monge and to Hermary. The last chapter treated skewed parallel projection, axonometry and Cavalière perspective. De Vries left Delft before his book was published, but it seems likely that he had in mind students of the TH as well as students of the university when writing his textbook. Clearly the teaching of descriptive geometry had expanded since 1870.

Hendrik van Veen, professor of pure and applied mathematics in Delft, published a textbook in two volumes on descriptive geometry, in 1925 and in 1929. He treated much the same topics as de Vries, with the addition of many problems, solved and unsolved. In the introduction van Veen discussed n-dimensional sets; he distinguished between algebraic curves and transcendental curves, symmetry was a concept used, as in "the top of a cone is a point of symmetry".

Meanwhile, the Technical faculties, such as Electrical Engineering, preferred a less theoretical propaedeutic exam, with more mathematics applicable to engineering, far less mathematics in a mainly formative role and they had no use for descriptive geometry. There were several attempts to modernize the program, e.g. in the late 1930s, with requests to abolish descriptive geometry and the elaborate drawing exercises in projection methods. However the majority of the mathematicians who taught the propaedeutic program at the TH in Delft continued to favour a rather traditional style of mathematics, with descriptive geometry as the main symbol of the propaedeutic function of mathematics (Alberts 1998).

But by 1955–1956 descriptive geometry was only obligatory for civil engineering and for aircraft technology. The program consisted of orthogonal and skew affinity; projection methods (orthogonal, skew parallel, orthogonal axonometry); principles of central projection; applications to planes, spheres, cylinders and cones; intersections and shadows. In 1960–1961 only the four projection methods and applications to planes were mentioned (*Programma der lessen*). This latter change may have had something to do with the disappearance of descriptive geometry from the curriculum of the HBS (see below), though there were important changes in the engineering courses of the TH as well. In 1968–1969 both projective and descriptive geometry were still mentioned in the program, but in 1969–1970 descriptive geometry was no longer mentioned.

From 1905, ongoing developments in different branches of geometry gradually found their way into the courses of the TH, though not in all faculties at the same rate. Methods of projection were eventually integrated in studies such as mechanical engineering, under different labels (Smid 1994).

4 The HBS, Preparation for "Delft"

In 1863 the HBS, the new type of secondary school, had two main aims:
- preparation for technical education, specifically the Polytechnic School in Delft;

- general education, in order to prepare sons of citizens for higher positions in industry and commerce.

Students with a diploma of the HBS were exempted from the propaedeutic exam (exam A) of the Polytechnic School. The law of 1863 specified for the HBS a broad range of 18 subjects, among them three modern languages, economic subjects, the sciences and mathematics. In order to receive a diploma the student had to pass a final examination which covered 16 subjects. One of these was mathematics, which in itself consisted of four topics: arithmetic/algebra, geometry, trigonometry and descriptive geometry (Krüger and van Maanen 2014).

When elucidating his proposal for this law, minister Thorbecke stated that the main role of mathematics in the curriculum of the HBS was to provide support for science. His main collaborator, from 1864 inspector of secondary education, Daniel Steyn Parvé, did not fully agree with him; Steyn Parvé considered both the supportive role of mathematics and the formative role, stimulation of thinking and reasoning, equally important. During the nineteenth and early twentieth century the formative aim of mathematics at the HBS would be more and more emphasized by mathematics teachers. This would influence the discussion about descriptive geometry as well, as gradually the formative value of descriptive geometry was emphasized as well.

In 1870 the program for descriptive geometry consisted of "the principles of descriptive geometry up to curved surfaces". A further description was not necessary; the books used in the Royal Military Academy, by Schmidt-Lacroix, by Strootman and by Badon Ghijben all treated the same topics more or less in the same order. Interestingly, in the examination program for mathematics of 1870 an optional topic was mentioned: new geometry (projective geometry). It was formulated as follows.

> Some knowledge of harmonic division, transversals and centres of similitude will be an advantage (Krüger 2014, p. 336).

It is not clear who initiated this rather modern topic, which would be introduced at the Polytechnic School in later years. In German speaking countries there was a discussion about the desirability of introducing this new mathematics in the Gymnasium. Professor Bierens de Haan (Leiden University) who in 1862 had advised on the proposed law for secondary education, had written an article about it and there were a few Dutch textbooks on some elements of projective geometry ('new' geometry) written by teachers of mathematics (Krüger 2014). The reason given at the time to treat this topic at the HBS was that many problems could be solved more conveniently through the methods of 'new' geometry (see Chap. 4, this volume). The topic was from the start taught at some schools, for example, at the HBS in Zwolle by Mr. Boxman, a mathematics teacher with a military background (Krüger 2014). Until the first decennium of the twentieth century, textbooks for the HBS on projective geometry had several reprints. So, even if it was an optional topic, during more than 40 years it was taught at least at some secondary schools. However, projective geometry did not have a lasting impact on the curriculum for mathematics at the HBS, though in the early twentieth century it would probably have been a good preparation for the Institute of Technology.

Teachers at the HBS were fairly autonomous regarding the content of their lessons, the order of treatment, the choice of textbooks, etc. They took into account the admission demands of institutes such as the Royal Military Academy, the Royal Marine Institute and the Polytechnic School and of course the requirements of the examination program. Quite a few mathematics teachers wrote textbooks for the HBS, including books on descriptive geometry. During the first years of the HBS the books by authors of the Royal Military Academy, Strootman and Badon Ghijben were often used; in some schools French (Catalan, Jullien) or German books (Brennecke) were used. From 1869 textbooks written specifically for the HBS appeared. One of the first books on descriptive geometry was by Adrianus van Pesch, an alumnus of the Royal Academy, who taught at the HBS in Deventer, before his appointment as professor at the Polytechnic School in Delft. The fourth edition of his book appeared in 1916; the content of the 1916 edition covered points, lines, planes, orthogonal projections, intersections of planes, rotation of lines and planes, changes of plane of projection, (trihedral) angles, distances, polyhedrons and sphere (Pesch 1916). The authors covered at least orthogonal projections of points, lines and planes, intersections, angles between planes, rotation of planes, trihedral angles and polyhedrons, with a great number of exercises; often a chapter on cylinders, cones and the sphere was added. In 1947 a textbook contained all the questions on descriptive geometry in the final exams from 1876 till 1940 (Thijn and Kobus 1947). Apparently at the HBS the subject had not changed very much in over 60 years. Most questions were about projections of polyhedrons and about intersection of a plane with a polyhedron; exam questions on a sphere or a cone were rare mentioned in an exam question. An example of a question in the exam of 1940 is the following.

> ABC is the base of the truncated prism $ABCDEF$, of which AD, BE and CF are the upright edges. The base is known: ABC is situated in the horizontal projection plane in front of the vertical projection plane; $AB = 7$cm and is parallel to the axis at a distance of 4cm; A is situated to the left of B, $AC = 8$cm; $BC = 6$cm. Furthermore: the angle of BE with the vertical projection plane is 300, E is situated in the vertical projection plane, 6cm above the axis, E is to the right of B; The distance of F to the horizontal projection plane is equal to its distance from the lateral face $ABED$; $DF = DE$. Construct the horizontal and the vertical projection of the truncated prism (Thijn and Kobus 1947).

In 1864 the goal of descriptive geometry at the HBS was to prepare students for a career in technology. It soon became clear that about 50% of the HBS students did not continue in a technical direction, so descriptive geometry was of little or no use for them from the point of view of career prospect. However, for mathematics as a whole and for descriptive geometry in particular, the emphasis of leading teachers during the first half of the twentieth century was very much on the value of mathematics education in the training of logical reasoning. In 1921 Eduard Dijksterhuis, teacher at a HBS and very dominant in the discussion on mathematics education, wrote in a journal for teachers about the aims of mathematics at the HBS (phrasing adapted by author).

The great value of pure mathematics is not in the first place in the attained results of extensive knowledge of properties; it is the style of mathematics and the mood of strict honesty evoked by an exact argument, which determines the high moral value of this subject (Dijksterhuis, in Groen 2000, p. 224).

This type of argument was repeated regularly in publications before the second world war (Groen 2000). Descriptive geometry was considered as "most suited to enhance the power of imagination and to promote logical thinking" (Nooten 1882). This was very much in accordance with the ideas expressed by professors of mathematics at the Polytechnic School, later the Institute for Technology. However, during the later part of the nineteenth and in the twentieth century descriptive geometry at the HBS became limited to a standard set of procedures and techniques, which the students could practice themselves. It is doubtful that this topic really promoted logical thinking for many students. As mixing topics, for example, using algebraic methods in geometry, was frowned upon, it also was an isolated topic (Hiele 2000). Descriptive geometry could thus hardly be considered a good preparation for future studies for the majority of the students of the HBS.

From very early on it was apparent that not everybody was happy with the position of descriptive geometry in the curriculum of the HBS. Already in 1874, when there was much discussion about the overloaded program of the HBS and especially the many subjects in the final examination, Steyn Parvé proposed as part of a solution to make descriptive geometry optional, so the approximately 50% of the students who did not plan to choose a technical profession could drop it (Krüger 2014). During the remainder of the nineteenth century several parties proposed diminishing the workload for students at the HBS, often mentioning abolishing descriptive geometry as an option. During the first half of the twentieth century the position of descriptive geometry was regularly questioned by teachers and others; nevertheless the review committees for the mathematics program always advised to maintain the topic, without major changes in the content.

One reason for keeping this isolated subject in the curriculum of mathematics was the strong presence in the review committees of the rather conservative mathematics professors of the TH in Delft. Another reason was the positive effect on the marks in the final examination. Descriptive geometry exams were predictable, so with a bit of practice students could gain high scores. This could balance lower scores for, e.g. algebra or trigonometry.

In the 1950s much discussion about the mathematics curriculum of the HBS took place, involving mathematics teachers, pedagogues, mathematicians and scientists, such as Tatjana Ehrenfest, Hans Freudenthal, Marcel Minnaert and Gerrit Mannoury (Wansink 1953). In 1958 a proposal for a new mathematics curriculum was accepted, in which the practical use of mathematics for other subjects was a decisive criterion to include topics, somewhat similar to the ideas of Thorbecke, a 100 years before. The formative value of mathematics was no longer mentioned. Descriptive geometry thus lost an important reason to be taught at secondary schools and the content was outdated. In 1958 it disappeared from the curriculum of the HBS. Some remnants, such as projection methods, were integrated with solid geometry (Bastide-van Gemert 2015; Groen 2000).

5 Conclusion

In the Netherlands mathematics was traditionally appreciated for its use by practitioners, as is evident from the success of the *Duytsche Mathematique* in Leiden in the seventeenth century and the many small private schools and teachers who taught mathematics to practitioners in the seventeenth and eighteenth century. Descriptive geometry fitted in that tradition (Barbin, Chap. 2, this volume).

As was the case in neighbouring countries the first institutes in which descriptive geometry became part of the curriculum were institutes for the training of (mainly military) engineers. Very soon drawing schools, providing lessons in the evening for craftsmen, also included some lessons in descriptive geometry. The curriculum for engineers was based on the French tradition, with a translation of Lacroix by Isaac Schmidt as the first textbook used in the lessons, from 1820. The deductive order used by Lacroix was thus introduced into the formation of (military) engineers in the Netherlands.

About 20 years later the first textbook on descriptive geometry by a Dutch author, Hendrik Strootman, appeared. It was based on the work of Monge and Lacroix, but there was also influence from the UK through the treatment of isometric perspective as used by Farish. The argument to introduce this projection method was again a practical one; it made the representation of instruments easier and made the relation between the parts and their function more transparent. This may be seen as an example of the growing tension between two points of view in (technical) education: should the emphasis in mathematics be on applicability, support for engineering and physics or should the emphasis be on pure mathematics, the general training of logical mathematical thinking? These two contradictory viewpoints would give rise to tension and conflicts around the programs for mathematics at institutes for the formation of engineers and in secondary schools, during the second half of the nineteenth and the first half of the twentieth century.

Strootman can be seen as a link between the two institutes for training of engineers; he first taught military engineers, at the Royal Military Academy and later on civilian engineers, at the Royal Academy in Delft. Two of his students became professor at the Polytechnic School, which replaced the Royal Academy in 1864. During the first years of the Polytechnic School there was still a strong influence of the teachers at the *École polytechnique*, as is noticeable from the textbooks used, but after some years, the influence of German mathematicians such as Fiedler and Wiener became stronger. The Royal Academy and the Polytechnic School provided both a preparatory and a final education for engineers, as was the case in other countries.

The lack of a solid system of secondary education posed severe problems. The Royal Academy, inspired by the model of the Polytechnic School in Karlsruhe, did not only receive inadequate funding, it also struggled with the insufficient knowledge of first-year students. These problems were remedied when the Polytechnic School replaced the Royal Academy. The introduction of the HBS, a secondary school with a curriculum of 5 years, in which modern languages, sciences and mathematics took about 65% of the teaching time and with final examinations on all subjects, proved very successful. One of the aims of the instruction at the HBS was preparation for technical studies, in particular the Polytechnic School. At the start the mathematics program of the HBS had a practical aim: to support physics instruction and to prepare for the Polytechnic School. About 50% of the students at the HBS who passed their final exams pursued a career in technology.

There was a strong alignment between the Polytechnic School and the HBS, originating in the law on secondary education and strengthened by the teachers. From 1864 all students who started at the Polytechnic School were supposed to have studied descriptive geometry up to curved surfaces, part of the mathematics program and of the final examinations of the HBS. Gradually the ideas about the role of mathematics at the Polytechnic School and at the HBS changed. The mathematicians at the Polytechnic School saw mathematics as mainly formative for the future engineers and descriptive geometry as the most important topic, the core of the program. A growing number of mathematics teachers at the HBS agreed with that point of view.

The large amount of mathematics in the propaedeutic years of the Polytechnic School and the emphasis on the formative function of mathematics as opposed to the supportive role became the cause of many complaints, of the Engineering faculties and of the students. Gradually the program of the Polytechnic School evolved, under influence of the developments in mathematics elsewhere, especially in Germany. In 1905 the Polytechnic School became the Institute of Technology, part of Higher Education, which enabled modernization of the programs and more specialization. However, the majority of the mathematics professors of the propaedeutic years did not budge as far as the position of descriptive geometry was concerned; notwithstanding complaints about the irrelevance of the topic for many specializations and about the amount of time students had to spend on it. There were gradual changes in the content of the program to make it more modern.

Something similar was going on at the HBS; descriptive geometry took a lot of time and only about 50% of the students would need it later on. Projective geometry remained an optional topic and as the final exams became more and more influential, it disappeared from the program. Proposals to make descriptive geometry optional or abolish it altogether from the curriculum were made regularly, starting in 1874, however the topic kept its position for more than 90 years.

In 1864 descriptive geometry was mainly seen as a useful preparation for technical studies, a very practical aim and as such it was in alignment with the program of the Polytechnic School. Towards the end of the nineteenth century the reasons to teach it had become more formative: the positive influence on the power of imagination and the promotion of logical thinking. A quite different argument was the positive influence on the results of the final examination. At the HBS the program for descriptive geometry and the questions in the exams did not change much through the years, students could get relatively high grades through practice. Thus descriptive geometry became for many students limited to the training of a restricted number of techniques for the final exam. The mathematics professors of the Institute of Technology and some leading teachers of the HBS were influential in maintaining descriptive geometry in the curriculum of the HBS and at the TH.

After the Second World War the need for usefulness of mathematics in the curriculum of secondary schools was again emphasized, by now a reason to remove descriptive geometry from the curriculum of the HBS, in 1958. At the TH descriptive geometry was for the last time mentioned in the program of 1968–1969.

References

Archives and Digital Libraries

DBNL. Digitale bibliotheek voor de Nederlandse letteren. http://www.dbnl.org/
NNBW. Nieuw Nederlandsch Biografisch Woordenboek. http://resources.huygens.knaw.nl/retroboeken/nnbw/
NL-HaNa. 2.13.22 Koninklijke Militaire Academie (KMA), 1816–1941, inv. 164, 261
NL-HaNA. 3.12.08.01 Technische Hogeschool Delft, 1841–1956, inv. 32, 39
SGD Staten Generaal Digitaal. www.statengeneraaldigitaal.nl/
Trésor, Library of Delft University.

Publications

Alberts, Gerard. 1998. *Jaren van berekeing*. Amsterdam: Amsterdam University Press.
Badon Ghijben, Jacob. 1858. *Gronden der beschrijvende meetkunst*. Breda: Broese & Comp.
Bastide-van Gemert, Sacha Ia. 2015. *All Positive Action Starts with Criticism*. Dordreccht: Springer.
Baudet, Ernest Henri Philippe. 1992. *De lange weg naar de Technische Universiteit Delft*. I De Delftse ingenieursschool en haar voorgeschiedenis. Den Haag: SDU.
Beckers, Danny. 1999. Lagrange in the Netherlands: Dutch attempts to obtain rigor in calculus: 1797–1840. *Historia Mathematica* 26: 224–238.
Bijl, Jacob. 1966. *Propedeutisch wiskundeonderwijs en ingenieursopleiding*. Delft: Delftsche Uitgevers Maatschappij N.V.
Cardinaal, Jacob. 1906. Eenige mededeelingen over de wiskunde aan de Polytechnische School. In *Gedenkschrift van de Koninklijke Akademie en de Polytechnische school 1842–1905*. Delft: J. Waltman jr.
Fiedler, Wilhelm. 1871. *Die darstellende Geometrie in organischer Verbindung mit der Geometrie der Lage*. Leipzig: B. G. Teubner.
Groen, Wim. 2000. Honderd jaar leerplanwijzigingen. In *Honderd jaar wiskundeonderwijs*, eds. Fred Goffree, Martinus van Hoorn and Bert Zwaneveld. Leusden: NVvW.
Hachette, Jean Nicolas. 1822. *Traité de géométrie descriptive*. Paris: Corby.
Hiele, Pierre van. 2000. De illusie van het streng redeneren. In *Honderd jaar wiskundeonderwijs*, eds. Fred Goffree, Martinus van Hoorn and Bert Zwaneveld. Leusden: NVvW.
Janssen, Jan. 1989. *Op weg naar Breda*. 's Gravenhage: Sectie Militaire geschiedenis Landmachtstaf.
Krüger, Jenneke. 2010. Lessons from the early seventeenth century for mathematics curriculum design. *BSHM Bulletin* 25: 144–161. https://doi.org/10.1080/17498430903584136.

Krüger, Jenneke. 2012. Mathematics education for poor orphans in the Dutch Republic, 1754–1810. In *"Dig where you stand"* 2. *Proceedings of the conference on the History of Mathematics Education*, eds. Kristin Bjarnadóttir, Fulvia Furinghetti, Jose Matos & Gert Schubring, pp. 263–280. Lisbon: Universidade Nova.

Krüger, Jenneke. 2014. *Actoren en factoren achter het wiskundecurriculum sinds 1600*. Utrecht: PhD thesis University Utrecht. http://dspace.library.uu.nl/handle/1874/301858.

Krüger, Jenneke, and Jan van Maanen. 2014. Evaluation and design of mathematics curricula: Lessons from three historical cases. In *Proceedings of CERME 8*, Antalya 2013.

Lacroix, Sylvestre-François. 1808. *Essais de géométrie sur les plans et les surfaces courbes (ou Éléments de géométrie descriptive)*. Paris: Fuchs.

Lintsen, Harry (ed.). 2009/1994. *Geschiedenis van de techniek in Nederland. De wording van een moderne samenleving 1800–1890*. Deel V. Zutphen: Walburg Press. (DBNL, 2009).

Lottman, Elizabeth. 1985. *Materiaal tot de geschiedenis van het ontstaan van tekenacademies en -scholen*. Zeist: Rijksdienst voor de Monumentenzorg.

Mannoury, Gerrit. 1907. *Systematische catalogus van de boekerij van het Wiskundig Genootschap*. Amsterdam: Delsman & Nolthenius.

Monge, Gaspard. 1811. *Géométrie descriptive par G. Monge avec un supplément par M. Hachette*. Paris: Klostermann fils.

Nooten, Sebastiaan van. 1882. *Vraagstukken over beschrijvende meetkunde*. Schoonhoven: Sebastiaan E. van Nooten & Zn. Onderwijsverslagen (1864–1877). SGD.

Pesch, Adrianus van. 1916. *Leerboek der beschrijvende meetkunde*. Groningen: Noordhoff. Programma der lessen, 1865–1970. Delft, Trésor.

Roberts, Lissa. 2012. Instruments of science and citizenship: science education for Dutch orphans during the late eighteenth century. *Science and Education* 21: 157–177.

Sakarovitch, Joël. 1993. La géométrie descriptive, une reine déchue. In: Bruno Belhoste, Amy Dahan Dalmedico and Antoine Picon (eds.). *La formation polytechnicienne 1794–1994*. Paris: Dunod.

Schmidt, Isaac. 1821. *Grondbeginselen der beschrijvende meetkunst*. 's Gravenhage: gebr. Van Cleef.

Smid, Harm Jan. 1994. Meetkunde en techniek. *Nieuwe Wiskrant* 13 (4): 13–15.

Smid, Harm Jan. 1997. *Een onbekookte nieuwigheid? Invoering, omvang, inhoud en betekenis van het wiskundeonderwijs op de Franse en Latijnse scholen 1815–1863*. Proefschrift: TU Delft.

Stamhuis, Ida. 1989. *'Cijfers en Aequaties' en 'Kennis der Staatskrachten'*. Amsterdam: Rodopi.

Strootman, Hendrik. 1841. *Gronden der beschrijvende meetkunst*. Breda: Broese & Comp.

Thijn, Abraham van, and Michel Kobus. 1947. *Inleiding tot de beschrijvende meetkunde*. Groningen: J.B. Wolters.

Veen, Hendrik van. 1925. *Leerboek der beschrijvende meetkunde*. Deel I. Groningen: P. Noordhoff.

Veen, Hendrik van. 1929. *Leerboek der beschrijvende meetkunde*. Deel II. Groningen: P. Noordhoff.

Vries, Hendrik de. 1908. *Leerboek der beschrijvende meetkunde, deel 1*. Delft: Waltman Jr.

Wansink, Johan. 1953. De wiskunde-werkgroep van de W.V.O. *Euclides* 28 (5): 197–205.

Well, Gerardus. 1899. *Oplossingen der wiskundige opgaven der examens B van de Polytechnische School te Delft*. Deventer: Kluwer.

Chapter 15
The Rise and Fall of Descriptive Geometry in Denmark

Jesper Lützen

Abstract The history of descriptive geometry in Denmark is primarily a story of teaching and its institutional setting. Only at the very end did a Danish mathematician contribute original research to the story. The subject was introduced in Denmark around 1830 in connection with the foundation of two new colleges, one civil and one military that were both inspired by the *École polytechnique*. The subject continued to be taught at the civil polytechnic college for about a century, after which descriptive geometry disappeared from Danish education. At the very end of the period, Hjelmslev's geometry of reality added an original approach to Danish descriptive geometry; otherwise Danish descriptive geometers limited themselves to importing new ideas from abroad, in particular from France and Germany. However, Danish textbooks and exam questions bear witness to a high theoretical level of the descriptive geometry education in Copenhagen.

Keywords Descriptive geometry · Denmark · *Polyteknisk Læreanstalt* · Ludvig Stephan Kellner · Carl Julius Ludvig Seidelin · Johannes Hjelmslev · Georg Frederik Ursin · Hans Christian Ørsted · University of Copenhagen · Projective geometry · History of mathematics

1 Mathematics in Denmark in the Nineteenth Century

During the nineteenth century, Danish mathematics rose from a well-informed but provincial level to an international level where Danish mathematicians contributed in a serious way to mathematical research. At the beginning of the century there were two universities in the lands ruled by the Danish King: The University of Copenhagen (founded in 1479) and the University of Kiel (founded in 1665). Since the level of mathematics was lower in Kiel than in the capital, and since Copenhagen

J. Lützen (✉)
University of Copenhagen, Copenhagen, Denmark
e-mail: lutzen@math.ku.dk

© Springer Nature Switzerland AG 2019
É. Barbin et al. (eds.), *Descriptive Geometry, The Spread of a Polytechnic Art*,
International Studies in the History of Mathematics and its Teaching,
https://doi.org/10.1007/978-3-030-14808-9_15

became the scene of the polytechnic movement, we shall consider mathematics only in Copenhagen in this chapter. For most of the nineteenth century there was only one chair of mathematics at the university. It was filled by the following professors: Carl Ferdinand Degen (1813–1825), Henrik Gerner von Schmidten (1825–1831), Christian Ramus (1831–1856), Christian Jürgensen (1857–1860), Adolph Steen (1860–1886), Hieronymus Georg Zeuthen (Docent 1871–1883, professor 1883–1910) and Julius Petersen (1886–1909). The two classmates Zeuthen and Petersen acquired world renown for their contributions to enumerative geometry, history of mathematics and graph theory; of their predecessors the most talented was probably von Schmidten who will play a role in our story. He studied in Paris and Göttingen and published papers in Crelle's Journal. Unfortunately, he died at a young age of tuberculosis (Andersen and Bang 1983, pp. 160–178).

During the first half of the nineteenth century the professor of mathematics at Copenhagen University had two main jobs: he examined the students leaving high school for the exam atrium, the entrance exam for the university, and he taught the mathematics course of the "second exam" that all university students had to pass after 1 year of study (the last remnant of the medieval studium generale). Occasionally the professor would also teach more advanced courses for a smaller audience. There was no specialized study of mathematics or natural science and the students could not take a degree in these subjects (Andersen and Bang 1983, pp. 160–161).

However, there were other institutions where mathematics was taught in a more systematic fashion, in particular the military schools: the Naval Cadet Academy (*Sø-Cadet-Academiet*)[1] founded in 1701, and the Land Cadet Academy (*Land-Cadet-Academiet*) founded in 1713. So, for example, von Schmidten joined the army in order to study mathematics at the latter institution (Rosenløv 1963). Moreover the Royal Academy of Arts (*Det Kongelige Kunstakademi*), founded in 1754, also offered classes in mathematics (see Fuchs and Salling 2004). In particular, architects could attend a class of linear perspective 1 h each week, taught by a specially appointed professor of perspective. However, these classes were not well attended until the 1820s, when Eckersberg began advocating for the importance of the discipline for painters as well. Christoffer Wilhelm Eckersberg (1783–1853) became professor at the Academy of Arts in 1818 and was its director from 1827 to 1829. He was the leading Danish painter of the time and wrote two books on linear perspective for painters (Eckersberg 1833, 1841). He also encouraged the architect and professor of perspective geometry, Gustav Friedrich Hetsch (appointed 1822) (Fuchs and Salling 2004, pp. 202–204), and the professor of mathematics, Georg Frederik Ursin, to develop the teaching of geometry at the Academy (Fuchs and Salling 2004, p. 60).

---

[1] All translations of titles, names, and quotes from Danish are the author's.

Ursin (1797–1849) had studied at the University of Copenhagen and then studied astronomy with Schumacher (1817–1818) and mathematics with Gauss (1818). He became doctor of philosophy in 1820 and taught at the high school Metropolitanskolen in Copenhagen until he was appointed professor of mathematics at the Royal Academy of Arts in 1827 (Zeuthen 1887–1905). The following year he and Hetsch published the book "Introduction to the geometric drawing science for use in particular in arts and craft schools" (*Begyndelsesgrunde af den geometriske Tegnelære til Brug for Kunst-og Haandverks-Skoler*) (Hetsch and Ursin 1828). It was a collection of 48 plates with many figures and short explanations on each plate. Plate 1–19 dealt with plane geometry, and the remaining plates dealt with stereometry beginning with "Elements of the science of projection". The plates showed the Monge'an type double rectangular projections of simple geometric figures such as lines, planes, and simple polygons, and explained simple constructions. This book amounted to the first introduction to descriptive geometry in Danish. However, the word descriptive geometry was not used in the book and there was no reference to Monge or the French tradition.

Ursin also founded and edited the first Danish polytechnic journal: *Magazin for Kunstnere og Haandværkere* (Magazine for Artists and Craftsmen), which included papers by Ursin and others on technical drawing (Wagner 1998, pp. 351–356).

The more advanced parts of descriptive geometry were imported into Denmark in connection with the establishment of two technical colleges that were both inspired by the *École polytechnique*: The civil school: The Polytechnic College (*Polyteknisk Læreanstalt*), and the military school: The Royal Military College (*Den Kongelige Militaire Højskole*). We shall now consider their different origins, beginning with the Polytechnic College.

2 The Establishment of the Polytechnic College

In 1827, the same year he was appointed professor of mathematics at the Academy of Arts, Ursin sent a letter to the Danish King in which he proposed to establish a new polytechnic school modelled on the German *Gewerbeschulen* (Steen 1879, p. 103). The King asked the academic council (*Konsistorium*) of the University to evaluate the proposal, and they appointed a committee to deal with the question. It consisted of professor of physics Hans Christian Ørsted, professor of astronomy R. G. F Thune, professor of Chemistry William Christopher Zeise, and professor of Mathematics Henrik Gerner von Schmidten. This group opted for an academically more ambitious solution, namely a polytechnic college modelled after the *École polytechnique* in Paris and connected to the university (Steen 1879; Wagner 1998).

This recommendation was a result of recent developments at the University. From its inception in 1479 and its reorganization in 1539 as a Lutheran University, there had been professorships in mathematics and astronomy. During the eighteenth century chairs in various natural sciences had been added. These professors were members of the faculty of philosophy or the faculty of medicine. In 1762 two professors, Oeder and Ascanius, proposed to establish a separate economical faculty for economy and natural science. However, the idea was rejected by the academic council. Half a century later in 1813, Ørsted proposed a slightly different plan, namely a division of the philosophical faculty into a historical-philological faculty and a mathematical-physical faculty. This proposal was also turned down, but Ørsted continued to work for the strengthening of the natural sciences in Denmark. After he had become world famous for his discovery of electromagnetism in 1820 he founded The Society for Dissemination of Science (*Selskabet for Naturlærens Udbredelse*). The society offered lectures for a wide circle of scientifically interested professional people, and it served as a substitute for the missing science faculty (Nielsen and Slottved 1983, p. 62).

When Ørsted, as part of the university committee, was asked to report on Ursin's proposal he saw it as a chance to create a kind of science faculty by the back door (Wagner 1998, p. 223). The commission recommended that the new school should be a higher scientific institution similar to the *École polytechnique*, which Ørsted knew from his stay in Paris in 1801. Moreover they recommended that the new polytechnic college should be connected to the University, and Ørsted suggested that his own professor's house and its laboratory, as well as another university building nearby could house the new institution. This time the academic council backed the idea, and in 1828 the King approved the plan. The Polytechnic College opened its doors to the first students in the fall of 1829 (Steen 1879, pp. 1–7).

As recommended by the committee the new Polytechnic College had complicated economical and organizational ties to the University, and 4 out of its first 7 professors were also professors at the University. This held true for the professors of mathematics (von Schmidten), chemistry (Zeise), chemistry and mineralogy (Forchhammer), and physics (Ørsted). The latter was also appointed director of the Polytechnic College. With its high academic ideals and its connection to the university the Danish Polytechnic College was unique on the European scene (see Schubring, Chap. 22, this volume).

According to the statutes, the aim of the college was "to teach young people with the necessary prerequisites such insights into mathematics and experimental science and such a proficiency in the use of these insights that they will be eminently useful for certain branches of the service of the state as well as for being in charge of industrial plants" (Steen 1879, p. 107).

The students at the Polytechnic College were divided into two classes, one for the mathematical sciences and one for applied science. When graduating the students of the former got the title Candidate of Mechanics and the latter got the title Candidate of Applied Science. Graduates from the college were called polytechnicians rather than engineers. The latter designation was, until 1857, reserved for military engineers.

3 Descriptive Geometry at the Polytechnic College: The First Failed Beginning

The original proposal from the university committee emphasized experimental work at the new school, in contrast to the purely theoretical teaching at the university. Ørsted stressed the need for laboratories where the students could themselves do experiments just as at the *École polytechnique*. It is well known that Ørsted was a strong advocate of the German romantic approach to physics as opposed to the French mathematical approach (Christensen 1995). Still, mathematics was a part of the planned curriculum from the start. In addition to physics, chemistry, machine science, technology and natural history students should learn algebra, trigonometry, analytic geometry, differential and integral calculus including solution of differential equations and mechanics, and they should follow practical exercises in applications of mathematics. The professor of mathematics was responsible for these mathematical disciplines. But already during the discussions in the university commission von Schmidten had emphasized the need for "geometric drawing or géométrie descriptive". And so this subject became part of the curriculum from the start (Steen 1879, p. 3).

In the original rules of the school (1829) descriptive geometry was mentioned only in passing:

> Every day two hours teaching is given in geometric drawing and machine drawing. Assisted by the teacher of machine science this teaching is conducted in such a way and is joined by such oral lectures that in this way a sufficient instruction is given in drawing-geometry, the géométrie descriptive of the French (Steen 1879, p. 116).

A special room was fitted out as a drawing laboratory where the professor of drawing and his assistant would teach the students to draw accurate geometric drawings. Descriptive geometry should then be taught in connection with the drawing classes by the professor of drawing collaborating with the professor of machine science. The two colleagues Ursin and Hetsch from the Academy of Arts were also appointed at the new Polytechnic College as professors of machine science and professor of drawing, respectively.

Though they had taught the elements of drawing geometry at the Academy of Arts and had composed a book on the subject, their teaching of descriptive geometry was strongly criticized by the first class of students. Students complained about the quality of Ursin's teaching and claimed that Hetsch was rarely present. And apparently the assistant, master carpenter Olsen, was not well versed in theoretical geometry. The student's complaints became an embarrassment for the new school, and so its director Ørsted suggested that a special teacher of descriptive geometry should be appointed. In theory Ursin and Hetsch agreed to the idea, but they did not want to surrender their part of the salary. In the end it was decided that Lieutenant Kellner would be employed as new teacher of descriptive geometry from October of 1830. For two different interpretations of this affair see Steen (1879, pp. 10–11) and Wagner (1998, pp. 309–322). Before we discuss Kellner's work at the school, we shall turn to the establishment of the other polytechnic school in which he was also involved.

4 The Establishment of the Royal Military College

The main initiative for the establishment of the Royal Military College came from Lieutennant-colonel Joseph Nicolai Benjamin Abrahamson (1789–1847). He had studied at the *École polytechnique* as well and wanted to implement this model. The new college founded in 1830, i.e. 1 year after the Polytechnic College, was aimed at educating officers for the general staff, the artillery corps, the engineering corps, the rocket corps, and the road corps that had been militarized in 1785.[2]

From the start, descriptive geometry was taught at the Military College by Ludvig Stephan Kellner (1796–1883) (Fig. 15.1). As a part of his military career as first lieutenant (1826) he was sent to Paris (1828–1829) with the explicit purpose of studying descriptive geometry.[3] After his return to Copenhagen he became a member of the committee planning the new Military College in 1830, and when it opened its doors later the same year he began to teach the subject to the new students. Kellner rose to the rank of Colonel in 1858 and became the leader of the Military College from 1860 to 1863.

5 Descriptive Geometry at the Two Colleges: Kellner's Era 1830–1861 and Its Aftermath

So from 1830 Lieutenant Kellner taught descriptive geometry at the two new polytechnic institutions in Copenhagen. He held the posts for 31 years until his retirement in 1861. He used the same textbooks at both colleges and the teaching seems to have been planned in parallel (Wagner 1998, p. 319).

At the Polytechnic College, students in the class of applied science were not taught descriptive geometry, but for students in the mechanics class and in the engineering class (opened in 1857) it was an important discipline. In 1845 they were taught the subject for 4 h/week during the first and second semesters and for 2 h/week during the third semester. During the last three semesters they did not follow lectures in descriptive geometry (Steen 1879, pp. 125–126). In 1864–1871 the number of hours was increased to 4 h/week during the first and third semesters

---

[2] For a survey of the many different earlier and later military schools in Denmark, Norway, and Holstein see Rosenløv (1963), in particular the schematic figure p. 185. See also Harnow (2005, pp. 40–42).

[3] However, Frédéric Brechenmacher has not been able to find Kellner's name in the archives of the *École polytechnique*.

15 Descriptive Geometry in Denmark

Fig. 15.1 Ludvig Stephan Kellner (1796–1883), The Historical Collection at the Technical University of Denmark

and 5 h during the second semester. The number of lectures in descriptive geometry was comparable to many other subjects, but less than the number of lectures of mathematics proper[4] (about 6 h weekly and continuing through most of the study), and also less than the drawing course that had about 6 h/week (Goos 1887, pp. 640–642). Still descriptive geometry constituted a substantial part of the final exam. There was one written exam and one oral exam, counting two marks out of a total of 9 marks for the mechanics line and 17 marks for the engineering line.

Kellner succeeded in solving the problems encountered during the first year of descriptive geometry teaching at the Polytechnic College. According to Wagner

---

[4]Descriptive geometry was taught as a separate course apart from the "mathematics" course.

(1998, p. 319) "his new appointment [at the Polytechnic College] allowed him to plan the education at the two institutions in parallel with such a successful result that his plans were followed in both places long after he had retired as a teacher".

Still, according to his former student Captain Bauditz, Kellner was not a great teacher:

> He was a tender-hearted and actually good humored, very conscientious man who probably always prepared himself with great diligence for his lectures. However, they were dragging and for me straight out soporific (inducing sleep) (Lundbye 1929, p. 360).

In the Danish Biographical Lexicon Zeuthen and Heegaard attributed to Kellner the honour of having introduced descriptive geometry in Denmark:

> The teaching of descriptive geometry that Monge and his school introduced at the École Polytechnique in Paris gradually also encroached on other European technical high schools and officer schools. Kellner has the honour of having introduced this education in Denmark. It was probably more as a conscientious officer who obeys an order than as a result of a special urge for research that he embarked on a study journey to France. Even though his work with the theories developed in particular by Poncelet and Dupin do not bear the impress of originality, and though his presentation shows some lack of precision, he has with great faithfulness completed the task that was expected of him. His textbook in descriptive geometry has thus at an early period disseminated the knowledge on this branch of mathematics that is so important for technicians (Tychsen et al. 1937, p. 391).

6 Kellner's Textbooks

In the Polytechnic College program the subject of descriptive geometry was described as follows: "The general representation of the geometric objects and applications to the science of shadows, perspective and gear wheel construction". From 1857 stone cutting was also mentioned explicitly (Steen 1879, pp. 124, 135, 141). As to textbooks the program stipulated: "All lectures are given after printed textbooks in Danish, German or French" (Steen 1879, p. 107). But instead of using one of the French or German textbooks on the market Kellner wrote his own material in Danish. First he composed a "Guide to the lectures on descriptive geometry at the Royal Military College", which came out in several installments in 1830–1831. The reason the Polytechnic College was not mentioned in the title is probably that the printing had begun before Kellner was appointed teacher there. In 1836 he published a real textbook "The theoretical part of the descriptive geometry, composed according to the best authors with particular view to the special aim of the book as a guide for lectures" (Kellner 1836). It was followed by "The applied part of the descriptive geometry" (Kellner 1840) published in installments from 1837 to 1839 and collected in one volume in 1840.

Who were the "best authors" mentioned in the title? Kellner was not very informative about this point. He only referred to Monge's lectures at the *École normale* and to G. Schreiber who published a German version of Monge's book, but according to Zeuthen he also drew on Poncelet and Dupin. Kellner did not claim any originality for himself, but his books were good and very complete compilations. In the preface Kellner briefly wrote about the history of descriptive geometry:

> The graphical methods have developed gradually as different cases in practice have given rise to them, and it would no doubt be useless to search for the first traces of them [...]. Not until the last decades of the previous century did the mathematically talented Monge collect the most important methods of construction and formed from them a systematic theoretical structure ordering the known elements and adding a large part of the missing elements. In this way Monge created a mathematical science that is not only the basis for the arts of drawing but also indispensable for the practitioners of a host of other technical subjects [...] (Kellner 1836, pp. I–II).

According to Kellner the aim of descriptive geometry was twofold, as it had been to Monge:

> The descriptive geometry teaches [us] to represent every object whose shape and extension can be given an exact definition whether the dimensions and position are given directly or they are only determined through certain conditions; moreover it provides us with methods to deduce the geometric properties that are conditioned by the extension and mutual position of the represented quantities (Kellner 1836, p. 1).

In order to explain the nature and importance of descriptive geometry Kellner ended the preface by a 4 page quotation in French of the "Programme" in Monge's *Géométrie Descriptive*.

To some extent Kellner followed Monge as far as structure, subjects, and constructions are concerned. However, in many details and in style he differed from Monge. Where Monge's book was composed as somewhat loosely structured lectures Kellner's books were more systematic and tightly structured. Monge included many philosophical, educational, or methodological reflections on the nature of space, the relation between synthetic and analytic methods, etc. Kellner did not deal with such matters, and he limited the discussion of the usefulness of descriptive geometry to the preface where Monge returned to this issue again and again. Kellner dealt with curves before surfaces and treated more specific curves and surfaces than Monge did. Where Monge in §95 dealt with a semi-practical surveying problem Kellner gave a general treatment of graphical solutions of trigonometric problems. Kellner also included methods and concepts from projective geometry, such as deductions using central projection, pole and polar, points and lines at infinity, and ideal cords, but he did not introduce the cross ratio and did not use Gergonne-style double column presentation. Moreover, Kellner went into much more detail with perspective drawing and applications to shadows, reflections, opaque bodies, tooth wheels, and stone building, and he included a special treatment of the cote method used in maps, where a point in space is given by its projection on the horizontal plane and a number denoting its height above this plane. Following Monge, Kellner dealt in great detail with synthetic differential geometry, including tangents, normals, tangent planes, osculating planes, curvature of curves, lines of curvature of surfaces, geodesics, developable surfaces, evolutes and involutes, etc. (see Fig. 15.2), but he did not include ideas from Gauss' *Disquisitiones Generales* (1828).

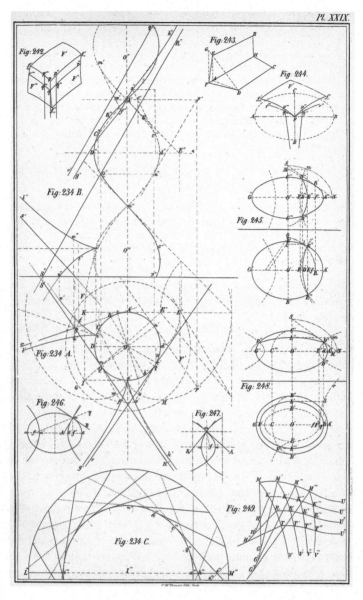

Fig. 15.2 On lines of curvature in Kellner's textbook. Pl. XXIX in Kellner (1836, second edition 1850)

As we saw above, Zeuthen and Heegaard later criticized the lack of precision in Kellner's books. This is not surprising. Indeed, later generations usually find mistakes in the presentation of their teachers. However, Kellner was very careful to define all the concepts used in his book (except those that were known from elementary geometry teaching) and he argued for all the results and methods that he presented. His books were not written in a Euclidean style; in particular they do not lay down any axioms, but they contain theorems and construction problems, some of them highlighted as statements in bold, others just embedded in the prose. The arguments are visual and intuitive and in general well presented. To be sure, many of the definitions and arguments are not rigorous, even given the standards of the day. In particular the differential geometric definitions and arguments use infinitesimally small quantities. For example, a curve is considered a polygon with infinitely small sides, and a tangent is defined as the straight line having one of these sides in common with the curve. In general two successive sides of the polygon do not lie on one line, but they lie in a plane called the osculating plane. But what Zeuthen and Hegaard objected to was most likely the fact that Kellner for the most part dealt with the general situation disregarding special configurations and singularities.

The intuitive style is also characteristic of the second and quite reorganized edition of Kellner's textbooks that were published in 1850.

7 The Collaboration and Competition Between the Colleges and the University

As we saw above, the Polytechnic College at its inception had strong economical and organizational ties with the University. It is highly probable that Ørsted from the beginning hoped that the new college would eventually be fused with the University and form a new Faculty of Science. A fusion was in fact suggested in 1840, but even though the plan was backed by Ørsted it fell through. Two years later Ørsted again proposed a fusion but failed again. In 1848/1849 the university introduced for the first time a separate final exam in mathematics and the natural sciences (*magisterkonferens*), and in 1850 a separate faculty of mathematical and natural sciences was finally created at the university (Nielsen and Slottved 1983, pp. 69–73). With these changes at the university it became less urgent to fuse the Polytechnic College and the University, and in 1863 the two institutions became financially independent.

Still, they continued to share many courses of mathematics, physics, and chemistry. In fact university students aiming for the new degree in mathematics, physics, and chemistry followed the first 2 years of mathematics courses at the Polytechnic College including the course of descriptive geometry. In 1924 the University introduced its own geometry course, but the analysis course continued to be shared by students of the two institutions until 1942, and even after that the curriculum at the courses of the University and the Polytechnic College differed only a little until the 1960s (Madsen 2008, p. 756). This was due to the fact that both institutions used Harald Bohr's and Johannes Mollerup's influential textbook (Bohr and Mollerup 1920–1923) from 1920 to 1923.

The Military College and the Polytechnic College also interacted in an uneasy manner. Historians do not agree about the wisdom of spreading the polytechnic education of a small country like Denmark across two colleges. Steen (1879, p. 11) expressed his strong disapproval whereas Essemann (see Rosenløv 1963, p. 89) believed that it was wise to keep the civil and the military education separate. At any rate, the result was a competition between the two—a competition for students and for the favor of the state. Initially, the Military College came out on top. Their students were better funded, and thus a larger percentage of their students passed their exams. Moreover, it was the only institution entitled to educate "engineers" (meaning constructional engineers) and so all engineering projects were planned by military engineers. The Polytechnic College many times (e.g. in 1848) tried to obtain the right to educate civil engineers, but the application was refused. Again in the 1850s there were long negotiations about the relation between the Polytechnic College, The Military College, and the University, but again they resulted in the status quo. As a result the polytechnic candidates led by Steen raised private funds that allowed the Polytechnic College to appoint a teacher of engineering. In 1857 the ministry finally gave in and granted the Polytechnic College the right to hold exams for engineers (Steen 1879, p. 44). After that civil engineers gradually replaced military engineers in civil projects.

Still, as late as 1855 the ministry of finance declared that "the Military College should be the Institute that was responsible for the higher teaching of mathematics and natural sciences" in Denmark (Steen 1879, p. 44). It is remarkable that the ministry expressed this point of view 7 years after the degree in mathematics and natural sciences had been introduced at the university and 5 years after the faculty of science had been created.

However, the Military College continued only for another decade. After the national disaster in 1864 when the Danish army lost the war against Prussia, and the country lost Holstein and Schleswig, the Military Academy was replaced in 1868 by a new institution: The Officer School of the Army (*Hærens Officerskole*). In the beginning this school also had a rather high mathematical level. For example, around 1870 all third year students were taught 150 h of descriptive geometry, and the artillery- and engineering officers were taught descriptive geometry for 270 h during their final fourth year at the school. The third year teaching of descriptive geometry was canceled in 1882 after which time Danish officers were taught descriptive geometry only during their final fourth year at the school. Also the mathematics course and the natural sciences were gradually reduced at the Officer school, leaving the Polytechnic College as the undisputed highest ranked polytechnic school in the country.

8 Carl Julius Ludvig Seidelin and His Textbooks

One year after Kellner had become the director of the Military College he resigned his teaching posts both at the Military College and at the Polytechnic College in 1861. Two different persons took over the courses. At the Military College, first lieutenant Lars Bache (1833–1903) applied for the position. Following the example of his teacher Kellner, he had prepared himself for the job by studying descriptive geometry in France (Nielsen 1910, p. 5). However, the senior captain Laurits Knudsen (1821–1904) was chosen for the job. He was transferred to the Officer School when it succeeded the Military College in 1868 and continued to teach there until 1893 when he was replaced by first lieutenant (later Captain) Anton Levinsen (1859–1920). Knudsen wrote three books for the students at the Military College, one on geometry for the first class (1874–1875) and two on descriptive geometry for the two last classes (1874 and 1882-1883, respectively). They were only autographed and never properly printed. So in order to follow descriptive geometry in Denmark after Kellner we should look to the Polytechnic College. Here Kellner's successor and former student Carl Julius Ludvig Seidelin (1833–1909) followed in the footsteps of his teacher. Seidelin's student Lundby did not think highly of his pedagogical talents:

> Considering how difficult many people find descriptive geometry, one could not call Seidelin a good teacher. But those few, who could envision the figures in space just as easily as he, could follow his always strictly correct presentation of even the most difficult subjects. However, if you lost the thread you could never pick it up again, for he never repeated what he had said or shown. [...] Therefore there were many who had to consult a coach because they could not follow the lectures, and usually it was only a small group who gathered around the small neat man with the old fashioned correct ways (Lundbye 1929, pp. 360–361).

Still, Seidelin taught the course for 42 years!

Seidelin published his own lectures first in autographed form (Steen 1879, p. 68) and then as a printed book: *Forelæsninger over Deskriptivgeometri* (Lectures on Descriptive geometry) (Seidelin 1873). It was updated twice during Seidelin's long term as lecturer, once in 1886–1887 and once in 1895–1896.[5] He dealt with much the same subjects as Kellner but many details were left out and projective geometrical methods became more central. For example, Seidelin considered orthogonal projections as a special case of central projection (from a point at infinity) and he introduced the cross ratio and duality and presented many dual results in the Gergonne double column style. It is not clear if Seidelin was influenced by the German mathematician Wilhelm Fiedler (1832–1912), who published a textbook at about the same time with a similar mix of descriptive and projective geometry (see Volkert, Chap. 10, this volume).

Seidelin gave fewer and more condensed arguments than Kellner, so that his book in one volume was much shorter that Kellner's. This condensed style was in fashion in Denmark at the time. In particular, Julius Petersen's school books were slim volumes written in an elegant but brief and intuitive style, famous (or infamous) for the often repeated phrase: "One can easily see that..." (Lützen et al. 1992). The

---

[5]Seidelin also published an elementary textbook in projection drawing (Seidelin 1879) (later eds. 1884, 1895).

condensed style of Seidelin's textbook combined with his lecturing style made the subject hard for the students.

9 Exam Problems

Both Kellner's and Seidelin's textbooks were quite demanding for polytechnic students. Of course that does not necessarily mean that the students learned, or were supposed to learn and master the entire curriculum. But fortunately we can quite easily infer what the students were supposed to learn because the exam problems from the written exams were published in the year books of the University which also included a section on the Polytechnic College. The first published year book contains the exam problems for the years 1864–1871. Here is the problem from the January-1864 exam:

> Two spheres and a luminous point are given by their two rectangular projections. Find a point of the shadow line that is produced on one sphere by the shadow of the other sphere, and the tangent at this point of the shadow line. Moreover, find the intersection point between this curve and the separation line between light and shadow by the illumination independent of the other sphere. (The given quantities may not have any special position mutually or with respect to the projection planes; moreover the required point may not be the result of any special facilitating circumstances) (Goos 1887, p. 689).

At the following January exam the students were posed the following problem:

> Given a sphere as well as the axis (in an oblique position to the projection plane) and the radius of a right circular cylinder. The intersection line (which it is not required to construct) of these two surfaces and the axis of the cylinder are directrices and the vertical projection plan is directing plane for a skew surface. Construct one rectilinear generator of the skew surface and the tangent plane at a point of this surface chosen arbitrarily on the constructed generating line (but outside the directrices) (Goos 1887, pp. 693–699).

Two years later the exam began with a plane problem:

> 1. Determine the intersection points between a straight line and a conic section given by 3 of its points and 2 of its tangents (all lying in the drawing plane) 2. Determine the intersection point between a straight line in the horizontal projection plane and a skew hyperboloid whose 3 straight directrices are given; one of them is horizontal; otherwise the lines have no special position mutually or with respect to the planes of projection (Goos 1887, p. 702).

Hyperboloids of revolution could also be the subject of the exam:

> Given two straight lines, one in each of the projection planes. These are generators of the same kind of a hyperboloid of revolution. Moreover a point is given on each of these lines, namely the vertical trace of the horizontal line and a point of the other line that is closest to the first mentioned point. These two points belong to the same parallel circle on the surface. Determine 1. another pair of points on the two given generators that belong to a parallel circle on the surface. 2. the axis and center of the surface, as well as the angle between the axis and the generator and 3. by construction show how one can for a given plane determine if its intersection with the surface is an ellipse, a parabola or a hyperbola (Goos 1887, p. 707).

It is obvious that these problems require a good knowledge of the techniques of descriptive geometry as well as the properties of quadric surfaces. More general aspects of curves and surfaces were probably subjects of the accompanying oral exam.

10 Theory and Practice

When the Polytechnic College was founded students were supposed to combine their theoretical scientific education with more practical work in workshops. However, the students did not attend the workshop training and so this part of their education was canceled after only 1 year. Only the geometric drawing workshop remained open. It was presided over by a series of teachers of drawing: Hetsch (1829–1838), Christian Gottfried Hummel (1838–1872), H. C. F. C. Schellerup (1872–1917), Erdmann Peter Bonnesen (1888–1917) (promoted to professor in 1902, and not to be confused with the professor of descriptive geometry Tommy Bonnesen), J. B. K. Gunner (from 1907).

In general it was difficult for the new college to find appropriate ways to combine theory and practice. We saw how the attempt at combining the geometric drawing class with the more theoretic teaching of descriptive geometry failed after 1 year. The ensuing, more theoretical course of descriptive geometry was a greater success, but it was also criticized by the students "not because it was characterized by incompetence but because the students claimed that they did not need the subject in their later work" (Harnow 2005, p. 36).

In order to connect descriptive geometry more closely to the practical needs of the engineering students, it was decided in 1903 to leave the introductory teaching to the professor of drawing E. P. Bonnesen. However, as had been the case in 1829, the experiment was not a success and was given up in 1908. By then a new professor had taken over the chair of descriptive geometry, a professor who directly addressed the problem of practical drawing at least in a theoretical way.

11 Johannes Trolle Hjelmslev's Geometry of Reality

With Hjelmslev's[6] appointment to docent in 1903 (promoted to professor 1905) the academic level of geometry at the Polytechnic College was raised. Where Kellner had never published anything but his textbooks and Seidelin had only published a few elementary mathematics papers in Danish, Hjelmslev had already published two papers in foreign languages (French and German) at the time he was appointed.

---

[6]1873–1950. Originally Hjelmslev was named Petersen, but since the name J. Petersen was already used by his colleague Julius Petersen he changed his last name in 1904.

Moreover, as the first geometry teacher at the college he had defended a doctoral thesis. It was on "Foundational principles for the infinitesimal descriptive geometry with applications to the science of variable figures" (defended 1897). He later became internationally known for his contributions to the foundations of geometry. In particular he contributed to Hilbert's axiomatic program and was quoted in later editions of *Grundlagen der Geometrie* (Hilbert 1930, pp. 50, 54 and 159).

Hjelmslev also modernized the course at the Polytechnic College. Only 1 year after his appointment he published a new textbook *Descriptive geometry, basis for Lectures at the Polytechnic College* (Hjelmslev 1904 (with a supplement in 1916)). In 1918 he completely rewrote it as *Textbook in Geometry for use at the Polytechnic College* (2. ed. 1923) (Hjelmslev 1918). In the first book Hjelmslev upgraded projective geometry and downplayed the Monge'an point of view with the vertical and horizontal projection of figures in space. In fact the separate volume of figures displays only an occasional double rectangular projection. The reason was that this part of the subject had been transferred to the course on drawing in 1903 or 1904 (Madsen 2008). The reversal of this decision may explain why the Monge'an point of view was slightly more visible in Hjelmslev's second book (1918), despite the noticeable change of title from "descriptive geometry" to "geometry". Hjelmslev also used the results of his doctoral thesis to replace the infinitesimal approach of his predecessors with a more rigorous approach using limits. This move is emphasized in the second book where the section "Theoretical Infinitesimal geometry" opens as follows: "In the theoretical infinitesimal geometry the foundation is exclusively arithmetic. The basic concept is the limit" (Hjelmslev 1918, p. 213).

However, Hjelmslev's most original contribution to geometry teaching, both at the Polytechnic College and in the primary and secondary schools, was his so-called geometry of reality. This approach to geometry focused on the accuracy with which one can execute geometrical constructions. In earlier books on descriptive geometry accuracy was emphasized. For example, Seidelin in §1 of his book wrote: "descriptive geometry teaches how to depict spatial quantities exactly [...]. It demonstrates the reliable way to construct exactly in space" (Seidelin 1873, 3rd ed. p. 1). Still, the books did not contain any theoretical reflections on how one should best obtain accuracy. It was left to the teacher of drawing to teach the students how to draw and construct precisely. Monge had argued that the two planes on which one projects, and thus the directions in which one projects the objects in space, should not make too small an angle with each other, because in that case the intersection between the two lines would not be well determined (Monge 1827, p. 11, § 8). But there were no such considerations in the Danish textbooks prior to Hjelmslev.

In order to deal with the question of accuracy in a systematic way Hjelmslev distinguished between two types of geometry: The geometry of reality, corresponding to the real geometry on the finite drawing paper, and the arithmetic geometry, corresponding to the Cartesian geometry of $\mathbb{R}^2$ and $\mathbb{R}^3$. A measurement of a length in the geometric plane cannot be determined exactly. There is a range of arithmetic values that equally well determine the length. When one chooses a specific value for the geometric length, Hjelmslev says that one fixes the length (Hjelmslev 1918, p. 9). A fixation is a mapping of the geometric plane into the arithmetic plane. There are many such mappings but according to Hjelmslev "every really existing figure can be fixed in such a way that the theorems of geometry are exactly valid for the fixed figure" (Hjelmslev 1918, p. 10). Conversely, there is one mapping that maps the arithmetic plane into the geometric plane at least if one restricts oneself to the part of the arithmetic plane corresponding to the finite limitations of the geometric drawing plane. But this mapping maps many arithmetic points into the same point. If, for example, one can draw with the accuracy of $\varepsilon = 1/25$ mm in the real drawing plane, then a line of this length in the arithmetic plane will be mapped into a point.

As a consequence of this point of view, two intersecting lines in the real geometry have a line segment in common. If, for example, the lines $y = 0$ and $y = 0.01x$ are mapped into the real geometric plane, they will have a line segment of length 8 mm in common, if the accuracy is $\varepsilon = 1/25$ mm. Similarly a circle has a whole line segment in common with its tangent, a segment that becomes larger when the radius of the circle increases. One can consider the geometric circle as consisting of these linear elements. Similarly, Hjelmslev introduced circular elements of a curve as the longest circular curves that are contained within the curve (within the uncertainty ε). The radius in this circle is the practical radius of curvature of the circle. This is in most cases different from the theoretical value (in the arithmetic geometry). For example, Hjelmslev could prove the theorem:

> The practical radius of curvature at the end points of the axes of an ellipse is somewhat larger than its theoretical value, namely so much larger as the half linear element on the theoretical osculating circle multiplied by the eccentricity of the ellipse (Hjelmslev 1918, p. 5).

Another result of Hjelmslev's focus on the practical execution of the geometric constructions was his insistence that it should be possible to make the constructions in the finite drawing plane. This was not a new idea, but Hjelmslev dealt with it in a more systematic way.

Hjelmslev published his ideas about geometry of reality in several papers and books, some in Danish and some in German (Hjelmslev 1923). Similar ideas had already been championed by Felix Klein in autographed lecture notes entitled *Anwendung der Differential- und Integralrechnung auf die Geometrie* (Application of Differential and Integral Calculus in Geometry) (1902 and 1907) published in book form as the third volume of his *Elementarmathematik vom höheren Standpunkte aus* (Elementary Mathematics Considered from a Higher Standpoint) with the title: *Präzisions- und Approximationsmathematik* (Precision and Approximation Mathematics) (Klein 1928). Klein also distinguished between a theoretical geometry (precision geometry) and a practical geometry (approximation geometry). In the former the quantities are accurately determined, but in the latter they are determined within a certain clearance (Spielraum). However, as the original title indicates, Klein mostly discussed subjects from analysis and did not deal with elementary geometry as Hjelmslev did.

Hjelmslev's geometry of reality was developed as a part of a lively discussion in the beginning of the twentieth century about the right way to teach geometry on all levels of the educational system (see Hansen 2002, in particular pp. 106–

125). Traditionalists defended a Euclidean axiomatic approach, whereas modernists advocated for a more empirical and intuitive approach. Hjelmslev and his successor as professor of descriptive geometry at the Polytechnic College, Tommy Bonnesen, were strong advocates for the latter point of view. They believed that geometric properties should be discovered by the pupils themselves and Hjelmslev stressed that elementary textbooks for schools should "not deal with abstractions but with things that belong to the practice of life". "Why do we need all these proofs of things that are often more evident than the axioms that the proofs are built on" (Hjelmslev 1913, p. 50).

In particular, Hjelmslev criticized the Euclidean paradigm of construction with ruler and straightedge. If such constructions involve many steps, they become so inaccurate in practice that they are useless. Instead Hjelmslev in his elementary textbooks taught the pupils to construct on squared paper, and he advocated for trial and adjustment methods. Even for a simple task like determining the midpoint of a line segment Hjelmslev recommended to try with a particular opening of the compass and adjust it until it fits twice on the segment. He argued that it was more important to know how to check if a proposed solution to a construction problem was accurate than knowing the ruler and compass construction of the solution.

12 The End of Descriptive Geometry in Denmark

In a sense Hjelmslev's era marked the high point of Danish descriptive geometry. For the first time a Danish mathematician introduced new original ideas. However, one can also consider his era as the beginning of the end of the discipline in Denmark. Indeed, the special Monge'an take on the subject was gradually weakened after Hjelmslev. When the mathematical and physical programs of the University got a new regulation in 1924 (authored primarily by Hjelmslev) descriptive geometry was still part of the curriculum, but here as well as at the Polytechnic College the geometry course gradually turned into a course of analytic and differential geometry (Madsen 2008, p. 779). Hjelmslev's successor at the Polytechnic College, Tommy Bonnesen (1873–1935) (professor 1917–1935) did not write his own textbook but continued Hjelmslev's line. When Bonnesen died in 1935 Børge Jessen (1907–1993) succeeded him. On that occasion the chair changed its name from "descriptive geometry" to "geometry" and at the summer exam of 1936 the course had also dropped the word "descriptive". Jessen's *Lærebog i Geometri* (Textbook in Geometry) (1939–1941) began with a 40 pages section on projections including the Monge'an method of double rectangular projections. Otherwise it bears little resemblance with the earlier books on descriptive geometry. In the earlier books one could find occasional algebraic formulas, but in general they were written in a purely synthetic style. Jessen abandoned this style and used vector notation and analytical formulas everywhere. And his successor Frederik Fabricius Bjerre (1903–1984) who stayed in the chair for 30 years (1942–1972) continued Jessen's line. So it is reasonable to say that descriptive geometry was history in Denmark from the middle of the 1930s.

The only vestige of the Monge'an double projection style descriptive geometry was taught as a part of the drawing course that continued until 1960. At the end of the 1940s the polytechnic students were still taught technical drawing for 3 h, three afternoons each week amounting to a total of 720 h. Gutmann Madsen who took part in these classes sarcastically recalls the outdated requirements including drawings of steam engines.

> The course ended with two exam drawings, in 1948 a perspective drawing of a bridge supported by a parabola and a technical drawing of an oil pump that we should in principle take apart and measure. (There was one oil pump and about 300 students) (Madsen 2008, p. 767).

By then the course was considered a relic from a bygone era.

13 Postscript on Geometry in Danish Schools

Contrary to education in many other countries, descriptive geometry seems never to have been taught in high schools or lower schools in Denmark. Synthetic plane geometry was a central subject in Danish schools through the first half of the twentieth century. Stereometry was also taught. In particular, construction of geometric problems with ruler and compass was a discipline that was developed to a high degree of perfection due in particular to Julius Petersen's textbook *Methoder og Theorier til Løsning af Geometriske Konstruktionsopgaver* (Methods and Theories) (Petersen 1866). This book was printed in many later editions until this day and it was translated into many languages.

References

Andersen, Kirsti, and Thøger Bang. 1983. Matematik. In *Københavns Universitet 1479–1979*, ed. Pihl, Mogens, 113–199. Bind XII, Det matematisk-naturvidenskabelige Fakultet, 1. del. København: Gad.

Bohr, Harald, and Johannes Mollerup. 1920–1923. *Lærebog i Matematisk Analyse*, vol. I–IV. København: Jul Gjellerup.

Christensen, Dan Charly. 1995. The Ørsted-Ritter Partnership and the Birth of Romantic Natural Philosophy. *Annals of Science* 52 (2): 153–185.

Eckersberg, Wilhelm Christoffer. 1833. *Forsøg til en Veiledning i Anvendelse af Perspektivlæren for unge Malere*. Kjøbenhavn: Thieles Bogtrykkeri.

———. 1841. *Linearperspektiven, anvendt paa Malerkunsten*. Kjøbenhavn: C.A. Reitzel.

Fuchs, Anneli, and Emma Salling (eds.). 2004. *Kunstakademiet 1754–2004*, Bind III. København: Det Kongelige Akademi for de Skønne Kunster & Arkitekternes Forlag.

Goos, Carl (ed.). 1887. *Kjøbenhavns Universitet, den polytekniske Læreanstalt og Kommunitetet indeholdende Meddelelser for de akademiske Aar 1864–71 med Sagregister*. Kjøbenhavn: Gyldendal.

Hansen, Hans Christian. 2002. *Fra forstandens slibesten til borgerens værktøj. Regning og matematik i folkets skole 1739–1958*. Aalborg: Papers from DCN No. 16.
Hansen, Hans Christian et al. 2008. *Matematikundervisningen i Danmark i 1900-tallet*, vol. 2, 7. ed. Odense: Syddansk Universitetsforlag.
Harnow, Henrik. 2005. *Den danske Ingeniørs Historie 1850–1920*. Herning: Systimes teknologihistorie.
Hetsch, Gustav Friedrich, and Georg Frederik Ursin. 1828. *Begyndelsesgrunde af den geometriske Tegnelære til Brug for Kunst- og Haandverks-Skoler*. Kjöbenhavn: Udgivernes Forlag.
Hilbert, David. 1930. *Grundlagen der Geometrie*, 7. ed. Berlin: Teubner.
Hjelmslev, Johannes. 1904. *Deskriptivgeometri. Grundlag for Forelæsninger paa Polyteknisk Læreanstalt. Separate volume of figures*. København: Jul. Gjellerup.
———. 1913. Om grundlaget for den praktiske geometri. *Matematisk Tidsskrift A* 2: 41–58.
———. 1918. *Lærebog i Geometri til Brug ved Den Polytekniske Læreanstalt*. København: Jul. Gjellerup.
———. 1923. *Die Natürliche Geometrie*. Hamburger Mathematische Einzelschriften, 1. Heft.
Jessen, Børge. 1939–1941. *Lærebog i Geometri. I Afbildninger og analytisk Geometri (1939)*, II Differentialgeometri (1941). København: Jul. Gjellerup.
Kellner, Ludvig Stephan. 1836. *Den beskrivende (descriptive) Geometris theoretiske Deel*. Figures in separate volume. Kjøbenhavn: Forfatterens Forlag. Second revised ed. 1850.
———. 1840. *Den beskrivende (descriptive) Geometris anvendte Deel*. Figures in separate volume. Kjøbenhavn: Forfatterens Forlag. Second revised ed. 1851.
Klein, Felix. 1928. *Elementarmathematik vom höheren Standpunkte aus Dritte Auflage, Dritter Band: Präzisions- und Approximationsmathematik*. 3. ed. Berlin: Springer.
Lundbye, Johan Thomas. 1929. *Den Polytekniske Læreanstalt 1829–1929*. København: Gad.
Lützen, Jesper, Gert Sabidussi, and Bjarne Toft. 1992. Julius Petersen 1839–1910. A Biography. *Discrete Mathematics* 100: 9–82.
Madsen, Tage Gutmann. 2008. Matematikundervisningen ved universiteter og højere læreranstalter. In *Fra forstandens slibesten til borgerens værktøj. Regning og matematik i folkets skole 1739–1958*, ed. Hansen, Hans Christian, 755–920. Aalborg: Papers from DCN No. 16.
Monge, Gaspard. 1827. *Géométrie descriptive*, 5. ed. Paris: Bachelier.
Nielsen, Jørgen Broberg, and Ejvind Slottved. 1983. Fakultetets almindelige historie. In *Københavns Universitet 1479–1979*, ed. Pihl, Mogens, 1–112. Bind XII, Det matematisk-naturvidenskabelige Fakultet, 1. del. København: Gad.
Nielsen, Niels. 1910. *Matematikken i Danmark 1801–1908. Bidrag til en Bibliografisk-historisk Oversigt*. København, Kristiania: Gyldendal.
Petersen, Julius. 1866. *Methoder og Theorier til Løsning af Geometriske Konstruktionsopgaver*. Kjøbenhavn: Schønberg. second enlarged ed. 1879. Translated 1879 into German and English.
Pihl, Mogens (ed.). 1983. *Københavns Universitet 1479–1979*, Bind XII, Det matematisk-naturvidenskabelige Fakultet, 1. del. København: Gad
Rosenløv, Mogens (ed.). 1963. *Uddannelsen af Hærens Linieofficerer 1713–1963*. Frederiksberg: Hærens Officerskole.
Seidelin, Carl Julius Ludvig. 1873. *Forelæsninger over deskriptivgeometri*. Figures in separate volume. Kjøbenhavn: Forfatterens Forlag. second ed. 1886–87, 3. ed. 1895–96.
———. 1879. *Elementær Lære i Projektionstegning*. Kjøbenhavn: Hauberg. 2. ed. 1884, 3. ed. 1895.
Steen, Adolph. 1879. *Polyteknisk Læreanstalts første halvhundrede Aar 1829–1879*. Kjøbenhavn: Bianco Luno.
Tychsen, Valentin Emil, Hieronymus Georg Zeuthen, and Poul Heegaard. 1937. Kellner, Ludvig Stephan. In *Dansk Biografisk Leksikon*, vol. XII, 390–391. København: Schultz.
Wagner, Michael F. 1998. *Det Polytekniske Gennembrud*. Århus: Aarhus Universitetsforlag.
Zeuthen, Hieronymus Georg. 1887–1905. Ursin, Georg Frederik Krüger. In *Dansk Biografisk Lexikon*, ed. Carl Frederik Bricka, vol. 18. Kjøbenhavn: Gyldendal.

Chapter 16
Descriptive Geometry in Czech Technical Universities Before 1939

Vlasta Moravcová

Abstract All branches of industry were developing rapidly in the Czech Lands in the first half of the nineteenth century. This caused an increasing demand for specialists with technical education. Descriptive geometry (as an important part of this education) first appeared as a subject in polytechnic schools, but soon also in secondary schools, especially in real-schools.[1] The greatest boom in Czech descriptive geometry came in the second half of the nineteenth century and it was still reverberating at the beginning of the twentieth century. Secondary schools, technical universities and other schools with the Czech teaching language were established and Czech textbooks and original scientific works were published. In this chapter, we give fundamental information about descriptive geometry education in Czech technical universities and about the most significant results, which were published by geometers who lectured at these universities.

Keywords Czech language · Descriptive geometry · Jan Sobotka · Karel Pelz · Polytechnic schools · Real-schools · Rudolf Skuherský · Technical universities

---

[1]The real-school (from German: *die Realschule*, in Czech: *reálka or reálná/reální škola*) was a special kind of secondary school with emphasis on mathematics and natural sciences as opposed to grammar school (gymnasium), where more lessons of Greek, Latin, etc. were provided. These real-schools were instituted in Austria in 1849 by the Exner-Bonitz reform.

V. Moravcová (✉)
Faculty of Mathematics and Physics, Department of Mathematics Education, Charles University, Prague, Czech Republic
e-mail: vlasta.moravcova@mff.cuni.cz

1 Beginnings of Descriptive Geometry Teaching in the Czech Lands

The first polytechnic school in the Czech Lands,[2] *Königlich-böhmische ständische Lehranstalt zu Prag* (royal Czech educational institution of the estates in Prague), was founded in Prague in 1806 in the manner of the *École polytechnique* in Paris. It originated from the *Ständische Ingenieurschule in Prag* (engineering school of the estates in Prague), which was established in 1707. At first, this polytechnic school[3] was a part of the *Karl-Ferdinand Universität in Prag* (Charles-Ferdinand university in Prague),[4] but it became an independent school in 1815. Its first principal[5] was František Josef Gerstner (1756–1832).

The main aim of the polytechnic school was the preparation of students for practice. Therefore the character of education was directed not only at theory, but particularly at applications of technical and natural sciences.

Descriptive geometry was introduced by Karel Wiesenfeld (1802–1870) as a part of civil engineering lectures at the polytechnic school in Prague in the 1830s. Students did not know this subject from secondary school, therefore Wiesenfeld at least provided the students with bases of Monge's projection and perspective. Moreover, orthogonal projection and construction of shadows were included in the syllabus for machine engineering students, thanks to professor Karel Wersin (1803–1880) in 1840. The education was provided according to César Nicolas Louis Leblanc's textbook *Choix de modéles appliqués à l'enseignement du dessin des machines, avec un texte descriptif* (selected models used for teaching of machine drawing with a descriptive text) (Leblanc 1830).

Inclusion of descriptive geometry in the lectures proved very useful and as a result, the department of descriptive geometry was established at the polytechnic school in Prague in 1850. The course *Beschreibende* (descriptive) *Geometrie* became obligatory for all first-year students, except for students of chemical engineering. At first, the lectures were provided by the professor of civil engineering Karel Wiesenfeld. The first professor of descriptive geometry, Rudolf (Rudolph) Skuherský, was appointed 2 years later. Rudolf Skuherský (1828–1863) studied at the polytechnic school in Prague and later in Vienna. There he published two

---

[2] The Czech Lands (also known as the Lands of the Bohemian Crown) were a constituent part of the Habsburg Monarchy, which was formally unified as the Austrian Empire (1804–1867) and later as the Austro-Hungarian Empire (1867–1918). The independent state Czechoslovak Republic (consisting of Bohemia, Moravia, Czech Silesia, Slovakia and Sub-Carpathian Ruthenia) was declared in 1918 and dissolved as a result of a German invasion in 1939 (more precisely, as a result of the Munich Agreement in 1938).

[3] This school had various official names throughout history (see Moravcová 2015, p. 159).

[4] The university in Prague, known as the Charles university in Prague, was named after Kaiser Ferdinand III between the years 1654–1918.

[5] This school did not gain a university statute immediately, it was approved in 1863. There was similar situation in other technical schools in the Austrian Empire.

original papers on descriptive geometry *Die orthographische Parallelperspective* (Skuherský 1851) and *Die Theorie der Theilungspunkte als Beitrag zur Lehre von der freien Perspektive* (theory of dividing points as a contribution to free perspective) (Skuherský 1851), that ensured him the post of a descriptive geometry assistant of professor Johann Hönig (1810–1886) in the school year 1851/1852 (see Velflík 1906, 1909, pp. 389–396).

Skuherský gave 15 h of lectures in descriptive geometry a week (including seminars) according to his own treatises, Johann Hönig's textbook *Anleitung zum Studium der darstellenden Geometrie* (introduction to the study of descriptive geometry) (Hönig 1845) and Charles François Antoine Leroy's work *Traité de Géométrie descriptive* (Leroy 1834). The lectures contained various methods of projections including Skuherský's own method (see Sect. 7), a projection of polyhedra with regard to crystallography, a theory of curves and surfaces, spherical geometry, illumination and anamorphosis. Moreover, Skuherský organized supplementary lectures on perspective and applications such as stereotomy or gnomonic projection.

The second polytechnic school in the Czech Lands was founded in Brno in 1849. Descriptive geometry was initially taught there by an assistant, Anton Mayssl (1826–1899), according to the second edition Georg Schaffnit's textbook *Geometrische Constructionlehre, oder darstellende Geometrie* (geometrical constructions or descriptive geometry) (Schaffnit 1837). Two years later Georg Beskiba was appointed professor. Georg Beskiba (1819–1882) studied at the polytechnic school in Vienna. He became professor of civil engineering at the polytechnic school in Lviv in 1846. There he also lectured in descriptive geometry (see Šišma 2002, pp. 36–38).

Descriptive geometry lectures at the polytechnic school in Brno were obligatory for students of engineering in the first year of studies and their number fluctuated between 3 and 13 h a week (including seminars) in the 1850s.

2 Origins of Czech Technical Universities

Although the Czech language was allowed by the law at the Charles-Ferdinand University in Prague since the revolutionary year 1848, the polytechnic school in Prague only provided German lectures. Skuherský was the first one who offered Czech lectures (in parallel with German lectures) in 1861. Owing to the great interest in these lectures they continued to be provided. This was one of the steps that resulted in the division of the technical university in Prague into the separated Czech technical university, *Český polytechnický ústav království českého* (Czech polytechnic school of the Kingdom of Bohemia) and the separated German technical university, *Deutsches polytechnisches Institut des Königreiches Böhmens* (German polytechnic school of the Kingdom of Bohemia), in 1869.[6] Both of these schools used in their official names the term technical university from 1879.

The polytechnic school in Brno was finally established as a German school and it was reorganized as a technical university in 1873. However, the *Česká vysoká škola technická v Brně* (Czech technical university in Brno) was established in 1899. Jan Sobotka was appointed professor of descriptive geometry in the same year. Jan Sobotka (1862–1931) studied at the Czech technical university in Prague. He then worked as an assistant in descriptive geometry there between the years 1886–1891. Later he studied in Zürich and Wrocław, taught at the real-school in Vienna (from 1894) and at the technical university in Vienna (from 1896). Sobotka influenced teaching of geometry in Brno from 1899 to 1904, then he was appointed professor of mathematics at the Charles-Ferdinand university in Prague. He is known for his textbook *Deskriptivní geometrie promítání parallelního* (descriptive geometry of parallel projection) (Sobotka 1906) and for the works on axonometry (especially oblique axonometry, see Sect. 9) and differential geometry (see Kašparová and Nádeník 2010). Sobotka created a syllabus[7] of descriptive geometry, organized a mathematical library and arranged a collection of geometrical models in Brno (Unknown Author 1911).

3 Descriptive Geometry in Czech Secondary Schools

Descriptive geometry also appeared in secondary schools, particularly real-schools, in the 1850s in connection with the development of technical studies.

Real-schools were instituted as a 6-year secondary school in 1849 by *Entwurf der Organisation der Gymnasien und Realschulen in Oesterreich* [outline of the organization of grammar schools and real-schools in Austria] (Unknown Author 1849). The number of classes increased to seven in about 1870. Czech and German real-schools were opened in the Czech Lands during the second half of the nineteenth century. At first, the education was provided mainly in the German language in Czech real-schools, but from the 1860s the Czech language began to be used more frequently.

Descriptive geometry was being integrated into education gradually and without a given curriculum in the 1850s and 1860s. The curriculum of descriptive geometry for Czech real-schools was first determined in 1874 and modified in 1898, 1909 and 1933 (see Moravcová 2015, pp. 29–38). Initially it contained orthogonal projection onto two planes (Monge's projection) and central projection (especially linear

---

[6]The period 1864–1869 (also called the utraquist period), when the school provided all lectures in German and Czech in parallel, preceded the division of the technical university in Prague (for history of the technical university in Prague see Lomič and Horská 1978).

[7]The syllabus contained Monge's orthogonal projection, oblique projection, central projection, axonometry and technical curves and surfaces.

perspective). The other projection methods like orthogonal axonometry and oblique projection were introduced in the first half of the twentieth century, but all parts were reduced and simplified as well.

The first graduation exams in some real-schools took place in 1869 (in connection with an extension to seven classes); from 1872 the exams were defined by law (before that the exams could be taken only in grammar schools). The descriptive geometry exam was one of the obligatory parts of graduation exams in real-schools, it had a written form and lasted 5 h. The exam was very difficult in comparison to current requirements in secondary education.

The advanced level of descriptive geometry in real-schools before World War II can be supported with several extant materials found in libraries or archives like graduation exam exercises, students' drawings, school notes, etc. (see Moravcová 2015).

The development of the Czech descriptive geometry curriculum can also be observed through the study of Czech textbooks for real-schools. The first one *Zobrazující měřictví* (descriptive geometry) (Ryšavý 1862, 1863) was written by Dominik Ryšavý (1830–1890). Even though it contained many errors, it was crucial for Czech descriptive geometry as it started the formation of the Czech terminology.[8] Problems with the terminology were worked out by Vincenc Jarolímek (1846–1921) in his textbook *Deskriptivní geometrie pro vyšší školy reálné* (descriptive geometry for high real-schools) (Jarolímek 1875, 1876, 1877). This one was supplemented with a German–French–Czech glossary of all the terms used. Moreover, the topics were organized very clearly and logically and the signage was similar to the contemporary one. The textbook was exceptional and there was a high demand for it abroad. It was translated into Bulgarian and published in Plovdiv[9] in 1895. Furthermore, Jarolímek published the first Czech collection of descriptive geometry exercises (Jarolímek 1873). It contained more than one thousand exercises on Monge's projection which Jarolímek created from his own experience. Before this collection (first published in 1873 and reprinted in 1880 and 1904), there had been no similar German collection in use in Czech schools.

---

[8]The first Czech textbooks together with the first Czech lectures were fundamental acts for a creation of Czech terminology. The main personalities who were involved in it were Rudolf Skuherský (he started the Czech lectures on descriptive geometry at the Czech polytechnic school in 1861) and Dominik Ryšavý (who started to teach in the Czech language at the *První česká reálka v Praze* (first Czech real-school in Prague) in the same year).

[9]Czech mathematicians had great merit in the introduction of descriptive geometry in secondary education in Croatia and Bulgaria in the second half of the nineteenth century (see Bečvářová 2009).

Descriptive geometry was also taught in real-grammar schools[10] and some secondary industrial schools before World War II. However, descriptive geometry teaching in these schools was not as significant as in real-schools.

Real-schools ceased to exist during World War II (most of them were transformed into grammar schools) and since then the importance of descriptive geometry in secondary schools has been constantly decreasing.

4 Descriptive Geometry at the Czech Technical University in Prague

The Czech lectures in descriptive geometry at the Czech technical university in Prague were given by professor František Tilšer from 1864. Tilšer (1825–1913) was a student of law at first, later he studied at military schools in Olomouc and Vienna. This school was moved from Vienna to Louka u Znojma and Tilšer obtained a post of descriptive geometry professor there in 1854. He became professor at the technical university in Prague in 1864 and after its division (1869) he lectured at the Czech technical university. He simultaneously worked as a member of provincial assembly from 1870, therefore his lectures were often delivered by supply teachers. Tilšer elaborated a theory of illumination in the work *Die Lehre der geometrischen Beleuchtungs-Constructionen und deren Anwendung auf das technische Zeichnen* (essay on geometrical constructions of illumination and their application in technical drawing) (Tilscher 1862), in which he made efforts to generalize a construction of parallel illumination isophotes by using systems of tangential planes, which have a constant angle with the illumination direction, of a surface. Equally significant is Tilšer's two-volume work *System der technisch-malerischen Perspective* (system of technical-painting perspective) (Tilscher 1865, 1866) (see Velflík 1910, 1925, pp. 45–54).

Tilšer divided the lectures into 2 years of studies between the years 1870 and 1874, but their total number was not changed. At first, Tilšer taught like Skuherský, but he transformed the concept of descriptive education in 1875. He perceived descriptive geometry as the one means of human cognition. He distinguished two parts of descriptive geometry—*morphognosy* (a science on objects which are projected) and *iconography* (a science on projection of objects). Moreover, Tilšer created a special system of signage and he called the whole scientific discipline iconognosy (Mikulášek 1924). His philosophical approach was correct but too complicated.

---

[10]The real-grammar school (from German: *das Realgymnasium*, in Czech: *reálné gymnasium*) was a special kind of 8-year secondary school that was instituted in Austria in 1908 (schools of this kind had already been established in the Czech Lands after 1862) as a compromise between real-schools and classical grammar schools.

Tilšer wrote several works on descriptive geometry, some of them were written in the Czech language. He also published the first Czech descriptive geometry textbook for university students titled *Soustava deskriptivní geometrie* (system of descriptive geometry) (Tilšer 1870); however, he finished only the first volume which contained basics of projective geometry, polar coordinates and selected curve properties.

Tilšer's philosophical ideas were elaborated in his works *Grundlagen der Ikonognosie* (basics of iconognosy) (Tilscher 1878) and *Gasparda Monge-a Géometrie descriptive po stoletém vývoji čili u východiště z labyrintu* (Gaspard Monge's descriptive geometry after one-hundred-years development, or near the exit from the labyrinth) (Tilšer 1898).

The next professor of descriptive geometry at the Czech technical university in Prague, Karel Pelz, was chosen during a selection in 1896. Karel Pelz (1845–1908) studied at the technical university in Prague, then he worked as an assistant in descriptive geometry at the German technical university in Prague from 1870 to 1875. After that he taught at the real-schools in Těšín and Graz. He became professor of descriptive geometry at the technical university in Graz in 1878. Pelz was interested mainly in a theory of curves and surfaces, axonometry (see Sect. 9) and central projection. He wrote more than thirty original scientific works, many of them were positively appreciated abroad (see Sklenáriková and Pémová 2007).

At the technical university in Prague, Pelz simplified the syllabus and returned to standard conception of descriptive geometry teaching. Although he did not publish in the Czech language and his Czech was allegedly not excellent, he was appreciated by students as a great teacher. We can see the high quality of his lectures, thanks to extant lithographic notes.[11]

The number of students of the technical university in Prague increased, therefore parallel descriptive geometry lectures were introduced at the beginning of the twentieth century. As a result, the second department of descriptive geometry was created in 1907 and Vincenc Jarolímek was appointed the next professor. Vincenc (Čeněk) Jarolímek (1846–1921) studied at the technical university in Prague. He worked at the real-school in Písek since 1868, subsequently he became the head at the real-school in Hradec Králové (1891–1893), Karlín (1893–1895) and at the first Czech real-school in Prague (1895–1904). He was a provincial school inspector from 1904, simultaneously he lectured at the Czech technical university in Brno from 1905. He moved to the Czech technical university in Prague in 1906. Jarolímek continued Tilšer's research into isophotes in central illumination in the work *Centrálné osvětlení* (central illumination) (Jarolímek 1871). Other fields of his activities were theory of curves and surfaces, especially of the second order, and imaginary objects in geometry (see Sobotka 1916).

From 1907, lectures in descriptive geometry were provided in three specialized sections. The first one (also called the first department of descriptive geometry) was intended for students of civil engineering; the second one (the second department

---

[11]Lithographic (lithography is a method of copying) notes were written by students as a replacement of missing Czech textbooks.

Table 16.1 Overview of the descriptive geometry lecturers for civil engineering students at the Czech technical university in Prague between the years 1907–1939 (Moravcová 2015, p. 420)

Deskriptivní geometrie (descriptive geometry)		
1907–1908	K. Pelz	5/6, 4/6
1908–1913	V. Jarolímek	
1913–1915		5/5, 4/5
1915–1921	F. Kadeřávek/B. Procházka	
1921–1939	F. Kadeřávek	5/4, 4/4

Table 16.2 Overview of the descriptive geometry lecturers for machine engineering students at the Czech technical university in Prague between the years 1907–1939 (Moravcová 2015, p. 421)

Deskriptivní geometrie (descriptive geometry)		
1907–1908	V. Jarolímek	5/6, 4/6
1908–1921	B. Procházka/F. Kadeřávek	
1921–1925	B. Procházka	6/3, 0/5
1925–1927	J. Kounovský	
1927–1932		4/4, 2/4
1932–1933		5/3, 0/4
1933–1939		5/3, 0/3

Table 16.3 Overview of the descriptive geometry lecturers for the third section at the Czech technical university in Prague between the years 1907–1939 (Moravcová 2015, p. 421)

Deskriptivní geometrie (descriptive geometry)		
1907–1921	B. Chalupníček	6/6, 4/6
1921–1922		4/6, 4/3
1922–1930	F. Kadeřávek	
1930–1932		4/6, 4/4
1932–1939		4/4, 4/4

of descriptive geometry) was for students of machine engineering and the third one for ground building students (see Tables 16.1, 16.2, and 16.3).[12] A third department was not formally created.

All lectures contained the most common kinds of projections. Moreover, the lectures for the first section included spot height projection,[13] projective geometry and illumination. Lectures for the second one put emphasis on kinematic geometry and the third one included perspective. Furthermore, the fourth section for students of forest engineering, which contained spot height projection, nomography and basics of photogrammetry extra, was opened in 1919. Descriptive geometry lectures were provided only in the first semester of the first year of studies.

---

[12] Tables 16.1, 16.2, 16.3, and 16.4 provide information on lectures in descriptive geometry including the lectures' names and the numbers of lectures. A symbol x/y in the last column refers to the number of lectures/seminars a week. The lectures were intended for first-year students (unless otherwise stated).

[13] Spot height projection is an orthogonal projection onto a single plane. Projections of points are annotated with numbers (spots) which give information about the distance between the point and the plane of projection. Minus indicates the negative semi-space. This method is used mainly in cartography and civil engineering.

Bedřich Procházka, Josef Kounovský and František Kadeřávek were three other important descriptive geometry professors at the Czech technical university in Prague.

Bedřich Procházka (1855–1934) studied at the Czech technical university in Prague and from 1876 worked as an assistant in descriptive geometry there. He passed teachers' competence exams in mathematics and descriptive geometry in 1879. After that he taught at secondary schools in Prague, Chrudim, Pardubice and Karlín. He was awarded the position of senior lecturer in geometrical illumination in 1884 and subsequently in kinematic geometry in 1895. He worked as the head at the real-school in Náchod from 1897. He became professor of descriptive geometry at the Czech technical university in Brno in 1904 and in Prague in 1908 (see Bydžovský 1934).

Josef Kounovský (1878–1949) studied at the Czech technical university in Prague and worked as an assistant in descriptive geometry there from 1902. He passed teachers' competence exams in mathematics and descriptive geometry in the same year. He then taught at the real-schools in Prague and Hradec Králové. He was awarded the position of senior lecturer in geometry in 1912 and then he lectured at the Czech technical university in Prague, where he became professor in 1927 (see Kadeřávek 1950).

František Kadeřávek (1885–1961) studied at the Czech technical university in Prague and worked as an assistant in descriptive geometry there from 1906. He passed teachers' competence exams in mathematics and descriptive geometry in 1908. He was awarded the position of senior lecturer in synthetic geometry in 1912. He became professor at the Czech technical university in Prague in 1917 but he had already lectured in descriptive geometry there from 1915. Together with Josef Kounovský and Josef Klíma (see below) he wrote an important two-volume Czech descriptive geometry textbook for all university students *Deskriptivní geometrie I, II* (Kadeřávek et al. 1929, 1932). The book contains all substantial topics of descriptive geometry and is used even by contemporary students from time to time. Kadeřávek was interested in the history of descriptive geometry and wrote several popular science books on it (see Kepr 1955).

At the Czech technical university in Prague, Procházka and Kadeřávek often swapped the first and second section of descriptive geometry lectures and also gave the lectures for the third and fourth section.

Jarolímek and Procházka wrote the first Czech descriptive geometry textbook for students of technical universities titled *Deskriptivní geometrie pro vysoké školy technické* (descriptive geometry for technical universities) (Jarolímek and Procházka 1909). The book was distinguished by its logical structure and was in great demand.

5 Descriptive Geometry at the Czech Technical University in Brno

The lectures in descriptive geometry provided at the Czech technical university in Brno had a similar form as in Prague.

Bedřich Procházka, who became professor in 1904, introduced additional topics from projective and kinematic geometry in 1906, when the number of descriptive geometry lectures was increased (see Table 16.4).

Miloslav Pelíšek was appointed professor of descriptive geometry in 1909. Pelíšek (1855–1940) studied at the German technical university in Prague. From 1881, he worked as an assistant in descriptive geometry there, after that he taught at the secondary industrial schools in Plzeň and Prague. He lectured at the Czech technical university in Brno from 1908. Pelíšek focused on synthetic, kinematic and analytic geometry. His determination method of trajectories curvature centres is known as *Pelíšek's construction* in Czech geometry textbooks (see Hlavatý 1941).

At the Czech technical university in Brno, Pelíšek modified the syllabus of the descriptive geometry lectures, in which he added spot height projection and its applications. At the same time, he reduced kinematic and projective geometry in them, however, he offered optional lectures on these topics. As the number of the descriptive geometry lectures was decreasing in 1921, Pelíšek wrote a textbook *Deskriptivní geometrie* (Pelíšek 1922) that was tailored for students.

The number of students in Brno was smaller than in Prague, therefore parallel lectures were provided later, from 1928. These lectures were organized by Josef Klíma, who was appointed professor of descriptive geometry at the Czech technical university in Brno in 1927. The lectures were common for all students in the first semester. In the second semester, they were divided into two sections (one for civil engineering students and the other for mechanical and electrical engineering students). Josef Klíma (1887–1943) studied at the Czech technical university in

Table 16.4 Overview of the descriptive geometry lecturers at the Czech technical university in Brno (Moravcová 2015, p. 424)

Deskriptivní geometrie (descriptive geometry)		
1899–1904	J. Sobotka	6/6, 4/6
1904	V. Jarolímek	
1904–1905	B. Procházka	
1905–1906		6/6, 4/6
Deskriptivní geometrie spojená s geometrií polohy (descriptive geometry associated with projective geometry)		
1906–1908	B. Procházka	6/6, 6/6
1908–1914	M. Pelíšek	6/6, 6/6
Deskriptivní geometrie (descriptive geometry)		
1914–1921	M. Pelíšek	6/6, 6/6
1921–1926		5/5, 5/5
1926–1927	J. Klapka	4/4, 4/4
1927–1939	J. Klíma	

Prague and became an assistant in descriptive geometry at the same school in 1909. He taught at the real-schools in Vinohrady from 1917, later in Vršovice and Karlín. He was awarded the position of senior lecturer in descriptive geometry in 1924 (see Seifert 1946).

6 Summary of Descriptive Geometry Teaching in the Czech Technical Universities

Descriptive geometry was one of the main subjects for all the first-year students except for students of chemistry at technical universities in the Czech Lands. Lectures continued in the curriculum of real-school, i.e., Monge's projection was not included, but axonometry, central, oblique and spot height projections were taught from basic principles. The emphasis was put on theories of curves and surfaces, collineation, projective and kinematic geometry were lectured on as well. Moreover, illumination was a significant topic in the nineteenth century, later it was reduced.

The number of lectures and seminars was very high in the nineteenth century, but we can see its decrease in the tables above, after 1900. On the other hand, syllabi were expanded on new topics such as photogrammetry and nomography, so that they adapted to requirements of the narrow specialization of graduates.

Lectured topics on descriptive geometry were similar at both Czech technical universities in the twentieth century, the slight difference was in projective geometry teaching. Its basics were a part of the descriptive geometry lectures in most cases. Moreover, advanced problems of projective geometry were taught during extra lectures in Prague; these kinds of lectures were offered in Brno only by Antonín Sucharda (1854–1907) between the years 1901–1906 and by Miloslav Pelíšek between the years 1911–1918.

Finally, special lectures for students who were preparing for teachers' competence exams should be mentioned. The exams were introduced as special exams for students who meant to become secondary school teachers in 1850. They were supervised by committees which were set up within classical universities,[14] but lectures suitable for candidates taking the exam in descriptive geometry were only provided in technical universities in the Czech Lands in the nineteenth century.[15] These lectures were modified every year and contained various advanced topics, for example, central axonometry, theories of high-order surfaces, etc. Bedřich

---

[14] In the nineteenth century there was only one classical university in the Czech Lands—the above mentioned Charles-Ferdinand university in Prague, which was divided into Czech and German universities in 1882. The German university in Prague was closed down in 1945. The second Czech classical university (and the last one to be found before World War II) was established in Brno in 1919.

[15] Descriptive geometry lectures at classical universities in the Czech Lands were not provided systematically in the nineteenth century. Jan Sobotka started these lectures at the Czech technical university in Prague in 1910 for the purpose of preparing student for teachers' competence exams.

Procházka prepared a special six-volume textbook for these students titled *Vybrané statě z deskriptivní geometrie* (selected topics of descriptive geometry) (Procházka 1912–1918), which contained less known information about orthogonal and oblique axonometry and surfaces of the second order, central axonometry, central illumination, spatial curves, translational surfaces and kinematic geometry (see Moravcová 2015, pp. 247–248).

7 Skuherský's Projection Method

The first attempts for a formation of graphic projection methods originated in England in the first half of the nineteenth century (see Lawrence, Chap. 18, this volume). Orthogonal axonometry was one of two approaches that were developed from these attempts. The other one was orthogonal parallel projection created by Rudolf Skuherský. Skuherský published his new theory that was based on transformations of an object in relation to the projection planes (Fig. 16.1) in the works *Die orthographische Parallelperspektive* (Skuherský 1850), *Die orthographische Parallel-Perspektive* (Skuherský 1858a) and *Die Methode der*

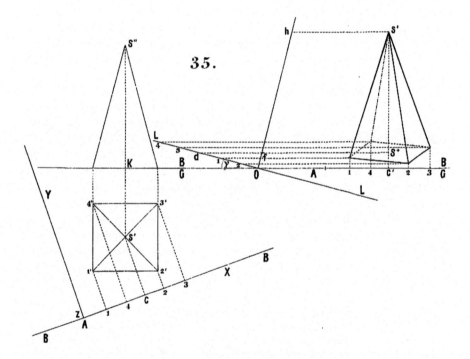

Fig. 16.1 Skuherský's projection of a pyramid from his work *Die Methode der orthogonalen Projekzion auf zwei Ebenen* (Skuherský 1858b, Tab. 2, Fig. 35)

orthogonalen Projekzion auf zwei Ebenen (method of orthogonal projection onto two planes) (Skuherský 1858b).

Skuherský's method can be explained easily through the projection of a cuboid $ABCDA'B'C'D'$, which is given in Monge's projection (Fig. 16.2a). We select

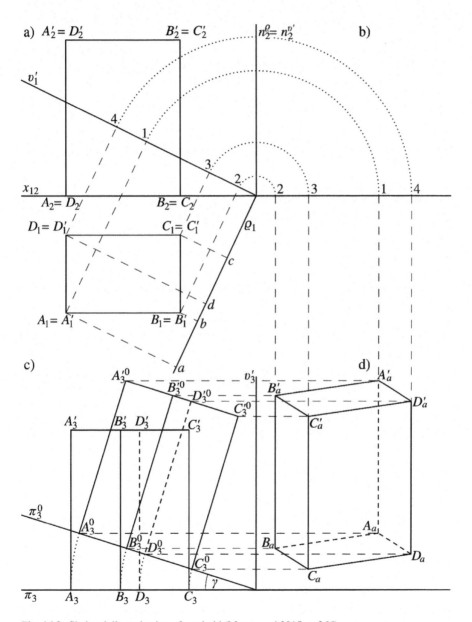

Fig. 16.2 Skuherský's projection of a cuboid (Moravcová 2015, p. 266)

a third projection plane ρ perpendicular to the first projection plane π. We then construct an orthogonal projection of the cuboid onto a plane ν' which is perpendicular to both π and ρ. We rotate ν'_1 to the axis x_{12} (see points 1, 2, 3, 4 in Fig. 16.2b) and also rotate the third projection onto ρ at a given angle γ around the line of intersection of π and ν' (Fig. 16.2c). Finally, we can project this transformed cuboid onto ν' again (using the auxiliary points 1, 2, 3, 4 from Fig. 16.2b) and obtain the orthogonal projection of the cuboid in Skuherský's method (Fig. 16.2d).

Skuherský's orthogonal projection enables us not only to display any object but also to solve spatial problems in a plane. Although this method was replaced by smart orthogonal axonometry in the second half of the nineteenth century, we can find it or its elements in many textbooks, for example, Gustav Adolf Viktor Peschka's *Darstellende und projective Geometrie* (descriptive and projective geometry) (Peschka 1883) or Emil Müller's *Lehrbuch der darstellenden Geometrie für technische Hochschule, zweiter Band* (descriptive geometry textbook for technical universities, 2nd vol.) (Müller 1923).

8 Pelz's Contribution to Axonometry

Karel Pelz had great merit in the development of orthogonal axonometry. He focused on it during his stay in Graz and published four original papers in which he described many constructions in orthogonal axonometry: *Zur wissenschaftlichen Behandlung der orthogonalen Axonometrie I, II, III* (on scientific conception of orthogonal axonometry I, II, III) (Pelz 1880, 1881, 1884), and *Beiträge zur wissenschaftlichen Behandlung der orthogonalen Axonometrie* (supplement to the scientific conception of orthogonal axonometry) (Pelz 1885). In the first and second, Pelz introduced, inter alia, constructions of perpendicular lines and planes and projections of a circle in various planes. He was the first one who proved that an orthogonal axonometry is unequivocally defined by an axonometric triangle (Sklenáriková and Pémová 2007). In his subsequent papers, Pelz dealt with illumination of cylinders and cones (Fig. 16.3). In the last paper, he focused on a sphere and its parallel and central illumination (Sobotka 1910, p. 458).

Pelz did not write a textbook on descriptive geometry, but his discoveries were published by his assistant Rudolf Schüssler (1865–1942) in a book *Orthogonale Axonometrie, ein Lehrbuch zum Selbststudium* (orthogonal axonometry, a self-study book) (Schüssler 1905).

Pelz also dealt with the basic theorem of oblique axonometry known as *Pohlke's theorem* after its author Karl Pohlke (1810–1876). Pelz published a new original proof in a paper *Über einen neuen Beweis des Fundamentalsatzes von Pohlke* (on a new proof of Pohlke's theorem) in 1877 (Pelz 1877). According to Pohlke's assistant Hermann Schwarz (1843–1921), Pelz's proof was similar to Pohlke's own, but Pohlke did not publish it. Principles of Pelz's proof, which was based on a system of confocal conics, are elaborated in (Sklenáriková and Pémová 2007).

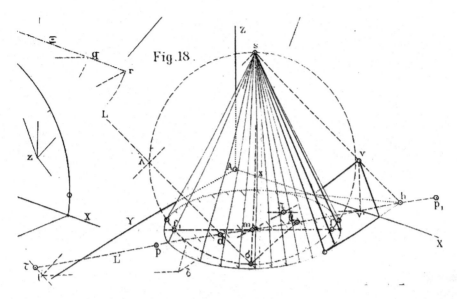

Fig. 16.3 Illumination of a cone in orthogonal axonometry from Pelz's paper *Zur wissenschaftlichen Behandlung der orthogonalen Axonometrie III* (Pelz 1884, Tab. 1, Fig. 18)

9 Sobotka's Constructions

The next Czech professor who claimed credit for the development of axonometry was Jan Sobotka. He devised three methods of transformation of oblique axonometry into orthogonal projection, which were published in the article *Axonometrische Darstellungen aus zwei Rissen und Koordinatentransformationen* (axonometric projection on the basis of two drawings and coordinate transformations) in 1901 (Sobotka 1901). These constructions are usually referred to as *Sobotka's constructions* in Czech literature.

The first construction transforms oblique axonometry into Monge's projection. It can be used in case the axonometry is defined by an axonometric triangle and a projection of the origin of coordinates. Oblique axonometry is first converted into orthogonal axonometry, then two of the auxiliary planes of projection are rotated to the axonometric plane and translated in the directions of the double-projection planes (see Fig. 16.4).

The second and third constructions are very similar to each other. They can be used if the axonometry is defined by oblique projections of the coordinates axes and the units of measurement on them. Sobotka excellently used the affine relation between the first and second auxiliary oblique projections and the first and second orthogonal projections (for detail see Moravcová 2015, pp. 273–275).

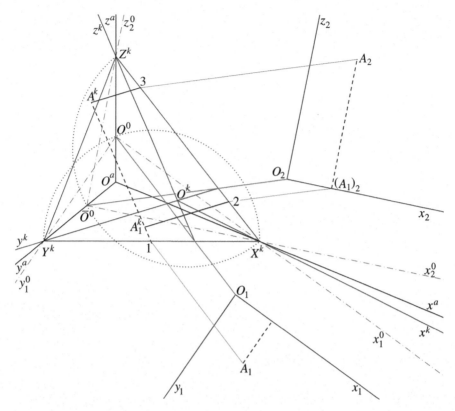

Fig. 16.4 Demonstration of Sobotka's first construction. A transformation of the oblique projection A^k of a point A into an orthogonal projection onto two perpendicular planes (Moravcová 2015, p. 272)

10 Conclusion

The most profound development of descriptive geometry in the Czech Lands was noticeable in the last third of the nineteenth century and its reverberations were still perceptible in the first half of the twentieth century. This science was integrated into the education of all engineers as an integral part of it. We have focused on Czech technical universities in this paper, but the situation and role of descriptive geometry at German technical universities in the Czech Lands was similar.

Origins of descriptive geometry in the Czech Lands were connected with its development in France, Germany and Austria; the greatest influence came from the polytechnic school in Vienna (see Stachel, Chap. 11 and Binder, Chap. 12, this volume). At first, German and French sources were studied until the 1860s. Nevertheless, in connection with Czech emancipation, we can observe the expansion of Czech science including the development of geometry (not only descriptive) as well. Many Czech geometers published works with new results on descriptive geometry but also on projective or differential geometry (besides those mentioned, we can refer to Czech mathematicians such as Eduard Weyr, Emil Weyr, Josef Šolín and others). The term *Czech geometric school* is taken for these activities (see Folta 1982).

The interest in technical branches gradually decreased in the period between the World Wars and Czech education system and science were definitively repressed by World War II (Czech universities were forced to close in 1939). Czech education was restored after 1945, but descriptive geometry never gained such importance again.

Acknowledgements The chapter has been supported by the project PROGRES Q17 (*Teacher preparation and teaching profession in the context of science and research*).

References

Bečvářová, Martina. 2009. *České kořeny bulharské matematiky*. Praha: Matfyzpress.
Bydžovský, Bohumil. 1934. *Bedřich Procházka*. Praha: Česká akademie věd a umění.
Folta, Jaroslav. 1982. *Česká geometrická škola: historická analýza*. Praha: Academia.
Hlavatý, Václav. 1941. *Miloslav Pelíšek*. Praha: Česká akademie věd a umění.
Hönig, Johann. 1845. *Anleitung zum Studium der darstellenden Geometrie*. Wien: Carl Gerold.
Jarolímek, Vincenc. 1871. *Centrálné osvětlení*. Praha: Eduard Grégr.
Jarolímek, Čeněk. 1873. *Deskriptivní geometrie v úlohách pro vyšší školy reálné*. Praha: Jednota českých mathematiků.
———. 1875/1876/1877. *Deskriptivní geometrie pro vyšší školy reálné. Část 1., 2., 3.* Praha: Jednota českých mathematiků.
Jarolímek, Vincenc, and Bedřich Procházka. 1909. *Deskriptivní geometrie pro vysoké školy technické*. Praha: Česká matice technická.
Kadeřávek, František. 1950. In memoriam techn. Dr. Josefa Kounovského. In *Časopis pro pěstování matematiky a fysiky* 75: D345–D349.
Kadeřávek, František, Josef Klíma, and Josef Kounovský. 1929/1932. *Deskriptivní geometrie I, II*. Praha: Jednota československých matematiků a fysiků.
Kašparová, Martina, and Zbyněk Nádeník. 2010. *Jan Sobotka (1862–1931)*. Praha: Matfyzpress.
Kepr, Bořivoj. 1955. Sedmdesát let prof. Ing. dr. Františka Kadeřávka. In *Časopis pro pěstování matematiky* 80: 375–382.
Leblanc, César Nicolas Louis. 1830. *Choix de modéles appliqués à l'enseignement du dessin des machines, avec un texte descriptif*. Paris: l'Auteur and Mahler.
Leroy, Charles François Antoine. 1834. *Traité de Géometrie descriptive*. Paris: Carilian-Goeury and Anselin.
Lomič, Václav, and Pavla Horská. 1978. *Dějiny Českého vysokého učení technického (1848–1918)*. Praha: SNTL.
Mikulášek, Artuš. 1924. *Dr. techn. František Tilšer, učenec, politik a vychovatel*. Brno: Moravský legionář.
Moravcová, Vlasta. 2015. *Výuka deskriptivní geometrie v našich zemích*. (*Descriptive geometry teaching in the Czech Lands*.) Doctoral thesis. Prague: Faculty of Mathematics and Physics, Charles University in Prague.
Müller, Emil. 1923. *Lehrbuch der darstellenden Geometrie für technische Hochschule, zweiter Band*. Leipzig: B. G. Teubner.
Pelíšek, Miloslav. 1922. *Deskriptivní geometrie*. Brno: Česká technika Brno.

Pelz, Karel. 1877. Über einen neuen Beweis des Fundamentalsatzes von Pohlke. In *Sitzungsberichte der kaiserlichen Akademie der Wissenschaften, Mathematisch-naturwisse Classe* 76, II. Abt.: 123–138.

———. 1880. Zur wissenschaftlichen Behandlung der orthogonalen Axonometrie I. In *Sitzungsberichte der kaiserlichen Akademie der Wissenschaften, Mathematisch-naturwissenschaftliche Classe* 81, II. Abt.: 300–330.

———. 1881. Zur wissenschaftlichen Behandlung der orthogonalen Axonometrie II. In *Sitzungsberichte der kaiserlichen Akademie der Wissenschaften, Mathematisch-naturwissenschaftliche Classe* 83: 375–384.

———. 1884. Zur wissenschaftlichen Behandlung der orthogonalen Axonometrie III. In *Sitzungsberichte der kaiserlichen Akademie der Wissenschaften, Mathematisch-naturwissenschaftliche Classe* 90: 1060–1075.

———. 1885. Beiträge zur wissenschaftlichen Behandlung der orthogonalen Axonometrie. In *Sitzungsberichte der königlichen böhmischen Gesellschaft der Wissenschaften in Prag, Mathematisch-naturwissenschaftliche Classe* 1885(1886): 648–661.

Peschka, Gustav Adolf Viktor. 1883. *Darstellende und projective Geometrie*. Wien: Carl Gerold's Sohn.

Procházka, Bedřich. 1912–1918. *Vybrané statě z deskriptivní geometrie*. vols. I–VI. Praha: Česká matice technická.

Ryšavý, Dominik. 1862/1863. *Zobrazující měřictví I, II*. Praha: I. L. Kober.

Schaffnit, Georg. 1837. *Geometrische Constructionlehre, oder darstellende Geometrie*. Darmstadt: J. Wilh. Heyer.

Schüssler, Rudolf. 1905. *Orthogonale Axonometrie, ein Lehrbuch zum Selbststudium*. Leipzig: B. G. Teubner.

Seifert, Ladislav. 1946. Prof. Dr. techn. Josef Klíma. In *Časopis pro pěstování matematiky a fysiky* 71: D35-D42.

Sklenáriková, Zita, and Marta Pémová. 2007. Zo života a diela Karla Pelza. In *Matematika v proměnách věků IV*, ed. E. Fuchs, 197–215. Brno: Cerm.

Skuherský, Rudolph. 1850. Die orthographische Parallelperspective. In *Sitzungsberichte der kaiserlichen Akademie der Wissenschaften, Matematisch-naturwissenschaftliche Classe* 5: 326–342.

———. 1851. Die Theorie der Theilungspunkte als Beitrag zur Lehre von der freien Perspektive. In *Sitzungsberichte der kaiserlichen Akademie der Wissenschaften, Matematischnaturwissenschaftliche Classe* 7: 471–477.

———. 1858. *Die orthographische Parallel-Perspective*. Prag: F. Tempsky.

———. 1858. *Die Methode der orthogonalen Projekzion auf zwei Ebenen*. Prag: Kath. Geržabek.

Sobotka, Jan. 1901. Axonometrische Darstellungen aus zwei Rissen und Koordinatentransformationen. In *Sitzungsberichte der königlichen Gesellschaft der Wissenschaften in Prag, Mathematisch-naturwissenschaftliche Classe* 35: 1–27.

———. 1906. *Deskriptivní geometrie promítání paralelního*. Praha: Jednota českých mathematiků and Česká matice technická.

———. 1910. O životě a činnosti Karla Pelze. In *Časopis pro pěstování matematiky a fysiky* 39: 433–460.

———. 1916. Vincenc Jarolímek. In *Časopis pro pěstování matematiky a fysiky* 45: 439–450.

Šišma, Pavel. 2002. *Matematika na německé technice v Brně*. Praha: Prometheus.

Tilscher, Franz. 1862. *Die Lehre der geometrischen Beleuchtungs-Constructionen und deren Anwendung auf das technische Zeichnen*. Wien: Carl Gerold's Sohn.

———. 1865/1866. *System der technisch-malerischen Perspective*. Prag: F. Tempsky.

———. 1878. *Grundlagen der Ikonognosie*. Prag: Verlag der königlichen böhmischen Gesellschaft der Wissenschaften.

Tilšer, František. 1870. *Soustava deskriptivní geometrie*. Praha: Privately printed.

———. 1898. *Gasparda Monge-a Géometrie descriptive po stoletém vývoji čili u východiště z labyrintu*. Praha: Výkonný výbor výstavy architektury a inženýrství.

Unknown Author. 1849. *Entwurf der Organisation der Gymnasien und Realschulen in Oesterreich.* Wien: Ministerium des Cultus und Unterrichts.

Unknown Author. 1911. *Památník českých vysokých škol technických Františka Josefa v Brně.* Brno: Česká vysoká škola technická Františka Josefa.

Velflík, Albert Vojtěch. 1906/1909. *Dějiny technického učení v Praze, I. díl.* Praha: Profesorský sbor c.k. české vysoké školy technické v Praze and Česká matice technická.

———. 1910/1925. *Dějiny technického učení v Praze, II. díl.* Praha: Česká matice technická.

Chapter 17
The Love Affair with Descriptive Geometry: Its History in Serbia

Katarina Jevtić-Novaković and Snezana Lawrence

Abstract While Serbia was part of the federation of South-Slav countries for majority of the twentieth century, the history of its mathematics, and in particular the history of descriptive geometry in the country is sufficiently independent from that of the twentieth century Yugoslavia to merit a separate historical analysis. We do this here, although we offer some remarks which include an overview of the 'Yugoslav' years as the twentieth century was in many ways different to the tradition we investigate. We conclude with remarks about the love affair with descriptive geometry, and its longevity in the Serbian educational system.

Keywords Descriptive geometry · Serbia · *Nacrtna geometrija* · Atanasije Nikolić · Emilijan Josimović · Dimitrije Stojanović · Stevan Davidović · Mihailo Petrović · Petar Anagnosti

1 Mathematics in Serbia in the Nineteenth Century: The First Mathematicians, Their Books, and Schools

In order to understand the importance of descriptive geometry for Serbia, and the importance of the Serbian story for this volume, we will first give an introduction by a way of short overview of the history of mathematics in the country prior to the appearance of this technique.

Serbia was, for main part of the five centuries, from fourteenth to the end of the nineteenth, a de facto divided country, most of which was under the rule of the Ottomans, but whose northern parts were under the rule and influence of the Habsburgs, and later Austro-Hungarian Empire. The situation changed after the Treaty of Adrianople (1829) which was forced upon the Ottoman Sultan by the

---

K. Jevtić-Novaković · S. Lawrence (✉)
Department of Design, Engineering, and Mathematics, Faculty of Science and Technology, Middlesex University, London, UK
e-mail: snezana@mathsisgoodforyou.com

© Springer Nature Switzerland AG 2019
É. Barbin et al. (eds.), *Descriptive Geometry, The Spread of a Polytechnic Art*, International Studies in the History of Mathematics and its Teaching, https://doi.org/10.1007/978-3-030-14808-9_17

Russians, who then granted Serbia an autonomy recognizing Miloš Obrenović (1780–1860) as hereditary prince. Except for a few garrisons in Belgrade, at this point the Ottomans evacuated Serbia, and the Metropolitante of Serbia was established in Belgrade autonomous from the Patriarch of Constantinople. This was the beginning not only of the modern Serbian nationhood and statehood but also of the pursuit of mathematics which could thus be described to be Serbian (Lawrence 2008).

Nothing is known or recorded as far as we know of Serbian mathematics before the arrival of the Ottomans. There are details of some activity, in particular in terms of mathematics education rather than any original mathematical activity, in the period before the autonomy from the Ottomans was proclaimed in the 1830s. There are also records of some mathematical activity in the Slav culture, for example, in the development of mathematical script (Dejić 2014).

The personalities and the books from which Serbian mathematics emerged begin to appear a century before the Serbian independence from the Ottomans (and later the Austro-Hungarians) was sought. We find, for example, that the first known mathematician, Vasilije Damjanović (1734–1792) published the first book on mathematics in Serbian language, *Nova Serbskaja Aritmetika* (The New Serbian Arithmetic), in 1767.

We also know that there were 16 town and several village schools prior to 1830. In the period of only 1 year from 1835 to 1836 though, 26 elementary schools were opened at the state expense. The Great School of Belgrade, founded during the First Serbian Uprising (against the Ottomans, 1804–1813), was opened in 1808 but closed only 5 years later, to be reopened in 1830. Reports from 1833 suggest that the school was at that time developing after the *gymnasium* model, although still having only one teacher for all the subjects. In 1835 the school transferred to Kragujevac, where it was further developed into a *gymnasium*-model school. It then gained in total four teachers, who all taught four grades, mathematics being a subject taught in the first three grades.

The northern part of Serbia known as Vojvodina was leading in developing the learning and teaching of mathematics in this period. Sombor's *Norma* (the 'Normal school'—school for the education of teachers) was founded in the town of Sombor on 1st May 1778 by Avram Mrazović (1756–1826) who was educated in Pesta and Vienna. Maria Teresa (1717–1780), the last Habsburg Empress, nominated Mrazović to lead the development of all educational institutions in Sombor. At the end of Mrazović's life, the then Austro-Hungary supported the foundation of the Matica Srpska, literally the 'Serbian Queenbee', founded in Pesta in 1826, to promote Serbian culture and science, an institution that survives to this day.

From the Great School of Belgrade (1808) and the *Matica Srpska* (1826), the Lyceum of Belgrade was founded in the capital, which grew into the Superior School (established in 1873). The first trained mathematician to teach at the Lyceum was Dimitriije Nešić (1836–1904). Nešić finished *gymnasium* in Belgrade and attended the Lyceum between 1853 and 1855, when he went to Vienna Polytechnic (founded in 1815) for 3 years, and then Karlsruhe Polytechnic (founded in 1825) for further 3 years of schooling. In 1861 he came back to Serbia and worked on introducing the metric system to the country. Josimović, about whom we will talk later, was also working on this and was part of the commission to formulate a law which would introduce metric system into Serbia. In November 1873 this law was proclaimed, and in 1879 Serbia joined the Metre Convention, an institution founded in the meantime (1875) in Paris to oversee and coordinate international metrology and the development of the metric system (McGreevy 1995). Nešić was a personal friend of Vuk Karadžić (1878–1864), the Serbian linguist who at the same time worked on reforming Serbian language, and Nešić worked with him on defining Serbian terminology to every and all mathematical concepts and process (Kastanis and Lawrence 2005).

In the period between 1863 and 1873, the Superior School of Belgrade founded a new section of Science and Mathematics within the Faculty of Philosophy. This was the decade during which Serbia finally won the full independence from the Ottomans and the international recognition of the fact in 1878.

Another prominent raising star in the sphere of mathematical sciences, in particular, as it will be shown in descriptive geometry, was Josimović (1823–1897). Josimović studied at Vienna Polytechnic; upon his return to the country he became professor of the Belgrade Lyceum in 1845; then professor of the Artillery School in 1854. Josimović became professor of the Superior School and rector of the same in 1874 (Jevtić-Novaković and Fontana-Giusti 2012; Maksimović 1957, 1967; Terzić 1994). We will look at his role in greater detail later on in this chapter.

There were others too, perhaps too many to number and describe here, but all of whom completed their doctoral studies in Jena, Budapest, Vienna, Berlin, and then increasingly Paris (Lawrence 2008). The pattern of the mathematicians schooled mainly in the countries of Western Europe continued to dominate the history of Serbian mathematics for some time. Dimitrije Danić (1862–1932) received his PhD in mathematics from the University of Jena in 1885, becoming professor at the Military Academy in Belgrade (of which more soon) upon his return. Bogdan Gavrilović (1864–1947) received his PhD in Budapest in 1887 and became professor in Belgrade upon his return. In 1894, the most prominent of Serbian mathematicians (probably of all time until the late twentieth century), Mihailo Petrović, also known as Alas (1868–1943), completed his doctorate studies in Paris at the *École Normale*. Upon return to Belgrade, he became the first professor of the newly founded University of Belgrade, and established the modern national mathematical school—virtually all the doctorates in mathematics between the two World Wars in Serbia were done under his supervision (Lawrence 2008; Božić 2002).

2 The Idea (or Not) of the Polytechnic School, and Descriptive Geometry in Serbia in the Nineteenth Century

Most of the protagonists of this story on the development of mathematical sciences in Serbia had some experience of the polytechnic schools, for example, Nešić

(Karlsruhe) and Josimović (Vienna), and hence had, what could be called 'polytechnic' experience. What amount of explicit and implicit values and knowledge that may have been embedded in those schools, and how much of that they were able or willing (or both) to bring back to Serbia is open to debate as we have no primary or secondary evidence to this effect. Let us therefore look now at the local military schools to see whether there could be some link made between these schools and the idea of a 'polytechnic' school as exemplified by the French model.

2.1 The Beginning of Descriptive Geometry in Serbia: Nikolić

For a short while, from December 1837 to June 1838, a Military Academy was established which operated initially in Požarevac, then Belgrade and finally Kragujevac. Stefan Krkalović, of whom there is no other information apart from being an officer of the Austrian army, was its initial director and its first teacher of mathematics. He was reportedly the first to teach advanced mathematics in Serbia, and this marked a triumph of the Austrian influence over the cultural strife and educational developments that began almost a century earlier (Kastanis and Lawrence 2005).

This academy was however short lived and it is unclear or unknown for what reason, but we may gather from some nineteenth century primary sources that its closure would have perhaps occurred because of the Turkish objection (Nikolić 2002). Although there is no direct proof, the diary of Nikolić with whom we will shortly meet describes in detail the actual situation and cultural and intellectual life in Serbia during his lifetime, in particular in terms of teaching and training army officers in mathematics. At some length he writes about learning in two areas of Serbia—that under the Ottomans, and the other under Habsburgs, giving some references to objections of the Ottomans to the development of learning in the parts of the country under their rule (Nikolić 2002, written in 1880).

Despite this setback, from this military academy a Lyceum in Kragujevac was established in 1838, and this marked the beginning of the higher study of mathematics in Serbia, i.e., beyond the elementary arithmetic, algebra, or geometry (Kastanis and Lawrence 2005). The curriculum for this first Lyceum in the country included philosophy, general history, mathematics, natural law, European statistics, drawing, German, French, and the Bible studies. The first teacher of mathematics was now appointed to be Atanasije Nikolić (1803–1882), who studied in the Austrian town of Dur (where he studied philosophy), Vienna (where he studied artillery), and Pesta (engineering). Nikolić completed his studies in Pesta in 1829 and had a number of engineering projects in Vojvodina (the northern part of Serbia, under Habsburg occupation until WWI). In 1838 Nikolić was appointed professor at the Lyceum (Kastanis and Lawrence 2005). While there, he wrote two textbooks for his courses, *Algebra*, published in 1839, and *Elementary geometry*, published in 1841. The latter, although entitled 'elementary', contained a few descriptions of descriptive geometry. It was of course published before the reform of the language and systems of measures, initiated by Vuk Karadžić and his friends, took place in the 1870s and was therefore difficult to read by students who took up the subject later on in the century.

Around this time the structure of mathematics education was solidified in Serbia. A law of 1844 prescribed the organizational structure of educational system, with the syllabus framework and the general programme of study: mathematics was to be taught at every stage of the lower first 4 years of gymnasium, the higher fifth and sixth grade to additionally contain algebra. In 1853 further reform introduced more mathematics, to include geometry and trigonometry in the main syllabus.

At this time also the new Artillery School was established in Belgrade (around 1850). The programme in mathematics included not only arithmetic, algebra, and geometry, but also descriptive geometry. This is an interesting fact as a unified or complete translation of descriptive geometry hadn't yet appeared by this time. Although a military school, and one at which modern mathematics was taught to include also analytic geometry, this was not a school which sought to resemble a 'polytechnic' either in its name or as an ideal.

2.2 The Non-polytechnic

While virtually all early mathematicians in Serbia were educated at some point in polytechnic schools of the German-speaking countries, there did not seem to be any attempt to establish a polytechnic school in Serbia. Various other schools were founded throughout the nineteenth century there, and so the lack or reference to 'polytechnic' is visible, noticeable, and may be significant. We can but speculate on the reasons: and one such speculation is that there had to be a critical mass of national culture, institutions, and infrastructure for this type of institution to be established. In the nineteenth century, Serbia did not achieve such status yet, nevertheless the further history will show us why the life of descriptive geometry in this country is important and perhaps unique.

3 The Official Descriptive Geometry Books, Courses, and Teachers

3.1 Emilijan Josimović: The Life

Emilijan was born in 1823 in the village of Stara Moldava in Romania. His family was Serbian and it is assumed that his grandfather fled to Romania during *Koča's frontier*: the time when the Habsburg Empire was in a war with the Ottomans

(1787–1791). At this time the Serbian Free Corps were organized and supported by the Habsburgs, in order to create the frontier with the aim of stopping the Ottoman advancement towards the west. The leader of the Serbian corps was one Koča Andjelković (hence the name of the uprising), but after the Habsburgs withdrew, the Ottomans reestablished themselves in central Serbia and members of the uprising fled to surrounding countries.

Emilijan enrolled in elementary school in Caransebes in 1831 and in Lugos finished the mathematical military school. He completed his studies in Vienna at the age of 22. By this time his family had returned to Serbia and was living on an island Poreč on Danube (now flooded since the erection of the damn on Danube, at Djerdap), on the border between Romania and Serbia.

Upon his return to Serbia, Emilijan gained employment at the Artillery School in Belgrade in 1845. He worked there until 1869, teaching mathematics and mechanics, and then moved to teach mathematics at the Great School of Belgrade, where he eventually became rector. Emilijan was a talented singer, won many awards for his educational work, and was an amateur photographer. He was a full member of the Serbian Academy of Sciences and Arts (founded 1841) and a keen Freemason.

3.2 Emilijan's Descriptive Geometry

Emilijan's book of descriptive geometry was published in 1874 in Belgrade as Book I, part of the planned series of three books on the basics of descriptive geometry and perspective, under the title *Basic descriptive geometry* (Fig. 17.1). The other two planned books never got published, and perhaps they were not written.

Josimović's book addresses two issues: that of the usefulness of the technique, and the translation of mathematical terms, related to conceptual understanding:

> In it [his book], there is no 'high' scientific nonsense to be found, which I, and many others with me, consider as an abuse of students' time. The one who would ask for such a thing, would show that he does not know what is the real goal of education ...

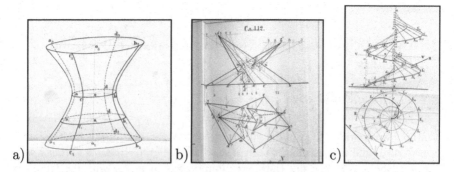

Fig. 17.1 Illustrations from Josimović's *Basic descriptive geometry*, 1874, The hyperboloid, the intersection of two pyramid, and the ruled surface

And

> Previous terminology [that described geometrical concepts] was of all sorts ...I had to recreate it from the first to the last word. That is how I did it using the best knowledge of our language (Josimović 1874, Preamble).

Emilijan's attention was especially turned towards developing a linguistic diversity in his explanation of geometric concepts and procedures. He of course added his own descriptions and explanations and set out to introduce the rules for geometric drawing to be different to the rules of perspective and of freehand drawing. He paid particular attention to the concept of rabatting: he explains the two projection planes as being flat mirrors, marking them as mirror planes, and their intersection as 'sample axis'. At several points Josimović gave rules for types of lines used in delineating different transformations and which are illustrated in the introduction to his book.

3.3 Other Contributions

Before Josimović, the translated terms and vocabulary were embedded in the main mathematical texts without clear reference to the meaning of the words to the concepts and their descriptions in other languages. Josimović introduced a new practice, giving at the end of the text, a 'registry of terminology' in which he gives phrases in Serbian, French, and German languages, (Fig. 17.2). This practice of providing a mathematical dictionary in mathematics textbooks was preserved in Serbia well into the twentieth century, and was evident also in the work of Peter Anagnosti (1909–1996), a professor of descriptive geometry at the University of Belgrade whose textbook is still in use.

Despite the importance of Josimović's work on writing the first book on the subject in Serbian language, his greatest contribution at the time, and possibly to national intellectual history, is not in this area. It was his architectural work that was of greater significance for the nation: he introduced urban planning to Serbia, and wrote several books on it, including Civil Architecture and the Building of Roads (*Gradjanska arhitektura i gradjenje puteva*), published in Belgrade in 1860 (Bogunović 2005; Djurić-Zamolo 2005; Jevtić-Novaković 2015). Josimović also published the first textbook on higher mathematics in Serbian language, *Načela vise matematike u tri časti* (the Basis of Higher Mathematics in three books), which was published in Belgrade between 1858 and 1872.

3.4 The Largest Book on Descriptive Geometry in Serbian Language, by Dimitrije Stojanović

Stojanović was born in 1841 (Fig. 17.3), the year in which the first introduction and mention of descriptive geometry appeared in the book by Nikolić. Stojanović's career was a cross between an academic and political and in that respect

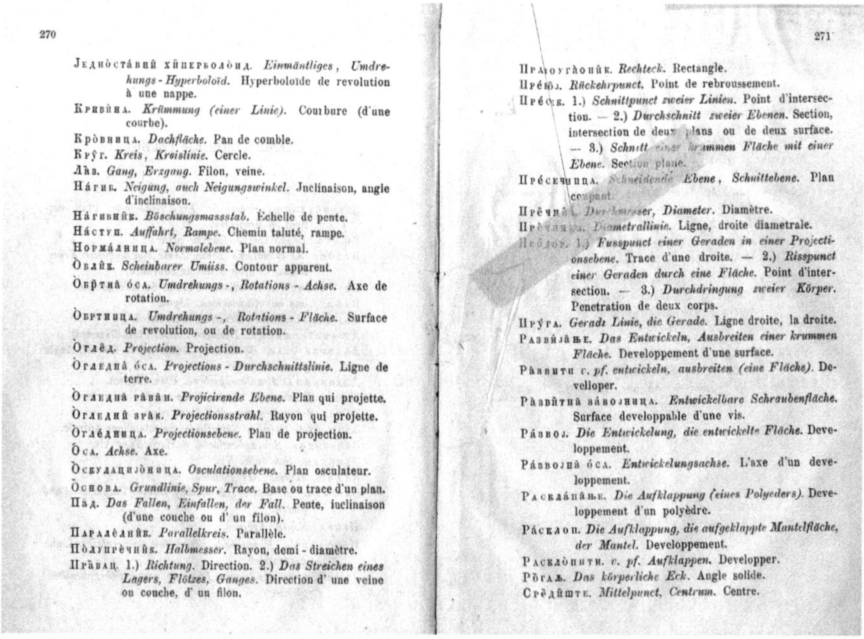

270

Једностáвнй хиперболòид. *Einmäntliges, Umdrehungs-Hyperboloïd.* Hyperboloïde de revolution à une nappe.

Кривѝна. *Krümmung (einer Linie).* Courbure (d'une courbe).

Кровница. *Dachfläche.* Pan de comble.

Круг. *Kreis, Kreislinie.* Cercle.

Лâз. *Gang, Erzgang.* Filon, veine.

Нáгиб. *Neigung, auch Neigungswinkel.* Inclinaison, angle d'inclinaison.

Нáгибнйк. *Böschungsmassstab.* Echelle de pente.

Нáступ. *Auffahrt, Rampe.* Chemin taluté, rampe.

Нормáлница. *Normalebene.* Plan normal.

Òблйк. *Scheinbarer Umiiss.* Contour apparent.

Òбртнâ óса. *Umdrehungs-, Rotations-Achse.* Axe de rotation.

Òбртница. *Umdrehungs-, Rotations-Fläche.* Surface de revolution, ou de rotation.

Òглêд. *Projection.* Projection.

Òгледнâ óса. *Projections-Durchschnittslinie.* Ligne de terre.

Òгледнâ рâвâн. *Projicirende Ebene.* Plan qui projette.

Òгледнй зрâк. *Projectionsstrahl.* Rayon qui projette.

Òглéдница. *Projectionsebene.* Plan de projection.

Òса. *Achse.* Axe.

Òскулацијòница. *Osculationsebene.* Plan osculateur.

Òснова. *Grundlinie, Spur, Trace.* Base ou trace d'un plan.

Пâд. *Das Fallen, Einfallen, der Fall.* Pente, inclinaison (d'une couche ou d'un filon).

Паралèлнйк. *Parallelkreis.* Parallèle.

Пòлупрèчнйк. *Halbmesser.* Rayon, demi-diamètre.

Прâвац. 1.) *Richtung.* Direction. 2.) *Das Streichen eines Lagers, Flötzes, Ganges.* Direction d'une veine ou couche, d'un filon.

271

Правоуглòнйк. *Rechteck.* Rectangle.

Прéвој. *Rückehrpunct.* Point de rebroussement.

Прéсек. 1.) *Schnittpunct zweier Linien.* Point d'intersection. — 2.) *Durchschnitt zweier Ebenen.* Section, intersection de deux plans ou de deux surface. — 3.) *Schnitt einer krummen Fläche mit einer Ebene.* Section plane.

Прéсечница. *Schneidende Ebene, Schnittebene.* Plan coupant.

Прèчнйк. *Durchmesser, Diameter.* Diamètre.

Пр... *Diametrallinie.* Ligne, droite diametrale.

Пр... 1.) *Fusspunct einer Geraden in einer Projectionsebene.* Trace d'une droite. — 2.) *Risspunct einer Geraden durch eine Fläche.* Point d'intersection. — 3.) *Durchdringung zweier Körper.* Penetration de deux corps.

Прўга. *Gerade Linie, die Gerade.* Ligne droite, la droite.

Развијâње. *Das Entwickeln, Ausbreiten einer krummen Fläche.* Developpement d'une surface.

Рàзвити *v. pf. entwickeln, ausbreiten (eine Fläche).* Develloper.

Рàзвйтнâ зáвојница. *Entwickelbare Schraubenfläche.* Surface developpable d'une vis.

Рáзвој. *Die Entwickelung, die entwickelte Fläche.* Developpement.

Рáзвојнâ óса. *Entwickelungsachse.* L'axe d'un developpement.

Расклáпâње. *Die Aufklappung (eines Polyeders).* Developpement d'un polyèdre.

Рáсклоп. *Die Aufklappung, die aufgeklappte Mantelfläche, der Mantel.* Developpement.

Расклòпити. *v. pf. Aufklappen.* Developper.

Рõгаљ. *Das körperliche Eck.* Angle solide.

Срèдйште. *Mittelpunct, Centrum.* Centre.

Fig. 17.2 Pages from Josimović's *Basic descriptive geometry*, 1874, giving the terminology relating to descriptive geometry in different languages

Fig. 17.3 Dimitrije Stojanović

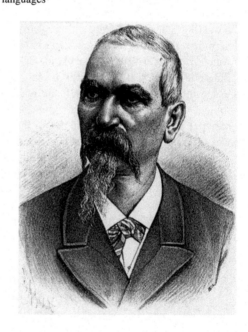

he resembled in no small measure the father of Descriptive Geometry, Monge (Lawrence and McCartney 2015). Dimitrije Stojanović (1841–1905) was educated in Požarevac and Vienna. He studied engineering at the Vienna Polytechnic and upon return to Serbia became a civil servant as county engineer based in Požarevac. In 1874, the year that the first book of descriptive geometry was published by Josimović, Stojanović became the professor of descriptive geometry at the Great School in Belgrade. He was a director and founder of the Railway Association of Serbia, a role in which he showed great integrity, working on maintaining national control over the development of railway system despite international corporations' attempts to privatize it. Stojanović was a minister of finances and of building works on several occasions, and was involved with the founding of many professional and national associations. His pastime was philology of Serbian language, writing articles for the *Serbian Learned Society* and the *Serbian Technical List*.

Stojanović wrote two books on descriptive geometry—the "Perspective" which appeared in 1871 and the "Methods of descriptive geometry" 1897, both of which were published in Belgrade (Stojanović 1897).

3.5 Stojanović's Method

In the introduction to his *Methods* (Stojanović 1897), the book which he wrote over a period of 20 years, Stojanović lists Monge, Hachette, Desargues, and Poncelet as mathematicians who established the science of descriptive geometry. The most interesting for him, and the one he suggests influenced him the most, is however Fiedler, and in particular in relation to his treatment of projective geometry. Stojanović doesn't give any credit to, or even mentions Josimović, and introduces new terminology, different to Josimović's. This book is indeed similar to the German interpretations. It is some 692 pages long and has 462 illustrations.

Stojanović's introduces the three rather than two projective planes, labeling them Π_1, Π_2, and Π_3. He gives the reasons for studying descriptive geometry which resonate with Monge's sentiments (Monge 1798, p. i):

> It is rare indeed, that there is a science which demands so much the development of ability of representation and description as it is so in descriptive geometry, but precisely because of that, (descriptive geometry) is the best way to study geometrical forms, as well as a good opportunity to develop sharp spirit and rigorous thinking (Stojanović 1897, p. 29).

In chapter five, Stojanović introduces projective space—he talks about points at infinite distances, the lines and planes at infinity and uses this to introduce perspective central projection. He gives a fascinating image of the space thus imagined (Fig. 17.4b).

This long treatise discusses various ways of presenting space—Stojanović, for example, shows the use of axonometric perspective, and the cavalier perspective (Fig. 17.5), showing his interest and knowledge of literature well beyond what was then available in the vernacular Serbian. He also looks at the application of

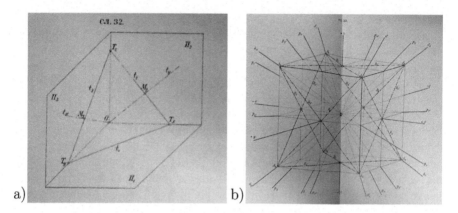

Fig. 17.4 Illustration from Stojanović's methods of descriptive geometry, 1899, Belgrade. Projecting planes (**a**), and 'projective hexahedron' (**b**) in which he shows the concepts of infinite lines and infinite planes

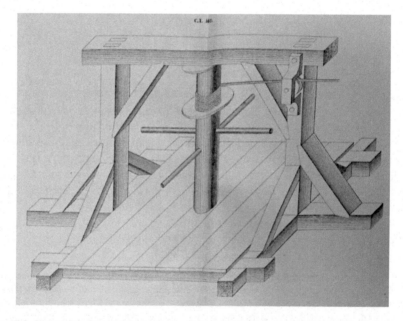

Fig. 17.5 Illustration from Stojanović's methods of descriptive geometry, 1899, showing the method of 'cavalier perspective'

this knowledge to various sciences—for example, to the study of crystals, or the mechanical engineering. Stojanović's approach is thorough and his book ends with the study of projective geometry. It is obvious from his work that he (a) spared no effort to make his work as complete as he could possibly achieve, and (b) that he had no immediate need to use this book as he published it when he was retired. This was a book through which he showed his love of the topic and in which he celebrated his long and productive career of teaching of the subject.

4 The Rest of the Nineteenth Century

Both Josimović and Stojanović established descriptive geometry as one of the important methods in the study of geometry and mathematics. By the end of the nineteenth century there were many more teachers of the subject, and so they were able to promote its acceptance and the perception that descriptive geometry is needed and indeed perhaps even necessary subject to be taught to a wide variety of students.

4.1 Stevan Davidović and Co

Davidović's contribution was not significant in brining any new insights, but it is certainly worthy of a mention. Stevan Davidović (dates unknown) was a professor of descriptive geometry at the Artillery School in Belgrade between 1883 and 1895. He published several books on the subject, his first two being translations from German. His first was a translation of a book on perspective by Berger, which he published in 1882 in Serbian. The second was his translation of Franc Močnik's (1814–1892) "Geometry", published in Belgrade in 1895. Močnik was a Slovene, but published in German language; he was schooled at the university of Graz, and later taught at the Polytechnic of Lemberg (now Lviv) in Ukraine.

Davidović published his textbook on descriptive geometry in 1896 (Davidović 1896, Fig. 17.6a and b), aimed specifically at the students of the Military Academies in Serbia. This textbook was republished a few more times until the WWII. In its second edition, we see two significant innovations—text and drawings are printed together (in the earlier book, the text is given separately, with drawings given in

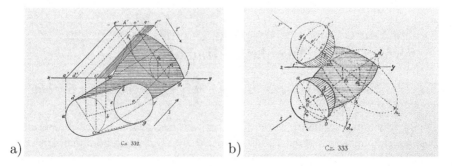

Fig. 17.6 Illustration from Stevan Davidović, Descriptive Geometry 1922, Shadows of round bodies—cylinder and sphere

a folio volume), and for the first time there is a chapter dedicated to the study of shadows.

4.2 New Mathematics

Serbian mathematicians from the end of the nineteenth century and until the WWI mainly showed interest in practical mathematics, mechanics included, applications to engineering, and the development of analytic geometry. It is from this background that the best known Serbian mathematician, Mihailo Petrović (1868–1943), came, and so it is interesting to see the sudden change in his work in the direction from predominantly practical mathematics and geometry to analysis.

Although, as we have shown so far, the influence in the learning and teaching of mathematics in general and descriptive geometry in particular had so far been predominantly German and Habsburg, then Austro-Hungarian, Petrović broke with this tradition and went to study mathematics at the *École Normale* in Paris. There he was by all accounts an exemplary student, and for his doctorate the examining commission consisted of Charles Hermite (1822–1901), Émile Picard (1856–1941), and Paul Painlevé (1863–1933). He made friends too, and remained in touch with Jules Tannery (1848–1910) and Painlevé, to whom he dedicated his thesis, *Sur les zéros et les infinis des intégrales des équations différentielles algébriques* (1894).

Upon his return to Belgrade, Petrović became professor at the Superior School in Belgrade, and, when the school was in 1905 transformed into the University of Belgrade, Petrović was given the first Chair in Mathematics. His main interest was in classical analysis, and he wrote papers on the properties of real and complex functions defined by power series (Petrović 2004, p. 100). Petrović established Serbian mathematical school based on his knowledge and interests and at this point the importance of descriptive geometry was somewhat diminished in favour of analysis (Lawrence 2008). However, as we will now see, the teaching of descriptive geometry, and its importance in the national system of education survived well into the twenty-first century in Serbia.

5 The Success and the Twentieth Century History: The Serbian and the Yugoslavian Traditions

5.1 A Few More Remarks on the Study of Geometry in Serbia

Euclid was not translated into Serbian until the late 1940s. It was then done in batches and published between 1949 and 1970, by a Ukrainian born mathematician Anton Bilimović (1879–1970). Bilimović fled Ukraine in the wake of the 1917 Revolution, and as did many other emigreés from the area, settled in Serbia.

Bilimović was educated at Kiev, Paris, and Göttingen, where he worked with David Hilbert (Djukić 2012).

During the WWII Bilimović turned to Euclid and translated it into the Serbian language. This meant that the modern mathematical techniques and works were known to Serbian mathematicians and students earlier than the ancient (Lawrence 2008). Perhaps it is within such a framework that we can find reasons for the 'whys' and 'hows' of the importance in which descriptive geometry remains to be held in the national education.

5.2 Yugoslavia

After WWI, the first Yugoslavian state was founded—the Kingdom of Serbs, Croats, and Slovenes. In 1929 this kingdom was renamed the Kingdom of Yugoslavia, and in 1943, after the Yugoslav Royal family fled to England (in 1941), it was renamed into the Federal Republic of Yugoslavia. Most of the readers will remember the horrors of civil war that befell the country in the 1990s, ending up with the break-up of Yugoslavia. It was therefore a relatively short period that Yugoslavia existed, and it did so outside of the main period of our interest which is mainly nineteenth century. Nevertheless, it is important to note a few developments, as they shed some light onto the reasons for the subject remaining to be a popular one in Serbia today.

Since Stojanović (1897), no other books were published on descriptive geometry until that by Stefan Adolf, in 1930. Adolf's textbook was aimed at secondary school pupils. With some 167 illustrations and 560 exercises, it was not a book for absolute beginners. This textbook was published in Belgrade and had the approval stamp of the Ministry of Education, to be used as a textbook in gymnasia across the country. We note a quote which shows the importance it was given in education at the time:

> It is important that, while learning descriptive geometry, one never overlooks detail and moves onto something else unless you are absolutely sure you understand what is going on. You will then see, that what you have learnt in descriptive geometry, you will need all the time while you learn more of it... If you adopt this approach, descriptive geometry will be your most easy and dearest subject of study, as there is nothing there that requires mechanical memory.
>
> Descriptive geometry is, as the beauty of nature, which, the longer you look at it, the more you love and know it. The beauty of nature you cannot learn but observe. That is how it is with descriptive geometry. So open your eyes well and observe (Adolf 1930, p. 1).

From this period we have another emigreé who became a professor of mathematics at the University of Belgrade, Peter Anagnosti (1909–1996). He became a professor of descriptive geometry at the Faculty of Architecture and wrote two famous textbooks during and after WWII on Descriptive Geometry (1942) and Perspective (1948). These books have been reprinted many times and used until the end of the twentieth century across the country's mathematical, architectural, and engineering schools (Fig. 17.7).

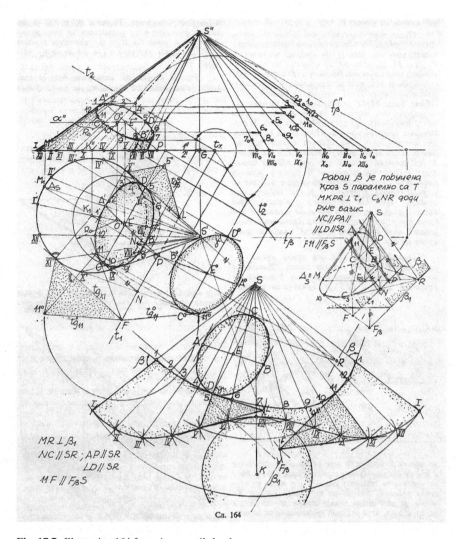

Fig. 17.7 Illustration 164 from Anagnosti's book

6 Conclusions: The Influences and the Wider Context

The nineteenth century was in Serbia a century of discovery—discovering intellectual and learning frameworks, traditions, and the nationhood. The reform of language, the adoption of the metric system (McGreevy 1995), and the rapid development of mathematical sciences were based on foreign influences, inventions, and developments. All of these, without exception, came from the west and the north, and first and mainly, from the German speaking countries. Situation changed at the end of the nineteenth century, when the Serbian mathematics with Alas at the lead turned towards France. It will therefore be interesting to see whether the origin of sources was noted in the books published by our main protagonists Josimović, Stojanović, and Davidović.

Josimović, the first to publish on descriptive geometry, does not give sources from which he drew, although he introduced the mathematical dictionary listing new Serbian words and their equivalents in German and French (Josimović 1874). This gives us some evidence that the first point of reference for him was German, and the second the French terminology. He mentions one book from which he learnt, and which was published in Serbian about geometric drawing—this book was in every other respect of negligible importance: "The science of geometrical drawing with theory to gain knowledge about geometry in general: for real-schools of the Dukedom of Serbia" (*Nauka o geometrijskom crtanju s teorijom kao priugotovno znanje za geometriju u opšte : za realku Kneževine Srbije*). This was written by Dragutin Jovanović and published in Belgrade in 1866. The book was aimed at secondary school students.

Stojanović (Stojanović 1897), on the other hand, mentions Henri Tresca (1814–1885), Théodore Olivier (1793–1853), and Gustav von Peschka (1830–1903) as well as Wilhelm Fiedler (1832–1912) as his teachers, although he does not give us the exact works from which he drew for his own work.

Finally, Davidović (Davidović 1896) was most precise in his description of the sources from which he drew. He says that, while he used many others, the main were the books by J. Kicies: *Traite élémentaire de Géométrie descriptive*, E. Lebon: *Géométrie descriptive*, B. Gugler: *Lehrbuch der descriptiven Geometrie*, and F. Smolik: *Lehrbuch der darstellenden Geometrie*. No dates of editions are given, and it is obvious that these are not the most important texts in the history of descriptive geometry. This in turn tells us perhaps about the scarcity of original sources in Serbia.

6.1 On the Periphery, Aiming for the Centre

The analysis of influences and sources used for writing Serbian textbooks tells us a few things about the learning and the spread of interpretations of the technique in nineteenth and twentieth century Serbia. Firstly, we saw that all the mathematicians involved with the writing of books on the subject had, without exception, some experience of education in German speaking lands, and some completed their education in polytechnic institutions. They used German and French as a language of learning, and published textbooks in Serbian. By the twentieth century native terminology for terms related to descriptive geometry solidified, so that Anagnosti's textbooks with ease use native words invented for the purpose, and the vocabulary is instead given to compare Serbian and Croatian expressions.

All of the authors we have mentioned were keen to mention at least some western sources in their editions. It may be that this practice gave some kind of credence and validation to the study of the subject, encouraging the effort to 'catch up' with the international centres of learning.

This of course, seemingly also changed in the early twentieth century with Petrović, and the success of both his school, and with the unique talents that arrived to Serbia following the Russian Revolution. But at this time too the focus changed from geometry and practical and applied mathematics to analysis.

7 The Post Script: Love Affair with Descriptive Geometry

Finally we cannot but conclude the chapter with some remarks about the love affair with descriptive geometry that may be evident from some quotes we have given here, and from the authors' own experiences in the educational system of the country. We conjecture on this sentiment in the following way.

Descriptive geometry was one of the first examples of higher mathematics that was introduced into the Serbian educational system, certainly the first taste of higher geometry that was given to learners who had no experience, until then, of an organized system of learning mathematics. With no other tradition to draw upon, and a complete lack of literature and therefore references to Euclidean tradition (Lawrence and McCartney 2015), descriptive geometry became the geometry that was the oldest and most traditionally taught subject in Serbian schools and higher institutions of learning.

It was also linked to some important people in the development of the Serbian society and its intellectual history: Josimović introduced urban planning and Stojanović developed the railway system in Serbia. Descriptive geometry was therefore seen to be at the core of national progress, prestige, and enlightenment that came after the centuries of Ottoman occupation.

For those reasons, we believe, descriptive geometry, although it may have become a subject of historical but very limited importance in many other locations, thrives in Serbia today. It is now mainly called *Mongeometrija*, after the name of an association which has bi-annual conferences and meetings of scholars and practitioners in the discipline. These meetings take place in universities all over the country to discuss the problems, methods, and advances and applications being developed in the teaching at secondary, technical, and university level institutions. To paraphrase Albers (1981, p. 226) we could easily say that 'descriptive geometry is alive and well, and living in Serbia under an assumed name'.

References

Publications

Adolf, Štefan. 1930. *Nacrtna Geometrija*. Beograd: Geca Kon.
Albers, Donald, J. 1981. Geometry is alive and well and living in Paris under an assumed name. *The Two-Year College Mathematics Journal* 12: 226.

Bogunović, Slobodan. 2005. *Architectural Encyclopedia of Belgrade XIX and XX & Centuries*. Belgrade: Belgrade Books.
Božić, Milan. 2002. *Pregled Istorije i Filozofije Matematike, Zavod za Udžbenike i Nastavna Sredstva*. Beograd.
Davidović, Stevan. 1896. *Nacrtna Geometrija za Pitomce Vojne Akademije*. Beograd
Dejić, Mirko. 2014. How the old Slavs (Serbs) wrote numbers. *BSHM Bulletin: Journal of the British Society for the History of Mathematics* 29(1): 2–17.
Djukić, Djordje. 2012. Anton Dimitrija Bilimovic. *Publications de L'Institut Mathématique*, Nouvelle série, tome 91(105): 3–17.
Djurić-Zamolo, Divna. 2005. *Contribution to the Biography Emilijan Josimovic*, vol. XXIII, 1976. Yearbook of Belgrade.
Jevtić-Novaković, Katarina and Gordana Fontana-Giusti. 2012. Emilijan Josimovic—the first Serbian urbanist. In *26th National and 2nd International Scientific Conference moNGeometrija*, 423–434.
Jevtić-Novaković, Katarina. 2015. Mirror images of professor Emilijan. *Journal Izgradnja* 69: 129–134.
Josimović, Emilijan. 1874. *Osnovi Nacrtne Geometrije i Perspective*. Belgrade.
Kastanis, Nicholas and Snezana Lawrence. 2005. Serbian Mathematics Culture of the 19th century. *HPM Newsletter* 15–24.
Lawrence, Snezana and Mark McCartney. 2015. *Mathematicians and Their Gods*. Oxford: Oxford University Press.
Lawrence, Snezana. 2008. *A Balkan Trilogy: Mathematics in the Balkans before World War I. The Oxford Handbook of the History of Mathematics*, ed. Elenor Robson and Jackie Stedall. Oxford: Oxford University Press.
Maksimović, Branko. 1967. *Emilijan Josimović, The First Serbian Urbanist*. Belgrade: Institute for Architecture and Urban Planning of Serbia.
———. 1957. *Josimovic Trench in the reconstruction of Belgrade*, vol. IV. Belgrade: Belgrade City Museum Yearbook.
McGreevy, Thomas. 1995. *The Basis of Measurement: Volume 1—Historical Aspects*. Chippenham, UK: Pitcon Publishing.
Nikolić, Atanasije. 2002. *Biografija verno Svojom rukom Napisana*. Belgrade: Serbian Society for the History of Science.
Petrović, Aleksandar. 2004. Development of the first hydraulic analog computer. *Archives Internationales d'Histoire des Sciences* 54(153): 97–110.
Stojanović, Dimitrije. 1897. *Methods of Descriptive Geometry*. Belgrade.
Terzić, Dušan. 1994. *Emilijan Josimović, Visionary Urban Belgrade*. Belgrade: Town Planners Association Belgrade.

Archives and Digital Libraries

University of Belgrade. www.bg.ac.rs
First Belgrade Singers' Society. www.pbpd.info
Serbian Academy of Arts and Sciences. www.sanu.ac.rs/
Big Masonic Lodge in Serbia. www.vmls.rs/
srpskaenciklopedija.org
www.mongeometrija.com

Chapter 18
Descriptive Geometry in England: Lost in Translation

Snezana Lawrence

Abstract This chapter looks at the history of descriptive geometry in England, and why here it had such a short, and not a very fulfilling life. Having arrived to England in the immediate aftermath of the wars between England and France, its translation and attempts to introduce it into the educational system happened only after the 1840s. The lack of direct, implicit knowledge of the original technique, and some aspects of mistranslation, meant that the technique was never properly understood. Descriptive geometry is still mainly regarded as a drawing, rather than a mathematical technique in England, and has not been practised since the end of the nineteenth century. Polytechnic schools in England were another short-lived phenomena, and only of any significance and showing similarity with the French model in the second half of the twentieth century.

Keywords Descriptive geometry · Gaspard Monge · William Farish · Peter Nicholson · Orthographic projection · Isometric perspective · Parallel oblique projection · Curvilinear perspective

1 Mathematics in England in the Eighteenth and Nineteenth Centuries

1.1 General Background

English mathematics in the eighteenth and nineteenth century was vibrant and diverse, not perhaps as spectacular as that of the seventeenth century, but bringing to the discipline some important developments and new concepts (Rice 2011). English

---

S. Lawrence (✉)
Department of Design, Engineering, and Mathematics, Faculty of Science and Technology, Middlesex University, London, UK
e-mail: snezana@mathsisgoodforyou.com

© Springer Nature Switzerland AG 2019
É. Barbin et al. (eds.), *Descriptive Geometry, The Spread of a Polytechnic Art*,
International Studies in the History of Mathematics and its Teaching,
https://doi.org/10.1007/978-3-030-14808-9_18

mathematical landscape was still under the heavy influence of Newton (1642–1726) and his work well into the eighteenth century. England of this time produced some mathematicians whose contributions were global, such as John Wallis (1616–1703), Edmund Halley (1656–1742), and Brook Taylor (1685–1731) to name but a few. But there was also, what Rice (2011, p. 3) called a certain stagnation, a 'lull' in producing original mathematics in the eighteenth century (Guicciardini 1989). And then suddenly the wheel turned and more mathematics was produced than at any other time.

So how did descriptive geometry fare amongst the locally made 'Maxwell's equations, Boolean algebra, histograms, and even the concept of standard deviation' (Rice 2011, p. 1), Venn diagrams (1880s), mechanization of mathematics in Babbage's Difference Engine (1832), the use of mathematics to make persuasive arguments in matters of changing policies (in Florence Nightingale's work, for example, 1850–1870s), to name but some of the most popular and widely known examples? This chapter will look at the place that descriptive geometry took in such a landscape of new mathematical developments.

1.2 Practical Mathematics and the Practicing Mathematician

In order to understand the context in which descriptive geometry was introduced in England, we will first give a short historical overview of the learning of geometry in this country. Until the sixteenth and seventeenth centuries, the learning of mathematics was confined to a small number of learning establishments to which no dissident, non-Christian (and after 1559 with the establishment of Anglican independent church) non-Anglican, or female, had access and certainly no one without the knowledge of the classics. Then, from mid-1500s, books on mathematics were published in vernacular English, like those of Recorde in 1543 and 1551, of Blagrave in 1585, and in 1570, the first edition of Euclid's Elements.

Several aspects of this emerging learning culture developed as, for example: the networks, associations of a kind, and schools for artisans, builders, and anyone who needed to know mathematics sprang around London and other big cities (Lawrence 2002). One such network grew into the first official mathematical society, which was formed in London in 1717—the Spitalfields' Mathematical Society (Cassels 1979). The early members of this society came from the lower middle classes of artisans and majority were weavers, but there were also brewers, bakers, and braziers.

> Soon after the accession of George II, in 1730, a small club of local men sat drinking in the snug parlour of a Westminster alehouse, gathered together to learn mathematics, so that "by their mutual assistance and indefatigable industry they are now become masters ... of logarithmetical arithmetic and some of them greatly advanced in algebra" (Clark 2000, p. 214).

The society's aim, along with drinking and socializing, was collective improvement—for it was a "fundamental rule of this society not to conceal any new improvement from another member ... ; before tackling mathematics they had taught themselves French" (Clark 2000, p. 1). One should not however make too much of this instance of learning French and mathematics together. This particular society arose in an area of London in which there was, at the time, a sizeable Huguenot population following their exodus from France more than a century earlier.

Nevertheless, this self-reliant at first, and then more organized movement to enlarge activities around the learning of mathematics, steadily grew and finally solidified itself in the first chairs of mathematics being established at the new modern institutions of learning, like that of the King's College and University College, both in London in the first half of the nineteenth century.

The two new universities were open to non-Anglicans, dissidents (more so University College London) and, by the end of the nineteenth century, women. The greater movement of mathematical texts, their translations and the branching of learning of mathematics needed more specifically for the newly founded professions (such as engineering and architecture), all ensued (Lawrence 2002).

2 English Tradition Pre-descriptive Geometry

2.1 Brook Taylor and His Linear Perspective

The work that had prepared the ground, or so it may seem, for the arrival of descriptive geometry to England would have been that of Brook Taylor (1685–1731). Taylor belonged to the tradition described above in more than one way. He was an artisan, musician, and a mathematician (Jopling 1835). He published two books on geometry which certainly paved the way to the study of graphical geometry in England before the arrival of descriptive geometry.

Having originated in a well to do family, Taylor had private tuition as a child and attended Cambridge as a young adult. By the time he graduated there in 1709, he had already written his first mathematical paper, and he published it a few years later, in the same year he was elected a Fellow of the Royal Society (Taylor 1712). Shortly after this, Taylor became the secretary to the Royal Society in 1714. At this time he was interested in several things: magnetism, calculus, and linear perspective. He published on all three subjects in 1715 (Jopling 1835), but the publication which we are most interested in was his *On the Principles of Linear Perspective* (Taylor 1715).

Linear perspective is a technique which was developed fully only during the Italian Renaissance, although there are instances of its appearance in art since the classical antiquity. The development of the technique can roughly be divided into three periods (Jones 1950): the first during which architects and artists made rediscovery of the technique (fifteenth to sixteenth centuries), the second in which geometrical study of the technique was presented more formally (seventeenth century), and finally the third in which the technique was presented in a more generalized form, as an abstract theory (eighteenth century). Taylor's work certainly falls into the third period not only because of the timing of his two treatises (Taylor 1715 and 1719) but because of the abstraction of thought that made this technique a mathematically sound one.

Taylor's work attracted some attention in England and on the continent—it was translated into Italian in 1755, and there gained important following. Luigi Cremona became interested in particular in the fundamental theorems from Taylor's second treatise and published new proofs of the same (Cremona 1865). This link between Cremona and Taylor remained in the Italian tradition and can be seen in the Gino Loria's contribution to Moritz Cantor's *Geschichte der Mathematick* (Loria 1908; Anderson 1992, p. 78).

2.2 Mathematical Catechisms

The number of practicing mathematicians or those who needed some mathematical knowledge as described earlier grew from there on and so did the need for resources from which to learn mathematics. In the eighteenth century, a new trend of publishing of popular geometrical books aimed particularly at artisans, builders, and workmen was born. The books such as these mainly resembled empirical recipes (Booker 1963, pp. 91–111; Lawrence 2002) like Joseph Moxon's (1627–1691) works: *Mathematics made easie* (Moxon 1700) and *Mechanick Exercises or Doctrine of handy-works* (Moxon 1703) or Batty Langley's (1696–1751) numerous books (Langley 1727, 1735, 1736a,b, 1738, 1739, 1740). These works invariably mixed the learning of geometry with freemasonic lore and mythology (Lawrence 2002; Lawrence and McCartney 2015).

An example is given below: it is a page from Langley's book giving an introduction to the science of Geometry:

> This Art was first invented by Jabal the Son of Lamech and Adam, by whom the first House with Sones and Trees was built.
>
> Jabal was also the first that wrote on this Subject, and which he performed, with his Brethren, Jubal, Tubal Cain, and Naamah, who together wrote on two Columns the Arts of Geometry, Music, working in Brass and Weaving, which were found (after the Flood of Noah) by Hermarines, a descendent from Noah, who was afterwards called Hermes the Father of Wisdom, and who taught those Sciences to other Men (Langley 1736b, p. 61).

As such, these books and manuals offered a very few underlying principles or encouraged geometrical understanding and thinking, or taught how to transfer principles from one case to another. Instead, they offered specific cases, and resembled catechism rather than an exact method. The need therefore, for a clearly defined geometrical technique to satisfy this particular application was discussed and entertained at the various levels of the engineering (both civil and military) and the architectural professions (Booker 1963; Lawrence 2002 and 2011) at the time when descriptive geometry was already taught in France.

3 Descriptive Geometry in England: Arrival, Translations, and Adaptations

Descriptive geometry's first public appearance in English language came around 1820s. Some records which we will soon explore in detail show that some copies of Gaspard Monge's books (whether the edition from an VIII – stenographic notes transformed into a book – or Hachette edition from 1811) were circulating around London at the time (Nicholson 1823).

When considering the reception, and the context of descriptive geometry in England in this light, one should also add that descriptive geometry was proclaimed a military secret (Booker 1963; Taton 1951; Lawrence 2002) when first conceived by Monge. Additionally when it was for the first time publicly taught in Paris on 20th January (1er Pluviôse III), Britain and France were at war with each other. This was of course during the War of the First Coalition (1792–1797); during the War of the Second Coalition (1798–1802), Britain led the coalition against France, and Monge was one of the savants that took part in the Egyptian Expedition where one of the biggest battles of this war took place—the battle of Alexandria (Lawrence 2015). Indeed the Coalition Wars meant something that was more important than the boundaries of countries or the prestige of the warring sides: they were about upholding the old or inventing a new social, political, and intellectual systems and frameworks (Fisher 2004).

In this light, it is clear that Monge's sentiments about the need for a technique such as descriptive geometry, and the role he imagined it play in building of the French national prestige (Fig. 18.1) did not exactly warm the hearts of the English (see Barbin, Chap. 1, this volume). Only following the death of Monge, different translations and adaptations of descriptive geometry appeared in England. There were three periods in this development, and we will examine them in the chronological order.

4 First Period: 1820–1840

4.1 Overview

This period saw an upsurge in publications considering techniques which would resemble descriptive geometry, all of which made references to it, but were not actually descriptive geometry. This period is defined by two publications, one from 1820 and another from 1840. It begins with the first alternative to descriptive geometry, Farish's *Treatise on Isometrical Perspective* (Farish 1820) and completes with Nicholsons' *A Treatise on Projection* (Nicholson 1840). The isometrical perspective of Farish is a system very different to that of descriptive geometry, and is de facto an alternative to parallel projection that was already in use and was, for example, described by Lambert in his *La perspective affranchie* (Lambert 1759).

PROGRAMME.

Pour tirer la nation française de la dépendance où elle a été jusqu'à présent de l'industrie étrangère, il faut, premièrement, diriger l'éducation nationale vers la connoissance des objets qui exigent de l'exactitude, ce qui a été totalement négligé jusqu'à ce jour, et accoutumer les mains de nos artistes au maniement des instruments de tous les genres, qui servent à porter la précision dans les travaux et à mesurer ses différens degrés : alors les consommateurs, devenus sensibles à l'exactitude, pourront l'exiger dans les divers ouvrages, y mettre le prix nécessaire ; et nos artistes, familiarisés avec elle dès l'âge le plus tendre, seront en état de l'atteindre.

Il faut, en second lieu, rendre populaire la connoissance d'un grand nombre de phénomènes naturels, indispensable aux progrès de l'industrie, et profiter, pour l'avancement de l'instruction générale de la nation, de cette circonstance heureuse dans laquelle elle se trouve, d'avoir à sa disposition les principales ressources qui lui sont nécessaires.

Il faut enfin répandre parmi nos artistes la connoissance des procédés des arts, et celle des machines qui ont pour objet, ou de diminuer la main-d'œuvre, ou de donner aux résultats des travaux plus d'uniformité et plus de précision ; et à cet égard, il faut l'avouer, nous avons beaucoup à puiser chez les nations étrangères.

On ne peut remplir toutes ces vues qu'en donnant à l'éducation nationale une direction nouvelle.

C'est, d'abord, en familiarisant avec l'usage de la géométrie descriptive tous les jeunes gens qui ont de l'intelligence, tant ceux qui ont une fortune acquise, afin qu'un jour ils soient en état de faire de leurs capitaux un emploi plus utile et pour eux et pour la nation,

A

Fig. 18.1 Introduction to *Geometrié Descriptive* ("In order to raise the French nation from the position of dependence on foreign industry, in which it has continued to the present time, it is necessary in the first place to direct national education towards an acquaintance with matters which demand exactness, a study which hitherto has been totally neglected; and to accustom the hands of our artificers to the handling of tools of all kinds, which serve to give precision to workmanship, and for estimating its different degrees of excellence. Then the consumer, appreciating exactness, will be able to insist upon it in the various types of workmanship and to fix its proper price; and our craftsmen, accustomed to it from an early age, will be capable of attaining it.

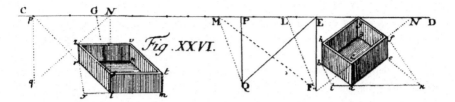

Fig. 18.2 Oblique projection as described by Lambert (1759, figure XXVI) prior to Farish's invention (1820, p. 1)

However it was portrayed as an English invention (Farish 1820) and remained to be considered as such through the nineteenth century (Heather 1851) (Fig. 18.2).

4.2 Farish and His Isometrical Perspective

William Farish (1759–1837) was Jacksonian professor of natural and experimental philosophy at the University of Cambridge from 1813 to 1836. We know very little of him, but do have some simple facts. Farish was an influential professor, one of the founders of the Philosophical Society of the University of Cambridge, and established its publication, *The Philosophical Transactions* in 1820.

In the first issue of the *Transactions*, in 1820, Farish published a treatise on his use of a graphical representation system which he called Isometrical Perspective (Farish 1820). Heather, of Royal Military Academy in Woolwich, made an interesting comparison of the Monge's and Farish's systems and explained the main difference between them, pointing to the reason why most British authors on the subject found descriptive geometry difficult to accept:

> Descriptive Geometry would require great accuracy in the construction of the shadows, and would frequently present great difficulties in practice … [on the other hand] a single projection, the construction of which is remarkable for its simplicity, forms a conventional picture, conveying at once to the eye the actual appearance of the objects, as in a perspective view, and also giving readily the dimensions of the objects represented, especially those dimensions which are situated in planes parallel to three principal planes at right angles to each other … This technique is Isometrical Perspective of William Farish … (Heather 1851, pp. ix–x).

Fig. 18.1 (continued) It is necessary in the second place to make popular a recognition of a number of natural phenomena indispensable for the progress of industry, and to exploit, through the advancement of the general instruction of the nation, the fortunate condition in which it finds itself of having at its command the principal resources which are necessary.

Finally, it is necessary to disseminate among our craftsmen the knowledge of the processes used in the crafts and in machines which have for their object either the diminution of manual labour or the imparting of more uniformity and precision to the results of workmanship; and in this respect it must be admitted that we have much to learn from foreign nations.")

Farish did not have the grand ambition that Monge did some 20 years earlier, in suggesting that it should be used throughout the whole territory defined by a national education system. As professor of Natural Philosophy at the University of Cambridge, Farish held lectures on mechanical principles of machinery used in manufacturing industries, and for these he often used models which exemplified particular principles. Storage of models and their transport from the store room to the lecture theatres posed a problem, which Farish solved by making the models from elements which were then assembled by his assistants. In order to communicate with his assistants, Farish devised this system and in drawings based on it he showed how the machinery was to be re-built.

> As these machines, thus constructed for a temporary purpose, have no permanent existence in themselves, it became necessary to make an accurate representation of them on paper, by which my assistants might know how to put them together, without the necessity of my continual superintendence. This might have been done by giving three orthographic plans of each; ...But such a method, though in [high] degree in use amongst artist, would be liable to great objections. It would be unintelligible to an inexperienced eye; and even to an artist, it shows but very imperfectly that which is most essential, the connection of the different parts of the engine with one another; though it has the advantage of exhibiting the lines parallel to the planes, on which the orthographic projections are taken, on a perfect scale ...(Farish 1820, p. 2).

Farish then published a short treatise describing this practice (Fig. 18.3): isometrical perspective, however, soon became very popular and was used in many other situations, from architecture to engineering (Sopwith 1834, 1854).

The difference between the circumstances in which the two techniques, that of descriptive geometry and that of isometrical perspective were invented, as well as the difference between their inventors, were to have major consequences for the ways the two were later adopted at the teaching institutions of the architectural profession. Peter Booker described these major differences between the two techniques:

> ...Whilst Monge had been a draughtsman and knew quite a lot about designing and manufacturing things, Farish was an academic and seems to have been concerned with assembling things which were already made or were familiar enough to be brought into existence by someone else, the craftsman ...Farish's interests lay entirely in their mechanics, the broad scientific principles. It is doubtful whether he understood the concept of accuracy as Monge did; and of course, being only concerned with broad principles, drawing did not find a place in his engineering lectures—quite the reverse of Monge's curriculum in which drawing was the key subject (Booker 1961, p. 73).

4.3 After Farish

Soon after this, a first edition of Descriptive Geometry in English language was published in the USA, by Claude Crozet (1821). And at almost the same time in England, Nicholson published his first book on perspective (Nicholson 1822), which was followed by his *A practical treatise on the art of masonry and stonecutting* (Nicholson 1823). In this latter book, Nicholson didn't only for the first time

18 Descriptive Geometry in England

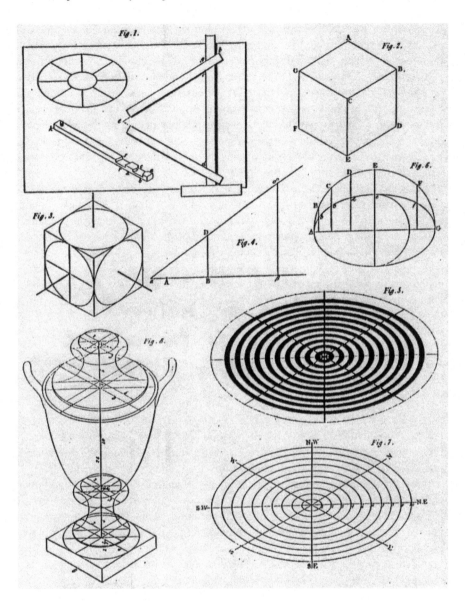

Fig. 18.3 The first page of Farish's treatise on *Isometrical Perspective* published in 1820

mention that he was working on descriptive geometry, but gave the first English translation of some of the basics of the technique published in England.

In 1823, Joseph Jopling, a civil engineer, published his *Septenary system of generating curves by continued motion*, which, although an interesting publication would not be of any interest to us but for the fact that it used the principle of motion for the description of geometrical objects, an underlying principle used in descriptive geometry (Lawrence 2011). Jopling later revised Farish's system, making it more applicable to engineering and architecture through giving concrete examples applicable to both disciplines (Jopling 1833). Thomas Sopwith, a friend and colleague of Jopling, a civil engineer from London, also published *A Treatise on Isometrical Drawing* (Sopwith 1834) but his was work aimed at mining and civil engineers. In the introduction to his book Sopwith stated that the drawback of isometrical perspective was that the real measurements could not be taken from a drawing, and suggested the method of 'crating'—an old technique used since the Renaissance for drawing objects by placing them in elementary reference box, from which the measurements could be extracted.

Some 5 years later, Thomas Bradley published a book on *Practical Geometry, Linear Perspective, and Projection* (Bradley 1834). Although this book contains no reference to descriptive geometry 'proper', we have the records (Lawrence 2002) that Bradley taught descriptive geometry at the time at the newly founded Engineering Department of King's College in London.

In 1840, Nicholson came back with a more comprehensive work on practical geometry, his *Treatise on Projection* (Nicholson 1840). Nicholson work transcended our two periods and is the most important contribution to the history of descriptive geometry as it lived, under an assumed name, in England.

4.4 Peter Nicholson and His Technique of Parallel Oblique Projection

A technique which most resembled the Monge's original came from the work of architect and mathematician Peter Nicholson (1765–1844). Nicholson was born in Prestonkirk, East Lothian on 20th July 1765, a son of a stonemason. He became interested in geometry and its applications to architecture, where he strove to develop an efficient system of graphical communication for the use of architects and craftsmen. Because he knew about descriptive geometry and was the first to apply its principles (Nicholson 1823), his translation of the practice of descriptive geometry for the architectural profession being founded at the time in Britain played an important part in leading research towards the establishment of a standardized graphical communication language. The character of his work may be seen as a mediating one between an architect and a craftsmen.

Nicholson was both an author and a practitioner, and, between 1805 and 1810 worked for Robert Smirke (1780–1867), the architect of the British Museum, as a superintendent of the building of the new court-houses at Carlisle. Both men were Freemasons and members of the same Masonic lodge (the Old Cumberland Lodge, united in 1818 with the Lodge of Fortitude, London).

Nicholson's family background is of great importance both for his career and our story. Freemasonry in Scotland was, in the late eighteenth and early nineteenth centuries, fundamentally different to that in England. Fully operative lodges, which nevertheless practised the ritual, or speculative Freemasonry, still persisted in Scotland up to the middle of the nineteenth century, although in England this ceased to be the case shortly after the founding of the Grand Lodge of England in 1717 (Lyon 1900; Lawrence 2002). In England, by 1717, the two concepts—that of operative masonry and speculative, or ritual, Freemasonry—were strictly defined, and masonic or building lodges were no longer involved with speculative or philosophical Freemasonry (Knoop 1935). In Scotland however, operative lodges admitted members through a variation of freemasonic rituals and customs, but maintained their status of the building trade organization into the 1800s (Lyon 1900).

Nicholson drew upon this practical knowledge he gained while a freemason, and as an apprentice to his uncle, a stonemason in Scotland. When he moved to London, Nicholson organized lectures for craftsmen in Berwick Street, Soho, in which he taught practical stonemasonry. These lectures served to provide learning opportunities for mechanics and workmen who were facing an open, post-lodge market, but where the lodge apprenticeship was non-existent and hence the practical instruction was lacking.

Nicholson's knowledge of projection techniques used within the stonemasons craft and the carpenters trade proved to be unquestionably important in this context. He became a well respected and well-known figure in this field, and published a number of books and treatises on practical geometry as well as on aspects of architecture (related to technical details and stonecutting). He obviously believed that his mathematics was sufficiently good, and in 1827 tried to get a professorship in mathematics at the newly established University College in London, but was of course passed over in favour of the much younger and better qualified Augustus De Morgan (1806–1871).

We can best trace the invention of Nicholson's system of projection through his own account of events:

> In 1794 I first attempted the Orthographical Projection of objects in any given position to the plane of projection; and, by means of a profile, succeeded in describing the iconography and elevation of a rectangular parallelpipedon: this was published in vol. ii of the Principles of Architecture (Nicholson 1822, pp. 46–47).

Nicholson also gave information on how and where he had become acquainted with descriptive geometry: "In 1812 Monge's treatise was lent to me by Mr Wilson Lowry, celebrated engraver ... (Nicholson 1822, p. 47)". Perhaps even more interesting is Nicholson's memory from an earlier period. Nicholson said (1822), that in 1796–1797 he had met Mr Webster, a drawing clerk for Mr Mitchell in Newman Street, who pointed out to him the similarity of his work with that of works from France. When in 1812 Nicholson reports to have been given Monge's treatise, he had it translated by Mr Aspin and considered publishing the major points in his *Architectural Dictionary* of 1813 (but in fact only mentioned it in passing). We do not unfortunately have that translation nor Nicholson's notes for the dictionary.

Nicholson's system undoubtedly rested on his knowledge and experience both of what was considered the necessary knowledge of builders' craft and of what was going on in this subject on the Continent. His account of the practical need for geometrical education appearing in his *Practical Treatise on the Art of Masonry and*

Fig. 18.4 Illustration showing the construction of stone arch using development of a surface into the flat plane (Nicholson 1839)

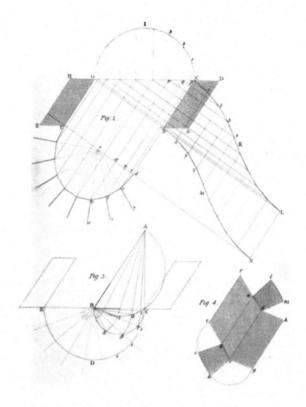

Stone-cutting (Nicholson 1839) described what he believed was the most important aspect of a new language of graphical communication (Fig. 18.4):

> To be able to direct the operations of Stone Masonry, taken in the full extent of the Art, requires the most profound mathematical researches, and a greater combination of scientific and practical knowledge, than all the other executive branches in the range of architectural science. [...] To enable the Workman to construct the plans and elevations of the various forms of arches or vaults, as much of Descriptive Geometry and Projection is introduced, as will be found necessary to conduct him through the most difficult undertaking (Nicholson 1839, p. 51).

Nicholson at this point called his system Parallel Oblique Projection; it was an orthographic system of projection which makes use of an oblique plane, so as to provide both the presentation of an object and the method by which such an object is to be executed at a building site. Nicholson's Treatise (1840) comes at the end of a series of his publications related to the topic that we already mentioned, and in which glimpses were given of the principles of his final technique. In this work (Nicholson 1840), he finally explains his system elaborately, together with numerous examples and listing of possible applications. Like most of the works in this genre, this book too was written for the engineer, architect, surveyor, builder, mechanic, and the like, suggesting that the technique should be used as a universal

18 Descriptive Geometry in England

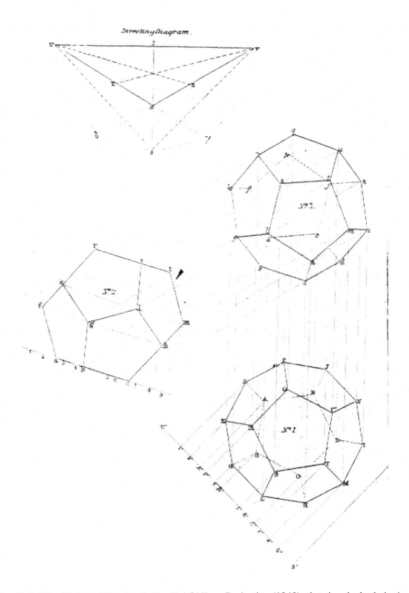

Fig. 18.5 Plate 20 from Nicholson's Parallel Oblique Projection (1840), showing the body in three views, the system is determined by the small diagram above, showing the position of the inclined plane and offers an easy method to obtain real measurements by manipulation of the object

language of graphical communication among the different parties involved in the building trade (Fig. 18.5).

The system of Parallel Oblique Projection is based on the principles of orthographic projection, and where there is a third, or oblique projection obtained through an auxiliary plane, which then enables the exhibition of a complex design. The third auxiliary plane of projection is seen in its trace through the two primary planes of projection.

5 Second Period: 1840–1864

Nicholson's technique, the parallel oblique projection, although it was based on descriptive geometry, was a heavily modified system of Monge. Not having the same title as Monge's technique should not then be surprising.

Following this book however, a couple of books on descriptive geometry, trying to introduce and present Monge's system in a straightforward but slightly simplified form, appeared. First was Thomas Grainger Hall's *The elements of Descriptive Geometry: chiefly intended for Students in Engineering* (Hall 1841). Hall's book is very interesting for two reasons: firstly it pays homage to the inventor and the invention of descriptive geometry and secondly, it suggests that descriptive geometry should be important in education and hence should be taught in England.

> The present work is intended for those students who are occupied in graphically representing the forms of bodies, and the delineations of machines. To such a class the advantage of having general methods by which the position of points, lines, and surfaces, may be determined with exactness and precision, is very obvious. Descriptive Geometry supplies that want. Invented by the genius of Monge, and pursued with ardour and success by the most eminent French Geometricians, it is now taught in almost all the universities and in the principal schools of the continent. In England it was unknown, as a branch of instruction, until lectures were given upon it by Mr. Bradley, in the Engineering Department of King's College; and the present work has been undertaken to supply the students with a text book, that by it they might the more profitably attend to what they heard in the lecture room: and as an elementary book was necessary for beginners, it has been thought expedient to place before the students, in an English dress, one which has stood the test of experience (Hall 1841, p. i).

Hall drew (Figs. 18.6 and 18.7) upon Lefébure de Fourcy—edition unknown—for the most part for his translation (on Lefébure de Fourcy, see Barbin, Chap. 2, this volume). The 'English dress' meant mainly the introduction to it went via inductive geometry.

Hall wrote when Monge was dead for some decades, and his revolutionary demeanour was not threatening any longer. A few further treatises on descriptive geometry were published in England. They were all given in the similar manner to Hall, and were aimed mainly at the engineers. First was by Joseph Wooley, *The Elements of Descriptive Geometry; being the first part of a treatise on descriptive geometry, and its application to ship building* (Wooley 1850), a very much simplified version of Hall. The second was by John Fry Heather, the *Elementary Treatise on Descriptive Geometry with a Theory of Shadows and Perspective* (Heather 1851), who gave Monge's an VII edition as the source for his book. Heather's book is divided into two sections: text at the front, with the illustrations at the back. The illustrations are given as drawings on the board, as can be seen below.

Proposition XX.

If two planes which intersect be each perpendicular to a third plane, their common section shall be perpendicular to the same plane.

Let the two planes AB, BC, be each ⊥ to the third plane, and let BD be the common section of AB, BC; BD shall be ⊥ to the third plane.

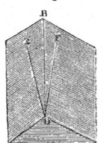

If it be not, from D in AB draw DE ⊥ AD, the common section of AB with the third plane; in BC draw DF ⊥ CD, the common section of BC with the same plane.

Then ∵ AB is ⊥ to plane ADC, and DE is ⊥ to AD, ∴ DE is ⊥ to ADC; similarly, DF is ⊥ to ADC; ∴ from the same point D, two ⊥^{s}, DE, DF, to the same plane ADC, have been drawn, which is impossible; ∴ the only line that can be drawn from D ⊥ to ADC is BD, the common section of AB and BC.

Fig. 18.6 Proposition XX from Hall's treatise (Hall 1841), showing the introduction to geometry to reiterate some basics before descriptive geometry principles are given below

Problem 6.

Given the projections of a straight line and a point, to find those of a line passing through the point, and parallel to the given line.

20. Let ab $a'b'$ be the projections of the line, cc' of the point; through c draw cd parallel to $a'b'$; through c' draw $c'd'$ parallel to $a'b'$: then cd and $c'd'$ are the projections required.

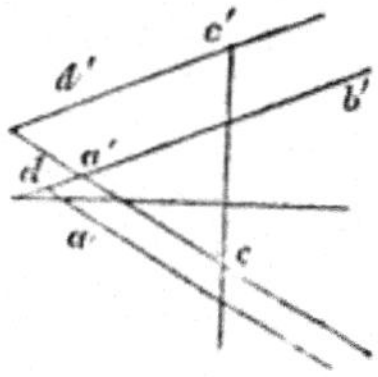

For when two lines are parallel, their projecting planes are also parallel, and ∴ the intersections of these planes with a third plane are also parallel; *i.e.*, the projections of two parallel lines are also parallel.

Fig. 18.7 Problem six from Hall's book (Hall 1841, p. 10) showing the remnants of same methodology used to teach geometry

While it is interesting to see that descriptive geometry 'proper' did briefly make appearance in England, it soon again vanished too. But let us first see the last few attempts and variations on the theme of descriptive geometry in the programmes of study and textbooks from the final period in its English history.

6 Third Period: 1851–1864

In this period the interest in inventing a new system of graphical communication or graphical geometry was waning, but some alternatives were further explored. Herdman's *A treatise on the Curvilinear Perspective of Nature* (Herdman 1853) explained geometry behind the technique used for centuries by various artists (see, for example, the detail of the convex mirror in Jan van Eyck's "Arnolfini Portrait" from 1434).

Various other treatises on descriptive geometry were published by professor of geometry William Binns, who taught at Putney College of Civil Engineers between 1846 and 1851. Binns' books at start have 'descriptive geometry' in titles, but slowly drop this in favour of 'orthographic' projection (Binns 1857, 1860, 1864). His books followed Heather's practice of providing the illustrations in contrast (white on black, Fig. 18.8), suggesting a pedagogical method, presumably to evoke images from the demonstrations on a blackboard. Further works by Bradley from this period did the same, and both incorporated Nicholson's system as a simplified method and one which aimed to give a final picture of the object (Bradley 1860, 1861) (Fig. 18.9).

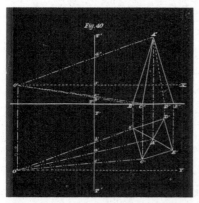

According to the definitions given above, it is easily perceived that *linear perspective* reduces itself to the construction of the section of a pyramid, the vertex and base of which are given, made by a determinate surface. The eye is the vertex; the base can be considered as spread over the surface of the objects to be placed in perspective, and the cutting surface is the picture.

The methods of descriptive geometry easily give the solution of this problem taken in all its generality, that is supposing the picture to be any curved surface whatever. However, keeping especially in view what is of constant utility in the arts, that only which concerns perspectives drawn on plane surfaces will be discussed with some detail, and a few observations will then be added respecting perspectives constructed upon curved surfaces.

The picture will be supposed to lie in a vertical plane, or perpendicular to what is considered the horizontal plane of projection; it could without difficulty be supposed to be inclined to these planes in any manner whatever; but the hypothesis proposed is more natural, and makes the constructions more simple.

Fig. 18.8 Heather's illustration of problem 40 (1849, p. 64), using descriptive geometry to demonstrate the principles of linear perspective

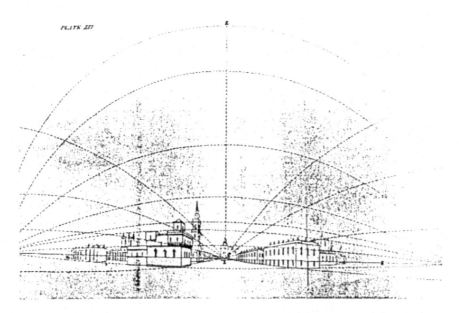

Fig. 18.9 Herdman's illustration of the principles behind curvilinear perspective (Herdman 1853)

6.1 Summary of a Life

Descriptive geometry had a short life in England. The simple, straightforward translations aimed at engineers, in particular those originating from Woolwich Military Academy and the King's College London, came following the first treatises which attempted to adapt descriptive geometry, in the period between 1820 and 1840s. While the Farish's and Nicholson's adaptations of old systems were interesting in their proposed use (and appeared throughout the nineteenth century as systems which were taught in engineering and art schools across England), the direct translations were more or less uninspired and not well received, and were never reprinted. From these came several further treatises by Binns and Bradley, mainly adopting Nicholson's approach—being more orthographic projection treatises than descriptive geometry 'proper'.

Nicholson's system combines in a way the methods of both Monge's and Farish's systems. It makes use of the processes of "rabatting" (bringing a plane of interest into the plane of projection to gain the real measurements) and at the same time offers an easy way of constructing an image of the object, without necessarily referring to its construction (generation).

The system which was developed by Nicholson subsequently even gained the name of 'British system of projection' (Grattan-Guinness and Andersen 1994). It became widely used in the schools of architecture and engineering that were established in the nineteenth century England, beginning from the ones in London.

7 The Wisdom of Hindsight

7.1 A Cunningham's Plan

In his paper on *Importance of Descriptive Geometry* in England did a comparison to its reception in France, Germany, and the UK. He concluded that the foci of national systems of education, and the perception of mathematics in relation to these, were most at fault for the technique never 'catching on' in England (Fig. 18.10).

But furthermore, the author argued that not only were the teaching practices different in the three countries, but that this was a consequence of the general differing predispositions in the way that space is perceived and taught. While descriptive geometry could be used, as indeed it was in many countries and national systems as we see in this volume, for very practical purposes, its strength was in the underlying mathematical principles, and not in the way the picture of an object was presented. In England descriptive geometry had a short and not all together fulfilling life; its main mathematical features were overlooked, and its benefits for education were considered sometimes negligible and sometimes undesirable. Nicholson's system was accepted as a method of solving practical geometrical problems in architecture and engineering, but gained no approval in relation to the fact that it can be applied to the learning and teaching of spatial mathematics, or as an introduction to projective geometry (Rogers 1995, pp. 401–412).

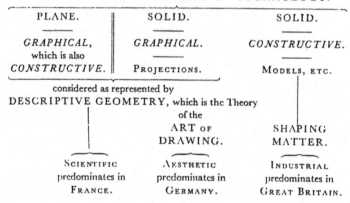

Fig. 18.10 Plate from Cunningham's *Notes on the History, Methods and Technological Importance of Descriptive Geometry* (1868) comparing the systems of graphical communication in France, Germany, and Great Britain

7.2 Lost in Translation

There was however, apart from the political context, another very important reason for descriptive geometry never being properly understood in England. It is the case of the lack of implicit knowledge: there was no one in England who had any personal experience of descriptive geometry as it was taught in France. The knowledge acquired seemed to have been entirely explicit, via a text or worse still, a translated text (Lawrence 2010).

This can be shown on the example of misunderstanding the simple and important terms and translating them slightly differently from their original meaning. Cunningham (1849–1919), for example, who wrote *On the History and Importance of Descriptive Geometry* in 1868, found that the terms 'plan and elevation' are erroneous in the case of Descriptive Geometry. Cunningham was an economist and a churchman and is largely credited with establishing the economic history as a scholarly discipline in British universities. He was a professor of Economics from 1891 to 1897 at King's College London, where he became interested in the technique, mainly because of his interest in its application to industry. His most important and complete work is *The Growth of English Industry and Commerce* (1882).

Cunningham's view was that

> [...] it is impossible to express the co-ordinate relation of the two planes of projection in such terms as Plan and Elevation, which involve the special ideas – 'horizontal' and 'vertical' (Cunningham 1868, p. 25).

Instead, Cunningham suggested the use of word 'rabatting' which refers to pulling the plane of projection to the plane of drawing. The orientation of planes in descriptive geometry was immaterial; the mathematical task at hand was to understand the ways objects are generated, in order to enable the practitioner to execute particular graphical operations and then perhaps to discover the precise measurements of the finished object. In English texts this was never understood because the generating principle was never explained, and instead the method of rabatting was taken to be the main principle in order to gain a view of the object.

7.3 To Translate or Not to Translate

The word 'rabatting' appeared for the first time in English language in Cunningham's paper (1868). Cunningham explained what was lost in translation from the original meaning of the method of rabatting (Fig. 18.11).

But the term 'rabatting', and even the mathematical importance of descriptive geometry, had by the time Cunningham's paper was published, become irrelevant in England. The technique had by then been cast aside as an abstract technique, the alternatives such as those of Farish and Nicholson were much easier to use. Its purpose and real nature misunderstood, descriptive geometry waned in both

Fig. 18.11 Cunningham's use of the term 'rabatting' and phrase 'to rabatt' is its first occurrence in the English language and directly derived from the French

> Now this is quite incorrect. For, assuming that an elevation means a projection on a vertical plane, how can we, with any propriety, talk of projecting the line A B upon a plane in which it actually lies—that is to say, its "plan-"projecting" plane ? No, the operation referred to is correctly described as "rabatting the plan-projecting plane of "the line," or simply "rabatting the line." By the use of this term the student sees at once that he can rabatt a line A B in space either upon the horizontal or vertical plane at pleasure, whereas, if he calls the former operation "making "an elevation," he is little likely ever to perform the latter, for the simple reason that he has no name to give to it.

importance and use, to completely disappear even as a passing reference, in the national educational system by the end of the nineteenth century.

The word rabatting is of little use today in English language, and if used, it is done almost exclusively in the literature relating to geometric manipulations of objects. As a curiosity, the author of this chapter noticed the first instance of the use of 'rabatting' in English language in Cunningham's paper (1868) and alerted the Oxford English Dictionary to the fact.

8 The Missing Link: The Polytechnic School

The English did not follow the example of the polytechnic school either. There was an attempt to establish such institutions, one of which was partially successful—it was the London Polytechnic, founded in 1838. But this institution did not resemble the French model in any way. It was a small institution, giving evening lectures to mainly workmen who wanted to advance themselves and did not have the resources to study at the universities. There were also polytechnic societies formed around the country, of which was, for example, The Royal Cornwall Polytechnic Society (founded in Cornwall in 1832). The purpose of this society was to offer a place for discussion, and organize a library.

Polytechnic schools perhaps more like the original developed in England in the 1960s, with the rise of the Labour government, which supported their image and role in education. But by this time the many iterations of the polytechnic school model meant that the main similarity between the French original and the English iterations was in the title rather than any of the principles that may have been embedded as the foundation stones of Monge's *École polytechnique*. In the 1992 though, and with the Conservative government gaining a new mandate, most of the polytechnic schools in England were given university status and their names changed. With this, the name polytechnic also disappeared as a school resembling the French original even in a minute detail, from the English educational landscape.

References

Anderson, Kirsty. 1992. *Brook Taylor's Work on Linear Perspective*. Berlin: Springer.
Binns, William. 1857. *An Elementary Treatise on Orthographic Projection; Being a New Method of Teaching the Science of Mechanical and Engineering Drawing, etc.* London: E. & F. N. Spon.
———. 1860. *A Course of Geometrical Drawing, Containing Practical Geometry, Including the Use of Drawing Instruments, etc.* London: John Weale.
———. 1864. *An Elementary Treatise on Orthographic Projection and Isometrical Drawing, etc.* London: Longman.
Booker, Peter Jeffrey. 1961. Gaspard Monge, The Father of Descriptive Geometry and Founder of the Polytechnic School. Paper presented to the Newcomen Society for the Study of the History of Engineering and Technology on 1st November 1961. *The Engineering Designer*. London.
———. 1963. *A History of Engineering Drawing*. London: Chatto & Windus.
———. 1834. *Practical Geometry, Linear Perspective, and Projection, etc.* London: Baldwin & Cradock.
———. 1860. *Lecture On Practical Plane and Descriptive Geometry, Mechanical and Machine Drawing, and Building Construction, etc.* London: G. E. Eyre & W. Spottiswoode.
———. 1861. *Elements of Geometrical Drawing or, Practical Geometry, Plane and Solid, Including Both Orthographic and Perspective Projection*. Published for the Committee of Council on Education. London: Chapman and Hall.
Cassels, John William Scott. 1979. *The Spitalfields Mathematical Society. Bulletin of the London Mathematical Society* 111:3'41-258. London.
Clark, Peter. 2000. *British Clubs and Societies 1580–1800: The Origins of an Associational World*. Oxford: Clarendon Press.
Cremona, Luigi. 1865. I principi della prospettiva lineare secondo Taylor. *Giornale di Matematiche* 3(14): 338–343.
Crozet, Claudius. 1821. *A Treatise on Descriptive Geometry, for the use of Cadets of the United States Military Academy*. New York: A. T. Goodrich.
Cunningham, Alexander W. 1868. *Notes on the History, Methods and Technological Importance of Descriptive Geometry, Compiled with Reference to Technical Education in France, Germany & Great Britain*. Edinburgh.
Farish, William. 1820. Treatise on Isometrical Perspective. In *The Transactions of the Cambridge Philosophical Society*, vol. i. Cambridge: Cambridge Philosophical Society.
Fisher, Todd, and Gregory Fremont-Barnes. 2004. *The Napoleonic Wars: The Rise and Fall of an Empire*. Oxford: Osprey Publishing.
Grattan-Guinness, Ivor, and Kirsty Andersen. 1994. Descriptive geometry. In *Companion Encyclopedia of the History and Philosophy of the Mathematical Sciences*, ed. Ivor Grattan-Guinness, 887–896. London: Routledge.
Guicciardini, Niccolo. 1989. *The Development of Newtonian Calculus in Britain 1700–1900*. Cambridge: Cambridge University Press.
Hall, Thomas Grainger. 1841. *The Elements of Descriptive Geometry: Chiefly Intended for Students in Engineering*. London: J. W. Parker.
Heather, John Fry. (1851). *Elementary Treatise on Descriptive Geometry with a Theory of Shadows and of Perspective. Extracted from the French of Gaspard Monge to which is added a description of the Principles and practice of Isometrical Projection the whole being intended as an introduction to the application of Descriptive Geometry to various branches of the Arts*. London: John Weale.
Herdman, William Gawin. 1853. *A Treatise on the Curvilinear Perspective of Nature; and Its Applicability to Art*. London: John Weale.
Jones, P.S. 1950. *Brook Taylor and the Mathematical Theory of Linear Perspective*. Presented on 1st September, 1950 to Section VII of the International Congress of Mathematicians, Cambridge Massachusetts, US.

Jopling, Joseph. 1833. *The Practice of Isometrical Perspective, etc.* London.

———. 1835. *Dr Brook Taylor's Principles of Linear Perspective etc.* London.

Knoop, Douglas. 1935. *On the Connection Between Operative and Speculative Freemasonry.* The Inaugural Address delivered to the Quatour Coronati Lodge, No 2076, on his Installation as Master. 8th November 1935. Sheffield.

Lambert, Johann Heinrich. 1759. *Perspective Affranchie de L'embarras du Plan Geometrical.* Zürich.

Langley, Batty. 1727. *The Builder's Chest-Book; or, a Complete Key to the Five Orders of Columns in Architecture, etc.* London: J. Wilcox.

———. 1735. *The Builder's vade-mecum or a Compleat key to the Five Orders of Columns in Architecture.* London and Dublin.

———. 1736a. *A Design for the Bridge at New Palace Yard, Westminster.... With Observations on the Several Designs Published to This Time,* etc. London.

———. 1736b. *Ancient Masonry, Both in Theory and Practice ... Illustrated by ... Examples Engraved on ... Copper Plates, etc.* vol. 2. London: Printed for the Author.

———. 1738. *The Builder's Compleat Assistant, or a Library of Arts and Sciences, Absolutely Necessary to be Understood by Builders and Workmen in General.* London: Richard Ware.

———. 1739. *The builder's chest-book; or, A complete key to the five orders in architecture ... The second edition, much improv'd. To which is added, Geometrical rules made easy for the use of mechanicks.* London: John Wilcox and James Hodges.

———. 1740. *The Builder's Complete Assistant; or, a Library of Arts and Sciences, Absolutely Necessary to be Understood by Builders and Workmen in General.* vol. 2 London: Printed for I. and J. Taylor.

Lawrence, S. 2010 Alternatives to Teaching Space Teaching of Geometry in 19th century England and France. In *Franco-British Interactions in Science Since the Seventeenth Century,* ed. R. Fox and B. Jolly. London: College Publications.

Lawrence, S. 2015. Geometry: Masonic lore and the history of geometry. In: Lawrence, Sand McCartney, M, eds. *Mathematicians and their gods: interactions between mathematics and religious beliefs.* Oxford University Press, Oxford, pp. 167–190. ISBN

Lawrence, Snezana. 2002. *Geometry of Architecture and Freemasonry in 19th Century England.* PhD Thesis presented to Open University, UK.

———. 2011. Developable surfaces, their history and application. *Nexus Network Journal* 13(3): 701–714.

Lawrence, Snezana and Mark McCartney. 2015. *Mathematicians and Their Gods.* Oxford: Oxford University Press.

Loria, Gino. 1908. Perspektive und darstellende Geometrie. *Moritz Cantor, Vorlesungen über Geschichte der Mathematick* 4: 577–637.

Lyon, David Murray. 1900. *History of the Lodge of Edinburgh, (Mary's Chapel) No. 1, Embracing the Account of the Rise and Progress of Freemasonry in Scotland.* London, Glasgow, and Dublin: The Gresham Publishing.

Moxon, Joseph. 1700. *Mathematicks Made Easie, or A Mathematical Dictionary Explaining the Tersm of Art and Difficult Phrases Used in Arithmetic, Geometry, Astronomy, Astrology, and Other Mathematical Sciences.* London.

———. 1703. Mechanick Exercises or Doctrine of handy-works (showing how to draw a true Sun-Dyal on any given Plane, however Situated; only with the help of a straight Ruler and a pair of Compasses, and without any Arithmetical Calculation). London, printed at St Paul's Church Yard.

Nicholson, Peter. 1822. *The Rudiments of Practical Perspective, in which the Representation of Objects is Described by Two Easy Methods, One Depending on the Plan of the Object, the Other on Its Dimensions and Position.* London, Oxford

———. 1823. *A Practical Treatise on the Art of Masonry and Stonecutting.* London: M. Taylor.

———. 1839. *Practical Treatise on the Art of Masonry and Stone-cutting.* London.

———. 1840. *A Treatise on Projection.* London: R. Groombridge [Newcastle printed].

Rice, Adrian. 2011. Introduction, in Raymond flood. In *Mathematics in Victorian Britain*, ed. Adrian Rice and Robin Wilson. Oxford: Oxford University Press.

Rogers, Leo. 1995. 'The mathematical curriculum and pedagogy in England 1780–1900: social and cultural origins', Histoire et épistémologie dans l'éducation mathématique, IREM de Montpellier (1995), 401–412.

Sopwith, Thomas. 1834. *A Treatise on Isometrical Drawing as Applicable to Geological and Minning Plans, Pictoresque Delineations of Ornamental Grounds Perspective Views and Working Plans of Buildings and Machinery and to General Purposes of Civil Engineering.* London: Taylor's Architectural Library; John Weale.

———. 1854. *On Models and Diagrams. In London.-III.-Royal Society for the Encouragement of Arts, Manufactures and Commerce. Lectures in connection with the Educational Exhibition of the Society of Arts, etc.* London.

Taton, René. 1951. *L'Œuvre Scientifique de Monge*. Paris: Presses Universitaires de France

Taylor, Brook. 1712. De Inventione Centri Oscillationis- Per Brook Taylor Armig. *Regal. Societat. Sodal. In Phil. Trans.* 28: 11–21.

———. 1715. *On the Principles of Linear Perspective*. London.

———. 1719. *New Principles of Linear Perspective*. London.

Wooley, Joseph. 1850. *The Elements of Descriptive Geometry; Being the First Part of a Treatise on Descriptive Geometry, and Its Application to Ship Building, etc.* London.

Part III
Descriptive Geometry in America and Africa

Chapter 19
Teaching Descriptive Geometry in the United States (1817–1915): Circulation Among Military Engineers, Scholars, and Draftsmen

Thomas Preveraud

Abstract In the United States, descriptive geometry was a subject very few mathematicians, teachers, or engineers knew about before 1820. Most of them were self-taught, as it was not introduced in any curriculum before 1817. Moreover, mathematics and science in general were not leading subjects in higher education, so their practice, teaching, and diffusion remained modest and in the making during the first half of the century. This chapter opens with the treatment of the first course on descriptive geometry taught in the United States. French polytechnician, Claude Crozet, was professor of civil engineering at West Point between 1817 and 1823, and introduced the subject into the West Point curriculum in 1817. Descriptive geometry soon became a subject taught in colleges, especially in those that had already started to offer their students elective courses, or special engineering-training programs. Thus, descriptive geometry gradually went from a subject limited in audience to one of general interest often taught as a sequel to the classical geometry course. Textbook authors introduced then new simplification of the method of projections in order to fit the changing readership and the changing place of the subject in the various curricula. After 1875, the practical role played by descriptive geometry remained crucial in emerging technical institutions, reassuming there its original mission as a graphic art for the training of engineers.

Keywords Teaching · Textbooks · Engineers · Draftsmen · West Point · Rensselaer Polytechnic Institute · Massachusetts Institute of Technology · Technical institutions · Military academies · United States · France

T. Preveraud (✉)
Department of Mathematics, Université d'Artois, Lens, France
e-mail: thomas.preveraud@espe-lnf.fr

1 The Introduction of Descriptive Geometry in the United States for the Training of Engineers (1817)

In the early years of the nineteenth century, American higher education was mostly provided in colleges where algebra, geometry, trigonometry, surveying, and some of the doctrine of fluxions were taught as subjects aimed at training students in the art of rigorous reasoning. Most students became teachers, lawyers, pastors, or traders, but some just intended to pursue intellectual, cultural, and moral training. Textbooks were imported to the colonies or were reprinted from English works, but later were written by American authors and contributed to the growth of the education publishing market for higher education.[1] Engineers were not trained in the colleges yet, but rather at West Point Military Academy, the only military engineering school in the country between 1802 and 1820.

1.1 West Point Inaugurated a Course of Descriptive Geometry in 1817

A few years after the declaration of Independence from England in 1776, Congress formally established the funding for a Military Academy at West Point to train officers and engineers for the American Army. The Academy officially opened in 1802 (Crackel 2002, pp. 29–35). In its early years, the Academy adopted basic standards for teaching contents and organization, so the cadets' training was deficient both in extent and structure. For instance, the curriculum included only very elementary contents of algebra and geometry (Cajori 1890, p. 84). The teaching of mathematics was mostly based on *A Course of Mathematics*[2] (Mansfield 1863, p. 29), a text characterized by contents of a deliberately general nature that made no mention of descriptive geometry, a subject that could have been useful for the training of the engineers (Preveraud 2011).

Between 1810 and 1815, the Secretary of War and the West Point administration planned to reform and modernize the scientific and theoretical training of the engineer-officers (Crackel 2002, p. 79; Preveraud 2014, p. 33) since West Point was then the only institution in North America where engineers could be properly trained. In 1815, General Sylvanus Thayer and Colonel William McRee were granted $5000 to travel to Europe, where they visited engineering schools (Preveraud 2014, pp. 40–51). In Paris, they met Charles François Antoine Leroy, professor of descriptive geometry at the *École polytechnique*, and studied the models

---

[1] For a general overview of the teaching and diffusion of the mathematical sciences in the beginning of the nineteenth century, see (Cajori 1890, pp. 57–83) for college curricula and (Kidwell et al. 2008, pp. 5–13) for textbooks.

[2] American edition of a British textbook. See (Preveraud 2011).

built by the students with the help of descriptive geometry. McRee reported on this meeting in a letter dated September 14, 1816 to General Joseph G. Swift, head of the Corps of Engineers: "I allude to the models of carpentry and stonecutting with the designs, and legends and explanations that accompany them; which constitute the system that is, (or was) taught at the *École polytechnique* on constructions in wood and stone [...]. General Bernard[3] has been so obliging as to make me acquainted with the Professor of descriptive geometry at the École" (Adams 1965). Back in the United States by 1817, Thayer was ordered by President James Monroe to become superintendent of the Military Academy. Under his stewardship, the Academy was thoroughly reorganized. Thayer established a new curriculum and numerous policies to reform and modernize the training of cadets, such as a strict mental and physical discipline, the demerit system and high academic standards, especially in mathematics (Preveraud 2014, pp. 51–64; Molloy 1975). Thayer was convinced that the French system of teaching based on high-level mathematics and strict and daily grading was the best way to train engineers.

Descriptive geometry was taught in the class of mathematics in the second year of study (third class), and introduced by Claude Crozet. Crozet was French and studied in Paris at the *École polytechnique*, graduating in 1807. After serving in the Napoleon's Army, he emigrated to the United States in 1816. He started working as an assistant professor before becoming professor of engineering at West Point Academy in 1817.[4] Engineering was a subject taught during the last 2 years of study, but initially descriptive geometry was not included in the curriculum. It was initiated by Crozet himself. When he started teaching the engineering course in the third year of study, he understood how poor the students' preparation was. Cadets were unable to produce the drawings and constructions demanded, were imperfectly trained basic reasoning, and lacked the mathematical tools to solve problems. According to Edward Mansfield, a former student of Crozet:

> [Crozet had] to supply these preliminary studies before he could commence in his own department. In other words, he must begin by becoming a teacher of mathematics, and drawing [...]. Among these preliminary studies was Descriptive Geometry, not an original and distinct science, but which by projecting geometrical figures and problems on co-ordinate planes, gave a more facile and practical mode of representing as well as solving many geometrical and practical problems (Mansfield 1863, p. 32).

Therefore, Crozet was forced to start his class 1 year before and introduce descriptive geometry in the second year of study so that his students would be ready to learn engineering in the third one. He was convinced that this was the only way to enable his students to follow a modern and high level course in engineering. He justified the introduction of the subject to third class students in a letter addressed to General

---

[3] In 1816, French officer Simon Bernard was about to become assistant in the Corps of Engineers.
[4] After his position at the U.S. Military Academy at West Point, he served as a state engineer in Louisiana and Virginia, and helped to found the Virginia Military Institute in Lexington, Virginia. Crozet was Principal Engineer for the Virginia Board of Public Works and oversaw the planning and construction of canals, turnpikes, bridges, and railroads in Virginia. See (Hunter and Dooley 1992) for his biography.

Swift in October 1818: "I've done all what was possible to give importance to this course [engineering course] and I deliberately added several related branches, since they were without any instructor; although they were not included in my teaching: such as descriptive geometry, the principles of artillery, great tactic and topography" (translation had been made by the author of this chapter) (Adams 1965, 5 October 1818). According to Crozet, descriptive geometry was a preparatory course that could help students in general mathematics and engineering just as it did at the *École polytechnique*. Descriptive geometry was a preliminary course, as much practical as theoretical, for the training of engineers (Sakarovitch 1994, pp. 77–83), which gave students a method to draw three-dimensional objects but also the theoretical tools to solve geometrical problems.

1.2 How to Teach Descriptive Geometry at West Point and the Need for a Textbook

For his early teaching, Crozet had no textbook, which soon became a problem. The need for the textbook was correlated with the way Crozet taught descriptive geometry: the Frenchman gave oral lessons, figures were drawn on the black board and the students were asked to reproduce them in their personal notebooks. This teaching method was time-consuming. Crozet complained about how difficult it was to monitor his students and assess their work: "the drawings [...] of descriptive geometry, the constructions require repeated explanations, an immediate monitoring" (translation had been by the author of this chapter) (Adams 1965, 5 October 1818). With a textbook given to all the cadets, he would not have to construct the figures himself but could work with the students. Also, he probably gave his course in very poor English. Mansfield reported how difficult it was to understand his teacher's explanations:

> [Crozet] does not more than half understand the American language. This difficulty is only to be overcome by practice. With extreme difficulty he makes himself understood. With extreme difficulty his class comprehends that two planes at right-angles with one another are to be understood on the same surface of the blackboard on which are represented two different projections of the same object (Mansfield 1863, p. 33).

Furthermore, around 1820, no more than 12 Americans knew what descriptive geometry dealt with, explained Mansfield and none of them was able to teach it. The teaching of descriptive geometry thus strongly relied on Crozet and only Crozet within the United States, and so the publication of an American book on the subject would help train not only future engineers but also future teachers.

Since his course was the first one ever taught on the subject in the United States, there was no existing American textbook on descriptive geometry. The domestic publishing market was emerging and the number of mathematical textbooks was very limited. Moreover, although Monge's *Géométrie descriptive* (1799) had been translated in London in 1809, it was produced in the political context of the war between France and England in the early years of the nineteenth century. The translation probably remained private and a few copies only circulated among the artisans (Lawrence 2003, pp. 1270–1271). It introduced a very different method for drawing objects, whereas English methods were practical and based upon observation. Descriptive geometry was more theoretical and mathematical (see Lawrence, Chap. 18, this volume). These were reasons why the translation of *Géométrie descriptive* was not used at the *Royal Military Academy* in Woolwich and why descriptive geometry was almost completely ignored in training English officers until 1840. This is probably why the work was unknown to the American audience. Consequently, Crozet had no textbook he could have given to his students, at least in English.

2 Claude Crozet and His *Treatise on Descriptive Geometry* (1821)

In 1821, Claude Crozet wrote and published *A Treatise on Descriptive Geometry* for his teaching, which turned out to be the first textbook ever published in the United States on that subject. It was initially designed with two volumes. The second volume that was never published would have contained "a complete Treatise on Shadows and Perspective" (Crozet 1821, p. ix). For Crozet, and most of the engineers, the drawing of perspective "requires nothing but some of the most simple problems of the cone and cylinder, which are laid down in the first pages of this work" (Crozet 1821, p. vi). With appendices dedicated to the applications of descriptive geometry, this articulation between drawing and descriptive geometry remained tangible during the nineteenth century in military academies' textbook publishing.

A Treatise on Descriptive Geometry was not a translation but Crozet gave credit and referred (Crozet 1821, pp. xi–xii) to the original lessons given by Monge at the *École normale* and assembled in *Géométrie descriptive* (1799). Monge's Lessons were not dedicated at first to the engineers trained at the *École polytechnique* but to future teachers. Indeed, the teaching of descriptive geometry came within the scope of the reorganization of teaching in France, a reorganization that relied on the simplification of the contents taught: this process allowed for the decomposition of the sciences, especially mathematics, in very small parts arranged and presented in such an order that they could easily be learned. This simplification thus determined the structure of textbooks published after the French Revolution. In Monge's lessons, it was correlated with the question of simplicity: the contents of descriptive geometry were presented to the learners from simple problems to more complex ones (Barbin 2015).

2.1 Tensions Between the Method, the Audience, and the Domestic Edition

What Monge called "simple" can be illustrated by the contents of the first pages of the first chapter of Crozet's book. Like Monge, Crozet started his textbook with an inaugural problem: how to determine the position of a point? (see Barbin, Chap. 1, this volume)

> The orthogonal projection is, therefore, almost exclusively employed in the operations of Descriptive Geometry; as, not only the most accurate, but the most simple method of determining the position of points, and consequently of lines and surfaces. This will appear evident by considering, that the position of points cannot be fixed otherwise than its relation to some known objects; and that these objects must be simple, in order to be an easy practical use. Hence the determination of a point by means of its distances from three known planes, seems to be the most convenient of the various methods which can be employed [...] (Crozet 1821, p. 2)

Thus, Crozet did not start his book with the definitions. Monge noticed how difficult they were to write. How to define a point, a line, or a plane in space geometry? Like Monge, Crozet preferred the order of invention: descriptive geometry was introduced from the point of view of its purpose, before the first principles and first problems were treated in Chap. 1 (Crozet 1821, pp. 3–10). His textbook offered a problematized geometry and methods to solve problems. As such, Crozet gave up the idea that figures should necessarily be introduced with axioms, definitions, and propositions, from the most simple to the more complex figures, in the logical order in which they appeared in classical Euclidean geometry textbooks then in use in America.

Knowing how geometry was taught in the United States during the first decades of the century is therefore crucial to understanding where and how descriptive geometry was introduced there. For the teaching of geometry,[5] American scholars, as their English peers, trusted Euclidean geometry, which referred to the Greek book, Euclid's *Elements*. In the United States, between 1800 and 1815, two textbooks were predominantly in use, namely new versions of Euclid's *Elements*, produced in Scotland in the eighteenth century: *Elements of Euclid*, by Robert Simson, was published in 1756 and offered a restored edition of the previous sixteenth century Latin versions of Euclid's text, while natural philosopher, John Playfair, also from Scotland, revised Simson's work in *Elements of Geometry* (Playfair 1795).[6] Simson's and, later on, particularly Playfair's textbooks were widely used in American colleges during the nineteenth century, for the same reason they were used in English colleges. Playfair's *Elements of Geometry* was even published in the United States with an American edition in 1806 by Francis Nichols (Playfair 1806). American scholars and professors appreciated the logical structure

---

[5]For a general overview and associated textbooks, see (Ackerberg-Hastings 2000).

[6]On the way Playfair restored and transformed Simson's translation, see (Ackerberg-Hastings 2002).

of Euclid, which helped students learn reasoning skills. Indeed, *The Elements* was characterized by an organized arrangement of geometrical propositions, proven through purely deductive reasoning. Each proposition was stated and proved in terms of the definitions, axioms, and previous propositions. In Euclid's *Elements*, most of the solutions to the proposed questions were first stated and then proven in order to emphasize the logical process of deductive demonstration. Those so-called synthetic proofs did not give hints as to how the solutions were found, but only as to why they were conclusive.

This was absolutely not how descriptive geometry was presented and taught; both Monge and Crozet began with three-dimensional considerations. For instance, the process of simplification borrowed from the idea of motion in geometry was nowhere to be found in Euclidean textbooks. One of its illustration was the crucial problem of the determination of the distance between two points whose projections were called a and b in the horizontal plane, and A and B in the vertical one. Crozet started his explanation using Monge's method:

> [...] It is then the hypotenuse AB of a triangle, the base of which AC is equal and parallel to its horizontal projection ab, and its altitude the difference BC between the projecting perpendiculars Bb, Aa of the extremities of the line. This triangle may be constructed by drawing through the lower point A the horizontal line AC; then will BC be the difference of the altitudes of the points B and A; because Aa' and Bb are respectively equal to Aa and Bb. If now $A'C$ be made equal to the horizontal projection ab, the hypotenuse $A'B$ will the length required (Crozet 1821, p. 9)

This method implicitly supposed the use of the movement because segment ab which is the horizontal projection has to be reported from point C, forming the original three-dimensional triangle before its projection (Fig. 19.1). But Crozet went further than Monge in the use of the movement, probably because his students were not as well-trained as their French counterparts. He added another method, which explicitly referred to the idea of motion:

> The following solution may sometimes be easier in practice. Knowing the point D where the line pierces the horizontal plane, draw to the horizontal projection ab the perpendiculars aA', Bb'; make one of them $B'b$ equal to Bb', draw DB'; the intercepted part $A'B$ is the true length of the line. This operation is similar to the construction of the preceding problem. While the projecting plane EDe is revolving round the horizontal projection De, the vertical lines Aa, Db remain, in every position of the plane, perpendicularly to the hinge De, and will yet be so when the plane coincides with the horizontal plane (Crozet 1821, p. 9).

The words "revolving" or "coincide" came with the idea of rotation. Crozet probably found this explanation in *Essais de géométrie sur les plans et les surfaces courbes* of Sylvestre-François Lacroix first published in 1795. In Lacroix's work, the rotation of planes was frequently implemented.[7]

---

[7] See (Lacroix 1802, pp. 11–17) and the French expressions "rotation" or "faire tourner".

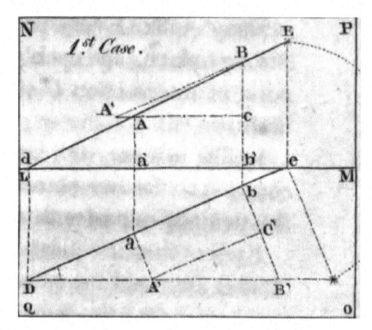

Fig. 19.1 A figure of *A Treatise on Descriptive Geometry* (Crozet 1821, Plate I, between pages 14 and 15)

2.2 Departing from the Original Purpose of Generalization

For Monge, the idea of simplicity came also with the aim of generalization. For instance, lesson 3 was dedicated to the generation of surfaces, but from a very general point of view. He explained:

> Thus in descriptive geometry, to express the form and the position of a curved surface, it is sufficient, for any point of any surface of which a projection can be taken, to give the way to construct the horizontal and vertical projections of two different generators passing through this point (translation had been made by the author of this chapter) (Monge 1799, p. 20).

He considered the particular case of the plane: "Let's apply these general considerations to the plane, the simplest surface, and which use is the most common" (translation had been made by the author of this chapter) (Monge 1799, p. 20). Crozet did not follow this organization and consistently started with particular situations: the plane as an introduction to the surfaces (Crozet 1821, pp. 10–11), the cone and the cylinder as the introduction of the intersection of a surface and a plane (Crozet 1821, pp. 30–35). This inversion may be explained by the way in which American publishers used textbooks. Many American authors and teachers believed that the presentation of several particular situations or so-called elementary situations was suitable to introduce the principles and rules of mathematics to beginners. In an article published in the *American Journal of Science and Arts*,

Brown College professor of mathematics Jasper Adams explained how American textbooks would benefit from using more general methods:

> It is time to distrust this predilection for particular methods, under the idea that they are more elementary than general methods; whereas the truth is, that they are preferred because more ancient, and more agreeable to habits previously acquired [...]. It is erroneous and contrary to established experience, to suppose that general methods must be preceded by an exposition of particular methods (Adams 1822, p. 311).

Nevertheless, the domestic publishing context influenced the writing: the taste for particular methods in American textbooks explained why Crozet moved away from generalization in structuring the contents in his textbook.

3 From Military to Civil Teaching (1822–1864)

After descriptive geometry entered the West Point curriculum, other military academies also taught the subject to their cadets: the American Literary, Scientific and Military Academy located in Norwich, Vermont (from 1819) or the Virginia Military Institute in Lexington, Virginia (from 1839) (Preveraud 2014, pp. 221–222, 2020).

The territorial expansion following the Louisiana Purchase of 1803 and the War of 1812 ultimately necessitated that other institutions beside West Point offered engineering training (Reynolds 1992). Colleges played a leading role in this training. After 1825, the curricula of some colleges included descriptive geometry (Table 19.1) and it seemed that it was pursued by students who did not necessarily plan to become engineers. Descriptive geometry was rather seen as a new subject that advanced students might be interested in. The creation of special programs

Table 19.1 Introduction of descriptive geometry in American higher education[a]

College/Institute	Dates of the introduction of descriptive geometry into the curriculum	Professor
Harvard	1824–1826	John Farrar
Rensselaer Institute	1830–1835	–
University of South Carolina	1836	Thomas Twiss
Kentucky University	1844	Robert T. P. Allen
Dartmouth	1851	James W. Patterson
University of Mississippi	1854	Albert T. Bledsoe
University of North Carolina	1854	James Philips

[a]This table, which is not comprehensive, was established with (Ricketts 1895), (Harvard College 1827), and (Cajori 1890)

for the training of civil engineers inside colleges after 1840[8] showed the need for specialization, but descriptive geometry, as part of the general-interest curriculum, was also disseminated to numerous and various American students.

3.1 Two Textbooks as Part of the General-Interest Curricula (1820–1830)

In the early 1820s, in France, descriptive geometry entered into the admission program of the *École polytechnique*.[9] As a consequence, several books dedicated to its candidates were published and used in the so-called lycées (French secondary schools) (Barbin 2015). Concurrently, however, the United States had only Crozet's work, until several authors wrote new textbooks on the subject to fill the needs of newly created descriptive geometry courses.

In 1822, Robert Adrain, who was professor of mathematics at Columbia College, published the fourth edition of his bestseller *A Course of Mathematics*, in which he added an appendix on descriptive geometry. It was the first time descriptive geometry had been mentioned in any general-content textbook in America, even though the subject was not yet explicitly mentioned in any college curriculum. Descriptive geometry was introduced through a 40-page supplement aimed at simplifying and clarifying its objects and principles for a general-interest audience, such as college students. There was no generalization and no intention to offer a complete treatise on the subject. Surfaces were only planes, and there was no careful consideration of the intersection of surfaces. The appendix had to be understandable by any curious student in mathematics, whether or not he wanted to become an engineer.

Adrain went further in the simplification of the method than had Crozet. He used simple drawing to sketch the projections of a point (Fig. 19.2).

Here, P' and P'' were the projections of a point P. Such elementary figures appeared in neither the French textbooks of the period 1795–1820 nor in Crozet's. Adrain drew the planes of projection on the two-dimensional figures so that the reader could easily imagine the rebatment ("rabattement") and the projections of the point. The rebatment was also made explicit with the direct use of the rotation of the vertical plane on the horizontal plane, as shown in Fig. 19.2. The word "rotation" was not mentioned, but hints indicated its use. Adrain introduced the rectangle $KLMP'$ with $KL = KP''$. He suggested that the rectangle "revolved" back to a vertical position with a 90° rotation around axis KP'. The reader could thus figure out the original position of point M by the "ascent" of the rectangle:

---

[8]At Dartmouth, the Chandler Scientific School offered elective and high-level courses for training scholars and engineers. Similar programs opened at Harvard and Columbia after 1860, and courses of descriptive geometry were also given. See (Preveraud 2014, pp. 152–155)

[9]A very detailed analysis of this process is given in Barbin (Chap. 2, this volume).

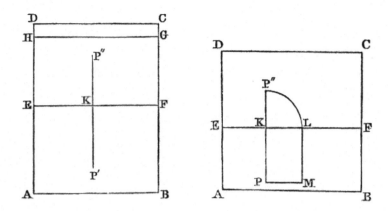

Fig. 19.2 Projection of a point in *A Course of Mathematics* (Adrain 1822, pp. 562–564)

> To conceive distinctly the place of the point P, [...] imagine now that the rectangle KLMP' revolves about its fixed side KP' from a horizontal to a vertical position by the ascent of the rectangle above the horizontal plane; and when the rectangle $KLMP'$ is in this vertical position, its angular point M will coincide with the point P of which the projections are P' and P''; and the angular point L, after having described a quadrant of a circle on the primitive vertical plane, will coincide with that point of it which is the vertical projection of P, and which is denoted by the point P'' (Adrain 1822, pp. 564–565).

The quarter circle clearly shows that Adrain wanted to visualize the rotation movement in space. This pedagogical technique gave a graphic representation of the mental process of the projection. In Lacroix's and Monge's treatises, this quarter circle appeared once, but only in the inaugural three-dimensional figures, and without any comment, without any use of the properties of the rotation Adrain explicitly referred to (conservation of the lengths). This pedagogical innovation must be linked to the level of the readers he wanted to reach: unprepared students with no teacher to explain what the method was.

For Adrain, descriptive geometry appeared to be a part and a sequel of the classical geometry curriculum. The subject came within an appendix on geometry. Moreover, it helped to solve problems that classical geometry can solve only with difficulty. Indeed, Adrain presented descriptive geometry with all its application to geometry:

> Besides the numerous corrections in this third American edition, there is added to the second volume an elementary treatise on Descriptive Geometry, in which the principles and fundamental problems are given in a simple and easy manner with a selected number of useful applications, in Spherics, Conics, Sections, &c (Adrain 1822, p. xii).

Therefore, according to Adrain, descriptive geometry was totally relevant as part of the collegiate mathematics curriculum, even for students who did not intend to become engineers. To highlight that point, Adrain made his appendix look like a Euclidean textbook. Indeed, Euclid's *Elements* were also known for a specific method of presentation. Each proposition was first stated in the most general way,

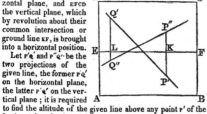

Fig. 19.3 Comparison of the presentation in *A Course of Mathematics* (Adrain 1822, p. 570) (LH side) and *Elements of Geometry* (Playfair 1806, p. 22) (RH side)

then followed the particular statement of the same proposition related to a particular diagram, and then came the proof and the conclusion. Adrain's presentation of descriptive geometry partly followed this structure as shown in Fig. 19.3.

The relationship between geometry and descriptive geometry as part of a general curriculum strengthened in Harvard assistant professor James Hayward's textbook. In *Elements of geometry upon the inductive method*, published in 1829 and intended for college students, Hayward added a short introduction to descriptive geometry. There, he explicitly articulated the relation between the teaching of geometry and descriptive geometry. In his opinion, the teaching of descriptive geometry helped students visualize the forms of geometry and their mutual relationship:

> By the study of Descriptive Geometry, the mind sees bodies and their parts in all their relations of position, magnitude, and figure; it becomes accustomed to the contemplation of forms, and acquires a certainty and readiness of the imagination which enables it to make with variety and skill, new combinations of the elements of form (Hayward 1829, p. xiii).

The table of contents also clearly indicated descriptive geometry as a sequel to the classical geometry course.

3.2 Mixed-Audience Textbooks (1825–1875)

In 1826, Charles Davies, professor of mathematics at West Point, published the second American textbook dedicated solely to the teaching of descriptive geometry. Though written at West Point and substituted there for Crozet's textbook in 1832 (Rickey and Shell-Gellash 2010), Davies' textbook was also aimed at students beyond the academy. He favored extending the teaching of descriptive geometry to the colleges at a time when their instruction in civil engineering was just beginning. He thought the subject should be part of the mathematics curriculum: "In France,

Descriptive Geometry is an important element of a scientific education; it is taught in most of the public schools" (Davies 1826, p. iii).

Davies' textbook started with the definitions, unlike the introduction of the inaugural problem found in the books of Monge and Crozet. The author studied curves before surfaces, and particular situations before generalizing. The generalization introduced by Monge appeared to be difficult for beginners, and it diverged from the purposes of domestic publishing. Moreover, Davies was a businessman (Ackerberg-Hastings 2000, Chap. 5) and wanted his textbook to be sold as widely as possible. The textbook included the simplification method initiated by Adrain to highlight the process of the rebatment:

> The part of the plane of the paper which lies above the ground line, will then represent that part of the vertical plane of projection which is above the horizontal plane, and also the part of the horizontal which is behind the vertical plane; and that part of the plane of the paper, which lies below the ground line, will represent that part of the vertical plane which is below the horizontal, and that part of the horizontal plane which is in front of the vertical (Davies 1826, pp. 2–3)

In Adrain's appendix, the movement of the plane was only indicated by a quarter circle, while Davies wrote the complete description of the process: "If a line be drawn perpendicularly from any point of a plane to the axis, and the plane be found in a circle, whose center is the point in which the perpendicular meets the axis, whose radius is this perpendicular" (Davies 1826, p. 3). Davies' textbook also shows strong evidence of links between descriptive geometry and the mathematics curriculum. First, Davies included a very general chapter about curved lines, their curvatures, and their tangents in plane geometry that one could have read in any analytical geometry textbook (Davies 1826, pp. 37–42). Moreover, Davies added three chapters about spherical projections (main principles, orthographic and spherical projections). He justified this triple introduction:

> To conceive of the whole surface of the earth, and the positions of objects situated on it, it is necessary to have recourse, either to artificial globes, or to drawings, which represent the earth and the different points of its surface. As it is quite difficult to construct artificial globes, and indicate on them the different places on the surface of the earth, as well as their relative positions, the method by drawings, or the representation on planes, has been generally adopted (Davies 1826, p. 146)

Davies established an analogy with the projections on a sphere and the projections used in descriptive geometry: both enabled to figure three-dimensional objects that had to be figured out. The author thus made connections with other aspects of mathematical knowledge, such as spherical geometry, navigation, astronomy, subjects taught in most American colleges.

Between 1826 and 1864, no other textbook was published on the subject at West Point; Davies' *Elements* seemed to serve the objectives of training students both in military academies and colleges. In 1864, the new West Point professor of mathematics, Albert E. Church, wrote a new specialized American textbook on descriptive geometry. The *Elements of Descriptive Geometry* was not that different from the book written by Davies almost 40 years before. But Church included some of the new pedagogical approaches the French had used in their books published

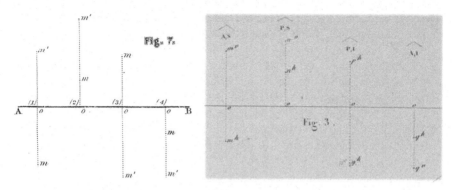

Fig. 19.4 A point in Church (1864b, Fig. 7) (LH side) and Olivier (1844), Plate 1 (RH side)

during the three previous decades. He explicitly (Church 1864a, p. iii) referred to the *Traité de géométrie descriptive*, written by Leroy and the *Cours de géométrie descriptive* by former *École centrale des arts et manufactures* professor Theodore Olivier. On the one hand, he followed the organization of Leroy's textbook. On the other, he transferred the new simplification Olivier had developed in his textbook: the "point, line, plane" method (Barbin 2015) (see Chap. 2, this volume). Olivier thought the method of the projections as well as the problems could be easily understood provided that the reader knew how to construct a point, a straight line, and a plane (Olivier 1844, p. vi). For instance, as shown in Fig. 19.4, and following Olivier, Church gave a detailed explanation of the different situations for the construction of a point with a degree of simplification never found before in any American textbook.[10]

Three years later, *Elements of Descriptive Geometry with its applications to Shades, Shadows and Perspective, and to Topography* was published by Virginia Military Institute professor Francis H. Smith (1867). Like West Point, VMI trained military and civil engineers, but Smith also explicitly intended his textbook for "architects, machinists and manufacturers [...] [and] to the general student" (Smith 1867, p. iv). The first part of the book was named "The point, the right line, the plane" and recalled the method exposed by Olivier. Smith began with some explanations about the projection of a point, a curve or a surface on a single plane with very elementary situations: "the projection of a point on a plane is the foot of the perpendicular let fall from this point on the plane" (Smith 1867, p. 16). He moved on to the double projection but still in three-dimensions: the rebatment came only on page 24, which showed a desire to introduce the subject with elementary considerations to potential readers whose mathematical background became more and more heterogeneous. Nevertheless, and although Monge's inaugural problem had been removed, Smith's textbook offered a problematized geometry as did

---

[10]Probably taken from Olivier's book, these figures had already been proposed in Vallée's book (see Barbin, Chap. 2, this volume).

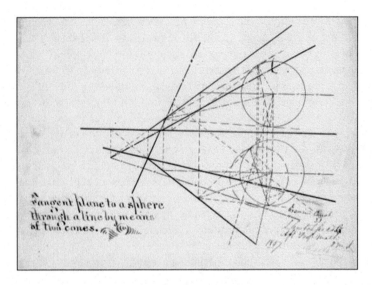

Fig. 19.5 Example of examination at VMI, 1857 (Cooke 1857)

Monge's: the chapters following the introduction of the main principles gathered the essential problems of the subject. Like in West Point, students had to complete the drawings on their notebooks. Examinations consisted on constructing such drawings. For instance, Cadet Giles B. Cooke was asked to draw "the projection of the tangent plane to a sphere through a line by means of two cones" (Cooke 1857). As shown in Fig. 19.5, the drawing was assessed by assistant professor of mathematics Stapleton Crutchfield.

4 Descriptive Geometry and the Rise of Technical Studies (1860–1915)

During the second half of the nineteenth century, the rise of technical and specialized schools (Brubacher and Rudy 1968, pp. 63–64)) for training not only engineers but also machinists, architects, masons, and carpenters led authors and teachers to write textbooks for this new audience. Descriptive geometry was still taught in some colleges, but was also removed from others curricula. For instance, Harvard did not teach the subject after 1860, while Samuel E. Warren taught it at the *Rensselaer Polytechnic Institute*. This first American institution to award a degree in civil engineering (1835) was refashioned in 1849. The organization and the contents of the studies were inspired by the curriculum of the *École centrale des arts et des manufactures*. Focused on applied sciences and practical studies (but also on mathematics), the Institute aimed at training civil engineers for industry as well as draftsmen, technicians, or architects (Preveraud 2014, p. 159).

In 1860, Warren published *General problems from the Orthographic Projections of Descriptive Geometry*. Explicitly dedicated to "draftsmen" (Warren 1860, p. xxiv), his textbook took issue with some of the previous American works published on the subject. Warren criticized his predecessors who had thought of "descriptive geometry as a science of rational (synthetic) geometry", explaining that they had confounded "the means of research with the results of the research" (Warren 1860, p. xxxiv). Whereas Adrain, Hayward, or even Davies had considered the descriptive geometry course as a sequel to classical geometry, Warren clearly made a distinction: "Descriptive Geometry is, then, the exact graphic art and not the science of geometry". Descriptive geometry had rather to be related to graphic arts and the graphic solving of problems. As a consequence, the book opened with projections and the laws of graphical elementary constructions. In *Elements of Descriptive Geometry* (1874), the same Warren, who had subsequently become professor at the Massachusetts Institute of Technology, wrote a new textbook on the subject and intended it for students at both technical schools and colleges. Still, he noted that the subject was crucial for the former and not necessarily for the latter:

> The rapidly increasing number of technical schools, and the consequently more manifest occasion for the wider diffusion of a general knowledge of the foundations of the higher industries taught in them, suggest the question whether the elements, simply, of a science which is so necessary to their members, should not be included among the mathematical studies in every course of liberal education (Warren 1874, p. xix).

In *Elements of Descriptive Geometry* (1914), Georges F. Blessing and Lewis A. Darling published a new method of presentation of descriptive geometry. Blessing and Darling had both been assistant professors of machine design at Cornell, and Darling had become an engineer in industry. In their introduction, the authors promised a "presentation of an experiment as a means of bringing out fundamental principles" (Blessing and Darling 1914, p. iv). According to them, experiments would help students "visualize" the problems of descriptive geometry and make its teaching as "practical" as possible. What they called an "experiment" was literally an illustrated handling of pieces of paper.

The inaugural experiment asked the reader to cut a piece of drawing paper as shown in Fig. 19.6. The authors gave then instructions to fold this so as to create solid angles, and planes they labelled "front wall", "side wall", or "floor" (Blessing and Darling 1914, p. 2). The student was asked to plot and trace points and lines with the help of plumb lines and wires as shown in the right-hand drawing of Fig. 19.6. The definitions and the principles were then introduced with respect to the walls of the room the readers had built. For instance, the rebatment described the deconstruction of the model:

> Remove the plumb line and wires, open the paper angle, spreading it out flat on the drawing board as it was originally [...]. The result is shown in Fig. 19.6 which is a graphic record or Descriptive Geometry drawing of all that has taken place (Blessing and Darling 1914, p. 4).

Here, one can find back traces of the original idea of Monge: descriptive geometry should also be practical.

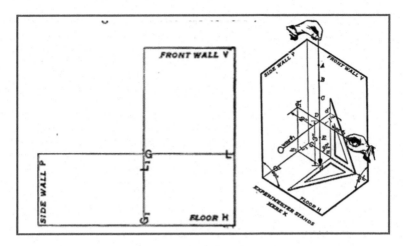

Fig. 19.6 Experiment 1 in (Blessing and Darling 1914, p. 2)

5 Concluding Remarks and the Absence of Translations

Unlike other mathematical subjects and other countries, no translation of any French textbook dedicated to descriptive geometry was published in America in the nineteenth century. Between 1815 and 1865, about 30 French textbooks were translated or adapted by American authors and professors, mostly for the teaching of algebra and geometry (Preveraud 2014). Also, Monge's lessons were translated into Greek in 1820, while the textbooks of Leroy and Olivier were published in the 1840s in a Greek version (Kastanis 2003, pp. 152–153) (for descriptive geometry in Greece see Phili, Chap. 8, this volume). Leroy's textbook was translated into German in 1845 (Morel 2013, p. 495) (for Germany, see also Benstein, Chap. 9, this volume). Olivier and other French textbooks were translated into Arabic (Crozet 2008, pp. 429–435). Like in many countries studied in this volume, those translations were dedicated to the teaching and the training of engineers in engineering schools or military academies (in Egypt, for instance, see Crozet, Chap. 20, this volume).

Instead, American authors chose not to translate a specific French author but rather to mix different sources. Those multiple and partial borrowings could be explained by the level of education in the domestic context, by the uses of the publishing market and by the textbook business. The authors faced students with very different levels of mathematical preparation, a preparation less homogenous than that of French students. With the multiplication of different institutions in which engineers were trained, the American context was completely different from the French. Moreover, in colleges, students did not all intend to become engineers and the subject ended up being taught as a part of the general curriculum. The audience the authors intended to reach was thus less well-defined, changing, and unsettled during the period under study there. Since future American engineers –

trained at West Point, at Rensselaer, at Harvard, and elsewhere – did not follow the same curricula, more individualized presentations of the subject like those discussed above were needed and so were produced.

References

Ackerberg-Hastings, Amy. 2000. *Mathematics is a Gentleman's Art: Analysis and Synthesis in American College Geometry Teaching*, 1790–1840. Ph.D. Dissertation. Iwo State University, Ames.
———. 2002. Analysis and synthesis in John Playfair's Elements of Geometry. *The British Journal for the History of Science* 35 (1): 43–72.
Adams, Cindy. 1965. *The West Point Thayer Papers (1808–1872)*, 82. West Point: Association of Graduates.
Adams, Jasper. 1822. Review of the Cambridge course of Mathematics. *The American Journal of Science and Arts* 5: 304–326.
Adrain, Robert. 1822. *A Course of Mathematics*, 2. New York: Campbell & Son, Duyokince, Sword, M'Dermut, Ronalds, Collins & Hannay, Long.
Barbin, Évelyne. 2015. Descriptive geometry in France: History of the simplification of a method (1795–1865). *International Journal for the History of Mathematics Education* 10 (2): 39–81.
Blessing, George F., and Lewis A. Darling. 1914. *Elements of Descriptive Geometry*. New York: Wiley and Sons.
Brubacher, John S., and Willis Rudy. 1968. *Higher Education in Transition, A History of American colleges and universities*, 1636–1968. New-York: Evanston. London: Harper & Row.
Cajori, Florian. 1890. *The Teaching and History of Mathematics in the United States*. Washington: Bureau of Education.
Church, Albert E. 1864a. *Elements of Descriptive Geometry*. New York: Barnes.
———. 1864b. *Plates of Descriptive Geometry*. New York: Barnes.
Cooke, Giles B. 1857. *Geometry Exercises. Tangent Plane*. 19th Century Cadet Drawing Exercises. Virginia Military Academy Archives Digital Collection. http://digitalcollections.vmi.edu/cdm/ref/collection/p15821coll15/id/12
Crackel, Theodore J. 2002. *West Point: A Bicentennial History*. Lawrence: University Press of Kansas.
Crozet, Claude. 1821. *A Treatise on Descriptive Geometry, for the use of the cadets of the United States Military Academy*. New York: Goodrich and Co.
Crozet, Pascal. 2008. *Les sciences modernes en Égypte, Transfert et appropriation (1805–1902)*. Paris: Geuthner.
Davies, Charles. 1826. *Elements of Descriptive Geometry, with their Applications to Spherical Trigonometry, Spherical Projections and Wrapped Surfaces*. Philadelphia: Carey and Lea.
Harvard College. 1827.*The Annual Report of the President of Harvard University to the Overseers on the State of the University for the Academical Year 1825–1826*. Cambridge, MA: University Press, Hilliard, Metcalf & Co.
Hayward, James. 1829. *Elements of Geometry upon the Inductive Method*. Cambridge, MA: Hilliard and Brown.
Hunter, Robert F. and Edwin L. Jr. Dooley. 1992. *Claudius Crozet: French Engineer in America, 1790–1864*. Charlottesville: University of Virginia Press.
Kastanis, Andreas. 2003. Descriptive Geometry in 19th Century Greece. In *Science, Technology and the 19th Century State: The Role of the Army*, ed. Konstantinos Chatzis, and Efthymios Nicolaidis, 147–162. Athens: Center for Neo-Hellenic Research/NHRF.
Kidwell, Peggy, Amy Ackerberg-Hastings, and David Roberts. 2008. *Tools of American Mathematics Teaching, 1800–2000*. Baltimore: The Johns Hopkins University Press.

Lacroix, Sylvestre-François. 1802. *Essai de géométrie sur les plans et les surfaces courbes*. Paris: Duprat.

Lawrence, Snezana. 2003. History of Descriptive Geometry in England. In *Proceedings of the First Internatinal Congress on Construction History, Madrid, 20th–24th January 2003*, ed. Santiago Huerta, 1269–1282. Madrid: Instituto Juan de Herrera, SEHC, COAC, CAATC.

Mansfield, Edward D. 1863. Military academy at West Point. *American Journal of Education* 13: 17–48.

Molloy, Peter M. 1975. *Technical Education and the Young Republic: West Point as America's École Polytechnique, 1802–1833*. Ph.D. Dissertation. Providence: Brown University Press.

Monge, Gaspard. 1799. Géométrie descriptive. Paris: Baudoin.

Morel, Thomas. 2013. *Mathématiques et politiques scientifiques en Saxe (1765–1851). Institutions, acteurs et enseignements*. Thèse de doctorat. Bordeaux: Université de Bordeaux 1.

Olivier, Theodore. 1844. *Cours de géométrie descriptive: Atlas*. Paris: Carilian-Goeury et Dalmont.

Playfair, John. 1795. *Elements of Geometry*. Edinburgh: Bell & Bradfute. London: G., G. & J. Robinson.

Playfair, John. Ed. 1806. *Elements of Geometry, Containing the First Six Books of Euclid*. Philadelphia: Nichols.

Preveraud, Thomas. 2011. A course of mathematics (1798–1841): The American story of a British textbook. In *Proceedings of the Second International Conference on the History of Mathematics Education*, October 2–5, 2011, ed. Kristin Bjarnadottir, Fulvia Furinghetti, José Manuel Matos and Gert Schubring, 383–399. Lisbon: UIED.

———. 2014. *Circulations mathématiques franco-américaines: transferts, réceptions, incorporations et sédimentations (1815–1876)*. Thése de doctorat. Nantes: université de Nantes.

———. 2020. La rationalisation des savoirs mathématiques français au sein des écoles militaires américaines avant la Guerre de Sécession. *Philosophia Scientiae* (to be published in early 2020).

Reynolds, Terry S. 1992. The education of engineers in America before the Morrill Act of 1862. *History of Education Quaterly* 32(4): 459–482.

Ricketts, Palmer C. 1895. *History of the Rensselaer Polytechnic Institute (1824–1894)*. New York: Wiley and Sons.

Rickey, V. Frederick, and Amy Shell-Gellash. 2010. *Mathematics Education at West Point: the First Hundred Years*. Convergence (on line at http://www.maa.org/publications/periodicals/convergence/mathematics-education-at-west-point-the-first-hundred-years)

Sakarovitch, Joël. 1994. La géométrie descriptive, une reine déchue. In *La formation polytechnicienne*, ed. Bruno Belhoste, Amy Dahan-Dalmedico and Antoine Picon, 77–93. Paris: Dunod.

Smith, Francis H. 1867. *Elements of Descriptive Geometry with Its Application to Shades, Shadows and Perspective, and to Topography*. Baltimore: Kelly, Piet & Co.

Warren, Samuel E. 1860. *General problems form the Orthographic Projections of Descriptive Geometry*. New York: Wiley.

———. 1874. *Elements of Descriptive Geometry in Three Parts*. New York: Wiley and Sons.

Chapter 20
The Teaching of Descriptive Geometry in Egypt

Pascal Crozet

Abstract In the period between 1837 with the translation of Emile Duchesne's *Éléments de géométrie descriptive* into Arabic, and the 1890s with the takeover of secondary and higher education by the British, the teaching of descriptive geometry enjoyed sustained development in Egyptian engineering schools and military academies. This paper will highlight the defining moments of French influence in education in Egypt. The protagonists' lives and the discipline's position in the curricula will be detailed, as well as the development of textbooks, from the translation of those by Olivier, Leroy and Gerono to the composition of original treatises by Aḥmad Naǧīb, Ṣābir Ṣabrī, etc.

Keywords École polytechnique · École centrale · Model function · Technical colleges · Orientalism · Arabic translation · Scientific language

1 Introduction

That one may chance upon a chapter concerning Egypt, or any other non-European country, in a book dealing with the dissemination of descriptive geometry, should come as no surprise. As a matter of fact, the very moment that this new discipline came into its own coincided with a massive wave of transmission and development of European sciences taking place *outside* Europe owing to territorial conquests as well as a variety of other forms of pressure applied by European powers. It is well worth recalling, in order to illustrate this concurrence, that Gaspard Monge himself was involved in one of the most notable instances of this expansion, namely Bonaparte's Egyptian campaign. Moreover both phenomena are linked, in as much as the training of engineers was the means by which European science was

P. Crozet (✉)
CNRS–Université Paris-Diderot, Paris, France
e-mail: crozet@univ-paris-diderot.fr

disseminated, and polytechnics were at the core the educational infrastructure that were established at this time.

Nonetheless, the Egyptian case presents many dissimilarities with its French, German or Dutch counterparts. Owing to the fact that the dissemination of descriptive geometry occurred in a radically different context, characterized above all by the problems raised by introducing a much wider scientific corpus into an environment that wasn't equipped to receive it. For example, how would this new knowledge relate to Egyptian society and its history? What kind of legitimacy could it claim? Or how to elaborate a scientific language capable of conveying it?

Thus the crossing of these cultural and linguistic barriers becomes an issue of paramount importance when considering these novel mathematical concepts and the models for technical education that accompany them. We will further expound on the subject at the end of this paper when an overview of treatises on descriptive geometry published in Egypt will enable us to gain a more precise view of the development of this discipline on the banks of the Nile.

2 The Introduction of Modern Science in Egypt (Nineteenth Century)

In order to gain a better understanding of our subject, we need to examine the political context in Egypt, and particularly during the reign of Muḥammad ʿAlī (1805–1848) which above all marks the beginning of the introduction of modern sciences. The extent of the Egyptian program in this regard, which in the 1830s far exceeded the claims of its Turkish or Persian contemporaries, must be emphasized here: it explains why the Egyptian experience constituted a kind of reference, if not a model, for all the region—and also why the introduction of descriptive geometry here was made earlier than elsewhere.

When Egypt's new ruler came to power in 1805, after 4 years of internal struggles following the French invasion, he was immediately confronted with the challenge posed by European expansionism. There was the imperative of modernizing his armed forces, in order to be able to stave off any surprise attacks. But also—and primarily—to modernize Egypt's economy in an attempt to bring it up to par with western European nations. For Muḥammad ʿAlī, building an independent state on the banks of the Nile was a way to counter increasingly overwhelming European economic power, and reduce imports and the influx of Western investments while rebuilding the Egyptian economy onto a sounder footing and providing additional outlets for exports.

The state assumed new prerogatives in the fields of manufacturing, public health and civil engineering, all of which required new skills based on scientific knowledge and techniques hitherto unknown on the banks of the Nile. To accomplish this, the master of Egypt summoned numerous European experts, and also made use of others who offered their services, such as the Saint-Simonians. He sent more than a hundred Egyptian students for training in Europe, notably in France, where several of them attended courses at the *École polytechnique* and the *Faculté de médecine* in Paris. Finally, graduate schools were founded in Egypt: a faculty of medicine and pharmacy, a faculty of engineering, a faculty of arts and crafts, military academies, etc. Although the organizational models adopted were French in origin (for example, the *École centrale des arts et manufactures* for the *Būlāq Engineering School*), courses were delivered in Arabic by a growing number of Egyptian teachers using textbooks translated for this purpose in all of these disciplines. The momentum created in this drive for translation was greatly favoured by the foundation of the Language Faculty by Rifā'a al-Ṭahṭāwī (1837) and resulted, 15 years later, in the elaboration of a modern Arabic scientific language, durable and respectful of its origins (Crozet 1996 and 2008, pp. 287–374). Lastly, one can note the appearance of new civil service echelons for these students (Alleaume 1987), which considerably institutionalized the scientific professions such as engineering and medicine.

All this, however, did not constitute the foundation of a genuine scientific community, since Egypt's rulers were much more interested in scientific applications and the resulting techniques than in scientific development per se: in a sense, European science was transferred essentially through its association with certain techniques, with no research being contemplated at the time. Thus no scientific institution emerged on the banks of the Nile that was comparable with the academies, which the European capitals had acquired; certain fields of technical expertise (currency, weights and measures, etc.) that could have been assigned to centralized institutions of this kind were left to short-lived boards.

The education system established by Muḥammad 'Alī alongside the traditional system was also entirely subject to the purpose for which it had been created: supplying experts for government departments, and nothing more. State-run primary and secondary schools were thus solely intended to educate pupils for the needs of further education, and not to dispense knowledge, scientific or otherwise, to a larger segment of the Egyptian population.

Not all of Muḥammad 'Alī's successors showed similar statesmanship, nor an interest in establishing their country's autonomy from European goals and projects. Throughout the century, this variety of attitudes, in addition to economic fluctuations, would subject the fields of sciences, techniques and their teaching to a wide-array of policies and institutional upheavals with each change of reign. Nonetheless, started as a means of thwarting European expansionism, this is the era during which Egyptians took command of their scientific destiny.

Heralded by the bombardment of Alexandria in 1882, the British invasion resulted in great change. Thereafter, scientific development gradually came under colonial management. Throughout the 1890s control over the educational system intensified. In 1902, the Egyptian director and teaching staff at the Engineering School were dismissed. Arabic was abandoned in favour of English for the teaching of sciences, etc. Additionally, references and educational models were aligned with those used by the British.

Thus there is a stark contrast between the nineteenth and twentieth centuries. Several elements which would be determinant in the development of descriptive geometry in Egypt can already be identified: firstly, the predominance of the French influence, which in this particular case is essential, followed by the drive to train high-level experts locally, and lastly the adoption of Arabic as the teaching language.

Descriptive geometry was relatively well represented in mathematical disciplines judging by nineteenth century Egyptian publications. One can thus list the following titles:

- Muḥammad Bayyūmī: *al-Lāzim min al-handasa al-waṣfiyya* (Būlāq 1837). Translation of: Émile Duchesne, *Éléments de géométrie descriptive*, 1ère éd., Malher (Paris 1828).[1]
- Ibrāhīm Ramaḍān: *al-La'āli' al-bahiyya fī al-handasa al-waṣfiyya* (Lithograph of the *Muhandisḫāna* 1842); reprint (Būlāq 1845). Translation of: Théodore Olivier, *Cours de géométrie descriptive*, cours autographié de *l'École centrale des arts et manufactures* (1840–1841); première partie («*du point, de la droite et du plan*»).
- Ibrāhīm Ramaḍān: *al-Rawḍa al-zahriyya fī al-handasa al-waṣfiyya*, a compilation of French textbooks translated by Ibrāhīm Ramaḍān, and Manṣūr ʿAzmī, in three parts (Lithograph 1852); the first part is printed in Būlāq in 1853 and again in 1873 with some additions.
- Aḥmad Naǧīb: *al-Tuḥfa al-bahiyya fī al-handasa al-waṣfiyya* (Būlāq 1873).
- Aḥmad Naǧīb: *Uṣūl al-handasa al-waṣfiyya li-istiʿmāl talāmiḏa al-madāris al-miṣriyya*, (Cairo 1873). A translation of Camille-Christophe Gerono and Eugène Cassanac's, *Éléments de géométrie descriptive* (Delagrave et Cie (Paris 1866)).
- Ṣābir Ṣabrī: *al-Barāʿa al-mašriqiyya fī ʿilm al-handasa al-waṣfiyya*, 2 vol., atlas (Būlāq 1881–1888).
- Maḥmūd Fahīm: *al-Nafḥa al-ʿurfiyya fī al-handasa al-waṣfiyya* (Cairo 1896).

We shall now peruse these textbooks (in the above order) as they embody the different stages through which the development of the teaching of this discipline unfolded in nineteenth century Egypt.

3 The Contribution of Muḥammad Bayyūmī (c.1810–1852)

The first descriptive geometry textbook published in Egypt is also one of the very first treatises published by the Būlāq Press, only preceded by an Arabic elementary

---

[1] We have chosen this unusual presentation owing to the absence of the original authors' name in most of these translations. The non-Arabic speaking, but astute reader, will have guessed that *al-handasa al-waṣfiyya* signifies 'descriptive geometry'. Beyond that, the Arabic titles generally bear no resemblance with the original textbook titles. Lastly, a meticulous comparison of all the editions of the French textbooks showed us that those we mention here are those that were actually translated.

arithmetic textbook written in the fourteenth century and several treatises written in Turkish. It is therefore a ground-breaking publication.[2]

The text is relatively basic. At the time it was published, its author Émile Duchesne (1793–1872) was a mathematics teacher in the *Collège de Vendôme*, and his work was explicitly written for "students wishing to pursue their studies at the *École polytechnique*, *École Militaire* or the *École de Marine*", as specified in the subtitle of the book.

The translator, Muḥammad Bayyūmī, was instrumental in the process of transferring European mathematical knowledge and adapting it to the Egyptian context. Born in Cairo around 1810, he was sent to Paris in 1826 to study, and remained there until 1835. Along with two Egyptian classmates, he attended courses at the *École polytechnique* (starting in 1830) and very probably thereafter at the *École des ponts et chaussées* (for Muḥammad Bayyūmī's biography, see Crozet (2008, pp. 337–340)). On his return to Egypt he was quickly assigned to the engineering school which had recently relocated to Būlāq, a suburb of Cairo. At this point, he was the only teacher at the institution to have followed such an extensive curriculum. It was in order to teach his classes that he undertook most of his translations, all completed before 1838: Duchesne's *Éléments de géométrie descriptive*, begun as soon as he returned to Cairo in 1835, Legendre's *Traité de trigonométrie*, Mayer's *Traité élémentaire d'algèbre*, and Terquem's *Manuel de mécanique*. Of all the French textbooks available to Bayyūmī it is probable that his choice of Duchesne's stemmed from his proficiency in that discipline. The momentum gained by the translation drive in Egypt led Bayyūmī, in 1842, to forgo his teaching responsibilities at the Engineering Faculty, in order to devote his energies completely to directing the mathematics section of the Bureau of Translation.[3] He even continued to allocate 3 days a week to this institution after he was reassigned in 1845 to the "Bureau of Schools" (an authority comparable to a ministry of education), probably in a supervisory capacity.

What were his duties at the "Bureau of Translation"? Undoubtedly, he did not translate much on his own. But he certainly tutored translators, supervised their work, and compiled and unified a lexicon of which he seems to have become the custodian. Consequently he is often credited, as a "specialist in mathematical terminology"? in the introductions of the writings he revised.

In 1850, he was exiled to Khartoum, for reasons that remain nebulous, but that are certainly in connection with the change of reigns, which had just taken place. He died prematurely in Sudan in 1851 or 1852, teaching in a grade school. And so, in a way that reveals the importance of restoring Arabic as a vehicle for scientific

---

[2] If this text is indeed the first Arabic translation of a French mathematical treatise to be printed, it is by no means the first translation to have been undertaken. One should mention, for example, Allaize Billy, Boudrot and Puissant's geometry textbook, whose translation was undertaken by Rifā'a al-Ṭahṭāwī around 1833 or 1834 for the Egyptian artillery academy, but only published in 1842. Or even, in the 1820s, the manuscript translation of a few relatively basic Italian textbooks.

[3] The Bureau of Translation comprised four sections: mathematical sciences, physical and medical sciences, humanities, history and geography, Turkish translations.

discourse—which Muḥammad ʿAlī as well as the educational establishment fully recognized—the reconstruction of such a language became the life—long endeavour of one of the major scientific authorities in the first half of that century.

In the introduction to his translation of Duchesne's *Éléments de géométrie descriptive*, Bayyūmī dwells on the complexity of the task ahead of him as he undertakes his first translation. He also expresses his debt of gratitude to Ibrāhīm Adham, a Turkish artillery officer stationed in Egypt from the 1810s and who would later head the *Bureau des Écoles*. Born in 1785, and most likely trained at Istanbul's Engineering School during Selim III's reign, Adham enjoyed considerable scientific prestige in that era. He had just published his Turkish translation of Legendre's *Eléments de géométrie* which are copiously annotated with his personal comments, thus demonstrating that he certainly was not a passive translator.[4]

What guidance did Adham provide for the fledgling Egyptian translator? It is hard to determine precisely. But he most likely induced him to research the Arabic versions of Euclid's *Elements*, and more particularly Naṣīr al-Dīn al-Ṭūsī's edition, which was then widely available: doesn't Bayyūmī end his introduction to Duchesne's book by commenting on the Greek geometer's "treasure", even as he underscores the distance that separates both books? Furthermore, isn't the vocabulary used in Adham's translation of Legendre essentially the same as that found in al-Ṭūsī's Arabic publication of the *Elements* in Constantinople in 1801, in the very circles frequented by the Turkish engineer during his studies?

Be that as it may, it should be noted that the lexicon is almost identical to that already used for centuries in classical Arabic geometry. The only differences are the relatively few terms which were coined long after Euclid, such as the constantly used projection, or those pertaining to descriptive geometry such as *ligne de terre* (ground line) and trace, or even the very term descriptive geometry. These last terms were not too difficult to translate since everyday language afforded natural equivalencies. For the more challenging projection, Bayyūmī resorts to the word *masqat*, from the root *sqt*, which quite appropriately refers to falling on something. For the even more challenging term *épure*, he uses the expression *rasm waṣfī*, which literally (and appropriately) means *descriptive drawing*.

The prevailing principals are thus, on the one hand to use the classical geometric lexicon as much as possible, and on the other to create neologisms in the spirit of the language in the rare cases where this lexicon is wanting. These rules would be applied in subsequent translations. In this way, when faced with the term *rabattement* while translating Olivier's treatise, Ibrāhīm Ramaḍān used the word *inṭibāq* which is found in ancient geometry treatises to signify a superposition (Ramaḍān 1845, p. 5). Later on, in order to render the words *développée, développante* and *surface développable*, he would follow Bayyūmī's lead who had already rendered *développement* as *inbisāṭ* and would respectively use the terms *mabsūṭ, bāsiṭ*, and

---

[4] The need for this translation arose when Adham was teaching geometry to Turkish-speaking officers who were still a majority in Muḥammad ʿAlī's army in the early 1830s. But by the end of the decade, this pre-eminence had disappeared.

saṭḥ qābil bi-l-inbisāṭ, relying on the *bsṭ* root as his predecessor had, which suggests in an even more direct manner than the French term, the idea of a plane and even more so of flattening (Ramaḍān 1852, p. 70ff). In comparison with other disciplines taught at the Būlāq Engineering School which did not have the benefit of a classical lexicon, as is the case with mechanics and its abundant technical vocabulary or with stereotomy (Crozet 2004), descriptive geometry was in a particularly favourable position. By determining the content of the lexicon from the outset, and providing a method for enriching it, Muḥammad Bayyūmī's contribution was decisive in rapidly endowing the discipline with a perfectly adapted language which would remain relatively unaltered till the end of the century.

4 Charles Lambert and the *École centrale* Model

Coincidently with the return in 1835 of the Egyptian scholarship students from the famous 1826 education mission, a French engineer who left a durable mark in Egypt's educational history during his tenure at the School in Būlāqentered the picture: the Saint-Simonian Charles Lambert.

Born in 1804, Lambert enrolled at the *École polytechnique* in Paris in 1822. Two years later he graduated top of his class, and for the next 4 years studied at the *École des Mines*. In 1829 he converted to Saint-Simonianism, and practised his calling as an engineer only briefly, before devoting his time exclusively to the activities of Prosper Enfantin's group within the new religion. Arriving in Egypt in 1833 with a group of eighty or so Saint-Simonians, Lambert was one of the few who stayed on after Enfantin's departure in 1836. There he led numerous projects spanning a variety of technical domains: dams on the Nile, railroad projects, irrigation, mining, topography, coin minting, gun powder and saltpetre manufacturing, paper mills, water distribution for Cairo, etc. Around 1836, he and Joseph Hekekyan—an Armenian engineer trained in England whose family was close to Muḥammad 'Alī's entourage—were appointed to reorganize the Engineering Faculty, which had just been transferred to Būlāq. He quickly became the schools' sole director, when Hekekyan left shortly afterwards to establish an arts and crafts school, and would pursue his tenure there until 1850 (Crozet 2008, p. 170 and thereafter).

As early as 1836, and stemming from his Saint-Simonian beliefs and his perceptions of the Orient, which were connected to the reasons for the arrival of Enfantin's group in Egypt, he opted for an organizational model based on Paris's *École centrale* (on the creation of *École centrale*, see Barbin, Chap. 2, this volume). He writes:

> Amongst Europe's deservedly celebrated schools, the one whose goals, whose industrial spirit, understanding of matter, the senses and practicality is closest to Boulac, is the École Centrale des Arts et Manufactures in Paris, a recently chartered establishment, full of vigour and promise for the future, and which is already dubbed the École Polytechnique industrielle.

> As in Boulac, the program takes three years to complete. The admission requirements are quite similar, scientifically speaking. For these reasons, we thought it wise to adopt the École centrale's curriculum as a base, modifying it according to Boulac's particular goals, location and history (Lambert 1836, fol. 7^v) (my translation).

In fact, choosing the *École centrale* as a blueprint calls into question the very structure of the engineer's acquisition of knowledge, i.e. the relationship between science and its application, the relationship between theory and practice in teaching, and the role and place of mathematics in the curriculum. In aiming to provide engineers for the public sector, which is to say for "Industry", the founders of the *École centrale*'s intentions turn out to be quite close to the Saint-Simonian worldview, so it seems quite natural that Lambert should have taken that route.

However in the 1837 program he co-signed with Hekekyan, references to the *École centrale*'s program are very scarce (more on this program in Crozet (2008, pp. 235–244)). There are two reasons for this: for one thing, the *École centrale*'s courses were not yet available in Egypt. And secondly, while he shared the same industrial goals as the other founding members, he believed—as did many of his contemporaries—that the qualitative level of mathematics at the *École centrale* was too wanting to impart the scientific background necessary to understand the physico-mathematical principles to which most of the courses referred (Lambert 1836, fol. 8^r–8^v).

Be that as it may, descriptive geometry emerged as a central part of these programs. As was the case during the first 2 decades of the fourteenth century, at the *École polytechnique* in Paris, for example, machine theory fell under the remit of descriptive geometry. Hachette's treatise on machines was recommended by Lambert and Hekekyan as well as Lanz and Betancourt's, whose Arabic translation was published in Būlāq in 1841 (Hachette 1819; Lanz et al. 1819). Both deal principally with the different descriptions and classifications of mechanisms that could be used to transform one type of movement into a different type of movement. In fact, they are more about the geometry of machines—where drawings (of cogs, for instance) are prevalent—than an actual study on mechanics per se (Dupont 2000). Even as far removed from descriptive geometry as a course in hydraulics may appear, they seem to entertain the same relation between theory and practice. In D'Aubuisson de Voisins' textbook, also recommended by Lambert, the author emphasizes in his preface the mainly descriptive nature of his book:

> Thus by its very nature, my work has more to do with the sciences of observation, and physics, than with mathematics; it is a treatise on applied and experimental hydraulics, and not rational hydraulics (Aubuisson de Voisins 1834, p. ix).

As for the required reading list for the descriptive geometry classes in 1837, Duchesne's textbook, which had just been translated by Bayyūmī, was to be used as a "textbook", and Leroy's "for reference" (probably the first edition of Leroy's *Traité de géométrie descriptive*, which was published in 1834; see Leroy (1842)).

But the 1837 program would undergo a thorough overhaul a few years later, when the *École centrale*'s textbooks became available in Egypt, and above all when the mathematical underpinnings in the Parisian courses had taken root. It was this new program dispensed in the early 1840s, and for years to come, that would characterize Lambert's school and even more crucially, constitute the bedrock for the organizational structure of the Egyptian Engineering Faculty for the remainder of the century.

As a matter of fact, when Bélanger took over from Liouville at the *École centrale* in 1838, the standards of admission had been raised, and he was able to—in the words of one of his successors—"frankly" reintroduce "the necessary notions of differential and integral calculus" in order to completely restructure the discipline's didactics (Comberousse 1879, p. 84). The ongoing changes occurring at the Parisian school, and which led to the creation in 1838–1839 of a complete and coherent mechanics class, were quick to spark Lambert's interest.

He promptly forsook Boucharlat's courses on differential and integral calculus, which he had at one time recommended, to adopt the first part of Bélanger's course, referred to by its author as Geometric Analysis. These hand-written courses were first printed in 1842 in Paris, then translated and printed in Arabic in 1844. The other parts of his mechanics (general mechanics, particular mechanics of solid or flexible bodies, hydraulics) would follow shortly afterwards (Crozet 2008, pp. 244–254).

Some disciplines such as machine theory, hydraulics or what would become the strength of materials would be directly encompassed by rational mechanics. As a result, descriptive geometry would gradually lose its preeminent position in Būlāq. Nonetheless it would remain an indispensable discipline until the British takeover at the end of the century.

Broadly following the *École centrale*'s curriculum would have the effect of introducing into Egypt the course taught by Theodore Olivier, who then held the chair of descriptive geometry within the Parisian school. A particularly remarkable sign of the translation movement's responsiveness is the translation and lithographic printing in 1842, in Egypt, of a version of his textbook dated from 1841 (Lambert just asked for all the new courses of the *École centrale*), whereas this textbook would be printed in Paris in letterpress only in 1843. The translator, Ibrāhīm Ramaḍān, would then teach this discipline for close to a decade.

5 The Contribution of Ibrāhīm Ramaḍān (c. 1815–1864/1865)

Hailing from a village in the vicinity of Zagazig, in the Nile delta, Ibrāhīm Ramaḍān was sent to study in France in the late 1820s. After completing his secondary and preparatory classes, he attended courses at the *École polytechnique* for several months as a non-matriculated student, along with three other members of the Egyptian mission. For reasons that remain elusive, all of them returned to Egypt around 1836 without having completed their studies.

Ibrāhīm RamaḍānRamaḍān was then dispatched to the Turā Artillery Academy where he was tutored by Muḥammad Maẓhar, who had completed his studies at the *École polytechnique* and *École des ponts et chaussées* alongside Muḥammad Bayyūmī. This arrangement enabled him to complete his instruction.

Towards 1837, he was appointed to teach descriptive geometry, stone-cutting, frameworks and topography at the Engineering School in Būlāq, where he furthermore performed with his students topographical surveys and levelling, some of which would later be used by the *Société du Canal de Suez*. Then in 1848 he was transferred to the Bureau of Translation, in order to supervise the translators' work. In 1854, he joined viceroy Sa'īd's staff, before dying in 1864 or 1865 (Crozet 2008, p. 469).

While teaching at the Engineering School in Būlāq, Ramaḍān was responsible for several publications bearing on the disciplines he taught. Aside from Olivier's text, and at a time when the first graduates of the Language Faculty founded by Rifā'a were not yet operational, he translated:

- *L'art de lever les plans, arpentage, nivellement et lavis des plans* by Thiollet, lithography, circa 1841;
- *L'art de lever les plans, appliqué à tout ce qui a rapport à la guerre, à la navigation et à l'architecture civile et rurale* by Verkaven (printed in Būlāq in 1844).

Later he composed—in French—a fairly comprehensive descriptive geometry, inspired by Olivier and Leroy's textbooks (on Olivier and Leroy's textbooks see Barbin, Chap. 2, this volume). He translated the first part and entrusted the remaining two to an alumnus of the Language Faculty, Manṣūr 'Azmī. Finally, he selected excerpts from Leroy's *Traité de stéréotomie* translated by Ṣāliḥ Maǧdī, from which two publications would be produced: one in 1852 dealing with the theory of shadows and perspective, the second in 1853/54 pertaining to stone-cutting and to frameworks (Crozet 2004).

Not much can be said about the translation of Olivier's text, except that it is a faithful and exhaustive translation of the first part of the treatise entitled "*Du point, de la droite et du plan*". The selection of this textbook resulting, as noted earlier, from Lambert's favouring of the *École centrale*, and with the exception of terms like "*rabattement*" most of the other terms are already used in Duchesne's textbook.

Let us dwell for a while on the second *Géométrie descriptive* published almost a decade later at the behest of the new director of the Engineering Faculty, 'Alī Mubārak, a former school-student of Lambert's who completed his education at the *École d'artillerie et du génie of Metz*.

The salient issue at hand is that it isn't a straightforward translation of French texts—as was the case with the two preceding publications. Ibrāhīm Ramaḍān's previous experience allowed him to make choices, and to deliver a textbook in keeping with the ideas he had concerning the teaching of this discipline at the Engineering School in Būlāq. Nonetheless, as we shall see, the mainstay is still very much akin to its French origins.

The treatise comprises three parts. The subject matter of the first deals with the same topics as the first part of Olivier's course, i.e. "*Le point, la droite et le plan*", and is quite obviously inspired by it. But on the whole, the result is considerably streamlined, as if to adapt itself to the framework of book I of Leroy's treatise, whose second edition is published in 1842 (Leroy 1842). As in Leroy's first book, the first part of Ramaḍān's textbook also includes 3 chapters, which in fact coincide with the first, third and fourth chapters in Olivier's course. The French authors' chapter titles are similar but not identical, and Ramaḍān blends both sources. Hence:

Olivier:	I. Preliminary Notions.
	III. Problems on the point, the line and the plane.
	IV. Of trihedral angles and pyramids.
Leroy:	I. Preliminary Notions.
	II. Problems on the line and the plane.
	III. Resolution of the trihedral angle.
Ramaḍān (literal translation):	I. Preliminary Notions.
	II. Problems on the point, the line and the plane.
	III. Resolution of the trihedral angle.

Fewer in number, the diagrams are almost all taken from Leroy's treatise, and some are also to be found in Olivier's Atlas. Concerning the text however, the situation is more complex. Hence, Chap. 1 is an almost literal translation of Olivier's Chapter 1, only slightly modified to accommodate Leroy's figures. Similarly, Chap. 2 is taken from Olivier's chapter III, albeit with only thirteen of the original forty problems included. In contrast, Chap. 3 is taken entirely from Leroy's treatise (Fig. 20.1).

The two remaining parts of Ibrāhīm Ramaḍān's textbook, which are more ambitious, are also inspired by Leroy's treatise. The second part is, in its quasi-

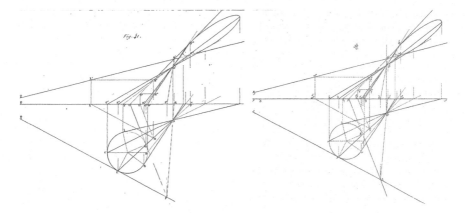

Fig. 20.1 Figure 41 from Leroy's treatise and figure 46 in Ibrāhīm Ramaḍān's textbook: By a given point on a conical surface, we propose to conduct a plane tangent to it (Leroy 1842, II plate 13; Ramaḍān 1852, II plate 13)

entirety, the French authors' book II (On surfaces and their tangent planes) and in an abridged form book III (On developable surfaces and on envelopes) and book VI (On surface intersections). As for the third part, it spans practically all of book V (On tangent planes whose point of contact is unknown) and in a condensed form, parts of books IV and VII (Miscellaneous questions; skew surfaces).

With this text, elaborated by such an obviously well-informed scholar, one who could meld his sources into a perfectly coherent whole, we behold what constitutes along with Ṣābir Ṣabrī's future textbook, one of the highpoints of what was published in Egypt concerning descriptive geometry. This textbook would have a great impact on the teaching of this discipline, both because of the scope and organization of the knowledge offered, and for the extension of the Arabic lexicon of the discipline which it allowed to fix (envelopes, skew surfaces, etc.).

However these were difficult times. To wit, 2 years later with the advent of Sa'īd, preparatory and secondary schools were shut down. And although, in 1857 new military academies, including a Military Engineering Academy were chartered, their aspirations and levels of science teaching were markedly inferior to those at the old Civil Engineering Faculty. So that it would not be before end the 1860s that this discipline would be taught again, and the dawn of the next decade before new textbooks were published.

6 The Contribution of Aḥmad Naǧīb

In 1866, 3 years after Sa'īd's successor Ismā'īl had become viceroy, the Egyptian Engineering School was re-established in Cairo under the leadership of an astronomer, Ismā'īl Muṣṭafā al-Falakī, who had spent a long time in Paris. After an initial few years during which the military academies' preponderance was much greater than it had been in the 1840s, an almost identical version of the previous educational system was recreated incrementally. A few teachers from the Lambert and 'Alī Mubārak schools regained their previously held positions in the faculty, and formerly abandoned textbooks were reclaimed.

Aḥmad Naǧīb, who published two descriptive geometry textbooks at this time, had never been a student of, or taught at the Engineering Faculty. He did however spend part of his teaching career in military academies where he himself had been a student.

Little is known about Aḥmad Naǧīb (Crozet 2008, p. 454). In March 1866, he enrolled at the Artillery School, where along with three other classmates, he was chosen to be sent to Paris to study at the *École des mines*. After a few months of preparatory classes at the Staff-officers' Academy, they were sent off to France before the end of that year. Owing to their lack of proficiency in both French and Science, their supervisory authority postponed their enrolment at the *École* until the end of 1867. By the end of 1869, two of his classmates had returned to Egypt, having failed their exams. Unlike Aḥmad Naǧīb, who was admitted for enrolment at the *École normale supérieure*, Joseph Serret, whom he had just met on this occasion, advised him to move towards pure mathematics.

Nevertheless, he returned to Cairo shortly afterwards, in all likelihood because of the turmoil wracking the besieged French capital. He was then quickly appointed as a teacher of mathematics at the Artillery and Staff-Officers Academies. These pursuits prompted him to publish two descriptive geometry textbooks in 1873 for students in military academies: a comprehensive translation of Gérono and Cassanac's *Élémens de géométrie descriptive*, and a small pocket textbook, a compendium of the discipline's major principles. Again in 1877 he translated a compilation of French textbooks for a preparatory school program, covering the same topics as those to be found in Legendre's *Élémens de géométrie*, together with texts on conics and other common curves (Šinān 1881). We lose sight of him shortly thereafter.

The motivations for translating Gérono's textbook are probably of a similar nature to those that drove Bayyūmī to translate Duchesne's textbook. In both cases these textbooks were designed for preparatory school students, and we can only surmise that Gérono's book was instrumental in Aḥmad Naǧīb's mathematical training. Conversely, Gérono's text is not as succinct and elementary as Duchesne's, and more in tune with the then currently available textbooks. The book comprises two parts. The first follows the same progression, although in a more detailed manner, as Leroy's and Ramaḍān's (preliminary notions, problems on the point, the line and the plane, trihedral angles). The second covers simple cases of tangential planes and surface intersections. Without explicitly introducing new subjects, this useful publication does however provide new exercises and viewpoints, which flesh out the available literature.

As for the pocket textbook, it adopts the form of a compendium (*mulaḫḫaṣ*). Had Aḥmad Naǧīb used the similar terms *talḫīṣ* or *muḫtaṣar*, he would have explicitly continued the tradition of a kind of textbook that was used quite widely in the instruction of many disciplines in the Arabic and Muslim world. Even though the result is the same, an ambiguity remains concerning his motivations. Of course the content has little to do with tradition, it spans the same program as Gérono's treatise and is divided into two parts, with identical chapter headings. But the similarities end there, as much for the text as for the figures, which are quite dissimilar. It is quite possible that the author used one or more French sources. But considering the fact that he made a point of explicitly mentioning Gérono as the author of his other publication, in this case he mentions no one (let us recall that neither Duchesne, Olivier or Leroy are credited by Bayyūmī or Ramaḍān); and even if compendiums had been compiled in French mathematical literature, they appear to have had little impact in the field of descriptive geometry; it would seem that we are most probably dealing here with an original composition, bearing witness to the naturalization of this discipline on the banks of the Nile.

7 The Contribution of Ṣābir Ṣabrī (c. 1853–1915)

At the same time as the publishing of Aḥmad Naǧīb's work in 1873, the first part of Ibrāhīm Ramaḍān's textbook including a few additional problems was republished for the benefit of civilian schools, and more particularly for preparatory schools. The unpublished second and third parts of the same textbook were probably used as course material at the Engineering Faculty. Be that as it may, it was at this school that the most advanced features of the discipline were taught, and that the next textbook written by Ṣābir Ṣabrī was published. He was one of the schools' foremost students in the early 1870s. After graduating, he seems to have taught for some time at the Faculty of Surveying and was then appointed, in the early 1880s, Professor of descriptive geometry and applications at Cairo's School of Engineering. Assigned to the institutions' sub-directorate, he remained there until 1892. He was then nominated chief engineer to the general directorate of the *waqfs* until his retirement, circa 1907. Relatively uninvolved as a member of the Egyptian Institute, he endeavoured to promote the teaching of sciences until his death in 1915; be it as a 23 year member of the executive board of the *Ǧama'iyya alḫayriyya alislāmiyya*, a charitable organization for the creation of elementary schools; or as a lecturer at the short-lived and privately owned University of Cairo in 1909–1910.

Among many other things, he wrote a treatise on geometrical curves initially intended for preparatory schools, and another quite sizeable treatise on descriptive geometry, which of course warrants perusal.

It comprehends two parts and an atlas. The first part, equally geared to preparatory schools, includes two chapters. The first spans the content of Ramaḍān's text and book I of Leroy's treatise and is structured in five sections: projection methods, various problems on lines and planes, changing projection planes and related problems, problems with the resolution of the trihedral angle, and regular polyhedrons. The second chapter concerns curved lines, surfaces and their generation, overall tangent planes, and the determining of tangent planes with cylinders and cones.

For his first part, Ṣābir Ṣabrī seems to have used a variety of different sources. The first to come to the fore are of course Leroy's and Ramaḍān's treatises (Fig. 20.2). Accordingly, the first illustration is identical in all three cases, as well as numerous passages in Chap. 2, with problems such as determining the tangential plane to a given point on a cone, where the three texts and illustrations are indistinguishable.

But the correlations end here. The very structure of the text is very dissimilar, especially in the much more detailed Chap. 1. Ṣābir Ṣabrī had obviously examined more recent French textbooks such as Gérono's, and many others as well, and he probably wrote a good number of passages himself, which lend a more personal touch—unlike any of the texts that preceded his in Egypt.

For the second part (Chaps. 3–7), which was written for the Engineering Faculty, the situation differs considerably. In this case Ṣabrī essentially retraces Leroy's treatise. Thus his Chaps. 3–5 (developable surfaces, surface intersections, tangential planes who's point of contact is unknown) coincide precisely with books III to IV of

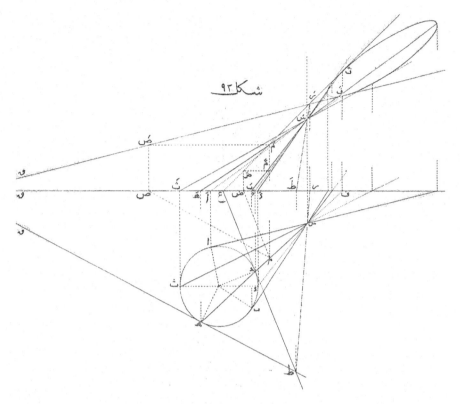

Fig. 20.2 Figure 92 of Ṣābir Ṣabrī's treatise to be compared with figures 41 from Leroy's treatise and 46 from Ibrāhīm Ramaḍān's textbook (cf. Fig. 20.1); note that the Arabic lettering is slightly different from Ramaḍān's (Ṣabrī (1881–88), II plate 22)

the French textbook, and thus even more so than in Ramaḍān's rendition. Chapter 6 is Leroy's book VII (of skew surfaces) and Chap. 7 blends the passage concerning epicycloids and part of book IX dealing with cogs. With the exception of book VIII on the curvature of lines and surfaces, almost all of Leroy's treatise is thus rendered in Arabic.

We are therefore in a similar context to Ibrāhīm Ramaḍān's: a skilful compiler who mastered the discipline at hand, and was capable of making sometimes challenging knowledge available in his country's language. However, one difference must be pointed out: unlike Muḥammad Bayyūmī, Ibrāhīm Ramaḍān and Aḥmad Naǧīb, Ṣābir Ṣabrī, although perfectly fluent in French, never travelled to France. He was entirely trained in Egypt, and kept largely abreast of the evolutions of the programs being taught in Europe. To a certain extent, this is an additional and decisive step in the transmission and appropriation of descriptive geometry that had been initiated several decades earlier.

Yet if an exception is made for Maḥmūd Fahīm's textbook, which was intended for the Arts and Crafts School, and which is much more elementary (and only mentioned earlier for the record) Ṣābir Ṣabrī's work would be the last treatise published in Egypt on the subject of descriptive geometry. The takeover of the educational system by the British would rapidly impose simpler and differently structured programs, where this discipline would be irrelevant.

8 Conclusion

From the intervention by Muḥammad Bayyūmī to that of Ṣābir Ṣabrī, most of the knowledge pertaining to descriptive geometry was introduced into Egypt and expounded upon in books written in Arabic. A perfectly stable language was elaborated, rendering all the subtleties of the discipline, which found its indisputable legitimacy in Egypt, thanks to its application (stone-cutting, frameworks, dimensioned drawings, machines, surveying, etc.).

It is a question of options: descriptive geometry's position in a given curriculum depended on the kind of educational model selected. And which textbook to translate also depended on the choices made by the French authors themselves, concerning the structure of their presentations and the methods they employed. How then can we reflect on the destiny of these models and understand how they operate? In this particular case, the crucial point, of course, is to understand how they managed to migrate across borders.

One should note that each model, each option or each bias, was often the fruit of observation, debate or more widely of historical developments particular to the advancement of sciences and general education in Europe, and therefore not immediately apprehensible elsewhere. And so the debate on the place of mathematical analysis in an Engineering curriculum did not cross the Mediterranean at the same time as the programs and textbooks that were spawned by them. In short, when European experts in Cairo recommended certain educational models, they were making choices whose premises were not readily perceivable on the banks of the Nile, but were all the more willingly accepted as the systems that were derived from them *worked*.

In fact, dislocated to another environment, these models mutate and take on another value. This is an important point to be taken into account, if one is to evaluate their future. Indeed, since Cairo's Engineering School was the only school where modern mathematical sciences were taught in Egypt, it seems plausible that it should have been looked upon, in certain respects, as a "temple of modern science". The Egyptians themselves often perceived it in this way—as did Ismāīl Muṣṭafā, the school's director from 1866 to 1887, when recalling his student days there (Crozet 2008, p. 193). In contrast, this appellation is scarcely applicable to Paris' *École centrale*, which could more fittingly be called, as Lambert did, an "*École polytechnique industrielle*". In addition, problems arose in Egypt that could not be resolved, or at least not in those terms, by European scientific authorities. These included questions surrounding the legitimacy of these novel disciplines, which were paramount in the grasping of scientific and technical endeavours in Egypt at that time, or the rebuilding of an Arabic scientific idiom.

The same holds true for the selection of textbooks to be translated. Of course, Lambert could have personally recommended Leroy's treatise, and the choice of Olivier's courses is a consequence of his opting for the *École centrale* model. But it is uncertain whether the nature of his choices was necessarily understood. What imports is rather that texts were made available for translation, such as Duchesne's and Gérono's textbooks, for more contingent reasons. Which is not to say of course, that Egyptian teachers of descriptive geometry did not have their own take on this discipline, Ibrāhīm Ramaḍān and Ṣābir Ṣabrī, for example, doubtlessly made choices in accordance with their own objectives. But with the exception of Aḥmad Nağīb's pocket textbook and perhaps the first chapter of Ṣābir Ṣabrī treatise, sufficiently original texts are not available in order to truly appreciate the characteristics of a nascent tradition, which British rule unfortunately nipped in the bud.

References

Alleaume, Ghislaine. 1987. La naissance du fonctionnaire. In *Égypte-Recompositions, Peuples méditerranéens*, vols. 41/42, 67–86.

Aubuisson de Voisins, Jean-François (d'). 1834. *Traité d'hydraulique, à l'usage des ingénieurs*. Paris: F. G. Levrault.

Comberousse, Charles (de). 1879. *Histoire de l'École Centrale des arts et manufactures, depuis sa fondation jusqu'à nos jours*. Paris: Gauthier-Villars.

Crozet, Pascal. 1996. Les mutations de la langue écrite au xixe siècle: le cas des manuels scientifiques et rechniques. In *Égypte/Monde arabe*, vols. 27–28, 185–211.

———. 2004. Entre science et art: la géométrie descriptive et ses applications à l'épreuve de la traduction (Égypte, xixe siècle). In *Traduire, transposer, naturaliser: la formation d'une langue scientifique moderne hors des frontières de l'Europe au xixe siècle*, ed. Pascal Crozet, Annick Horiuchi, 171–200. Paris: L'Harmattan.

———. 2008. *Les sciences modernes en Égypte. Transfert et appropriation (1805–1902)*. Paris: Geuthner.

Duchesne, Émile. 1828. *Éléments de géométrie descriptive*, 1ère ed, Paris: Mahler.

Dupont, Jean-Yves. 2000. Les cours de Machines de l'École polytechnique, de sa création jusqu'en 1850. *Bulletin de la Sabix* 25: 3–5.

Gérono, Camille-Christophe, and Eugène Cassanac. 1866. *Élémens de géométrie descriptive*. Paris: Delagrave et Cie.

Hachette, Jean-Nicolas-Pierre. 1819. *Traité élémentaire des machines*, 2nd ed. Paris: Vve Courcier.

Lambert, Charles. 1836. *Rapport de Ch. Lambert au vice-roi d'Égypte sur l'École Polytechnique de Boulac*. Paris, Bibliothèque de l'Arsenal, Fonds Enfantin, MS 7746/2.

Lanz, José María, and Agustín de Betancourt. 1819. *Essai sur la composition des machines*, 2nd ed. Paris: Bachelier.

Leroy, Charles-François-Antoine. 1842. *Cours de géométrie descriptive*, 2nd ed. Paris: Bachelier.

Olivier, Théodore. *Cours de géométrie descriptive, cours autographié de l'École centrale des arts et manufactures, rédigé par P.-A*. Casimir de Paul (1840–1841), no indication of place and date.

Ramaḍān, Ibrāhīm. 1845. *Al-la'ālī' al-bahiyya fī al-handasa al-waṣfiyya*. Būlāq: Maṭba'a Būlāq.

———. 1852. *Al-rawḍa al-zahriyya fī al-handasa al-waṣfiyya*. Būlāq: Maṭba'a al-Muhandisḫāna al-ḫidīwiyya.

Ṣabrī, Ṣābir. 1881–88. *Al-barā'a al-mašriqiyya fī 'ilm al-handasa al-waṣfiyya*. Būlāq: Maṭba'a Būlāq.

Šinān, Ṣādiq Sālim. 1881. *Al-nuḫba al-saniyya fī uṣūl al-handasa*, translation Aḥmad Naǧīb, 2nd ed. Būlāq: Maṭba'a Būlāq.

Chapter 21
The Dissemination of Descriptive Geometry in Latin America

Gert Schubring, Vinicius Mendes, and Thiago Oliveira

Abstract The intention of this chapter was to analyse the dissemination of descriptive geometry within the entire region of Latin America in the broad sense, thus of all the countries in the Americas with either Spanish or Portuguese as their main language. Yet, despite the decisive difference between the former Spanish colonies and Brazil as a former Portuguese colony, in that higher education became established in the Spanish colonies in the sixteenth century—in contrast to Brazil, without such structures—the history of this dissemination is poorly researched for the former region while there is pertinent research for the latter.

The chapter will therefore describe this situation for the former Spanish colonies and analyse in more detail the development in Brazil from 1810, where a net of institutions teaching descriptive geometry emerged.

Keywords Argentina · Brazil · Colombia · Descriptive geometry · Monge · Latin America · Polytechnic school · Professional training · Secondary school

1 Introduction

"Latin America" is used here in two different meanings: firstly, the entire region of Central and South America, with both Spanish and Portuguese speaking countries; secondly, the more restricted meaning of just the Spanish speaking countries in this region.

G. Schubring (✉)
Universidade Federal do Rio de Janeiro, Instituto de Matemática, Rio de Janeiro, Brazil

Universität Bielefeld, Fakultät für Mathematik, Bielefeld, Germany
e-mail: gert.schubring@uni-bielefeld.de

V. Mendes
Universidade Federal Fluminense, Niterói, Brazil

T. Oliveira
Colégio Militar de Rio de Janeiro, Rio de Janeiro, Brazil

© Springer Nature Switzerland AG 2019
É. Barbin et al. (eds.), *Descriptive Geometry, The Spread of a Polytechnic Art*, International Studies in the History of Mathematics and its Teaching, https://doi.org/10.1007/978-3-030-14808-9_21

Using the restricted meaning, one has to emphasise a decisive difference between Brazil and Latin America regarding the development of mathematics in general, and a particular difference regarding the reception of descriptive geometry. While various regions in Latin America were endowed with universities from the early colonial period on, universities in Brazil were founded only from the 1930s. There, from the end of the colonial period, in 1808, and particularly after independence in 1822, higher education adopted the French model of *écoles spéciales*. The Military Academy and the later polytechnic schools were of higher education level and taught descriptive geometry.

2 The Countries of Spanish Latin America

Unfortunately, there exist very few studies dedicated to the history of mathematics in the many countries of Latin-Spanish America. In fact, Juan José Saldaña, the editor of the first volume studying the history of science as a whole in Latin America, deplores that this history had "remained hidden" and wonders "why have historians, as a rule, not studied it?", characterising the few "historical studies about local science" basically as "laudatory histories, chronologies of events, and commemorative accounts" (Saldaña 2006, p. 5).[1] Therefore, some hints can be given for only three countries: Colombia, Argentina and Chile. In Bogotá, the capital of **Colombia**, one of the greatest countries in Latin America, a *Colégio Militar* was founded in 1848. The first 3 years of studies were devoted to mathematics. It comprised a broad program, in particular:

– geometría descriptiva y sus aplicaciones a las sombras. [shadows]

To open the school, an enormous number of mainly French textbooks were bought from Europe. For descriptive geometry, three textbooks were favoured:

– Gaspard Monge, *Géométrie Descriptive* (Monge 1799)
– Louis-Léger Vallée, *Traité de géométrie descriptive* (Vallée 1821), and
– Mariano de Zorraquin, *Geometria analitica-descriptiva*. Alcalá (1819).[2]

In 1867, the school was renamed *Colégio Militar y Escuela Politechnica*, and in 1868 it became the *Faculdad de Ingenieria* of the newly founded Universidad Nacional in Colombia. Thus, the so far missing function of mathematics in higher education became now attributed to providing the foundations for studies of engineering.

---

[1] Carvalho, in his chapter on mathematics education in Latin America in the nineteenth and twentieth centuries, invested enormous efforts to detect local histories for the various countries; it proved to be enormously difficult to obtain relevant information (Carvalho 2014).

[2] Regarding Zorraquin's textbook (see Ausejo, Chap. 5, this volume).

Interestingly enough, at the end of the 1880s heated debates emerged about this function of mathematics. In an analogous anti-mathematical movement,[3] the amount of theoretical mathematics to be taught was questioned and required to be restricted to what should be strictly necessary for engineers (Sanchez 2007).

In **Argentina**, the *Universidad de Buenos Aires* was founded in 1821. It had six departments, one of which, the "Estudios Preparatorios", included Latin, modern languages, philosophy and "fisico-matemáticas". Within its department of "Ciencias Exatas" there were the disciplines of *cálculo y mecanica* and, in particular, *geometria descriptiva con sus aplicaciones*. The teaching of descriptive geometry began there in 1822. Its professor from 1827 to 1830 was Romano Chauvet who is said to have been a student of Lacroix and of Cauchy. His successor in 1830 was Avelino Diaz (Venturini 2011). More information is unfortunately so far unavailable.

The third country is **Chile**, where it was also a Frenchman who tried to innovate education: Charles Ambroise Lozier (1784–1864), hired in 1822 to establish a map of Chile. He was an engineer who had studied at the *École polytechnique*. In 1813, an Instituto Nacional had been founded for secondary and higher education. Lozier became its director in 1826 and tried to modernise it. It was modelled according to a French *lycée*; among others, non-elementary mathematics was taught. Lozier was very active in introducing modern French textbooks to Chile and commissioned translations of them. One of these translations was that of Leroy's *Traité de géométrie descriptive* (Carvalho 2014, p. 340). Lozier had to resign after a few years, however, due to a rebellion by students and parents against the high standards of learning he wanted to realise.[4] The translation extant in the *Biblioteca Nacional de Chile* dates of 1845:

> Tratado de jeometria descriptiva: acompañado del metodo de los planos de acotacion de la teoria de los encargantes cilindricos y conicos con una colección de depurados compuesta de 69 láminas; escrita en francés por C.F.A. Leroy; traducida de la segunda edición por D. Andrés Antonio de Garbea. Santiago: Impr. del Progreso.

3 Brazil

The institutional development in Brazil shows the same pattern as in Colombia: the first modernising institution was mainly for military engineers; the institution later on splits into a military and a civil part; and the part for forming civil engineers becomes one of the germs of university institutions.

---

[3]For anti-mathematical movements in other countries, see for Germany (Schubring, Chap. 22, this volume), and for Spain the paper by Ausejo (2006).

[4]Communication by Pitombeira Carvalho.

3.1 The Military Academy

While Brazil had been kept by the Portuguese government in a state of underdevelopment since the sixteenth century, serving almost exclusively for extracting raw materials (like *pau Brasil*, a Brazilian typical redwood), a decisive change and rupture occurred in 1808 at the arrival of the Portuguese court, which had fled an imminent invasion of Portugal by Napoleonic troops (see Pinho et al., Chap. 7, this volume). For the first time, a printing press became established and institutions for higher learning were created. The first one was the *Academia Real Militar*, created on 4 December 1810, and teaching military and civil engineers from 1811. The Royal Decree defined not only the teaching subjects, based on "a complete course of Mathematical Sciences" and named the professors who should give these courses, but it also stipulated the textbooks to be used. And since none of them were in Portuguese, the decree required translating them. For mathematics, the following textbooks were used: *Éléments de géométrie* by Adrien-Marie Legendre,[5] *Traité élémentaire de calcul différentiel et intégral* by Sylvestre-François Lacroix and the *Géométrie descriptive* by Gaspard Monge. At the same time, the professors were urged to publish their own lecture courses ("compendios") (Saraiva 2007).

Almost all the professors of the new Academia had been formed in Portugal, at the University of Coimbra or at the *Academia Real dos Guardas Marinhas*. Descriptive geometry had to be taught in the second year of the study course. The first professor of descriptive geometry at the Academy was José Vitorino dos Santos e Sousa (?–1852), who had graduated from the Mathematics Faculty of the University of Coimbra and was an officer of the Corps of Engineers. His translation *Elementos de Geometria Descriptiva* is based on the 1799 first-edition of Monge's textbook and was published in 1812 by the newly created Royal Press.[6]

About his contribution to such a text, Vitorino Souza said that his desire was

> to cooperate to raise the empire of the sciences and of the finearts in this new world, offering many natural resources for their application to industry, and for improvements of the arts, which are the springing of the great social *machina*. (Sousa 1812, p. xix; transl. by Oliveira)

Vitorino's table of content is a literal translation of Monge's 1799 table (Sousa 1812, pp. 240–244). The translation of Monge by Vitorino was made very accurately; he added an extensive commentary. Although there are some problems in rendering the new French terminology into Portuguese, almost all of the text is a literal translation, and no part of the original text was omitted. The decree establishing the creation of the Academy determined not only the translation of the textbooks, but required their modification as new discoveries would be made and so Vitorino added 27

---

[5]Legendre's Geometry was already translated in 1809 by Manoel Ferreira de Araújo Guimarães. This same year, Legendre's appendix was published as *Tratado de Trigonometria*, and Euler's algebra as *Elementos d'Algebra*.

[6]Strangely, Saraiva in his excellent paper on the early history of mathematics at the Academy (Saraiva 2007) does not mention the discipline and the professor of descriptive geometry—as if he would not count this discipline as part of mathematics.

pages of notes to his translation of Monge. In their introductory statement, Vitorino exposes the rationale for the use of double projection both from a theoretical point of view, due to the possibility of studying properties of geometric figures, and from a practical point of view, for its use by artists and architects. Then, the author discusses the insufficient notion of distance for two fixed points, or of the distance for two straight lines when determining the position of a point in the plane. He argues, after the previous discussion, "it will be very easy now to apply these considerations to space" (Sousa 1812, p. 211), as does Monge in his introduction. Vitorino includes a discussion about the generation of cylindrical and conical surfaces, as surfaces generated by a straight line. Vitorino also extends to treating the generation of surfaces, including those of revolution, as a way to characterise them.

The notes and additions by José Vitorino did not include new information to Monge's text from a theoretical point of view. They were a commentary intending to overcome difficulties, which a reader less familiar with geometry might have to understand Monge's text.[7]

3.1.1 Monge versus Lacroix

There is a strange element in the history of teaching descriptive geometry in Rio de Janeiro: although Monge's textbook was prescribed in the royal document, and although it was in fact translated and published, the translator, himself the professor of descriptive geometry, changed at a yet unknown point to the textbook by Lacroix:

Essais de géometrie sur les plans et les surfaces courbes (Éléments de géométrie descriptive) (Lacroix 1795).

It is not known whether Lacroix's book was translated—in fact, no copy of Lacroix's book has been found so far in libraries in Rio de Janeiro. It seems that not only during Vitorino's time but also beyond, Lacroix's book was the official textbook. There is evidence for this, from 1831, 1836 and 1837. In 1836, the Minister of War, to whose operational area the school belonged, asked the mathematics professor José Saturnino da Costa Pereira to give a report on the textbooks in use at the Academy; Saturnino emphasised that Lacroix's textbook on descriptive geometry was "the most difficult of the textbooks written by its author, and the least suited for elementary teaching" (Mormêllo 2010, p. 121). And for 1837, one has the list of textbooks in use, in particular for the second year—there still figures Lacroix's textbook. One can therefore assume that Lacroix's book was used for teaching in Rio de Janeiro for an extended period (Fig. 21.1).

In general, descriptive geometry seems not to have had an easy status at the school. In 1831, the military junta, responsible for administering the Academy, said that the professorship of descriptive geometry was vacant, and that of the thirteen

---

[7]Besides Monge, Vitorino also translated Lacroix's *Application de l'Algèbre à la Géométrie* and published in 1832 an own textbook: *Geometria e Mecânica das Artes, dos Ofícios e das Belas Artes*. He retired at the Academia, some years before 1840.

Fig. 21.1 Textbooks used by the second professor of the Academia Real Militar in 1837. AN 1

teaching subjects, only four had properly qualified professors. Moreover, Miranda (2001) and Mormêllo (2010) indicate another problem that brought consequences for geometry: it was one of the tasks of the professor of descriptive geometry to substitute any of the so-called mathematical courses in case of impediment of their professors.

3.2 An Interplay: For Architects and Surveyors

There was a revealing effect, notwithstanding, of the descriptive geometry courses at the *Academia Militar*. A graduate of the Academia, Pedro d'Alcântara Niemeyer Bellegarde (1807–1864), published in 1840 the first textbook in Brazil on descriptive geometry: *Noções de geometria descriptiva para uso da escola de architectos medidores*. It was a booklet of only 27 pages; unfortunately, no copy of it has been be traced so far. The book would be telling regarding the practice at the Academia, in particular whether Monge's textbook was used by Bellegarde as source, or that by Lacroix. The book is, moreover, noteworthy because of its author: a very influential engineer, administrator and politician; a topographer having established a great number of maps for various regions in Brazil; author of many textbooks including topography, architecture, mechanics, history, etc.; director of the *Escola Central*; minister of the Empire: for War, for the Navy, and for Agriculture and Industry.[8] But the booklet is also noteworthy both due to the school, which functioned from 1837 to 1844 in Niteroi near Rio de Janeiro, and the task for which it was destined. The graduates—25 in the end—were taught surveying in the first year and engineering formation in the second and third year.[9] Hence, surveyors and architects were trained there. Remarkably, the first textbook on descriptive geometry in Germany was destined for architecture students, too (see Benstein, Chap. 9, this volume).

---

[8] Site with the history of the Niemeyer family: http://www.cbg.org.br/novo/niemeyer-na-engenharia/.

[9] "Escola de Architetos Medidores", Diario do Rio de Janeiro, Saturday, 4 March 1837. We are grateful to Bruno Dassie who informed us of this newspaper publication.

3.3 Descriptive Geometry at the Main Institution of Higher Education in Rio de Janeiro

After 1831, it is not yet known who the professor lecturing descriptive geometry at the Military Academy was, and which textbook was used after 1837. Moreover, the institution experienced various structural changes. Firstly, in 1839, it was renamed *Escola Militar*. Then, in 1855, the separation of training civil and military engineers occurred. The school for the military was also named *Escola de Aplicações* and changed its name various times, while maintaining descriptive geometry (Mormêllo 2010, p. 132). We will follow the civil school, renamed *Escola Central* in 1858 and *Escola Politécnica* in 1874. With these changes, the school moved from the War Ministry to the *Ministério do Império*, thus documenting its civil status.

Though throughout almost all of these changes, descriptive geometry was kept as a major discipline, its position in the curriculum varied somewhat. In 1833, it was taught in the second year (Mormêllo 2010, p. 113). In 1839, it was put into the third year, and renamed *geometria descriptiva e analítica* (Mormêllo 2010, p. 114). A new organisation, in 1842, implied that exceptionally there was no course of descriptive geometry. In 1845, it returned, however, and now in the second year—now named, *geometria descriptiva, e suas aplicações à Estereometria, e à Prespectiva* (Mormêllo 2010, p. 116). In the curriculum of the Escola Central of 1858, descriptive geometry continued in the second year (Mormêllo 2010, p. 129).

The next, more concrete, information on the teaching of descriptive geometry dates from 1861, in line with the structural changes: the teaching programme for descriptive geometry, found in the files of the school in the National Archives. It is rather succinct, giving just the main topics of the lecture course, but constitutes the first proper development for descriptive geometry in Brazil:

- Preliminares
- Método das projeções. Representação de um ponto.
- Representação de uma reta.
- Representação de um plano.
- Achar o comprimento [lenghth] de uma reta.
- Traços de uma reta. Ângulos que ela faz com os planos de projeção.
- Método dos rebatimentos.
- Projeção de uma curva. Exemplo do circulo.
- Representação de figuras quaisquer [whatsoever].
- Projeção de um prisma, de uma pirâmide.
- Projeção de um cilindro vertical ou inclinado, projeção de um cone.

This programme was probably established by Ignácio da Cunha Galvão, the next known professor of descriptive geometry. He studied first in Paris, obtaining the baccalauréat-ès lettres, and continued to study at the *Escola Militar* in Rio, obtaining the doctoral degree in mathematics there in 1848 with a thesis on enveloping surfaces. He taught descriptive geometry until 1882, when he became the director of the school until his retirement in 1889. After the first teaching programme of

1861, Galvão established more extensive programmes; his programme of 1878, with 120 items (Escola Polytechnica 1879),[10] will be analysed in a further section in comparison with those of the other two schools having emerged in the meantime.

His successor was João Baptista Ortiz Monteiro who had studied at the *Escola Central* from 1872, and graduated there in 1877 as a civil engineer with a doctoral degree in mathematics and the physical sciences. In 1879 he became substitute professor at the Polytechnic School and in 1882 Galvão's successor. With the permission of the Emperor, who appreciated his intellect, Ortiz Monteiro travelled abroad in order to specialise in Monge's science. According to Pardal, Ortiz Monteiro assisted higher mathematics courses in Vienna, Paris and Leipzig, returning to Brazil in 1885 (Pardal 1984). He was director of the *Escola* between 1905 and 1913. He delivered talks at scientific conferences in Latin America, was a member of the *Sociedade de Ciências*, honorary professor at the University of Zurich and a member of the *Instituto Historico* (Pardal 1984). Monteiro continued at first with Galvão's teaching programme, but changed it slightly from 1884 by introducing the study of polygons and polyhedra.

In this period, the textbooks in use were: *Elementos de Geometria Descritiva* by the collective French author F.I.C.,[11] translated by Eugênio de Barros Raja Gabaglia, mathematics teacher at the *Colégio Pedro II*, with an appendix by Ortiz Monteiro; and the French textbooks *Traité de géométrie descriptive* by C.F.A. Leroy and *Traité de géométrie descriptive* by A. Javary.

Ortiz' successor in 1913 was Henrique Cesar de Oliveira Costa (also called Costinha):

> He effected a true revolution in the processes hitherto applied to teaching Monge's science: the textbooks by F.I.C., Leroy and Javary were eliminated; the period of Roubaudi began now; the crutches ('muleta'), as he called the ground line, disappeared due to this textbook (Pardal 1984, p.139; transl. by Oliveira).[12]

3.4 More Publications and the Escola Naval in Rio de Janeiro

During the nineteenth century, there were more activities regarding descriptive geometry, besides the courses at the *Academia Militar/Escola Central*. In particular, there were a number of doctoral theses defended. And some dissertations, which were elaborated for competitions for a professor position, are extant. As mentioned previously, Ignácio da Cunha Galvão wrote a thesis on surfaces and obtained a doctoral degree in mathematics in 1848. Another example is the *These de concurso*

---

[10] The programme for the year 1876 is documented completely in Miranda (2001). For this research, the teaching programmes for descriptive geometry between 1878 and 1899 have been analysed. They are preserved in the *Biblioteca de Obras Raras of the Universidade Federal do Rio de Janeiro*.

[11] Frères de l'Instruction Chrétienne.

[12] C. Roubaudi was the author of a descriptive geometry textbook, published in Paris in 1916.

para a vaga de lente da cadeira de geometria descritiva e topografia da Escola Naval, written by João da Costa Pinto in 1898 (Pinto 1898). For this exam, the author had written 115 pages in his dissertation discussing the classification and general properties of surfaces, based on Monge's concept of generation of surfaces. The text presents no figures, no *épure*, just a long exposition of several examples of surfaces and their generation.[13]

During the first half of the twentieth century, the number of Brazilian textbooks on descriptive geometry increased. Carlos Süssekind was a teacher of descriptive geometry at the *Escola Naval* and the Military School of Rio de Janeiro. Süssekind published a textbook entitled *Geometria Descriptiva*, in 1924, and this text was based on the author's lecture notes at the *Escola Naval* (Süssekind 1933). This textbook constitutes the second textbook on descriptive geometry published in Brazil, after the first publication in 1840. A second edition was released in 1933, including the solutions of several problems that usually appeared in the exams of the institution.

Süssekind followed the conception previously discussed for the discipline. The author represents points, straight lines and planes in several particular positions. And he discusses the construction of *épures* without the *ligne de terre*. The descriptive methods presented in Süssekind's text are the change of planes, rotations and *rabattements*. Thereafter, the text discusses problems that focus on metric issues; these problems involve distances and measures of angles. Regarding the construction of polyhedra, the author represents the platonic solids and pyramids. Regarding the space curves, Süssekind presents the *épures* of the circle and the helix. After a brief presentation of basic concepts of generation of surfaces, the author describes the construction of conic sections and cylinders of revolution, as well as spheres, ellipsoids and hyperboloids of revolution. Süssekind's text also includes the solution of problems by means of the *projection cotées* and a presentation of basic concepts of perspective, shadows and drawing of projections. The edition of 1933 includes the solution of 21 problems selected from previous exams at the *Escola Naval*. Although Süssekind's text presents a huge number of topics, the book comprises 248 pages and focuses on a pragmatic approach of solutions to problems, emphasising "how" to solve but not "why" that's the solution.

3.5 Escola de Minas in Ouro Preto (Minas Gerais)[14]

While higher education in mathematics had been concentrated in the capital of Rio de Janeiro since 1810, a certain diversification occurred by the second half of

---

[13]The thesis is extant in the BOR, the Biblioteca das Obras Raras, Instituto de Matemática, UFRJ.

[14]The development of mathematics at this school and also that at the *Escola Politécnica* de São Paulo (see below) are subjects of the PhD thesis of Vinicius Mendes Couta Pereira, supervised by Gert Schubring.

the nineteenth century. The first school with courses in higher mathematics was the *Escola de Minas* in Ouro Preto, in the federal state Minas Gerais, for training mining engineers. The initiative for its foundation was taken by the Emperor Pedro II himself, who also searched and contracted the director of the school, the chemist Henri Gorceix. The school, the second to train engineers, was opened in 1876. In the following year, 1877, because of the insufficient preparation of the students by secondary schools, particularly in mathematics, a *Curso Preparatório* was introduced. In this preparatory year, mathematics courses were taught: elementary geometry, trigonometry and descriptive geometry, graphical exercises, algebra, notions of the derivative calculus, analytic geometry, and, moreover, notions of mechanics, elementary physics, chemistry and biology (Pereira and Schubring 2017).

It is noteworthy that descriptive geometry was not only taught within the higher education sections of the school, but also at the basic level and—even more remarkably—knowledge of it was even required in the exams for admission to the school (Figs. 21.2, 21.3 and 21.4).

Initially, when the study course for mining engineers lasted 2 years, descriptive geometry was taught in the first year. This continued at first when the study course was extended to 3 years in 1882, but in 1885 descriptive geometry was transferred to the second year.

The first lecturer was the French Armand de Bovet, from 1876 to 1882. He had studied at the *École des Mines* in Paris. He gave the structure of his lecture course in 1876 in French [EMOP 1]:

- Hyperboloide de révolution
- Intersection de Surfaces
- Théorie des Ombres
- Plans cotés

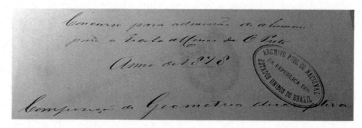

Fig. 21.2 Cover page of the files for the entrance exam in descriptive geometry at the EM in 1878 (AN 3). "Concurso para admissão de alunos para a Escola de Minas de O Preto. Anno de 1878. Composição de Geometria descriptiva"

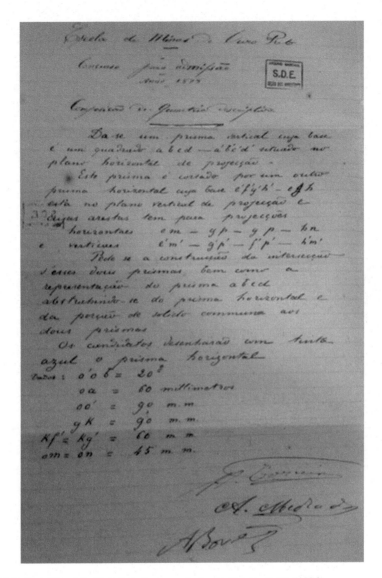

Fig. 21.3 Bovet's task for the admission exam in descriptive geometry (AN 3)

Bovet's successor was Domingos da Silva Porto (1856–?), with a degree as a mining engineer from the *Escola de Minas* itself, became professor of *Geometria Descriptiva, Estereotomia e Topografia* in 1883, and remained in this function until 1913. His teaching programme of 1885 [AN 3] will be analysed in the later section.

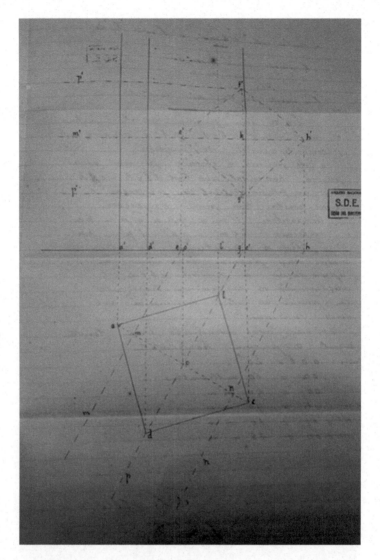

Fig. 21.4 A solution by one of the candidates (AN 3)

3.6 Teaching of Descriptive Geometry at Secondary Schools

Since descriptive geometry was required for the entrance examinations at the Mining School in Ouro Preto, one wonders where aspirants might have learned such knowledge. There is evidence that it was in fact taught in secondary schools, if only from 1894. The *Colégio Pedro II* in Rio de Janeiro, the leading secondary school in Brazil—its curriculum had to be applied by all other *colégios* that wanted to be recognised by the government—had introduced this teaching subject in 1895 (Beltrame 2000, p. 67), and its teacher was Eugênio Raja Gabaglia who had translated the textbook of the French collective F.I.C. This textbook was prescribed in the first programme for 1895 (Beltrame 2000, p. 185). The teaching programme of 1897 listed the following topics:

- Objeto e utilidade da Geometria. Definição. Método de Monge.
- Rotação de um ponto em torno de um eixo.
- Rebatimento de um plano.
- Aplicação do método de rebatimento à determinação de distâncias e de ângulos.
- Representação dos sólidos.
- Seções planas dos poliedros, do cone e do cilindro de revolução.

The following textbook was used: A. Julien, *Cours élémentaire de géométrie descriptive, conforme aux programme du baccalauréat ès sciences*. Paris 1875 (Beltrame 2000, p. 190). Descriptive geometry was maintained in the curriculum until 1930 (Beltrame 2000, p. 246). In the curricular reform of 1931 it did no longer appear (Beltrame 2000, p. 247).

3.7 The Escola Politécnica in Sao Paulo

The next important regional institution to be founded was the *Escola Politécnica* in the economically important city of São Paulo. It was founded in 1893 and began its teaching in 1894. Its director, Antônio Francisco de Paula Souza, had studied at the Karlsruhe polytechnic school in the 1860s and applied the structure of the Karlsruhe school[15]: a general course, followed by professionalised courses for the various engineering professions. Mathematics, and hence also descriptive geometry, was taught in the general course. After a change in the school structure of 1897, the discipline was called *Geometria Descriptiva e Geometria Superior*.

Its first, and almost perennial, professor was Carlos Shalders (1863–1963). He had studied at the *Escola Politécnica* of Rio de Janeiro. Shalders taught the subject there for 40 years, and from 1931 to 1933 he served as the director of the school. Among others, he published: "Elementos de geometria projectiva; versão portuguesa da tradução francesa por Ed. Dewulf, com a colaboração do autor Luigi Cremona, por C. G. S. Shalders". His teaching programme of 1899 (Annuario 1900) will be analysed in the following section.

---

[15] See Schubring (Chap. 22, this volume).

3.8 Comparing the Teaching Programmes of Rio de Janeiro, Ouro Preto and São Paulo

3.8.1 The Programme of the Rio School in 1878

In the period when the discipline in Rio de Janeiro was under the responsibility of Inácio Galvão, his programme considered more than 120 topics and detailed all theorems and problems that should be solved by the students. The course was divided into three parts: the first was the representation of the point, of the straight line, of the plane, of the intersection between straight lines and planes, of straight lines and planes determined by various conditions, of straight lines and perpendicular planes, of angles between straight lines and planes, of solving problems involving the trihedron; in the second part of the course, surfaces and tangent planes, the cylinder and the cone, the surfaces of revolution, the contact curves between two surfaces, the properties of the skew surfaces of revolution were studied; and finally, in the third part, the plane sections, the intersection of two curved surfaces, the envelopes, the helix and the epicycloid were studied.

The topics of the programme were structured similarly to Lacroix's *Essais*' table of contents. For each part, there is a list of problems and theorems that the Academy's students should learn. The part dedicated to the representation of points, straight lines and planes contains several theorems that deal with the determination of these elements by their projection on two perpendicular planes.

The programme explicitly mentions the Monge'an conception of surface that is its classification by its generation, and considers two classes of surfaces: those generated by the straight line and those generated by other curves. This theme is explored by Lacroix's *Essais* in its second part and in Vallée's *Traité* in the first chapter of *Livre II*. Yet, instead of a general approach to the theme, the Academy's programme presented theorems and problems involving particular surfaces like cylinder and cone.

The Academy's programmes presented the properties of surfaces before those of curves. Curves are studied by means of the intersections of surfaces. According to Barbin (2015), "as in the second *élémentation* introduced by Vallée, Adhémar introduced a study of curves before coming to surfaces". The Academy followed a different conception for curves, more similar to Lacroix's *Essais* that focus on the representation of surfaces and their curves of intersection.

By analysing the programmes of the discipline during the period in which Galvão was at the Academy, one observes a specific change that deserves attention: item 123 of the 1878 programme (construct the intersection curve between the torus and the plane) no longer appears in subsequent programmes. The torus is a special case, since only in some particular cases can intersection curves be constructed by ruler and compass. Since no documents were found justifying such a change, one can only infer a reason for its exclusion. The difficulty in treating such a problem from a constructive point of view lies in the choice of particular cases that can be solved by ruler and compass. The lack of generality in the processes for the solution of this problem imposes a difficulty that might be inappropriate for a basic course on the subject.

3.8.2 Comparing the Three Schools

The other schools for training engineers considered here enable to study comparatively, in the Brazilian context of the second half of the nineteenth century, the approaches and conceptions to teach descriptive geometry. The programme of the *Ouro Preto Mining School* for the year 1885, for example, was divided into eight parts. In the first part, entitled "Preliminary Notions", the representations of points, straight lines and planes, as well as the theorems on the determination of these geometric entities, were presented. In addition, straight lines and planes are studied in different positions.

The second part deals with "The problems of the straight line and the plan", including the solution of the problems of intersection between straight lines and planes. The third part is dedicated to the problems of distances. In the study of straight lines and perpendicular planes, the problems of distances between point and plane, between parallel planes, between point and straight line, and between two straight lines are solved. In this third part of the programme, "The theorem of the shortest distance between two straight lines" is presented. Although the statement of the theorem is not included in the programme, it can be inferred that it is the problem of determining the perpendicular common to two skew straight lines, treated by Monge (1799) to present the study of planes touching the curved surfaces and their normal.

> Two straight lines being given by their horizontal projections AB, CD, and by their vertical projections ab, cd; construct the projections PN, pn of their shortest distance, that is, of the straight line, which is at the same time perpendicular to both, and to find the magnitude of this distance? (Monge 1799, p. 37)

In Part IV of the programme, the method of *rabattement* is presented. In the Mining School programme, the general solution of the problem is presented, as well as the *rabattement* of a point and a line. The inverse problem and its applications are presented. In Part V, various problems are analysed in which the change of projection plane facilitates the implementation of the *épures*.

In the sixth part of the programme, various problems are analysed whose solutions involve the method of rotations. Also, various practical exercises are made, and 12 *épures* chosen containing applications of the topics covered in the course.

The seventh part, entitled "Sphere", deals with its representation as well as various problems. Theorems are shown about planes tangent to cones and cylinders of revolution, the contact curves between a sphere and a cylinder and a circumscribed cone.

The eighth and final part is dedicated to trihedrons with six cases to be resolved, a fundamental problem, direct solution of the six cases, three cases of solution considering the additional trihedron, and the reduction of an angle to the horizon.

The programme of the *Polytechnic School* of São Paulo for the years 1899 and 1900 was divided into two parts—the first devoted entirely to the problems linked to points, straight lines and planes, and the second dealing with curved surfaces. There is also a final section devoted to projective geometry.

In the first part, the programme exposes, besides conventions and fundamental theorems, a set of 53 items that must be dealt with by involving points, straight lines and planes. It gives problems of distances (including, in item 25, the problem of the perpendicular common to skew straight lines) and the determination of angles. Already suggesting at the beginning of the programme the use of the methods of *rabattement* and rotations, it includes the application to solving problems involving plane figures (polygon and circle) and polyhedra. Cubes, tetrahedrons and pyramids should be represented "in the plane and by elevation". Finally, the plane sections of polyhedra are studied and analysed as oblique sections of straight prisms, pyramids, tetrahedron.

The second part begins with the classification and generation of surfaces, their representation and determination of tangent planes. Then, the "developable" surfaces are studied. The planes tangent to cylindrical surfaces and to conical surfaces and to surfaces of revolution are studied, then the contact curves between some surfaces, the projections of the helix and developable helicoide, and the skew surfaces. Plans côtés and tracing contour lines in the landscape were treated, too.

Finally, under the heading "Higher Geometry", some topics of projective geometry are studied. It should be noted that the professor of descriptive geometry, Carlos Shalders, was the translator of Luigi Cremona's textbook on projective geometry into Portuguese. Thus, the fundamental forms are studied, the homological figures, the principle of duality, projective forms, harmonic forms, the anharmonic relation, projective forms in the circle and the conic sections, including Pascal's theorems, Brianchon, Möbius, MacLaurin and Apollonius, the theorem of Desargues, poles and polar axes. The introduction of projective geometry in the lecture course of descriptive geometry of the Polytechnic School of São Paulo was due to Shalders' conception for his teaching that was, in turn, influenced by the Italian tradition stemming from the work of Cremona.

The programme of the *Polytechnic School* of Rio de Janeiro for the year 1898 was divided into three parts. The first, following the same design from the other schools already presented, was the representation of point, straight line and plane. The main theorems on the determination of these subjects were presented, as conventions and the study of different positions. Two types of problems were considered: descriptive problems and metrical problems. In the first, the construction of points, lines and planes under various conditions were treated. In the second type, problems were solved regarding the determination of measures of segments and angles. The methods mentioned in the programme were rotations and *rabattements*. Changes in the ground line parallel to its original position and its effect on *épures* were discussed as well as their suppression. The study of polygons and polyhedra was made at the end of the first part, in particular for prisms and pyramids. The plane sections of these solids were studied according to the concept of plane homology.

The second part of the programme dealt with curves and surfaces. In the study of curves, the programme required their representation, the tracing of tangents and normals, and the tracing of tangents by means of the *error curves* and of the remarkable and singular points in the projections of the curves. The curves that should be studied were not specified. For surfaces, the programme described their classification according to the nature of the generating element. Theorems about tangents planes had to be taught, as well as notions on surfaces considered as envelopes, problems about planes touching cylindrical and conical surfaces, hyperboloids of one sheet and the hyperbolic paraboloid, and eventually problems on planes touching surfaces of revolution and contact curves between some surfaces.

Finally, the third part presented the intersections of surfaces, plane sections of cylinders and conics, notions of enveloped surfaces, helices, helicoides, geodetics and osculating and epicicloidal planes. Furthermore, the *plans côtés* and contour lines were studied.

Some important aspects appear in the analysis of the teaching programmes of these three scientific institutions. First, the initial approach to descriptive geometry is based on the study of particular positions of points, straight lines and planes. Problems involving these three geometrical elements are found in all analysed programmes and the order: point–straight-line–plane is adopted in all three institutions.

Secondly, the study of curves and surfaces also followed the approach of particular cases. Different from the notion of generality that characterises Monge's textbook, problems involving these elements are presented in the programmes. The choice of certain families of curves and surfaces and the particular sequence in which they should be studied stood in relation to the kind of professional instruction that the institutions offered. Problems involving certain polyhedra were solved with the previously developed knowledge on points, straight lines and planes. The programmes analysed also indicate that only some particular polyhedra were studied, and that the emphasis on the intersection problems reinforced the kind of knowledge considered relevant to the professional instruction provided by the Brazilian institutions. According to Barbin, this conception can be found in Lacroix (1795) since "Lacroix followed an order of simplicity of figures: points and straight lines in a plane, then in a space and then spheres" (Barbin 2015, p. 47). She also says that, contrary to Monge, he began the second part of his *Essais* "with particular kinds of surfaces (conical surfaces, cylindrical surfaces, etc.) before coming to the general conception of surfaces". From her analysis it also follows that the already mentioned characteristics of the Brazilians institutions' programmes appear in Vallée (1819) and Adhémar (1832), since both authors propose a study based on certain particular positions of points, lines, planes and surfaces.

Moreover, descriptive methods like rotations and *rebattements* were presented as basic skills that should be acquired by the students. This was reinforced by the analysis of several particular positions for the elements that should be transformed by means of the two mentioned methods. The key idea is that, by means of rotations or *rabattements*, figures could be represented in a convenient position so that it should be easy to determine the measurements of its angles and sides. Barbin shows that this is an *élémentation* introduced by Oliviers' *Leçons* (see Barbin 2015). Even in Brazilian publications on descriptive geometry during the twentieth century, one can find chapters dedicated to descriptive methods like *rabbattements*, rotations and change of planes.

A remarkable distinction is found at the Polytechnic School of São Paulo's programme. Its "Higher Geometry" described an approach that related descriptive geometry to projective geometry.

3.9 The Academia Imperial de Belas Artes in Rio de Janeiro[16]

Brazil shows hence a number of institutions for training engineers, where descriptive geometry constituted since the beginnings one of the key teaching disciplines. Moreover, there existed also a school for a quite different public, for training artists and architects, the *Academia Imperial de Belas Artes* (AIBA). Founded officially in 1816, it began to function, however, only in 1826.

Although the teaching of descriptive geometry had been introduced there formally only in 1890, from 1831 students were required to have taken courses of elementary geometry and of descriptive geometry in the *Academia Militar* in order to be admitted. In 1855, the school became structured into five sections: architecture, sculpture, painting, additional sciences and music. Among these sciences there was perspective and theory of shadows as well as linear drawing, so that some elements of descriptive geometry were already taught. In 1890, the school was renamed *Escola Nacional de Belas Artes* and restructured to provide a general course, of 3 years, and subsequently special courses of 3 years, too. Among the disciplines taught in the general course were:

- Desenho Linear, in the first year;
- Geometria Descritiva e seus respectivos trabalhos gráficos, in the second year;
- Perspectiva e Sombras e seus respectivos trabalhos gráficos, in the third year.

As a consequence, a professor for descriptive geometry, perspective and shadows became contracted. The main task was the training of architects, but his duties also included the training of teachers for drawing.

---

[16]This school and its professor Alvaro Rodrigues is the subject of the PhD thesis of Thiago Oliveira, supervised by Luiz Carlos Guimarães (Oliveira 2016).

3.9.1 Prof. Alvaro José Rodrigues (1882–1966) and His Textbook of Descriptive Geometry

Alvaro José Rodrigues was the perennial professor of descriptive geometry at the *Escola das Belas Artes*. He had studied at the *Escola Politécnica* in Rio de Janeiro, and made a study voyage to Germany in 1909 and 1910. He is said to have studied descriptive geometry at the Technical College in Berlin. Soon thereafter, in 1911, he became professor of descriptive geometry at the *Escola de Belas Artes*. Rodrigues was the third Brazilian scientist to publish a textbook on descriptive geometry.

His textbook comprised two volumes. The first one was published as *Geometria Descritiva—Operações Fundamentais e Poliedros*. It had an enormous success. Published in 1941, it had six more editions—in 1945, 1950, 1951, 1961, 1963 and 1964—and was even reprinted three times after the author's death, until 1973. It was based on his lectures in the first year of the ENBA, on painting, sculpture and engraving, and on the first year of his lecture course for teachers of drawing. The second volume *Geometria Descritiva—Projetividades, Curvas e Superfícies*, published in 1945, had further editions in 1953, 1960 and 1964, and two more reprints after the author's death. It was based on his lectures on architecture in the first year of the ENBA and the second year of his lecture course of forming teachers for drawing.

The work is remarkable not only for including historical notes, but also for presenting approaches to the teaching of descriptive geometry in use in various countries. In the introduction, he gives a short biography of Monge. Then, Rodrigues discusses the theoretical foundations necessary for the entire work. He classifies the projection of a point in the plane by two systems: as cylindrical and as conical projections. And he presents a set of definitions and theorems about straight lines perpendicular to planes, angles and dihedrals. His didactical strategy for the study of the *épures* was based on the presentation of various particular cases to represent points, straight lines and planes—thus adapting the post-Monge approach typical for France as identified by Barbin (2015). In Vallée's textbook one encounters the concept of the study of the elements via their les plus *remarquables* positions. This approach can be found in Rodrigues' textbook, too. And it is similar to the F.I.C. textbook where nine different positions are exposed for studying the point (see below).

The various chapters of the textbook are accompanied by exercises for the readers (students). Typical for them is the practice to work with carton, i.e. building three-dimensional models of *épures* in cardboard to develop the ability to view and create a mental image for *épures*. This concept relates to the mode for teaching the discipline, whose influences were presented in the book's introduction.

In the second chapter, Rodrigues presents the method of *rebattements*. To apply it to a plane figure on a plane, he establishes a homological correspondence between the projection of the figure and its *rabattement* (Rodrigues 1964, p. 102). This constitutes a revealing pattern of Rodrigues' work who applied here Poncelet's *Traité des propriétés projectives des figures* (1865), in its reception by Cremona (1893) (on Cremona, see Menghini, Chap. 4, this volume). Hence, Rodrigues has been influenced by Shalders' conception from the São Paulo Polytechnic. Rodrigues was aware that the method of *rabattement* had been made explicit by Théodore Olivier in 1843.

While Chap. 3 deals with metrical problems, the Chaps. 4–6, the main body of volume I—present, respectively, trihedron, irregular polyhedron and regular polyhedron. In the following parts, on *plans côtés*, Rodrigues refers to surveying methods for engineers developed already in France before Monge. Various metrical problems are solved for this approach.

The Chaps. 9 and 10 are inserted from the third edition (1950) onwards and deal with central projection as method for representing three-dimensional objects. Rodrigues based this on a broad spectrum of historical authors, from Taylor (1749), Cousinery (1828), Bergery (1835) to Fiedler (1871) (on Fiedler, see Volkert, Chap. 10, this volume).

Rodrigues' second volume is distinguished by presenting projective geometry, principally based on Poncelet. While emphasising projections and sections as Poncelet's fundamental operations, he refers also to Jakob Steiner's fundamental forms (Rodrigues 1960). The concept of homology, already exposed in volume I, is treated here more extensively. Likewise, Cremona's book is used more extensively. No other teaching programme or textbook used such an approach of connecting descriptive geometry with projective geometry. One can see this as an outcome of Rodrigues' studies in Germany. His two-volume work documents a profound knowledge of descriptive geometry and proper approaches for its teaching.

3.10 An Assessment for Brazil

In the Brazilian context, we note the predominance of approaches to studying descriptive geometry by using French textbooks. This incorporated the changes prepared before 1843 and led to the new level of conceptualisation introduced by Olivier's textbook of 1843: *Cours de géométrie descriptive*. In the period between 1812 and 1843, characteristic French textbooks were, according to Barbin (2015, p. 63), those by Lacroix and Jean Nicolas Hachette (Hachette 1822); different from Monge, the preliminary notions are extended, so that "more and more considerations are introduced to help students solve problems" (Barbin 2015, p. 63). Authors such as Louis Leger Vallée and Joseph Adhémar proposed a complete decomposition of the projections of points, straight lines and planes. Their changes constituted different *élémentations* in Barbin's terms: to decompose descriptive geometry into teachable and ordered elements (Barbin 2015, p. 41). One characteristic pattern is to study the objects of descriptive geometry by means of particular cases, and another one is that the methods of *rabattement*, of rotations and of changes of projection planes have to be acquired by the students as a "basic ability". Likewise, the study of surfaces, via Monge's classification, has to start from particular cases.

Considering the teaching programmes of the three institutions, but also Rodrigues' textbook, they address, in their introductory parts, the same topics

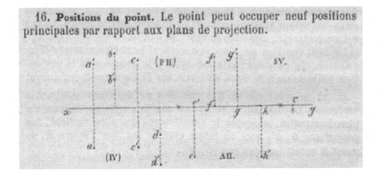

Fig. 21.5 The nine positions for a point, in the F.I.C. textbook (F.I.C. 1876, p. 6)

(representation of points, straight lines and planes, considering several special positions). In Lacroix's textbook, different manners to represent points and straight lines had been studied, as well as the theorems concerning the determination of these geometric entities. In Monge's textbook, however, there is no presentation at all of these geometric entities in various particular positions.

In Vallée's textbook, before resolving various problems, one finds the basic geometric entities represented in their "les plus remarquables positions" (Vallée 1825, p. 10). The author shows, for example, how vertical and horizontal straight lines are represented parallel and perpendicular to the ground line. This study, by means of various particular positions, is later on to be found in the textbook *Éléments de Géométrie Descriptive avec de Nombreux Exercices* by the author group F.I.C. Figure 21.5 shows, for example, an *épure* for representing a point, considering nine different positions:

One can say that, in general, all programmes analysed here required in their last parts the study of curved surfaces, of tangent planes and of contact curves between certain surfaces. These topics were initiated by discussing a classification of surfaces. There, classifying by analytic geometry via degrees of equations was considered as impractical and the application Monge's classification via generation of surfaces was preferred. These topics constitute parts II and III of Monge's text, which favour the generality of presented results and leave the discussion of particular cases to the end. For surfaces, for example, Monge shows that it is possible to construct a tangent plane to a surface at a given point from the tangent line to two generatrices of the surface containing that point. Then, by solving various problems, the author shows how the general method is applicable to the specific cases, such as cylinders, cones and spheres.

In Lacroix's textbook, as one sees from the programme of the Ouro Preto School of Mines, there is a chapter called *De la Sphère*, which deals with the representation and the construction of tangent planes and sections of the sphere. Only in the second part of his textbook, does Lacroix deal with the generation of surfaces to include the cylindrical and conical surfaces, and surfaces of double curvature and of revolution. For the surfaces, Lacroix proposes a set of problems concerning intersection curves between surfaces, discusses the development (flattening) of the cylinder and of the cone, the construction of the tangent plane and the normal line to a cylinder, and to a cone and a surface of revolution.

The problem of determining the perpendicular common to two skew straight lines is noteworthy because, depending on the analysed programme, it appears at different instances, which require therefore different theoretical frameworks for their solution. Monge discussed this issue after presenting the study of surfaces. In the programme of the Ouro Preto School it is analysed in its first part, dedicated to the study of straight lines and planes, assuming a different conceptual framework than that appearing in Monge's book. While this book is characterised by the notion of generality in its constructions, later textbooks apply the framework of complete decomposition of the figures, as is the case of Vallée and Adhémar (see Barbin 2015).

Another fact, which is characteristic for the Brazilian context is the influence of *Elementos de Geometria Descriptiva* by the French collective F.I.C. In the late nineteenth century and early twentieth century, this text was one of the references for the lecture course at the Polytechnic School of Rio de Janeiro and other institutions such as the *Colégio Pedro II*. The structure of F.I.C. book resembles the analysed programmes.

In the Brazilian edition of the F.I.C. book, one finds terms that do not appear in the original French text, but which are used until today in the Brazilian books on descriptive geometry. Terms such as "reta de frente" (straight line parallel to the vertical plane), "reta de topo" (straight line perpendicular to the vertical plane), "reta de perfil" (straight line situated in a plane parallel to the planes of projection), "plano do topo" (plane perpendicular to the vertical plane), "reta de maior declive de um plano" (straight line being the horizontal projection perpendicular to the horizontal trace of the plane or the same for the vertical plane) appear in the Brazilian translation of the F.I.C. book without having corresponding terms in the original French text. These and other adaptations apparent in the textbook can be understood as introduced by didactic purposes, aiming to thereby facilitate the student in understanding the concepts of descriptive geometry, starting from the analysis of many particular cases.

References

Sources

Arquivo Nacional, Rio de Janeiro

- Série Educação
- Academia Militar. Textbooks used in 1837. Código IG35. [AN 1]
- Programme of the Escola Central, 1861. Código IE278. [AN 2]
- Provas de admissão à Escola de Minas, 1878. Código IE3265. [AN 3]
- Bovet's programme for teaching descriptive geometry 1876, in: Officios Escola de Minas 1876 à 1878—Código IE3265. [AN 3]

Arquivo da Universidade Federal do Ouro Preto (UFOP) Arquivo Permanente [Archives of the Escola de Minas]

- Programme for teaching descriptive geometry 1885, in: Annaes da Escola de Minas de Ouro Preto. No. 4—Rio de Janeiro—1885. [EMOP 1]

Arquivos da USP (Universidade de São Paulo) Arquivos da Escola Politécnica

Publications

Adhémar, Joseph. 1832. *Cours de mathématiques à l'usage de l'ingénieur civil. Géométrie descriptive*. Paris: Carillan-Gœury.
Annuario da Escola Polytechnica de Sâo Paulo para o anno de 1900—São Paulo. Typografia do «Diario Official» – 1900.
Ausejo, Elena. 2006. Quarrels of a marriage of convenience: On the history of mathematics education for engineers in Spain. *International Journal for the History of Mathematics Education* 2 (1): 1–13.
Barbin, Évelyne. 2015. Descriptive geometry in France: History of Élémentation of a method (1795–1865). *International Journal for the History of Mathematics Education* 10 (2): 39–81.
Beltrame, Josilene. 2000. *Os programas de ensino de matemática do Colégio Pedro II: 1837–1932*. Dissertação (Mestrado em Matemática) - Departamento de Matemática, Pontifícia Universidade Católica do Rio de Janeiro: Rio de Janeiro.
Bergery, Claude Lucien. 1835. *Géométrie descriptive appliquée à l'industrie, à l'usage des artistes et des ouvriers*. 3. ed. Metz: Thiel.
Carvalho, João Bosco Pitombeira de. 2014. Mathematics education in Latin America. In *Handbook on the history of mathematics education*, ed. A. Karp, and G. Schubring, 335–359. New York: Springer.
Cousinery, Barthélémy Édouard. 1828. *Géométrie perspective, ou principes de projection polaire appliqués à la description des corps*. Paris: Carilian-Gœury.
Cremona, Luigi. 1893. *Elements of Projective Geometry*. 2nd ed. Oxford: The Clarendon Press.
Escola Polytechnica. 1879. *Programmas da Escola Polytechnica referentes ao ano de 1879*. Rio de Janeiro: Imprensa Nacional.
F.I.C. 1876. *Éléments de géométrie descriptive avec de nombreux exercices*. Paris: Alfred Mames et Fils.
Fiedler, Otto Wilhelm. 1871. *Die dastellende Geometrie. Ein Grundriß für Vorlesungen an Technischen Hochschulen und zum Selbststudium*. Leipzig: Teubner.
Gabaglia, Eugênio de Barros Raja. 1946. *Elementos de geometria descritiva com numerosos exercícios*. 11 ed. [Translation of F.I.C.]. Rio de Janeiro: F. Briguiet.
Hachette, Jean Nicolas Pierre. 1822. *Traité de géométrie descriptive*. Paris: Corby.
Javary, Adrien. 1882. *Traité de géométrie descriptive*. Paris: Ch. Delagrave.
Lacroix, Silvestre-François. 1795. *Essais de géométrie sur les plans et les surfaces courbes (Élements de géométrie descriptive)*. Paris: Courcier.
Legendre, Adrien Marie. 2009 [1809] *Elementos de Geometria, tradução Manoel Ferreira de Araújo Guimarães, adaptação e organização Luiz Carlos Guimarães*. Rio de Janeiro: Editora LIMC.
Leroy, Charles François Antoine. 1834. *Traité de géométrie descriptive, avec une collection d'épures composée de 60 planches*. Vol. 1 Texte; vol. 2 Planches. Paris: Carillan- Gœury and Anselin. [Many reeditions and translations all over the 19th century].
Miranda, Hernani de Oliveira. *O Ensino da Geometria Descritiva no Brasil: Da Academia Real Militar à Escola Politécnica do Rio de Janeiro*. São Paulo: Pontifícia Universidade Católica, 2001. Dissertação de mestrado.

Monge, Gaspard. 1799. *Géométrie descriptive, leçons données aux écoles normales de l'an III de la République*. Paris: Baudouin.
Mormêllo, Ben Hur. 2010. *O Ensino de Matemática na Academia Real Militar do Rio de Janeiro, de 1811 a 1874*. Campinas: Universidade Estadual de Campinas. Dissertação de mestrado.
Oliveira, Thiago Maciel de. 2016. *A obra de Alvaro José Rodrigues*. Tese de doutorado. Programa HCTE, Universidade Federal do Rio de Janeiro.
Olivier, Theodore. 1843. *Cours de géométrie descriptive*. Paris: Carilian-Gœury & Dalmont.
Pardal, Paulo. 1984. *Memórias da Escola Politécnica*. Rio de Janeiro: Xerox do Brasil.
Pereira, Vinicius Mendes Couto, and Gert Schubring. 2017. A Matemática Desconhecida da Escola de Minas de Ouro Preto. *Anais do 70 Encontro Luso-Brasileiro de História da Matemática*. [forthcoming]
Pinto, Joao da Costa. 1898. *These de concurso para a vaga de lente da cadeira de geometria descritiva e topografia da Escola Naval*. Rio de Janeiro: Typ. Leuzinger.
Poncelet, Jean Victor. 1865. *Traité des propriétés projectives des figures*. Paris: Gauthier-Villars.
Rodrigues, Alvaro José. 1960. *Geometria Descritiva – Projetividades, Curvas e Superfícies*. 3. ed. Rio de Janeiro: Ao Livro Técnico S.A.
———. 1964. *Geometria Descritiva – Operações Fundamentais e Poliedros*. 6. ed. Rio de Janeiro: Ao Livro Técnico S.A.
Roubaudi, C. 1916. *Traité de Géométrie Descriptive*. Paris: Masson.
Saldaña, Juan José (ed.). 2006. *Science in Latin America. A History*. Translated by Bernabé Madrigal. Austin: University of Texas Press.
Sanchez, Clara Helena. 2007. *Los ingeniero-matemáticos colombianos del siglo XIX y comienzos del siglo XX*. Bogotà: Universidad Nacional de Colombia.
Saraiva, Luis. 2007. The Beginnings of the Royal Military Academy of Rio de Janeiro. *Revista Brasileira de História da Matemática* 7 (13): 19–41.
Sousa, José Vitorino dos Santos e. 1812. *Elementos de Geometria Descriptiva com aplicações às artes. Exrahidos das obras de Monge*. De ordem de sua alteza real o Principe Regente N.S. Para uso dos alunos da Real Academia Militar. Rio de Janeiro: Imprensa Régia.
Süssekind, Carlos. 1933. *Geometria Descritiva: Geometria Descritiva, Perspectiva, Sombras e Desenho de Projeções*. Rio de Janeiro: Freitas Bastos.
Taylor, Brook. 1749. *New principles of linear perspective: or the art of designing on a plane, the representations of objects all sorts of objects, in a more general and simple method than has been hitherto done*. 3. ed. London: John Ward.
Vallée, Louis-Léger. 1821. *Traité de géométrie descriptive*. Paris: Courcier.
Venturini, Alejandro E. Garcia. 2011. *Historia de la Matemática en la Argentina*. Buenos Aires: Ediciones cooperativas.

Part IV
Epilogue

Chapter 22
The Myth of the Polytechnic School

Gert Schubring

Abstract From 1806 on, polytechnic schools were founded in various European states. According to traditional historiography, and in particular to the *Festschriften* of these institutions published at some anniversary, these schools took the *École polytechnique* in Paris as a model, with descriptive geometry as a key teaching discipline. A closer investigation shows, however, that these schools began at a rather low educational level, often as commercial schools and dependent on the ministry of commerce, and taught quite elementary mathematics. The only institution projected at the level of higher education in the first half of nineteenth century Europe was a Polytechnic Institute in Berlin: its rationale was to be mathematics teacher education. It was only in the second half of the nineteenth century that these institutions, after steady rise in the status and level of formation provided, succeeded in attaining the level of higher education and in offering demanding mathematical courses. The question is hence why a structure, which proved so successful in France, was not viable in other countries—at least in the first half of the nineteenth century.

What is crucial proves to be the character of applications of mathematics as demanded and as practiced in the respective countries. The paper will study these differences in the level and degree of development of applications of mathematics in these countries on the one hand and, on the other hand, the differing demands in these countries for such applications and, in particular, of the requisite geometrical knowledge and the functions being realised by descriptive geometry. This investigation reveals a very specific and telling pattern that enabled the functioning of the *École polytechnique* in a unique way.

Keywords *École polytechnique* · Model function · Polytechnic characteristics · Technical colleges · Department structure

---

G. Schubring (✉)
Universidade Federal do Rio de Janeiro, Instituto de Matemática, Rio de Janeiro, Brazil

Universität Bielefeld, Fakultät für Mathematik, Bielefeld, Germany
e-mail: gert.schubring@uni-bielefeld.de

1 Introduction

One refers often to the *École polytechnique* (EP) as a model—without, however, specifying in which respects it should exert a model function. To analyse more specifically, the respects understood here as relevant for the model function are specified as three patterns:

- descriptive geometry should constitute a major discipline within the curriculum;
- the institution should be of higher education level;
- the institution should not provide a formation for just one technical branch or engineering profession, but either provide a formation for the then relevant branches of engineering and technology, within a net of institutions, or provide such a range of professional formations within its proper scope;
- and this formation should be based on mathematics: the polytechnic character of mathematics meant its foundational status for study courses of military and civil engineering.

It is with regard to these three basic constituents of the original conception of the EP that institutions claimed to have been modelled according to the Paris *École polytechnique* will be analysed—and where a myth of the model function will be shown. As I want to show in particular, there is not just one myth of the *École polytechnique*—there is even a double myth. And one of them applies to the *École polytechnique* itself.

The first myth is the one mentioned in the abstract, namely that the École polytechnique at Paris served as a model for numerous analogous foundations in other countries. I will discuss this soon. First I intend, however, to analyse the second myth. In the context of the spread of the *École polytechnique* as model for technical education, the model is conceived of as being intimately tied to descriptive geometry as one of the key disciplines of such an institution.

2 Rise and Decline of Descriptive Geometry at the École Polytechnique

On analysing the curricular development of the *École polytechnique* in its details, this essential role of descriptive geometry in its curriculum proves to be a myth. Clearly, there is no doubt at all that descriptive geometry had been conceived of by Monge, indeed, as one of the key teaching disciplines at the newly founded school, together with chemistry—both representing the methodological approach of *analyse*—in the sense of Condillac. In fact, there is a key document among the founding texts for the *École Centrale des Travaux Publiques*: a type of Manifesto written by Monge that explains the key function of descriptive geometry (*Développemens sur l'enseignement adopté pour l'École centrale des travaux publiques*, 1794). This key role was actually established in the teaching practice, by the lecture courses and the accompanying exercises of the students to elaborate the sketches (*épures*). The high percentage of the teaching ascribed to descriptive geometry and to the exercises has been analysed in the PhD thesis of Paul (1980).

But this thesis showed at the same time the marked decline of this percentage after 1800: descriptive geometry was then concentrated in the first year of the studies; while its lessons, exercises and repetitions still comprised 40% of the teaching time in 1801, this percentage was reduced to 26% in 1806 and to 23% in 1812 (Paul (1980), p. 124, see also Sakarovitch (1994)). In fact, it has been agreed for many years in the research into the development of the EP, that one distinguishes between the original *École polytechnique* as the École de Monge and the later EP as the *École de Laplace*. This radical and abrupt rupture is a consequence of the years 1799–1800, when—on the one hand—the EP survived the imminent danger of loosing all its characteristic features (see Fourcy 1828), but—on the other hand and at the same time—Laplace, the former enemy of the EP, managed to redress its curricular conception: "analyse" in the sense of the analytic methodological programme of Condillac became replaced by analysis in the sense of the proper mathematical discipline. The ensuing reduction in the teaching descriptive geometry at the *École polytechnique* turned out to become one of the issues that legitimated the decisive crisis of the EP in 1810/1811, provoked by the so-called *plaintes de Metz*: one of the main criticisms in the list of faults of the formation of engineers at the EP were the poor quality and scant knowledge of drawing and of elaborating the *Épures* of descriptive geometry.

I have studied these *plaintes* extensively, because they were the means to reduce the role of theory even further in the teaching at the EP (Schubring 2004). The *École du génie* at Metz, which aimed to become independent of the institutional net consisting of the preparatory school—the EP—and of the following schools of applications, had criticised, among others that:

– Almost all students could draw only very little; there was no precision in the execution of their sketches of the mining industry, buildings and fortifications.

If one looks at the points of criticism raised in Metz, in fact only the weakness of the graduates in graphic works remains as a strong point. This weakness can be seen as the natural consequence of the fundamental change in the conception of the curriculum of the year 1800, since from this time on descriptive geometry had been pushed into the background. The *Conseil d'instruction* and the *Conseil de perfectionnement* had indeed recognised this weakness; after 1809 it was discussed several times, and attempts had been made to implement remedial measures (Schubring 2004, p. 108).

The fragility of descriptive geometry at the *École polytechnique* is confirmed by a document elaborated in the context of the restructuring the curriculum of the EP after its closure in 1816. This document, presented to the *Conseil de Perfectionnement* on 12 December 1816, discusses whether it should be maintained as one of the disciplines of formation at the EP. The rapporteur was Gilles de Laumont, who was external to the *École polytechnique*—he was the head inspector of the corps of mining and member of the *Académie des Sciences*. He argued for maintaining descriptive geometry and the *Conseil* approved his memoir "*Quelques reflexions sur la géométrie descriptive et sur les professeurs de cette partie*".

These reasons were given:

- "This science is indispensable for all the services to which the students of the *École polytechnique* are destined" (AEP I)[1]
- "it is essential that this science does not degenerate at the new school" (Schubring 2004, p. 108)[2] and it was therefore concluded:
- "the choice of a professor of descriptive geometry is of the highest importance for the prosperity of the school, but the choice is extremely difficult to be made" (Schubring 2004, p. 108).[3]

Given the reasons for the closure of the *École polytechnique*, besides scientific qualities of the candidate ("force majeure en théorie et en pratique") moral, religious and political qualities were required. Three candidates were discussed: Coriolis, Lefebure de Fourcy and Charles Leroy who was finally selected (Schubring 2004, p. 108). He taught there descriptive geometry for more than 30 years, according to Dhombres without any innovations (Dhombres 1987, p. 177).

It should be added that, despite the high esteem descriptive geometry is enjoying in the general public, its flourishing as a research discipline had ended already with the first generation of Monge's disciples (Sakarovitch 1998).

3 Polytechnic Schools in Europe: Modelled According to the Paris School?

3.1 Germany

It is true that numerous polytechnic schools were founded throughout Europe during the first half of the nineteenth century—and particularly in Germany. But were they really modelled on the Paris original?

Let us at first look at such schools in Germany; their number reflects the number of independent states within Germany throughout the nineteenth century. A list of the foundations is as follows:

- Berlin 1821
- Karlsruhe 1825
- Darmstadt 1826

---

[1] "Cette science est indispensable à tous les services auquels on destine les élèves de l'ecole polytechnique."

[2] "Il est donc essentielle que cette science ne dégénere pas dans la nouvelle Ecole."

[3] "le choix d'un professeur de Géometrie descriptive est de la plus haute importance pour la prospérité de l'école, mais qu'il est extrémement difficile à faire."

- Dresden 1828
- Stuttgart 1829
- Hannover 1831
- Brunswick 1835
- Aachen 1870

A revealing case is presented by the school in Munich, the capital of Bavaria: in the official histories of the present Technical University at Munich, 1868 is given as year of its foundation. In reality, a polytechnic school was founded there in 1833, attached to the university (the successor of a *Polytechnische Zentralschule*, which had existed since 1827)—as second level for a system of trade and agricultural schools. But it failed to flourish, in the context of the then-dominant spirit of humanities in Bavaria, and it had to be closed in 1842; such a school became viable only much later. Thus, one has to add:

- Munich 1827/1833–1842; 1868 (Schubring 1989, pp. 176 f.).

In fact, all these schools founded in the first half of the nineteenth century differed in a decisive sense from the Paris original: none of them was of the level of higher education. Some would have been even difficult to be classified as of secondary school level: they used to be of much lower status than the classical gymnasia and used not to belong to the administration of the ministry of education, but rather to that of the ministry of commerce.

Almost all these schools began at a very low level. This is evident from the usual age at which students entered—between 12 and 15 years old. Sometimes they functioned primarily as evening schools for artisans, as was the case at Dresden. For the first period of the Karlsruhe school, Hoepke gives a realistic report of its badly organised state, its incoherent conceptions and the miserable state of a school that used some rooms of the Gymnasium and some in a private building. The mathematics teaching, although nominally dominant, was very elementary—destined for students entering at an age of 13 years. The first positive point was that Guido Schreiber was hired in 1828 to teach descriptive geometry—he was the author of the first German textbook of this discipline (Hoepke 2007, pp. 31 ff.). Apart from the obvious exceptions—the foundations that took place after 1860, i.e. already in the new age of rising toward college status—only the schools at Berlin and Hannover operated at a somewhat higher level, since they accepted only graduates from lower-ranking technical schools.

A pattern common to all these technical schools, and which marks them as schools rather than as institutions of higher education, was the lack of *Lernfreiheit* (academic freedom)—the classical pattern of German universities: students were organised in yearly grades and had to follow a strictly prescribed curriculum (Schubring 1989, pp. 179 f.).

These schools did not even necessarily have a "polytechnic" in their name. For instance, the Berlin foundation was called "Gewerbe-Institut". This means an institution for trade and commerce. The school in Hannover was opened in 1831 named "Höhere Gewerbe-Schule".

This objective—*Gewerbe*—provides an essential keyword for historical analysis. Some histories of these schools emphasise as a difference from the Paris original, that they prepared students not only for civil (and military) service like the École polytechnique, but also for industry (see Manegold (1970, pp. 19 ff.)). This is in fact an important issue to discuss. On the one hand, Monge—in his 1794 document—had emphasised the importance of descriptive geometry for "industry". But one has to warn against a naive and unreflecting use and understanding of "industry" for this period. Too easily "industry" is associated with "industrial revolution" and thus with industry in the modern sense of factory production. In reality, before, say at least the 1840s, one could not understand industry in this sense, at least for German states—rather it meant manufacture and commerce. Therefore, the term *Gewerbe* reveals exactly a non-industrial—in modern terms—orientation of these schools—characterising them as a type of professional schools. When they adopted "polytechnic" in their name this was a programmatic step to secure them a certain status—alluding to the key role of mathematics in the Paris original—thus using 'mathematics' and 'polytechnic' as a means to improve their status in relation with the higher ranking genuine secondary schools—the gymnasia, often dependent of another ministry.

And on the other hand, as already the name for the *École polytechnique* in its first name suggests, the EP was destined to prepare for "Travaux Publics"—i.e., for public service. In fact, the mission of the EP remained to train people for public service. Formation for what could be later called "industry" was provided by another school founded somewhat later, in 1829: the *École Centrale des Arts et Manufactures*, which developed to become one of the *Grandes Écoles*.

3.2 "Europe"

Let us now look beyond the realm of what later constituted the (second) German Empire: principally the Austrian-Hungarian Empire with the foundations in Prague 1806 and Vienna 1815. A foundation in the Netherlands is rather late: a Royal Academy in 1842, in Delft, for the education of civil engineers, serving both nation and industry, and for the training of apprentices for trade—thus a quite mixed institution. It became, in 1864, the Polytechnic School of Delft, being at first at secondary school level. In Switzerland, where there had been no prior structures of military or technical formation, the *Eidgenössische Technische Hochschule* Zürich was founded in 1855, first under the name *Eidgenössische polytechnische Schule* (for sake of brevity: ETH)—immediately at higher education level. As regards England, the Military Academy at Woolwich was founded already in 1741 and remained its major institution for technical formation.[4]

---

[4]The *Royal Naval College* at Greenwich was founded as late as 1873.

The school in Prague was a reorganisation of an engineering school founded in 1786, with the new name *Polytechnikum*. It provided training for commercial professions and for engineering. In 1863 it achieved the status of a *Hochschule*, thus of higher education (Stark 1906). The school in *Vienna* was founded in 1815 as "polytechnisches Institut in Wien". Although the term "Institut" should imply a higher status than a secondary school, it had two departments—a technical one, for forming engineers, and a commercial one, to prepare for commercial professions; a 2-year realistic[5] school was associated, as preparatory school. This implies that the Polytechnic Institute was not at a coherent higher education level. It achieved this level only after a profound reform in 1865, when the commercial department was closed (Neuwirth 1915).

It seems therefore that one can conclude that within Europe the foundations of polytechnic schools in the first half of the nineteenth century, although they claimed to take the Paris school as their model, were of rather low status and not at all compatible with the French original.[6] The apparently only exception is given by Denmark, where—although military schools already existed there—a polytechnic school of a somewhat higher level was founded: but actually, due to very specific circumstances, because the original plan of a low-level school became replaced by a higher ranking one (see Lützen, Chap. 15, this volume). Yet, the parallel existence of two polytechnic schools there—one for civil professions and the other for military careers—proves a model functioning differently from the Paris one.

This is further confirmed by the only attempt to found a Polytechnic Institute of a truly academic level: these were the plans for a Polytechnic Institute in Prussia, in Berlin. I have studied these plans, which were never realised, intensively and they demonstrate the remarkable fact that it was intended to produce teachers of mathematics—and not engineers!

There were four periods when the project was promoted:

- The first project was elaborated by the mathematics professor at the newly created Berlin university, Johann G. Tralles, in 1817, by transforming an existing institution—the *Bau-Akademie*, School of Architecture—into a polytechnic institute. The high standard of formation intended for this reformed institution by the ministry of education met strong resistance in the ministry of commerce who aimed only to train low-level civil engineers. In the end, the ministry of education had to hand over responsibility for vocational and trades training to the ministry of commerce, in 1820 (Schubring 1981, pp. 166 ff.).
- The next stage was occasioned by the attempt to call Gauß to Berlin. Due to Gauß's two demands—a high salary and no teaching obligation at the

---

[5] "Realistic" as translation for the not translatable German "Real-".

[6] It should be noted that schools founded in several countries, for which one might claim a polytechnic character, were in reality "mono"-technic: the schools founded in Spain and in Russia were schools for forming "ponts et chaussées" engineers (see Ausejo, Chap. 5 and Gouzevitch et al., Chap. 13, this volume)—thus not having the characteristic pattern of providing a broad mathematical basis for various professions.

university—the solution seemed to be the creation of a proper institute. The report of the ministry of education in 1823, to demand from the king the authorisation to deliberate with Gauß and to allow the necessary funds, was really a memorandum for creating a polytechnic institute. It argued that training specialised mathematicians was of considerable practical importance for the whole state. Mathematics and physics formed the basis for practical knowledge, it argued, and in particular for "several of the most important administrative units concerned with public security and public welfare; the whole military system and some of its parts particularly, the entire building industry, several trades, smelting and mining". The proposed new centralised institution, called *Centralseminarium*, should be independent of universities. The report concluded by saying that a need existed for a "so-called *polytechnic institute*". Yet, the ministry of education did not obtain the necessary support from the other ministries who, given that training institutions for pertinent lower-level professions already existing in their administrative domains, were not keen to have a competing higher-grade institution (Schubring 1981, pp. 169ff.).

- The third stage of the project began in 1828, after the return of Alexander von Humboldt from Paris. Humboldt was completely aware of the structure of scientific institutions in France and in particular of the *École polytechnique*. One of his goals was to create "a seminary of chemistry and mathematics at the University". It was August Leopold Crelle who was charged by the education ministry with elaborating such a project. Crelle proposed an institute to train Gymnasium teaches of mathematics. As professors, he thought of Dirichlet, Jacobi, Abel and Plücker. He extolled pure mathematics, as the best means to provide for applications:

    > So it is also important that pure mathematics should be explained in the first instance without regard to its applications and without being interrupted by them. It should develop purely from within itself and for itself. For only in this way can it be free to move and evolve in all directions. In teaching the applications of mathematics it is results in particular that people look for. They will be extremely easy for the person who is trained in the science itself and who has adopted its spirit (quoted from Schubring 1989, pp. 180f.).

- Characteristically, Crelle claimed as the function of the EP exactly what he proposed for the new institute: "an institution having as its essential task the training of mathematics teachers" (Schubring 1981, p. 182).

- Crelle's plan seemed too extended for the ministry and remained therefore for 2 years without anything being done. It was revived, however, after the visit in 1831 of the French education minister Victor Cousin. Now the ministry together with Dirichlet elaborated a more concrete plan. It took quite a time that the project was seriously discussed, but eventually in 1835 there seemed to be real chances to get the project approved by the King. It was exactly in this "hot" phase that Jacobi gave a public lecture on the Paris *École Polytechnique*: actually, speaking about the Paris school to promote the Berlin project. Interestingly enough, he gave the same wrong appraisal of the Paris original as Crelle earlier on: declaring it to be "the seminar of all teachers of mathematics, physics, and chemistry of France, having formed all members of the present Academy of Sciences in these disciplines" (Jacobi 1835/1891, p. 356). Eventually, also this time, resistance from the ministries of war and of commerce made the project fail (Schubring 1981, pp. 184 ff.).

- There was a last revival in 1844, occasioned by Jacobi's move from Königsberg to Berlin. Here, the project definitely failed—it became too obvious that the formation of highly qualified mathematics teachers was already carried out by the universities (Schubring 1981, pp. 186ff.).

3.3 Outside Europe

It is therefore highly remarkable that the first institutions that adapted the conception of the *École polytechnique* at its level of higher education and not being "mono"-technic but preparing for a range of professions, were founded outside Europe:

- the *Academia Militar* 1811 in Rio de Janeiro, in Brazil,
- the Military Academy in Westpoint (USA), founded in 1802 but profoundly revised in 1818, after a careful study of the *École polytechnique* by a Westpoint administrator,
- and the school in Bulaq in Egypt 1837.

The basic reason for this discrepancy seems to be that all the countries in Europe were operating in the tradition of their "Ancien Régime". That means that all the structures of education and technical formation were determined by the weight of traditions, or to put it better, by the social weight and influence of the respective corporations, of the corps of engineers, and the impact of the military leaders on the state government. Given the net of already existing institutions dependent on different and competing authorities, no completely new structure was viable that would have been in conflict with so many well-established structures.

Such a radical innovation was thus only viable in a "new world". This is in fact the case for these three foundations: they constituted a type of "tabula rasa" at least as regards technical education.

- Brazil as a colony had intentionally been kept undeveloped by the Portuguese crown. Only in 1792 had a minor course for technical training been established: at the Real Academia de *Artilharia, Fortificação e Desenho*. It was therefore a radical change when the Royal family arrived in 1808, fleeing from the Napoleonic invasion of Portugal, accompanied by an enormous fleet, carrying libraries, printing presses, scientists, etc. Immediately, in 1810, a Military Academy was founded, for the formation of military and civil engineers. The royal decree required

    that from these courses there will graduate skillful Artillery and Engineering Officers, and also Officers from the Class of Geographic and Topographic Engineers, who can have the useful task of directing administrative matters in mining, roads, ports, channels, bridges, waterworks and pavements; I hereby establish in my present Court and city of Rio de Janeiro a Royal Military Academy with a complete course of Mathematical Sciences, of Observational Sciences, such as Physics, Chemistry, Mineralogy, Metallurgy, Natural History, which will include the plant and animal kingdoms, and of Military Sciences in all its range, as well as of Tactics and Fortification and Gunnery (Saraiva 2007, pp. 25–26).

- The United States, after its independence had to establish a proper system of training engineers. This was brought about through the foundation of the United States Military Academy in 1802 in Westpoint. The first organisation was rather chaotic, however, and it was only due to an intensive visit of the *École polytechnique* in 1816 by Major Thayer ("the father of the USMA") that eventually a definite structure became established in 1818, adapting many characteristic features of the Paris model. In fact, Westpoint is probably the only case of a creation, resp. reform of an institution where the institution in Paris had been visited for assessing its functioning (see Preveraud, Chap. 19, this volume).
- Muhanmed Ali, the semi-independent governor of Egypt from 1805, applied a policy of strong modernisation, despite the resistance of the ulemas and at first of the Mameluks, too. He paid special attention to education, to mathematics and to science. An engineering school established in 1825, was transferred, in 1837 to Bulāq and became known as *Mühendeskhāne (Bulāq Polytechnic School)* (see Crozet (2008, p. 118 and p. 120)). In this school Egyptian youth received high quality training which enabled them, once they graduated, to modernise Egypt despite the resistance of large sections of the society, as well as the hesitations and hostility to reforms of some of the successors of Muhammad Ali, in addition to the disastrous financial conditions and strong pressures exerted by the imperialist powers seeking to settle in the country. Indeed England was to impose its protectorate on Egypt no later than 1882; it put an end to the development of high-level modern science in Egypt (Abdeljaouad 2012, 490).

4 A Decisive Difference: Department Structure Instead of an Institutional Net

I should like now to analyse and comment on a profound structural difference between the typical polytechnic schools since about the 1830s/1840s and the Paris original—a difference apparently neither well known hitherto nor studied. And this difference reveals a decisive and pertinent difference with regard to the function of mathematics in these two types of institution.

The characteristic structural pattern of the EP since its reorganisation in its second year, in 1795, was to constitute the basis of a "net" of institutions: the EP as fundamental step in training was to be complemented by the group of *écoles d'applications*, providing the specifics of the basis for the particular future professions.

In contrast to this, none of the institutions founded under the label "polytechnic" belonged to such a net. In fact, all these schools provided a formation complete in itself, i.e. its graduates were immediately able to enter the respective profession. How then was there the relation of foundations to applications organised? It turns out that the foundational part, and hence the function of mathematics, became ever more weakened in these schools.

The school in Karlsruhe, founded in 1825, was reorganised in 1832, by Carl Friedrich Nebenius (1784–1857), and displayed there the first time the structure that was to become adopted by other schools later on: first, a preparatory year had to be passed, called *Klassen für die mathematische Grundausbildung* and thereafter students would continue in one of five departments:

- the engineering school (*Ingenieurschule*)
- the architecture school (*Bauschule*)
- the higher commerce school (*höhere Gewerbeschule*)
- the school of forestry (*Forstschule*)
- the trade school (*Handelsschule*) (Hoepke 2007, 35).

A further step of specialising and of upgrading was taken in 1847, when the *Höhere Gewerbe-Schule* became transformed into two new departments: the mechanical-technical *Fachschule* (professional school) and a chemical-technical *Fachschule*. Mechanical Engineering (*Maschinenbau*) and chemistry turned now to constitute separate study courses. The Karlsruhe school therefore now offered separate study courses for architects, construction engineers, chemists and mechanical engineers. Mathematics was reduced to providing a "prior mathematical propaedeutic (*Grundlagen-Ausbildung*)" (Hoepke 2007, p. 55). This structure was adopted by the ETH when it was founded in Zürich in 1855—thus the first technical college of higher education status in Europe outside France.

4.1 An Example of Polytechnic Studies

I should like to show this curricular structure in the studies of a Brazilian student, Antônio Francisco Paula Souza (1843–1917), who first studied at the ETH, from 1861 to 1863, and then in Karlsruhe, from 1864 to 1867, and who became later the founder of the *Escola Politecnica* in São Paulo.[7]

The structure of the studies at the ETH at this time was the following. It might begin with 1 year in a *mathematische Vorbereitungsklasse* (mathematical preparatory course), with focus on two major disciplines: mathematics and mechanics: "Candidates who came from the practice or could not be admitted immediately to one of the *Fachschulen* due to insufficient previous knowledge or language problems

---

[7] Antônio Paulo Souza is a subject of research by Vinicius Mendes, one of my doctoral student at the Universidade Federal do Rio de Janeiro.

should be qualified in an 1 year course to enter the school" (Fig. 22.1). The proper ETH was divided in five schools and a general department, for the sciences and the humanities (AETH I):

- *Bauschule* (architecture school),
- *Ingenieurschule* (engineering school)
- *Mechanisch-technische Schule* (mechanical-technical school),
- *Chemisch-technische Schule* (chemical-technical-school)
- *Forstschule* (school of forestry)
- *Sechste Abteilung* (sixth department)

The mathematical courses taught in these four departments reveal a remarkable difference in their level. In the architecture school, the infinitesimal calculus is taught by the same teacher who taught the general mathematics course in the mathematical preparatory year. For engineering and for mechanics, the calculus is taught by a genuine mathematician, by Richard Dedekind (Fig. 22.2).

For chemistry, however, no such course of higher level mathematics was taught (Fig. 22.3).

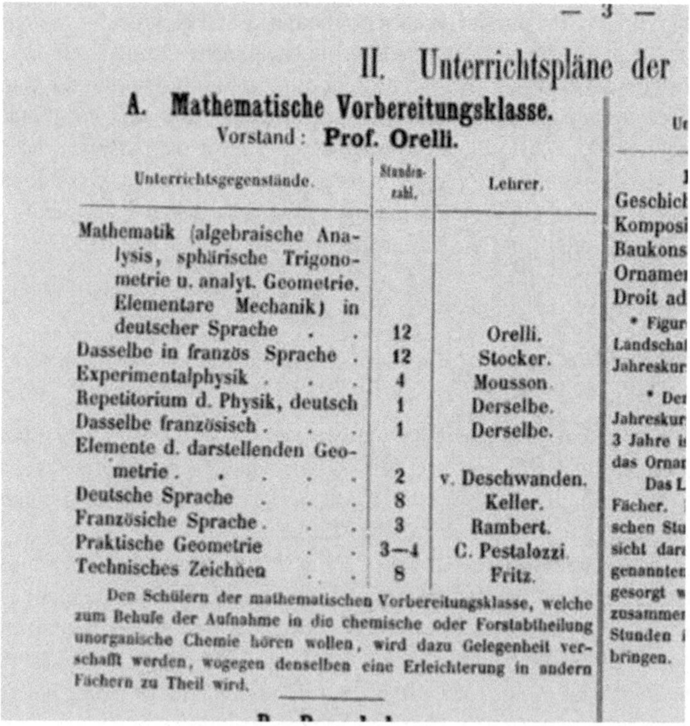
— 3 —

II. Unterrichtspläne der

A. Mathematische Vorbereitungsklasse.

Vorstand: Prof. Orelli.

Unterrichtsgegenstände.	Stunden-zahl.	Lehrer.
Mathematik (algebraische Analysis, sphärische Trigonometrie u. analyt. Geometrie. Elementare Mechanik) in deutscher Sprache	12	Orelli.
Dasselbe in französ Sprache	12	Stocker.
Experimentalphysik	4	Mousson.
Repetitorium d. Physik, deutsch	1	Derselbe.
Dasselbe französisch	1	Derselbe.
Elemente d. darstellenden Geometrie	2	v. Deschwanden.
Deutsche Sprache	8	Keller.
Französiche Sprache	3	Rambert.
Praktische Geometrie	3—4	C. Pestalozzi.
Technisches Zeichnen	8	Fritz.

Den Schülern der mathematischen Vorbereitungsklasse, welche zum Behufe der Aufnahme in die chemische oder Forstabtheilung unorganische Chemie hören wollen, wird dazu Gelegenheit verschafft werden, wogegen denselben eine Erleichterung in andern Fächern zu Theil wird.

Fig. 22.1 AETH I: course structure in the mathematics preparatory course ihnen gefordert wird.

B. Bauschule.
Vorstand: Prof. Semper.

Unterrichtsgegenstände	Stundenzahl	Lehrer.
I. Jahreskurs.		
Differential- u. Integralrechnung, deutsch	5	Orelli.
Dasselbe französisch	9	Méquet.
Darstellende Geometrie	4	v. Deschwanden.
Archäologie und Geschichte der antiken Kunst	4	Lübke.
Baukonstruktionen	4	Gladbach.
Baukonstruktionszeichnen	8	Derselbe.
Architektonisches Zeichnen	4	Semper.
Ornamentenzeichnen	4	Stadler.
* Figurenzeichnen	9	Werdmüller.
* Landschaftzeichnen	4	Ulrich.
* Modelliren	9	Keiser.
Experimentalchemie (für solche, welche noch keinen Unterricht in der Chemie genossen haben)	5	Städeler.
II. Jahreskurs.		
Vergleichende Baulehre	4	Semper.
Mechanik (Fortsetzung)	6	Zeuner.
Schattenlehre	3	v. Deschwanden.
Petrographie	3	Kenngott.
Strassen- und Wasserbau	3	Pestalozzi.
Kompositionsübungen im Hochbauwesen	4	Semper.
Baukonstruktionszeichnen	4	Gladbach.
Ornamentenzeichnen	4	Stadler.
*) Figurenzeichnen, Landschaftszeichnen und Modelliren wie im ersten Jahreskurs.		

sicht darauf genommen, dass die Schüler dieser Abtheilung den genannten Unterricht benutzen werden. Es soll im weitern dafür gesorgt werden, dass die Schüler der Bauschule täglich einige zusammenhängende, von der Abtheilungskonferenz bezeichnete Stunden in den Zeichnungssälen unter angemessener Aufsicht zubringen.

C. Ingenieurschule.
Vorstand: Prof. Culmann.

Unterrichtsgegenstände.	Stundenzahl.	Lehrer.
I. Jahreskurs.		
Differential- u. Integralrechnung	6	Dedekind.
Repetitor. in deutscher Sprache (Die Schüler in Gruppen abgetheilt)	3	Ders. u. Durège.
Calcul différentiel et intégral	6	Méquet.
Repetit. in französischer Sprache	3	Derselbe.
Darstellende Geometrie	4	v. Deschwanden.
Topographie	3	Wild.
Baukonstruktionen	4	Gladbach.
Baukonstruktionszeichnen	4	Derselbe.
Maschinenzeichnen	4	Fritz.
Planzeichnen	3	Wild.
Petrographie	3	Kenngott.
Experimentalchemie (für solche, die noch keinen chemischen Unterricht genossen haben)	5	Städeler.
II. Jahreskurs.		
Anwendung des Differential- u. Integralrechnung	2	Dedekind.
Experimentalphysik für die technische Richtung	4	Clausius.
Technische Mechanik, Fortsetz	6	Zeuner.

Fig. 22.2 AETH I: course structure in the architecture and the engineering departments

Unterrichtsgegenstände	Stundenzahl	Lehrer.
Graphische Statik	?	Derselbe.
* Konstruktionsübungen	6	Derselbe.
Kartenzeichnen	3	Wild.
III. Jahreskurs.		
Theoretische Maschinenlehre	6	Zeuner.
Ausgewählte Parthien aus der höhern Astronomie	4	Wolf.
Geodäsie	3	Wild.
Eiserne Brücken, Strassen- u. Eisenbahnbau	4	Culmann.
Konstruktionsübungen	6	Derselbe.
Droit administratif	1	Dufraisse.
Kartenzeichnen	3	Wild.

(In das Sommersemester fallen an der Ingenieurschule ferner die Vorträge chemische Technologie der Baumaterialien, die Feldmessübungen und die Uebungen auf der Sternwarte.)

D. Mechanisch-technische Schule.
Vorstand: Prof. Zeuner.

Unterrichtsgegenstände.	Stundenzahl.	Lehrer.
I. Jahreskurs.		
Differential- u. Integralrechnung	6	Dedekind.
Repetitorium deutsch, die Schüler in Gruppen getheilt	3	Ders. u Durège.
Calcul différentiel et intégral	6	Méquet.
Repetitorium französisch	3	Derselbe
Darstellende Geometrie	4	v. Deschwanden.
Maschinenzeichnen	8	Fritz.
Civilbau	2	Gladbach.
Mechan. Technologie, I. Thl.	4	Kronauer.
* Experimentalchemie (für solche, welche nicht früher Vorträge dieses Faches hörten.)	5	Städeler.

der Baumaterialien und Civilbau.)

E. Chemisch-technische Schule.
Vorstand: Prof. Städeler.

Unterrichtsgegenstände.	Stundenzahl.	Lehrer.
I. Jahreskurs.		
*Ausgewählte Kapitel der unorganischen Chemie	3	Städeler.
Experimentalphysik für die chemische Richtung	4	Clausius.
Chemische Technologie: a) das Wasser und Fabrikation chemischer Produkte 3 b) Thonwaaren und Glas 1	4	Bolley.
Repetitorium u. Conversatorium	1	Derselbe.
*Mineralogie mit Repetitorium	6	Kenngott.
Allgemeine Botanik	2–3	Cramer.
÷Zoologie	6	Frey.
Technisches Zeichnen	4	Fritz.
Analytisches Praktikum	9	Städeler.
**Experimentalchemie mit einer Uebersicht der organischen Chemie	5	Derselbe.
II. Jahreskurs.		
*Chem. Technologie (Bleicherei, Färberei, Zeugdruck)	3	Bolley.
Repetitorium u. Conservatorium	1	Derselbe.
*Mechanische Technologie	4	Kronauer.
Angewandte Krystallographie	3	Kenngott.
Geologie	3	Escher v. d. Linth.
Technisches Praktikum	12	Bolley.
Technisches Zeichnen	4	Fritz.
÷Pharmazeutische Chemie	4	Gastell.
÷Pharmazeutische Botanik	3	Heer.

Die mit * bezeichneten Fächer sind nur für Studirende der

Fig. 22.3 AETH I: course structure in the mechanical and chemical school

The Brazilian student had to pass an entrance examination. The exam proved that he had sufficient previous knowledge and likewise no language problems so he was admitted to the engineering school. The marks obtained are telling (Fig. 22.4); they ranged them from 1 (lowest) to 5 (highest):

While the result in arithmetic was poor, he obtained the highest mark in mathematics and a medium mark in descriptive geometry. In the following term he was lucky to have Dedekind as teacher of the calculus, with best results; they weakened when Dedekind had left and was substituted by Durège. The results in descriptive geometry were impressing (Fig. 22.5):

In Karlsruhe, Paula Souza continued to study in the engineering school, with like success. The last year, 1866/67, shown here, attests his assiduity (Fig. 22.6).

The courses he attended "assiduously" were on road constructions, drawing, practical geometry, higher geodesy and analytical mechanics. The final exam of the same year, 1867, relates the subjects of the questions in the numerous disciplines. Regarding descriptive geometry, the examiner Christian Wiener stated: "Die Fragen betreffend die Construktion des Winkels von Geraden und Ebenen. Urtheil: gut"— the questions regarding the construction of the angle of straight lines and planes. Judgment: good (Fig. 22.7).

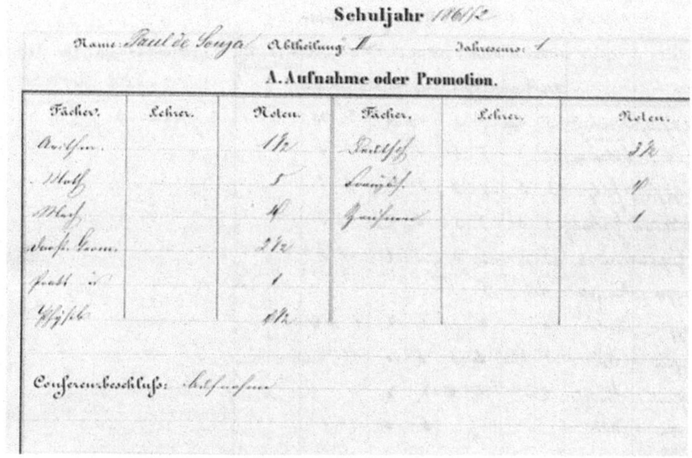

Fig. 22.4 AETH II: Results of the entrance examination in 1861

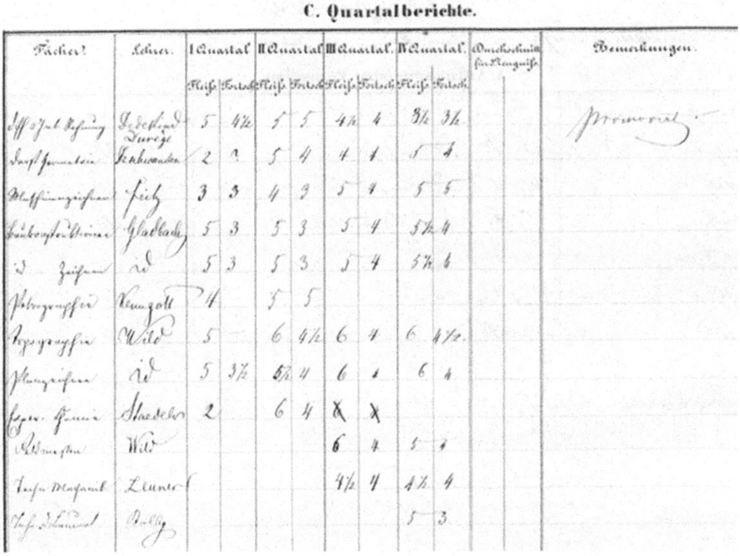

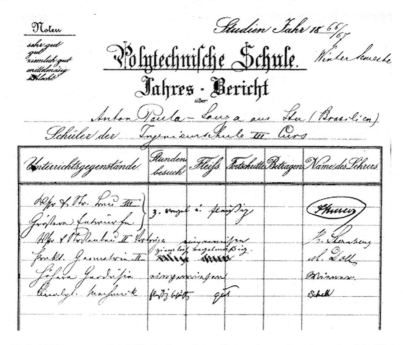

Fig. 22.5 AETH II: Paula Souza's results studying in the engineering school 1861/62

Fig. 22.6 AKIT: report of A. F. Paula Souza's studies at the engineering school in Karlsruhe 1866/67

Fig. 22.7 Excerpt from Paula Souza's exam in 1867 (Padilha 2009, p. 79)

5 Polytechnic Schools Becoming Technical Colleges

After their elevation in status to achieve—at least partially—a level of higher education, polytechnic schools were structured in general into five departments, *Fachschulen*, or study courses:

- architecture,
- mechanical engineering,
- chemistry,
- construction engineering,
- mathematics

The claim that mathematics constituted now a parallel department, hence having the "status of a Fachschule" (Hoepke 2007, p. 61), and no longer a propaedeutic one, is somewhat misleading. On the one hand, given the higher education status and the establishment of schools—like *Realgymnasien*—that provided now the propaedeutic teaching, a preparatory course could no longer exist. On the other hand, there was no independent study of mathematics at polytechnic schools having now raised their level, being called *polytechnische Hochschulen*, which can be translated as polytechnic colleges, the mathematics lectures now were service lectures for the other study courses.

But this demonstrates exactly the basic conflict about the function of mathematics within these colleges. Mathematics was no longer understood as the foundation for the various sciences and for the applications. The various departments for engineering etc. understood themselves as independent. At best they would call mathematics for service lectures.

In the evident conflict over how to design the mathematics lessons so as to be of service for these applications lies the reason for a rather unusual movement: the anti-mathematical movement of the engineers in the last third of the nineteenth century—thus exactly a conflict about the polytechnic nature of mathematics.

Being no longer preparatory courses of a secondary school character, the colleges did no longer hire mathematics teachers but mathematics professors (Hoepke 2007, p. 60)—and in general young graduates, who had freshly obtained their doctoral diploma and who understood the apparently independent status of the mathematics department as the liberty to give calculus lectures according to the new demands of rigour. The reaction against such lectures was strong: professors of engineering claimed to know better what mathematics their students needed and intended to give such service lectures themselves (see Schubring 1989, p. 181).

The conflicts were very strong, and threatening the status of mathematics at the technical colleges. A mathematics professor thought to have found a compromise formula:

> For technical education, mathematics constitutes a foundational discipline, it is a necessary part. For later practice, mathematics changes, however, to become just an auxiliary discipline (Papperitz 1899, 45–46; my translation, G. S.).

Yet this did not really present the solution: The anti-mathematical movement of the engineers denied mathematics a foundational character, they accepted mathematics only as service for them, as auxiliary (see Hensel 1987).

6 Conclusion

In short: the very notion of "polytechnic" as launched and propagated by the Paris school was no longer accepted in these Polytechnic Colleges. The only connection to the conception of the EP was that descriptive geometry had been from the beginning, even in low ranking schools for instigating *Gewerbe* assiduity, a strong component of the curriculum throughout the nineteenth century—while paradoxically descriptive geometry had suffered a marked decline at the EP. The further development of the Polytechnic Colleges to Technical Colleges and eventually to Technical Universities led to the extension of the Mathematics Department to a General Department, embracing humanities as well, such as history of arts. Thus, the General Department provided something like a higher-level general culture—mathematics being included in such a conception meant that the polytechnic character was in fact lost. One can thus sum up that the polytechnic character of mathematics, i.e. its foundational status for study courses of sciences and engineering, was maintained only in France—even if the internal role of descriptive geometry within this function had decisively weakened. One might attribute this to the strong tradition of rationalism, assuring mathematics a key function in scientific epistemology and in cultural views of science and technology.

References

Sources

Archives de l'École polytechnique, Palaiseau (France) [AEP] AEP I: X2C 29—[1816] Conseil de perfectionement da l'école polytechnique le 12 décembre. Quelques reflexions sur la Géométrie Déscriptive et sur le professorat de cette partie.

Archiv der Eidgenössischen Technischen Hochschule, Zürich (Switzerland) [AETH] AETH I: ETH-Bibliothek, Archive, SR2: Anhang 1861, Programm der eidgenössischen polytechnischen Schule für das Schuljahr 1861/ 62. beziehungsweise für das erste Halbjahr. AETH II: EZ_Rek_1/1/387 Paul de Souza, Ant. Francisco.

Archiv der Technischen Universität Karlsruhe [AKIT]. AKIT I: Technische Hochschule Karlsruhe, "Einschreibebücher der Studierenden", Band 2: Studienjahre 1860/61 bis 1865/66. Signatur 10001, 1644 (for the year 1866/67).

Publications

Abdeljaouad, Mahdi. 2012. Teaching European mathematics in the Ottoman Empire during the 18th and 19th centuries: Between admiration and rejection. *ZDM - The International Journal on Mathematics Education* 44 (4): 483–498.

Crozet, Pascal. 2008. *Les sciences modernes en Egypte. Transfert et appropriation*, 1805–1902. Paris: Geuthner.

Dhombres, Jean. 1987. *Histoire de l'École Polytechnique – Réédition de Fourcy 1828*, avec une introduction [et quelques additifs]. Paris: Belin.

Fourcy, Ambroise. 1828. *Histoire de l'École Polytechnique*. Paris: chez l'auteur.

Hensel, Susann. 1987. *Die Auseinandersetzungen um die mathematische Ausbildung der Ingenieure an den Technischen Hochschulen Deutschlands Ende des 19. Jahrhunderts*. Dissertation A, Universität Leipzig.

Hoepke, Klaus-Peter. 2007. *Geschichte der Fridericiana. Stationen in der Geschichte der Universität Karlsruhe (TH) von der Gründung 1825 bis zum Jahr 2000*. Karlsruhe: Universitätsverlag.

Jacobi, Carl Gustav Jacob. 1835/1891. Über die Pariser polytechnische Schule. In C. G. J. Jacobi, *Gesammelte Werke*, Band 7, ed. by Carl Weierstraß, 355–370. Berlin: Reimer.

Manegold, Karl-Heinz. 1970. *Universität, Technische Hochschule und Industrie: ein Beitrag zur Emanzipation der Technik im 19. Jahrhundert unter besonderer Berücksichtigung der Bestrebungen Felix Kleins*. Berlin: Duncker und Humblot.

Neuwirth, Joseph. 1915. *Die K.K. Technische Hochschule in Wien 1815–1915. Gedenkschrift*. Wien: Gerold.

Padilha, Rodrigo Bastos. 2009. *A formação científica e humanística de Antônio Francisco de Paula Souza, o fundador da Escola Politécnica de São Paulo*. Dissertação de mestrado em história da ciéncia, Pontifícia Universidade Católica de São Paulo.

Papperitz, Erwin. 1899. *Die Mathematik an den deutschen technischen Hochschulen: Beitrag zur Beurteilung einer schwebenden Frage des hoeheren Unterrichtswesens*. Leipzig: Veit.

Paul, Matthias. 1980. *Gaspard Monges "Géométrie Descriptive" und die Ecole Polytechnique. Eine Fallstudie über den Zusammenhang von Wissenschafts- und Bildungsprozeß. Materialien und Studien des IDM der Universität Bielefeld*, Band 17. Bielefeld: IDM.

Sakarovitch, Joël. 1994. La géométrie descriptive, une reine déchue. In *La formation polytechnicienne: 1794–1994*, ed. Bruno Belhoste, Amy Dahan Dalmedico, Antoine Picon, 77–93. Paris: Dunod.

———. 1998. *Épures d'architecture: de la coupe des pierres à la géométrie descriptive XVIe - XIXe siècles*. Basel: Birkhäuser.

Saraiva, Luis. 2007. The beginnings of the Royal Military Academy of Rio de Janeiro. *Revista Brasileira de História da Matemática* 7 (13): 19–41.

Schubring, Gert. 1981. Mathematics and teacher training: Plans for a polytechnic in Berlin. *Historical Studies in the Physical Sciences* 12/1: 161194.

———. 1989. Pure and applied mathematics in divergent institutional settings in Germany: The role and impact of Felix Klein. In *The History of Modern Mathematics. Volume II: Institutions and Applications*, ed. David Rowe, John McCleary, 171–220. Boston: Academic Press.

———. 2004. *Le Retour du Refoulé. Der Wiederaufstieg der synthetischen Methode an der École Polytechnique*. Reihe Algorismus, No. 46. Augsburg: Erwin Rauner.

Stark, Franz. 1906. *Die k.k. Deutsche Technische Hochschule in Prag, 1806-190. Festschrift zur Hundertjahrfeier*. Prag: Deutsche Technische Hochschule.

Author Index

A

Abad, Urbano Mas, 75
Abel, Niels, 410
Abrahamson, Joseph Nicolai Benjamin, 260
Adham, Ibrāhīm, 364
Adhémar, Joseph-Alphonse, 23, 25, 27, 28, 33, 35, 44, 76, 78, 87, 390, 393, 396, 398
Adolf, Stefan, 307
Adrain, Robert, 348–351, 354
Alberti, Leon Battista, 172
al-Falakī, Ismā'īl Muṣṭfā, 370
Ali, Muhammad, 412
Alix, Luis Felipe, 75, 76
Allen, Robert T.P, 347
Almeida, Carlos Augusto Morais de, 107
al-Ṭahṭāwī, Rifā'a, 361, 363
al-Ṭūsī, Naṣīr al-Dīn, 361, 363
Álvarez Ude, José Gabriel, 90, 92
Amadieu, M.F., 27
Amaldi, Ugo, 66
Amiot, Antoine, viii, 31, 50
Ampère, André-Marie, 213
Anagnosti, Petar, 301, 307
Antonopoulos, Dimitrios, 121, 132
Apollonius, 392
Apostolou, Apostolos, 130, 131
Arillaga y de Garro, Francisco de Paula, 90
Arroquía, 75, 76
    See also Rodríguez Arroquía
Ascanius, 258
Ascarza, 76
    See also Lozano y Ascarza
Aschieri, Ferdinando, 49
Azmī, Mansūr, 362, 368

B

Babinet, Jacques, 87
Bache, Lars, 267
Badon Ghijben, Jacob, 237–240, 243, 247
Baehr, George, 243
Bailly-Baillière, Carlos, 76
Barbery, Manuel María, 75, 76
Bardin, Libre-Irmand, 87
Bathrust, Henry, 115
Battaglini, Giuseppe, 62, 63
Bauche de la Neuville, Jean-Nicholas, 43
Bauditz, 262
Bayyūmī, Muhammad, 362–365, 368, 373, 374
Bazaine, Pierre-Dominique, 213, 215, 218, 224
Baz, Jiménez, 74, 76
Bélanger, Jean-Baptiste, 367
Bellavitis, Giusto, 45, 86
Bellegarde, Pedro d'Alcântara, 382
Bergery, Claude Lucien, 396
Bernard, Simon, 341
Bertaux-Levillain, C., 30, 31
Berthollet, Claude-Louis, 43
Beskiba, Georg, 277
Betancourt, Agustín de, vii, 70, 213, 224
Bézout, Étienne, 22
Bielsa y Ciprián, José, 75, 76, 87
Binns, William, 328, 329
Biot, Jean-Baptiste, 21, 72
Biris, Konstantine, 126, 129–131
Bjerre, Frederik Fabricius, 272
Blagrave, John, 314
Blaschke, Wilhelm, 189, 191

Bledsoe, Albert T., 347
Blessing, Georges F., 354, 355
Bompiani, Enrico, 66, 67
Bonitz, Hermann, 275
Bonnesen, Erdmann Peter, 269
Boucharlat, Jean-Louis, 72, 367
Bouquet, Jean-Claude, 243
Bourdon, Pierre-Louis-Marie, 119
Bouris, George, 125, 126
Bourlet, Carlo, 35
Bovet, Armand de, 386
Boxman, A., 247
Bradley, Thomas, 322, 328
Brennecke, Wilhelm H., 160
Brianchon, Charles, 245, 392
Brioschi, Francesco, 48
Briot, Charles, 31, 32, 243
Brisson, Bernabé, 4, 7, 10, 23, 44
Brugg, David Rytz von, 185
Bruyère, Léon, 23
Burmester, Ludwig, 146, 188, 189
Butz, Wilhelm, 160

C

Caballé, 82
   See also Torroja Caballé
Cabrera, 76
   See also Pedraza y Cabrera
Cajal, 90
   See also Ramón y Cajal
Campomanes, 84
   See also Rodriguez de Campomanes
Cantor, Moritz, 316
Carandinos, Ioannis, 115, 116, 123
Carboni, 50
   See also Ortu Carboni
Cardinaal, Jacob, 242–245
Cardona y Escarrabill, Baltasar, 85
Carnot, Lazare N.M., 72
Carvalho, Rómulo de, 99, 105
Casati, Gabrio, 46
Cassanac, Eugène, 362, 371
Castel-Branco, José Frederico d'Assa, xv, 99, 103
Cauchy, Augustin Louis, 213
Cavallero, Agostino, 89
Cellar, Joseph, 131
Cerruti, Valentino, 59
Chasles, Michel, 58, 87, 104, 189
Chastillon, Nicolas de, 4
Chauvet, Romano, 379
Chizhov, Fedor, 221
Chomé, Félix, 104

Christoffel, Elwin, 171
Church, Albert E., 351, 352
Cioci, Antonio, 46
Ciprián
   See also Bielsa y Ciprián
Cirodde, Paul-Louis, viii, 23, 75, 76
Clebsch, Alfred, 58, 173
Cloquet, Jean-Baptiste, 75
Cohen Stuart, Lewis, 241–243
Colombo, Giuseppe, 42
Comberousse, Charles de, 27, 28, 32
Condillac, Bonnot de, 404, 405
Conte, Antonio, 41
Cooke, Giles B., 353
Coriolis, Gustave de, 406
Corridi, Filippo, 43, 44
Cortázar Albásolo, Juan de, 87
Coste, M., 16
Cousinery, Barthélémy Édouard, 396
Creizenach, Michael, 141, 142
Crelle, August Leopold, 410
Cremona, Luigi, vii, viii, 48, 57–67, 83, 86, 104, 167, 178, 316, 389, 392, 395
Crozet, Claude, xiv, 320, 341, 343–347
Crutchfield, Stapleton, 353
Culmann, Carl, 48, 58, 86, 170, 171
Cunha Galvão, Ignacio da, 383, 384
Cunningham, William, 331, 332

D

Danić, Dimitrije, 297
Darboux, Gaston, 87
Darling, Lewis A., 354, 355
Daubenton, Louis Jean-Marie, 43
d'Aubuisson de Voisins, Jean-François, 366
d'Audebard, André E.J., 222
Davidović, Stevan, 305, 309
Davies, Charles, 350, 351
D'Ayala, Salvatore, 45
Dedekind, Richard, 414, 416
Degen, Carl Ferdinand, 256
Delgado, Joaquim Filipe Nery, 100
Delprat, Isaac, 237
De Morgan, Augustus, 323
Denisov, Vesevolod, 226
Desargues, Girard, 36, 58, 63, 303, 392
Descartes, René, 58
Despotopoulos, Dimitrios, 120
Destrem, Maurice, 213, 224
Devjatnin, Aleksandr, 226
Diaz, Avelino, 379
Diesterweg, Adolph, 117
Dijksterhuis, Eduard, 248, 249

Dinet, Charles-Louis, 21, 213
Dirichlet, Johann Gustav Lejeune, 410
Disteli, Martin, 189
Doležal, Eduard, 189, 203
Domènech i Estapà, Josep, 88
Douglas, Howard (Lord), 116
Duchesne, Émile, xiv, 21, 362–364, 371, 375
Duhamel, Jean-Marie, 22
Dupin, François Pierre Charles, 4, 10, 46, 84, 115, 186, 187
Durège, Heinrich, 416
Dyck, Walter F.A. von, 155, 171, 189

E

Echegaray y Eizaguirre, José de, 85–87
Eckersberg, Christoffer Wilhelm, 256
Eckhart, Ludwig, 204–207
Ehrenfest, Tatjana, 249
Einstein, Albert, 189
Eizaguirre, 85
    See also Echegaray y Eizaguirre
Elizalde, José Antonio, 75, 78, 87, 89
Emilijan, Josimovic, 299–300
Enfantin, Prosper, 365
Enriques, Federigo, 48–50, 53, 65
Escarrabill, 85
    See also Cardona y Escarrabill
Essemann, 266
Estapà, 88
    See also Domenech y Estapà
Euclid, 216, 306, 344, 345, 364
Exner, Franz Serafin, 275

F

Fabre, Alexandre, 212, 213, 224
Fahīm, Maḥmūd, 362, 373
Fano, Gino, 48, 49
Farish, William, 45, 239, 240, 250, 317, 319–320, 329, 331
Farrar, John, 347
Favaro, Antonio, 88
Ferdinand III, Kaiser, 276
Ferdinand VII, King of Spain, 71, 73, 74
F.I.C., 384, 389, 395, 397, 398
Fiedler, Wilhelm, vi, xxviii, 57–67, 86, 89, 140, 167–179, 186, 189, 244, 267, 309, 396
Finsterwalder, Sebastian, 189
Flauti, Vincenzo, xiv, 43
Forchammer, Johan Georg, 258
Fourcy, Ambroise, 21, 213, 405

Fourcy, Louis Lefebure de, 21, 75, 99, 100, 104, 119, 243, 326, 406
Francoeur, Louis Benjamin, 21, 70, 84, 117
Freire, Francisco de Castro, 99
Frenet, Jean Frédéric, 243
Freudenthal, Hans, 249
Frobenius, Ferdinand G., 171
Fromholt, F., 35, 36

G

Gabaglia, Eugênio de Barros Raja, 384, 389
Galdeano, see Garcia de Galdeano
Galvão, Ignácio da Cunha, 383, 384, 390
Garat, Dominique-Joseph, 43
García de Galdeano, Zoel, 82, 88
García Otero, José, 73
Garnier, Jean Guillaume, 22, 72
Garro, 90
    See also Arillaga y de Garro
Gärtner, Friedrich von, 125
Gascheau, Gabriel, 23, 99
Gauß, Carl Friedrich, 188, 409, 410
Gauthier-Villars, Jean-Albert, 87
Gavrilović, Bogan, 297
Gelder, Jacob de, 236
Georgiades, Michael, 126
Gerakis, George, 119
Gercken, Wilhelm, 161
Gergonne, Joseph Diaz, 16, 22, 267
Gérono, Camille-Christophe, 371, 372, 375
Ghyben, Jacob Badon, 233
Giró, see Mundi i Giró
Gonçalves, António Augusto, 107
Gorceix, Henri, 386
Gournerie, Jules de La, 30, 32, 34, 76, 87, 89, 104, 188, 243
Grossmann, Marcel, 171
Grünwald, Josef, 191
Gugler, Bernhard, 140, 145, 146
Guilford, Frederic North 5th Earl of, 114
Guimarães, Manoel Ferreira de Araújo, 380
Gunner, J.B.K., 269

H

Haan, Bierens de, 247
Hachette, Jean Nicolas, 3–17, 20, 22, 124, 213, 218, 223, 224, 238, 317, 366, 396
Halley, Edmund, 314
Hall, Thomas Grainger, xiv, 326, 327
Hansen, Christian, 125, 126, 271
Hansen, Theofile, 125, 126

Hans, Robert, 189
Harpe, Jean-François de la, 43
Hassenfratz, Henri, 213
Hast, Adolph, 120
Hauck, Hermann Guido, 178
Haussner, R., 140, 144
Haüy, René-Just, 43
Hayward, James, 350, 354
Heather, John Fry, 319, 326, 328
Heegaard, Poul, 262, 264
Heffter, L., 154
Hekekyan, Joseph, 365, 366
Heller, Karl, 126
Henkel, Otto, 92
Henry, André Guillaume, 215
Hetsch, Gustav Friedrich, 256, 257, 259, 269
Hetzel, Pierre Jules, 84
Hiele, Pierre van, 249
Hjelmslev, Johannes, 269–272
Hoepke Klaus-Peter, 407, 413, 418, 419
Hohenberg, Fritz, 191
Hönig, Johann, xv, 146, 182, 184–186, 198, 199, 277
Hreczyna, Gregor, 221, 222
Humboldt, Alexander von, 410
Hummel, Christian Gottfried, 269
Hurwitz, Adolf, 171

J
Jacobi, Carl Gustav Jacob, 411
Janni, Vincenzo, 62
Janssen, Jan, 234–236
Jarolímek, Vincenc, 279, 281–284
Javary, Adrien, 384
Jessen, Børge, 272
Jiménez y Baz, José, 74, 76
Jopling, Joseph, 315, 321, 322
Josimović, Emilijan, xv, 297–301, 303, 305, 309, 310
Julien, A., 389
Jürgensen, Christian, 256

K
Kadeřávek, František, 282, 283
Kanellopoulos, Miltiadis, xv, 123
Kapodistrias, Ioannis, 116–119
Karakatsanidis, Nikolaos, 131
Kavtanzoglou, Lyssandros, 126–128
Kellner, Ludwig Stephan, xiv, 259–269
Kiprijanov, Valerian, 219

Klapka, Jiří, 284
Kleanthis, Stamatis, 117
Klein, Felix, xxv, 58, 148, 155–157, 160, 171, 178, 185, 271
Klíma, Josef, 283, 284
Knudsen, Laurits, 267
Kobus, Michel, 248
Kokkinakis, Konstantinos, 117
Kokkonis, Ioannis Komninos, Theodore, 117
Konovalov, Ivan, 221
Kontouris, Ioannis, 116
Koppe, Karl, 119
Kötter, Ernst, 82
Kounovský, Josef, 282, 283
Králové, Hradec, 281, 283
Krames, Josef L., 187, 189, 191, 192, 204–207
Kriezis, Andreas, 128, 129
Kruppa, Erwin, 188, 190–192, 201, 204–207
Kurdjumov, Valerian, 218

L
Labey, Jean-Baptiste, 213
Lacroix, Sylvestre-François, xiv, 3–17, 22, 45, 72, 74, 98, 213, 236, 345, 380, 381
Laderchi, Giacomo, 46
Lafrémoire, Henri Charette de, 30
La Gournerie, Jules de, 30, 32, 34, 76, 243
Lagrange, Joseph-Louis, 22, 28, 72, 236
Lambert, Johann H., 58, 172, 317, 319
Langley, Batty, 316
Lanz, José María, 70, 366
Laplace, Pierre-Simon, 43, 72
Lauberg, Carlo, 43
Laumont, Gilles de, 405
Laurent, Charles, 84, 126
Lavallée, Alphonse, 27
Leblanc, César Nicolas Louis, 276
Lebon, Ernst, 34, 130, 309
Legendre, Adrien-Marie, 116, 243, 380
Leroy, Charles François Antoine, xv, 20, 30, 33, 45, 75, 99, 100, 120, 123, 277, 340, 355, 366, 369, 379, 384, 406
Levi, Beppo, 48, 92
Levinsen, Anton, 267
Libritis, Theodore, 123
Lie, Sophus, 191
Lietzmann, Walther, 161
Liouville, Joseph, xxv, 367
Lipkens, Antoine, 241
Lobačevskij, Nikolaj, 221
Lobatto, Rehuel, 241, 243

López de Peñalver, Juan, 74
Loria, Gino, 44, 48, 49, 64, 99, 182, 316
Louis XVIII, King of France, 73
Lowry, Wilson, 323
Lozano y Ascarza, Antonio, 76
Lozier, Charles Ambroise, 379

M

Machado, Achilles Alfredo da Silveira, 107, 108
MacLaurin, Colin, 392
Mağdī, Ṣāliḥ, 368
Majocchi, Giovanni Alessandro, 46
Majurov, Aleksej, 212, 224
Makarov, Nikolaj, 218
Mannheim, Amédée, 34, 104
Mannoury, Gerrit, 243, 249
Mansfield, Edward, 340–342
Martelet Émile, 20, 30
Mas Abad, 75
   See also Abad
Matos, Ana Cardoso de, 100
Mavroyannis, Gerasimos
Mayssl, Anton, 277
Mazzitelli, Domenico, 50
McRee, William, 340, 341
Mechnikov, Evgraf, 220
Mehler, Ferdinand, 162
Mentelle, Edme, 43
Mindler, Joseph, 127
Minkowski, Hermann, 171, 189
Minnaert, Marcel, 249
Mitchell, 323
Möbius, August Ferdinand, 58, 63, 168, 174, 176, 392
Monge, Gaspard, v, vii, x, xi, xiv, xv, 3–17, 20, 21, 58, 59, 69–78, 98, 113–132, 139, 140, 182, 184, 185, 193, 212–214, 227, 238, 270, 281, 317, 346, 359, 378, 380, 391
Močnik, Franc, 305
Monroe, James, 341
Monteiro, João Baptista Ortiz, 384
Monteiro, José, 103
Moschopoulos, Timoleon, 123
Mossotti, Ottaviano Fabrizio, 116
Moxon, Joseph, 316
Mubārak, ʿAlī, 368, 370
Muhanmed Ali, *see* ʿAlī, Muhammad
Müller, Emil, 184, 187, 201–204, 288
Müller, Hans Robert, 189

Mundi i Giró, Santiago, 88
Mustafā, Ismāʿīl, 370, 374

N

Nagīb, Aḥmad, 362, 370–373, 375
Napoleon (Bonaparte, Napoléon), 40, 71, 213, 214, 234, 341, 359
Nebenius, Carl Friedrich, 413
Negris, Konstantine, 124, 125, 127
Nešić, Dimitriije, 296, 297
Neuville, *see* Bauche de la Neuville
Neveu, François-Marie, 213
Newton, Isaac, 314
Nichols, Francis, 344
Nicholson, Peter, 317, 320, 322–326, 329, 331
Niemeyer Bellegarde, *see* Bellegarde
Niemtschik, Rudolf, 199
Nightingale, Florence, 314
Nikolić, Atanasije, 298–299, 301
Nooten, Sebastiaan van, 249
Nordio, Attilio, 53
Nordštejn, Alexandr, 218

O

Oeder, J.C., 258
Olivier, Théodore, 4, 5, 27–29, 31, 75, 99, 100, 243, 309, 352, 367, 396
Ørsted, Hans Christian, 257–259, 265
Ortu Carboni, Salvatore, 50
Otero, *see* Garcia Otero
Otto, King of Greece, 119–121, 123
Overduyn, Willem Lodewijk, 241

P

Papadakis, Ioannis, vii, 125, 127, 129
Papperitz, Erwin, 140, 145, 146, 152, 154, 159, 419
Parvé, Daniel Stein, 247, 249
Pasch, Moritz, 92, 173, 177
Pasi, Carlo, 45
Patterson, James W., 347
Pauzié, Jean-Pierre-Augustin, 118
Péclet, Eugène, 27
Pedraza y Cabrera, Pedro, 76
Pegado, Luiz Porfírio da Mota, 103, 104, 108
Pelíšek, Miloslav, 284, 285
Pelz, Karel, 281, 282, 288–289
Peri, Giuseppe, 50, 51
Pery, Gerardo A., 100

Pesch, Adrianus van, 248
Peschka, Gustav Adolf Viktor, 186, 187, 189, 200–202, 288
Pestalozzi, Johann, 84
Petersen, Julius, 256, 267, 269, 273
Petridis, Michail, 128
Petrović, Mihailo, 297, 306, 310
Philips, James, 347
Pilotos, Sotirios, 128
Pilotto, Salvatore, 128
Pinto, Rodrigo Ribeiro de Sousa, 99
Placci, Giuseppe, 43
Playfair, John, 344, 350
Pleskas, Panos, 128
Pluchart, A., 222, 223
Plücker, Julius, 58, 410
Pohlke, Karl Wilhelm, 89, 146, 188, 288
Poisson, Siméon-Denis, 213
Poletti, Luigi, 46
Poncelet, Jean-Victor, 22, 28, 32, 33, 36, 58, 60, 63, 104, 173, 174, 233, 262, 303, 395, 396
Porto, Domingos da Silva, 387
Potier, Charles–Marie, 25, 114, 142, 211–227
Potrér, Charles–Marie, *see* Potier, Charles–Marie
Presler, Otto, 161
Procházka, Bedřich, 282–284, 286
Prony, Gaspard de, 25
Puissant, Louis, 72, 363

Q

Quérard, Joseph Marie, 222, 223
Quetelet, Adolphe, 241

R

Rahmanov, Petr, 212, 214, 224
Ramaḍān, Ibrāhīm, xv, xii, 362, 364, 365, 367–373, 375
Ramón y Cajal, Santiago, 90
Ramus, Christian, 256
Recorde, Robert, 125, 314
Reder, Aleksandr, 218
Rerberg, Fedor, 225, 226
Rey Pastor, Julio, 82, 90, 92
Reye, Theodor, 49, 58, 61, 86, 88, 171, 173, 189
Reynaud, Antoine-André-Louis, 22
Rheineck, Eduard von, 119
Riemann, Bernhard, 188
Rittinger, Peter, 183, 184

Roberval, Gilles P., 144
Rodrigues, Alvaro José, 394–396
Rodrigues, Carlos Telo, 107
Rodrigues, Manuel Augusto, 107
Rodríguez Arroquía, Ángel, 75, 76
Rodríguez de Campomanes, Pedro, 84
Romas, Vassilios, 121
Rouché, Eugène, 32–35
Rumbovich, Hipolit, 221
Rynin, Nickolaj, 218, 219, 223
Ryšavý, Dominik, xv, 279
Rytz von Brugg, David, 185

S

Ṣabrī, Ṣābir, 362, 370, 372–375
Sakarovitch, Joël, 3, 28, 35, 240, 243, 342, 405, 406
Saldaña, Juan José, 378
Salmon, George, 58, 168–170
Santos, Cândido dos, 100
Sarazin, Louis-Charles, 117
Schaffnit, Georg, 45, 277
Schaubert, Eduard, 117
Schellerup, H.C.F.C., 269
Scherling, Christian, 160, 167
Schiappa Monteiro de Carvalho, Alfredo Augusto, 104
Schlesinger, Josef, 145, 167, 173
Schmid, Theodor, 191, 201–205
Schmidt, Isaac, 233, 236–238, 250
Schmidten, Henrik Gerner von, 256–259
Schottky, Friedrich, 171
Schreiber, Guido, xiv, 140, 142–145, 153, 262, 407
Schumacher, Heinrich C., 257
Schüssler, Rudolf, 288
Schwarz, Hermann, 171, 288
Scorza, Gaetano, 53
Seidelin, Carl Julius Ludvig, 266–270
Sella, Quintino, 45, 89
Sendim, Maurício José, 106
Senovert, Etienne-François, 224, 225
Sequeira, Luiz Guilherme Borges de, 102–104
Sereni, Carlo, 45
Serret, Joseph, 371
Sevastianov, Jakov, xii, 214, 215, 217–222, 224, 226, 227
Severi, Francesco, 48, 50, 53, 66
Sganzin, Joseph-Matieu, 23, 213
Shalders, Carlos, 389, 392, 396
Shcheglov, Nikolaj, 221
Shelejhovskij, Kondratij, 221

Author Index

Sicard, Roch-Ambroise, 43
Silva Porto, Domingos da, 387
Simson, Robert, 344
Šinân, Sâdiq Sâlim, 371
Sivers, Egor, 220
Skalistiris, Dimitrios, 129, 132
Skuherský, Rudolf, viii, 276, 277, 279, 280, 286–288
Smirke, Robert, 322
Smith, Francis H., 352
Snell, Friedrich, 119
Sobotka, Jan, 201, 278, 281, 284, 285, 288–290
Šolín, Josef, 291
Solomos, Nikolaos, 130
Sonnet, Hippolite, 23, 83
Sopwith, Thomas, 320, 322
Sousa, José Victorino, 380, 381
    See also Souza
Souza, Antônio Francisco de Paula, 389, 413, 416–418
Souza, José Victorino dos Santos e, 98, 103
    See also Sousa
Spyromilios, Spyridon, 119, 121
Stäckel, Paul, 140, 142, 144, 148–152, 155–159, 167, 178, 179
Staudigl, Rudolf, 184–186, 188, 199–201
Staudt, Christian von, 58, 82, 88, 89, 92, 145, 173, 174, 177, 189
Stavridis, Dimitrios, 120, 123
Steen, Adolph, 256–260, 262, 266, 267
Steiner, Jakob, 58, 172, 174, 175, 396
Stephanos, Cyparissos, xxvi, 124, 125
Stojanović, Dimitrije, xi, 301–305, 307, 309
Stroganov, Sergej, 225, 226
Strootman, Hendrik, xiv, 237–243, 247, 248, 250
Stuart, Lewis Cohen, 241–243
Sturm, Jacques, 243
Sucharda, Antonín, 285
Suppantschitsch, Richard, 162
Süssekind, Carlos, 385
Swift, Joseph G., 341, 342

T

Taylor, Brook, 58, 59, 314–316, 396
Terquem, Olry, 22
Thayer, Sylvanus, 340, 341, 412
Thénot, Jean-Pierre, 25, 128
Theofilas, Anastasios, 130, 132
Thijn, Abraham van, 248
Thiollet, François, 368
Thomaidis, Dimitrios, 131

Thorbecke, Johann Rudolf, 235, 244, 247, 249
Thune, R.G.F., 257
Tilšer, František, 280, 281
Torelli, Gabriele, 44
Torroja, Eduardo Caballé, 78, 82, 87
Tournakis, Dimitrios, 121, 129
Tralles, Johann G., 409
Tramontini, Giuseppe, 43, 44
Tresca, Henri, 31, 309
Trudi, Nicola, 62
Tucci, Francesco Paolo, 44
Tuzzi, Vincenzo, 45
Twiss, Thomas, 347

U

Ude, José Gabriel Alvarez, 90, 92
Uglieni, Marco, 59
Ursin, Georg Frederik, 256–259

V

Vafas, Christos, 119
Vallée, Louis Léger, 23–28, 30, 33, 35, 75, 84, 87, 212, 218, 352, 378, 390, 393, 395–398
Vallejo, José Mariano y Ortega, 87
van den Berg, Franciscus, 243
van Eyck, Jan, 328
van Goens, Ryklof, 242, 243
van Hiele, see Hiele
Vandermonde, Alexandre-Théophile, 43
Veen, Hendrik van, 246
Verkaven, Jean-Jacques, 368
Veronese, Giuseppe, 171
Vidal, António Augusto de Pina, 107
Vidal Abascal, Enrique, 82, 83
Viollet-le-Duc, Eugène, 84
Vitorino, see Souza
Volney, Constantin-François de Chassebœuf de, 43
von Zentner, Friedrich, 125, 126, 132
Vries, Hendrik de, 171, 244–246

W

Wallis, John, 314
Wansink, Johan, 249
Warren, Samuel E., 353, 354
Weber, Heinrich, 171
Webster, 323
Weinbrenner, Friedrich, 140, 141, 150
Weisbach, Julius, 45, 168, 172

Well, Gerardus, 244
Wersin, Karel, 276
Weyr, Eduard, 291
Weyr, Emil, 171, 291
Wiener, Christian, 140–142, 144–146, 153, 163, 167, 178, 244, 250, 416
Wiesenfeld, Karel, 276
Willers, H.F., 170
Wooley, Joseph, 326
Wunderlich, Walter, 189, 191, 206, 207

Z

Zablockij, Evgenij, 220
Zavadsky, Joseph, 222
Zeise, Christofer, 257, 258
Zelenoj, Aleksandr, 221
Zentner, Friedrich (von), 125, 126, 132
Zeuthen, Hieronymus Georg, 256, 257, 262, 264, 265
Zinopoulos, Andreas, 130
Zorraquin, Mariano de, 71, 74, 378
Zühlke, Paul, 159–161

Subject Index

A
Abstract theory, 316
Abstract *vs.* applied mathematics, 152
Academia Polytechnica do Porto, 100, 105
Academic Council, 101
'Academic' exercises, 186
Academic model, 46–47
Academy of Sciences, 173, 225
Advanced algebra, 242
Aerial perspective, 5
"Affine" formulation, 61
Agricultural engineering, 70
Algebra, 117, 169, 189, 242, 340
Algebraic elimination, 9
Analytical-descriptive geometry, 72
Analytic geometry, 71, 88, 145, 242
Analytic reasoning, 192
Arabic elementary arithmetic textbook, 362–363
Arabic scientific language, 361
Arabic translation, 366
Architectural Dictionary, 323
Architectural drawing, 182
Argentina, 378, 379
Arithmetics, 117
Artillery school, 236–240, 299, 305, 370
Artists, 23–27
Austria, evolution, 181–193
Austro-Hungarian Empire, 198, 199, 202, 295, 408
Auxiliary planes, 26
Axonometric perspective, 76, 77, 243, 303
Axonometric projections, 244

B
Babbage's Difference Engine, 314
Badon Ghijben, Jacob, 237–240, 243, 247, 248
Basic geometry, 85
"Basis of elevation," 4
Biblioteca do Povo e das Escolas, 107–108
Bildung neohumanistic ideal, 86
Board of Ways of Communication, 220
Boolean algebra, 314
Bourbon Monarchic Restoration, 83
Brazil, 97, 98, 108, 378–398, 411
British system of projection, 329
Bulletin des sciences technologiques, 222
Bundesamt für Eich-und Vermessungswesen, 206

C
Cadet Corps of Mines, 220
Candidate of Applied Science, 258
Carpentry drawings, 5
Cartesian geometry, 72
Casati law, 46, 47, 49, 52
Cavalier perspective, 303, 304
Central collineation, 176, 177
Central projection, 45, 49, 59, 61–63, 65, 66, 140, 143, 145, 146, 160–162, 168, 172, 175–177, 245, 246, 263, 267, 278, 281, 303, 396
Centralseminarium, 410
Charles-Ferdinand University, 278
Cisalpine Republic, 40
Citoyen Legendre, 12

Civil engineers, 23–28, 41, 70, 73, 78, 116, 124, 130, 131, 221, 242, 266, 321, 322, 328, 348, 352, 353, 379, 380, 384, 408, 409, 411
Colégio Militar, 378
Colombia, 378, 379
Conseil de perfectionnement, 21
Constructive theory, 177, 178
Council for Advanced Education and Scientific Research, 90
Council of Studies, 120
Course of Mathematics, 348–350
Cube shadow, 6
Cultural and intellectual life, 298
Cultural channels, 40
Cultural-political framework, 84
Cunningham's Notes, 330
Curs der darstellenden Geometrie in ihren Anwendungen, 145
Curved surfaces, 10, 20, 23, 27, 144
Curvilinear perspective, 328, 329
Cyclography, 189, 191
Czech Lands, 276–277
Czech language, 277, 278, 281
Czech Technical Universities, 275–291

D

Danish mathematics, 255
Darstellende Geometrie, 146
Delft, 246–249
Denmark, 255–273, 409
Department structure, 412–418
Desargues theorem, 61
Descriptive geometry, 407, 408, 416, 419, 420
   Czech Technical Universities, 275–291
   Denmark, 255–273
   École normale de l'an III, Leçons, 7–11
   École polytechnique, 404–406
   Educational Institutions of Greece, 113–132
   Egypt, 359–375
   England, 313–332
   evolution, Austria, 181–193
   in France, 19–36
   in Italy, 39–53
   Latin America, 377–398
   method of projections, 10, 12–16
   method to teaching, 3–7
   Netherlands, 233–252
   nineteenth-century Spain, 69–78
   and polytechnic, 139–163
   Portuguese textbooks, 97–109
   Russia, 211–227
   Serbia, 295–310
   Spain, late modern outlook, 81–92
   synthesis, 167–179
   technical instruction, 57–67
   United States, 339–356
   Vienna school, 197–207
Deskriptivní geometrie I, II, 283
Didactic approach, 215–217
Differential calculus, 237, 242, 367
Differential geometry, 185–189, 191, 193, 206, 207, 263, 265, 272, 278, 291
Dimension drawing, 76
Directrix, 13, 31
Disquisitiones Generales, 263
Draftsmen, 339–356
Drawing course, 261
Drawing notebook, 29, 30
Duality principle, 60

E

Echegaray, José de, 85–87
École centrale, 32, 33, 35, 100, 109, 365–368, 374, 375
École centrale des arts et manufactures, 27, 28, 128, 130, 352, 361–362, 365
École centrale des Quatre-Nations, 12
École Centrale des travaux publics, 5, 20, 40
École des arts et métiers, 23
École des Beaux Arts, 127
École des ponts et chaussées, 4, 7, 23, 24, 30, 74, 100, 129, 213, 223, 224, 237, 363, 368
École normale de l' an III, 5, 7–11, 20, 43
École normale de l'an III, Leçons, 7–11
École polytechnique, 3, 5, 7, 10, 12–14, 16, 20–24, 27, 28, 30–32, 34, 35, 70, 72, 109, 116, 118, 121, 124, 127, 129, 147, 148, 150, 160, 197, 198, 212–216, 222–224, 241, 243, 257–260, 276, 332, 340, 341, 343, 348, 361, 363, 365–368, 379, 404–406, 408, 410–412
École préparatoire du dessin, 127
École royale du génie de Mézières, 4
Educational Institutions of Greece, 113–132
Educational models, 46–47, 361
Educational planning, 117
Education of engineers, 154–155
Egypt, 359–375
Electromagnetism, 258
Elementary Descriptive geometry, 123
Elementary geometry, 242
Elementary industrial education, 84–85
Elementary mathematics, 31, 34, 125, 269

Elementary technical teaching, 46
Employment system, 151
Engineering school, Rome, 48
Engineers, 3–5, 7, 14, 16, 23–28, 34–36, 41, 42, 46, 49, 59, 60, 64, 70, 71, 73, 74, 76, 86, 99, 100, 105, 131, 152, 154–155, 170–171, 175, 178, 200, 213, 221, 234, 236, 241–243, 250, 266, 322, 339–356, 366, 379, 380, 386, 411–412
England, 313–332
English tradition pre-descriptive geometry, 315
Entrance program, 20
Escola Politécnica de Lisboa, 100, 102, 103, 105, 107
Essais de géométrie sur les plans et les surfaces courbes, 12
Essential theorems, 142
Euclidean axiomatic approach, 272
Euclidean space, 174
Euclidean style, 265
Euclid's Elements, 12, 83, 314, 344, 345, 349, 364
Europe, 407–412
Europeanization, 82
European Union scientific research policy, 92
Europe of sciences, 87
Exercises, 21, 128

F
Fachschulen, 150, 153
Faculté des sciences de Paris, 13, 14
Faculty of Architecture, 66
Faculty of Philosophy, 116
Faculty of Sciences, 66, 85
Fine Arts Academy, 126
Forestry engineering, 70, 74
France, 7, 19–36, 40, 45, 58, 71, 73, 83, 87, 89, 97, 109, 123, 129, 163, 172, 193, 213, 215, 217, 225, 227, 241, 243, 267, 309, 316, 317, 323, 330, 331, 343, 348, 350, 361, 367, 395, 396, 410, 411, 413
French educational models, 41
French models, 146, 298
French polytechnicians, 224
French Revolution, 12, 342
Freye Perspective, 172
Fundatie van Renswoude, 234

G
General-interest curricula, 348–350
Generalization, 346–347, 351
General Law on Public Education, 74

General State Archives, 118, 130
Generation of surfaces, 9, 12, 14, 17
Gennadius Library, 116
Geometrical transformations, 90
Geometric drawing, 257, 259
Geometrisches Port-Folio, 145
Geometrisches Zeichnen, 193
Geometry applications, 23
Geometry of mappings, 191
German education, 167
German model, 131
German technical universities, 150, 151, 281
German translation, 139–140, 144
Gerwerbeinstitute, 147–150, 407
Gewerbeschule, 158, 168, 172
Giornale di Matematiche, 63
Gournerie, Jules de la, 30, 32, 34, 76, 104
Graphical methods, 88, 184
Grassmannian methods, 201
Great School of Belgrade, 296
Great schools, 35, 36
Grinding-engine model, 240
Gugler, Bernhard, 140, 145–146, 309
Gymnasien, 155–157, 162, 163
Gymnasium, 119, 129, 149, 155–159, 243, 296

H
Hall's treatise, 327
Higher education, 40–41, 47
Higher technical education, 59, 148, 149, 158, 226
Hinge line, 184
Histograms, 314
History of mathematics, 295, 378
Hjelmslev, Johannes, 269–272
Hogere Burger School (HBS), 235, 237, 242, 243, 245–252
Horizontal projections, 4
Humanistic education, 52
Hyperboloid, 13, 28, 268

I
Indeterminate analysis, 72
Institute of Technology (TH), 245–246, 249
Institute of the Corps of Engineers of Ways of Communication (ICEWC), 213, 217
Institute of the Corps of Ways of Communication, 212
Institutional net, 412–418
Instituto Profissional de NovaGoa, 100
Integral calculus, 237, 242, 367
International languages of science, 227

Ionian Academy, 114–116
Isometric perspective, 238–240, 243, 250
Istituti tecnici, 58, 60
Istituto tecnico, 47, 50
Istituto tecnico superiore, 48
Italian secondary schools, 51–53
Italian translation, 62
Italian treatises, 43–46
Italy, 39–53

J
Jornal de Sciencias Mathematicas e Astronomicas, 104
Josimović, Emilijan, 297–303, 305, 309, 310

K
Kinematics, 32, 34, 187, 189, 207

L
Laplace system, 115
Latin America, 82, 377–398
Law of Education, 85, 88
Legitimisation, 151
Leroy's treatise, 369
Liberal Triennium, 73
Library and the School of Drawing and Architecture, 59
Linear drawing, 83, 85, 117, 126
Linear perspective, 5
Line segment, 271
Line-sphere-transformation, 191

M
Manchester of the East, 168
Massachusetts Institute of Technology, 354
Mathematical catechisms, 316
Mathematical courses, Italian Universities, 47
Mathematical education, 243
Mathematical exactness, 84
Mathematical Laboratory and Seminar of the Council, 90
Mathematical language of technology, 83
Mathematical modernity, 83
Mathematical school, 59
Mathematical teaching organization, 5, 7
Mathematical textbooks, 113
Mathematics curriculum, 119
Mathematische Hauptsätze, 161
Mathesis, 53
Maxwell equation, 314

Mechanical engineering, 70, 184
Mémoires de l'Académie des sciences de Turin, 10
Method of changes, 28, 30
Method of orthogonal projections, 105–106
Method of projections, 8–10, 12–16, 31, 105–107, 132, 238
Method of projections with elevation, 105, 106
Method to teaching, 3–7
Military academies, 71, 73, 343, 351, 355, 361, 370, 371
Military School, 21, 31, 40, 42–44, 116–123, 125, 129, 130, 132, 223, 234, 256, 257, 280, 298–300, 385, 409
Military School Colonel Spyridon Spyromilios, 119
Military school of Evelpides, 116–123
Mining engineering, 70, 74
Ministry of Internal Affairs, 127–129
Mixed-audience textbooks, 350–353
Model function, 404, 409
Modern geometry, 32–35
Modern science, 360–362
Moebius' barycentric calculus, 61
Mongeometrija, 310
Müller, Emil, 159, 161, 184, 185, 187–193, 201–206, 288
Municipal Technical Schools, 42

N
National Astronomic Observatory, 88
National education system, 10
Netherlands, 233–252
Nineteenth century, 32, 35, 39–53, 59, 65, 69–78, 82–84, 86–87, 90, 92, 99–100, 106, 107, 113–132, 141, 147–150, 152–158, 160, 163, 168, 172, 182, 187, 189, 193, 198–201, 212, 245, 249, 251, 255–257, 278, 285, 286, 295–299, 305–306, 308, 315, 319, 329, 332, 340, 343, 353, 360–362, 386, 391, 398, 406, 407, 409, 419
Non-polytechnic school, 299

O
Oberrealschulen, 157, 161, 162, 199
Oblique projection, 319, 322–326
On-the-job training, 84
Oral teaching, 16
"Order of invention," 14, 20
Order of knowledge, 9
Ordinary elements of geometry, 9

Organizational model, 365
Organizational structure, 299
Orthogonal projection, 34, 42, 50, 66, 141, 143, 216, 278
Orthographic projection, 320, 323, 324, 328, 329, 351, 354
Osnovaniâ naèertatel'noj geometrii, 219
Ottoman Empire, 114

P

Parallel oblique projection, 140, 161, 162, 322–326
Perpendicular planes, 4, 22
Perspective, 5, 7, 10, 14, 20, 23, 25, 30, 33, 34, 49, 61, 63, 65, 72, 75–77, 89, 90, 117, 120–124, 126, 127, 129, 140, 141, 144–146, 155, 160–162, 172, 184, 188, 202, 216–220, 235–240, 243–246, 250, 256, 262, 277, 300, 301, 303–305, 315–317, 319–322, 326, 328, 329, 343, 394
Photogrammetry, 189, 203
Physicomathematical department, 127
Physics-mathematics courses, 64
Picture plane, 61
Plagiarism, 12
Planar geometry, 98
Plane geometries, 12
Planes of projection, 8, 24–26, 30, 34
Plane trigonometry, 74
Pohlke, Karl, 89, 140, 146, 188, 288
Pohlke's theorem, 188, 288
Political turbulence, 99
Politics and education, 234–236
Polytechnic at Zürich, 49, 170, 186
Polytechnic characteristics, 404, 420
Polytechnic school, 59, 100–106, 120, 121, 123, 125–132, 147–155, 157, 158, 163, 178, 182, 235, 240–251, 257, 260, 266, 276–278, 290, 297–299, 332, 378, 384, 389, 392, 394, 398, 403–420
Polytechnicum, 182, 186, 198–200
Polytechnikum, 409
Polyteknisk Læreanstalt, 257
Poncelet's geometry, 32
Portugal, 97–100, 105, 107–109, 380, 411
Portuguese textbooks, 97–109
Preliminary technical drawing, 198
Preparatory grades, 20–21, 30–32
Preparatory institutions, 157–158
Preparatory School for Engineers and Architects, 74
Privatdozent, 199

Professional careers, 151
Professional education, 42–43, 148, 156
Professional mathematics, 82
Professional training, 147, 151
Projective geometry, 36, 48–50, 57–67, 82, 83, 86, 88–90, 120, 140, 144–146, 155, 160, 167, 168, 170, 172–175, 177, 186, 188, 191, 199, 202, 203, 206, 207, 223, 244, 245, 247, 251, 263, 267, 270, 281, 282, 284, 285, 303, 304, 330, 392, 394, 396
Prussian educational reforms, 86

Q

Quoted geometry, 30

R

Raumanschauung development, 62
Realanstalten, 156–157, 159
Realgymnasien, 157, 161, 162
Real-schools, 278–281, 283, 285, 309
Realschulen, 198
Rebatimento, 107
Rebatment, 23, 25–27, 29–35, 348, 351, 352, 354
Rectilinear trigonometry, 129
Relief perspective, 188, 189
Rensselaer Polytechnic Institute, 353
Representation systems, 89
Restoration, 40–41, 59, 213
Revista Matemática Hispanoamericana, 82
Revolution, 306
Rio de Janeiro, 383–385, 390–396
Roman mathematics community, 66
Rotation, 26–35, 49, 76, 102, 107, 143–144, 248, 345, 348, 349, 385, 391, 392, 394, 396
Royal Academy, 90, 223, 235, 241–246, 248, 250, 256, 257
Royal Academy of Sciences, 90
Royal Conservatory of Arts, 73
Royal Cornwall Polytechnic Society, 332
Royal Military Academy, 223, 235–240, 247, 248, 250, 319, 343, 412
Royal Technical Institute, 42, 45
Ruled surfaces, 191
Russia, 211–227
Russian Biographical Dictionary, 222
Russian–French diglossia, 225
Russian Revolution, 310

S

Sao Paulo, 389–394
Sapienza, 58
School courses, 44
School for engineers, 4, 23, 41, 46, 49, 59, 64, 74
School of Industrial Arts, 131
School of Mathematics, 59
School of Roads and Waterways, 71, 73, 75
School subject, 159–162
Schrägrissverfahren, 206
Schulprogramme, 199
Scientific language, 360, 361
Scientific school, 227
Scuola sintetica napoletana, 43
Scuola tecnica, 47
Secondary school, 4, 10, 22, 30–34, 50–53, 58–60, 62, 85, 86, 106–109, 117, 119, 148–150, 153, 155–157, 159, 221, 235, 238, 243, 246, 247, 249, 250, 252, 270, 276, 278–280, 283, 285, 307, 309, 361, 370, 386, 388–389, 408–409, 419
Secondary technical education, 49–51
Serbia, 295–310
Serbian Academy of Sciences and Arts, 300
Serbian Language, 301–303
Serbian mathematics, 308
Serbian Queenbee, 296
Sezione fisico matematica (SFM), 47, 49–51, 58
Shadows and shading, 187
Shadow theory, 99
Skew affinity, 246
Skuherský's projection method, 286–288
Sobotka, Jan, 201, 278, 289–290
Social turbulence, 99
Spain, 69–78, 81–92, 117
Spain, late modern outlook, 81–92
Spatial curves, 10, 11
Spatial geometries, 12, 22
Special mathematics, 21
Special secondary school teaching, 32
Spherical trigonometry, 245
Spitalfields' Mathematical Society, 314
State education national council, 90
Stereometry elements, 60
Stonemasonry drawings, 5
Surfaces' intersections, 120
Surfaces of revolution, 14, 16, 21, 31
Surveying, 85, 340
Symmetrische Schrotungen, 207
Synthetic geometry, 283, 354

T

Tangents, 14, 15, 25, 27, 32, 186
Teacher education, 153–154
Teaching, 3–17, 20, 22–23, 28, 31–36, 39–53, 58–60, 62–66, 70, 71, 73, 75, 82–83, 85, 90, 99, 100, 103–109, 117, 120, 123–132, 140, 141, 155, 156, 174, 178, 206, 212–214, 218–227, 244, 259–261, 266, 276–277, 285–286, 330, 339–356, 359–375, 380–384, 388–396, 404–405, 407, 410
Teaching of geometry, 16, 28, 32, 59, 60, 63, 66, 256, 278, 344, 350
Teaching of mechanics, 28
Teaching of methods, 32, 35, 117, 342
Technical colleges, 257, 395, 413, 418–419
Technical drawing, 179
Technical education, 42–43, 49–50, 52, 58, 86–90, 131, 132, 147–149, 152, 155, 158, 163, 167, 168, 226, 246, 360, 411, 419
Technical education institutions, 163
Technical faculties, 246
Technical institutions, 42, 45, 47, 49, 50, 52, 53, 58, 61, 62, 65, 66, 147, 148, 150, 152, 158, 162, 193
Technical instruction, 47, 57–67
Technical sciences, 150, 152, 154, 155, 163, 168
Technical training, 235
Technical universities, 151, 155, 261, 275–291
Technicians education, 83–85
Technische Hochschule, 62, 147, 148, 150–155, 157–159, 162, 190, 288, 408, 409
Technische Hogeschool, 235, 244
Textbooks, 4, 10, 12–14, 20–23, 25, 27–31, 33–35, 49, 50, 63, 65, 66, 70–76, 87, 97–109, 113, 140–144, 161, 182, 184–188, 191, 206, 214–220, 222, 225, 236–238, 240, 243–248, 250, 260, 262–270, 272, 273, 276–279, 281, 283, 284, 286, 288, 298, 301, 305, 307, 309, 310, 326, 340, 342–355, 361–363, 366–375, 378–385, 389, 392, 393, 395–398, 407
Thénot, Jean-Pierre, 25, 128
Theory of involution, 60
Theory of projections, 60
Theory of shadows, 75, 76, 120
Three-dimensional analytic geometry, 124
Topography, 74
Traces, *Traité de géometrie descriptive*, 24, 25

Trading schools, 149
Training of engineers, 340–343
Traité du calcul différentiel et du calcul intégral, 12
Traité élémentaire des machines, 12
Transformation of coordinates, 26
Transformation of projections, 31
Treatise on Isometrical Perspective, 317
Treatise on Projection, 322
Trigonometry, 242, 340
Trimetric projection, 202
Turā Artillery Academy, 367–368
Twentieth century, 53, 65, 66, 82, 150, 153, 155, 157, 163, 191, 193, 219, 238, 245, 247–250, 271, 273, 279, 281, 285, 290, 297, 301, 306–310, 385, 394, 398

U
Umklappung, 175
United States, 339–356, 412
Universidade de Coimbra, 107
University of Athens, 123–125, 127
University of Copenhagen, 257
University of Naples, 41
Ursin, Georg Frederik, 256–259

V
Vaterländische Front, 206
Venn diagrams, 314
Vertical projections, 4
Vidal Abascal, Enrique, 82–83
Vienna, 40, 114, 123, 125, 148, 173, 182, 184, 186, 187, 189–191, 197–207, 276–278, 280, 290, 296–298, 300, 303, 384, 408, 409
Vienna school, 197–207

W
War of Independence, 116
Waterstaat, 234
Weinbrenner, Friedrich, 140, 141, 150
Werkmeisterschule, 168
West Point, 223, 340–343, 350–353, 356, 411, 412
Wisdom of Hindsight, 330–332
Workshop training, 269

Y
Yugoslavia, 307–308
Yugoslavian traditions, 306–308